More Than Just a Textbook

Log on to **ca.algebrareadiness.com** *to…*

- access your Online Student Edition from home so you don't need to bring your textbook home each night.
- link to Student Workbooks and Online Study Tools.

See mathematical concepts come to life

- Personal Tutor
- Concepts in Motion: Interactive Labs
- Concepts in Motion: Animations

Practice what you've learned

- Chapter Readiness
- Extra Examples
- Self-Check Quizzes
- Reading in the Content Area
- Vocabulary Review
- Chapter Tests
- California Standards Practice

Try these other fun activities

- Real-World Careers
- GameZone Games

Glencoe McGraw-Hill

California

Algebra Readiness

Concepts, Skills, and Problem Solving

JOHN ADAMS MIDDLE SCHOOL

JOHN ADAMS MIDDLE SCHOOL

Author
Jack Price, Ed.D.

Glencoe

New York, New York   Columbus, Ohio   Chicago, Illinois   Woodland Hills, California

About the Cover

Reaching a dizzying height of 146 feet and a top speed of 55 miles per hour, the Knott's Silver Bullet at Knott's Berry Farm in Buena Park, California, is one of the most exciting roller coasters in the world. Riders of the Silver Bullet zoom through six inversions, including the cobra roll you see on the cover. In Chapter 6, you will learn about rates of change, such as speeds.

About the Graphics

Twisted torus. Created with *Mathematica*.
A torus with rose-shaped cross section is constructed. Then the cross section is rotated around its center as it moves along a circle to form a twisted torus. For more information, and for programs to construct such graphics, see: www.wolfram.com/r/textbook.

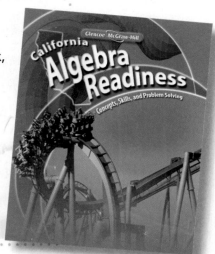

 **Glencoe**

The *McGraw-Hill* Companies

Send all inquiries to:
Glencoe/McGraw-Hill
8787 Orion Place
Columbus, OH 43240-4027

ISBN: 9780078777370
MHID: 0-07-877737-2

Printed in the United States of America.

3 4 5 6 7 8 9 10 043/079 16 15 14 13 12 11 10 09 08 07

Contents in Brief

Authors

Jack Price, Ed.D.
Professor Emeritus
California State
   Polytechnic University
Pomona, California

Contributing Authors

Viken Hovsepian
Professor of Mathematics
Rio Hondo College
Whittier, California

FOLDABLES Dinah Zike
Educational Consultant
Dinah-Might Activities, Inc.
San Antonio, Texas

Consultants

Content

Les Winters
Instructor
California State University, Northridge
Northridge, California

Assessment

Donna M. Kopenski, Ed.D.
K–5 Math Coordinator
Rosa Parks Elementary
San Diego, California

ELL Support and Vocabulary

ReLeah Cossett Lent
Author/Educational Consultant
Alford, Florida

California Advisory Board

Glencoe wishes to thank the following professionals for their invaluable feedback during the development of the program. They reviewed the table of contents, the prototype of the Student Edition, the prototype of the Teacher Wraparound Edition, and the professional development plan.

Linda Anderson
4th/5th Grade Teacher
Oliveira Elementary School
Fremont, California

Cheryl L. Avalos
Mathematics Consultant
Retired Teacher
Hacienda Heights, California

Bonnie Awes
Teacher, 6th Grade Math
Monroe Clark Middle School
San Diego, California

Kathleen M. Brown
Math Curriculum Staff
    Developer
Washington Middle School
Long Beach, California

Carol Cronk
Mathematics Program
    Specialist
San Bernardino City Unified
    School District
San Bernardino, California

Audrey M. Day
Classroom Teacher
Rosa Parks Elementary
    School
San Diego, California

Jill Fetters
Math Teacher
Tevis Jr. High School
Bakersfield, California

Grant A. Fraser, Ph.D.
Professor of Mathematics
California State University,
    Los Angeles
Los Angeles, California

Eric Kimmel
Mathematics Department
    Chair
Frontier High School
Bakersfield, California

Donna M. Kopenski, Ed.D.
K–5 Math Coordinator
Rosa Parks Elementary
San Diego, California

Michael A. Pease
Instructional Math Coach
Aspire Public Schools
Oakland, California

Chuck Podhorsky, Ph.D.
Math Director
City Heights Educational
   Collaborative
San Diego, California

Arthur K. Wayman, Ph.D.
Professor Emeritus
California State University,
   Long Beach
Long Beach, California

Frances Basich Whitney
Project Director,
   Mathematics K–12
Santa Cruz County Office of
   Education
Capitola, California

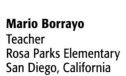

Mario Borrayo
Teacher
Rosa Parks Elementary
San Diego, California

Melissa Bray
K–8 Math Resource Teacher
Modesto City Schools
Modesto, California

California Reviewers

 Each California Reviewer reviewed at least two chapters of the Student Edition, providing feedback and suggestions for improving the effectiveness of the mathematics instruction.

Alma Casey Allen
Retired
ABC Unified School District
Cerritos, California

Marilyn Armas Allotta
Mathematics Teacher
Rancho Bernardo High School
San Diego, California

Dr. Hal Anderson
Professor Mathematics Education
California State University, Long
    Beach
Long Beach, California

Lee W. Ault
Mathematics Instructor
Selma High School
Selma, California

Wendy L. DenBesten
Mathematics Teacher
Hoover High School
Fresno, California

Stella Kumagai
Mathematics Coach
Monterey Peninsula Unified
    School District
Monterey, California

Rebecca Newburn
Mathematics Teacher
Mill Valley Middle School
Mill Valley, California

Dr. Donald R. Price
Teacher, Professor, Motivational
    Speaker
Alvarado Intermediate School
Rowland Heights, California

James D. Sherman
Assistant Principal/Math Academies
    Coordinator
Quimby Oak Middle School
San Jose, California

Larry R. Waters
Mathematics Teacher
Eastlake High School
Chula Vista, California

Prerequisite Skills

H.O.T. Problems
Higher Order Thinking

- Talk Math   **7, 8, 11, 12, 15, 16, 19, 20, 23, 24, 27, 28, 31, 32, 35, 36, 40, 44, 47, 48, 51, 52**
- Writing in Math   **8, 12, 16, 20, 24, 28, 32, 36, 40, 44, 48**
- Reflect   **9, 13, 17, 21, 25, 29, 33, 37, 41, 45, 49, 53**

CHAPTER 1
Variables, Expressions, and Properties

CHAPTER 2 Integers and Equations

CHAPTER 4 Decimals

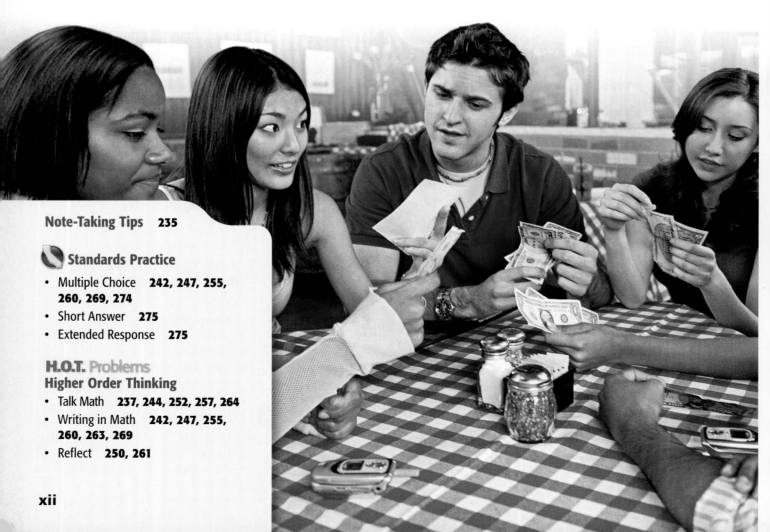

CHAPTER 5 Exponents and Roots

Note-Taking Tips **277**

CHAPTER 6

Ratios, Rates, Proportion, and Percent

CHAPTER 7 Algebra on the Coordinate Plane

Table of Contents

Note-Taking Tips   365

Standards Practice
- Multiple Choice   **370, 379, 386, 392, 398, 405, 410**
- Short Answer   **411**
- Extended Response   **411**

H.O.T. Problems
Higher Order Thinking
- Talk Math   **368, 375, 384, 390, 395, 400**
- Writing in Math   **370, 373, 379, 386, 392, 398, 405**
- Reflect   **371, 387, 399**

CHAPTER 8 Geometry Basics

Note-Taking Tips **413**

 Standards Practice

- Multiple Choice **419, 425, 433, 438, 447, 456, 461, 468**
- Short Answer **469**
- Extended Response **469**

H.O.T. Problems
Higher Order Thinking

- Talk Math **417, 423, 430, 436, 441, 454, 457**
- Writing in Math **419, 424, 433, 438, 446, 449, 456, 461**
- Reflect **427, 428, 439, 450**

Student Handbook

California Mathematics Correlation Standards

California Content Standards, Algebra Readiness, Correlated to *California Algebra Readiness*, ©2008

California Content Standards for Algebra Readiness

🔑 = **Key Standards defined by Mathematics Framework for California Public Schools**
* = **Standard assessed on the California High School Exit Exam (CAHSEE)**

Part A: Grade 7 and Standards

Standard	Text of Standard	Primary Citations	Supporting Citations
NUMBER SENSE			
🔑 **7NS1.2*** Topic 4	Add, subtract, multiply, and divide rational numbers (integers, fractions, and terminating decimals) and take positive rational numbers to whole-number powers.	22–24, 70–75, 126–147, 186–218, 243–247, 251–263, 291–292	220–226, 264–269, 278–283, 330–334
7NS1.3* Topic 4	Convert fractions to decimals and percents and use these representations in estimations, computations, and applications.	236–247, 251–260, 318–328, 336–340	304–309
🔑 **7NS1.5** Topic 4	Know that every rational number is either a terminating or a repeating decimal and be able to convert terminating decimals into reduced fractions.	34–36, 236–242	304–309
7NS2.1* Topic 4	Understand negative whole-number exponents. Multiply and divide expressions involving exponents with a common base.	285–290	299–303
ALGEBRA AND FUNCTION			
7AF1.1* Topic 6	Use variables and appropriate operations to write an expression, an equation, an inequality, or a system of equations or inequalities that represents a verbal description (e.g., three less than a number, half as large as area A).	60–69, 91–92	114–118, 336–340
🔑 **7AF1.3** Topic 6	Simplify numerical expressions by applying properties of rational numbers (e.g., identify, inverse, distributive, associative, commutative) and justify the process used.	77–89, 93–105, 128–145, 243–247, 251–255	91–92, 264–269
7AF2.1* Topic 4	Interpret positive whole-number powers as repeated multiplication and negative whole-number powers as repeated division or multiplication by the multiplicative inverse. Simplify and evaluate expressions that include exponents.	22–24, 70–75, 278–292	299–303
🔑 **7AF3.3*** Topic 8	Graph linear functions, noting that the vertical change (change in y-value) per unit of horizontal change (change in x-value) is always the same and know that the ratio ("rise over run") is called the slope of a graph.	388–398	494–495
🔑 **7AF3.4*** Topic 8	Plot the values of quantities whose ratios are always the same (e.g., cost to the number of an item, feet to inches, circumference to diameter of a circle). Fit a line to the plot and understand that the slope of the line equals the ratio of the quantities.	372–381	382–386
🔑 **7AF4.1*** Topic 6	Solve two-step linear equations and inequalities in one variable over the rational numbers, interpret the solution or solutions in the context from which they arose, and verify the reasonableness of the results.	149–153, 220–226, 264–269	330, 334, 336–348
🔑 **7AF4.2*** Topic 6	Solve multi-step problems involving rate, average speed, distance, and time or a direct variation.	330–356	382–386

Standard	Text of Standard	Primary Citations	Supporting Citations
MEASUREMENT AND GEOMETRY			
⚷ **7MG1.3*** Topic 8	Use measures expressed as rates (e.g., speed, density) and measures expressed as products (e.g., person-days) to solve problems; check the units of the solutions; and use dimensional analysis to check the reasonableness of the answer.	382–386	316–322, 330–335
⚷ **7MG3.3*** Topic 7	Know and understand the Pythagorean theorem and its converse and use it to find the length of the missing side of a right triangle and the lengths of other line segments and, in some situations, empirically verify the Pythagorean theorem by direct measurement.	400–405	420–425
MATHEMATICAL REASONING			
7MR1.0	Students make decisions about how to approach problems:	60–64, 91–92, 146–147, 184–185, 262–263, 291–292, 355–356, 371–372, 420–427, 451	496
7MR1.1*	Analyze problems by identifying relationships, distinguishing relevant from irrelevant information, identifying missing information, sequencing and prioritizing information, and observing patterns.	38–40, 146–147, 262–263, 414–427, 434–438, 451	349–353, 374–379
7MR1.2*	Formulate and justify mathematical conjectures based on a general description of the mathematical question or problem posed.	60–64	146–147, 335–336
7MR1.3	Determine when and how to break a problem into simpler parts.	291–292	262–263
7MR2.0	Students use strategies, skills, and concepts in finding solutions:	60–64, 184–185, 262–263, 291–292, 440–449, 457–461	91–92, 146–147
7MR2.1*	Use estimation to verify the reasonableness of calculated results.	42–44, 206–212, 335–336	342–348, 400–405, 434–438, 440–447, 457–461
7MR2.2	Apply strategies and results from simpler problems to more complex problems.	291–292	262–263,
7MR2.3*	Estimate unknown quantities graphically and solve for them by using logical reasoning and arithmetic and algebraic techniques.	429–433, 440–447, 457–461	335–336
7MR2.4*	Make and test conjectures by using both inductive and deductive reasoning.	91–92	335–336
7MR2.5	Use a variety of methods, such as words, numbers, symbols, charts, graphs, tables, diagrams, and models, to explain mathematical reasoning.	98–99, 126–127, 162–163, 184–185, 248–249, 262–263, 304–305, 355–356, 372–373, 448–449	60–64, 91–92, 146–147, 184–185, 291–292, 448–449
7MR2.6	Express the solution clearly and logically by using the appropriate mathematical notation and terms and clear language; support solutions with evidence in both verbal and symbolic work.	146–147, 434–438	374–379
7MR2.7	Indicate the relative advantages of exact and approximate solutions to problems and give answers to a specified degree of accuracy.	42–44	264–269, 429–433
7MR2.8	Make precise calculations and check the validity of the results from the context of the problem.	91–92, 248–249	60–64, 146–147, 184–185, 262–263, 291–292, 355–356, 371–372, 448–449

Standard	Text of Standard	Primary Citations	Supporting Citations
7MR3.0	Students determine a solution that is complete and move beyond a particular problem by generalizing to other solutions:	380–381	60–64, 91–92, 146–147, 184–185, 262–263, 291–292, 371–372
7MR3.1	Evaluate the reasonableness of the solution in the context of the original situation.	380–381	91–92
7MR3.2	Note the method of deriving the solution and demonstrate a conceptual understanding of the derivation by solving similar problems.	60–64, 128–133	91–92, 146–147, 184–185, 262–263, 291–292, 355–356, 371–372, 448–449
7MR3.3*	Develop generalizations of the results obtained and the strategies used and apply them to new problem situations.	60–64, 126–127	91–92, 146–147, 184–185, 262–263, 291–292, 355–356, 371–372, 448–449

Part A: Algebra 1 Standards

Standard	Text of Standard	Primary Citations	Supporting Citations
ALGEBRA I (INTRODUCTORY EXAMPLES)			
ALG2.0 Topic 9	Students understand and use such operations as taking the opposite, finding the reciprocal, taking a root, and raising to a fractional power. They understand and use the rules of exponents [*excluding fractional powers*].	119–124, 149–153, 278–283, 293–303	220–226, 264–269
ALG4.0 Topic 9	Students simplify expressions before solving linear equations and inequalities in one variable, such as $3(2x - 5) + 4(x - 2) = 12$ [*excluding inequalities*].	70–75	98–105, 112–145
ALG5.0 Topic 9	Students solve multistep problems, including word problems, involving linear equations and linear inequalities in one variable and provide justification for each step [*excluding inequalities*].	149–153, 220–226, 262–269, 330–334, 336–353, 400–405	98–105

Part B: Referenced Standards from Foundational Skills and Concepts

Standard	Text of Standard	Primary Citations	Supporting Citations
Topic 1: Whole Numbers			
NUMBER SENSE			
3NS1.3	Identify the place value for each digit in numbers to 10,000.	10–12	34–36
3NS1.5	Use expanded notation to represent numbers.	10–12	34–36
Topic 2: Operations on Whole Numbers			
NUMBER SENSE			
4NS3.1	Demonstrate an understanding of, and the ability to use, standard algorithms for the addition and subtraction of multidigit numbers.	14–16	128–133, 243–247
4NS3.2	Demonstrate and understanding of, and the ability to use, standard algorithms for multiplying a multidigit number by a two-digit number and for dividing a multidigit number by a one-digit number; use relationships between them to simplify computations and to check results.	18–20	141–145, 251–260

Standard	Text of Standard	Primary Citations	Supporting Citations
ALGEBRA AND FUNCTIONS			
🔑 **2AF1.1**	Use the commutative and associative rules to simplify mental calculations and to check results.	77–83	100–105
3AF1.5	Recognize and use the commutative and associative properties of multiplication	77–83	100–105
5AF1.3	Know and use the distributive property in equations and expressions with variables.	84–89	100–105
Topic 3: Rational Numbers			
NUMBER SENSE			
🔑 **5NS1.4**	Determine the prime factors of all numbers through 50 and write the numbers as the product of their prime factors by using exponents to show multiples of a factor (e.g., $24 = 2 \times 2 \times 2 \times 3 = 2^3 \times 3$).	26–28, 177–182	482
🔑 **6NS1.1**	Compare and order positive and negative fractions, decimals, and mixed numbers and place them on a number line.	164–175, 304–309	323–328
Topic 4: Operations on Rational Numbers			
NUMBER SENSE			
🔑 **6NS1.4**	Calculate given percentages of quantities and solve problems involving discounts at sales, interest earned, and tips.	342–348	355–356
🔑 **6NS2.0**	Students calculate and solve problems involving addition, subtraction, multiplication, and division.	126–127, 323–328, 342–348	14–20, 42–44, 243–263, 342–348
6NS2.1	Solve problems involving addition, subtraction, multiplication, and division of positive fractions and explain why a particular operation was used for a given situation.	186–218	220–226
6NS2.2	Explain the meaning of multiplication and division of positive fractions and perform the calculations $\left(\text{e.g., } \frac{5}{8} \div \frac{15}{16} = \frac{5}{8} \times \frac{16}{15} = \frac{2}{3}\right).$	186–197	220–226
Topic 5: Symbolic Notation			
ALGEBRA AND FUNCTIONS			
🔑 **4AF1.2**	Interpret and evaluate mathematical expressions that now use parentheses.	70–75	77–89, 100–105
🔑 **4AF1.3**	Use parentheses to indicate which operation to perform first when writing expressions containing more than two terms and different operations.	70–75	77–92, 100–105
5AF1.0	Students use variables in simple expressions, compute the value of the expression for specific values of the variable, and plot and interpret the results.	65–69, 293–303, 342–348	70–75, 100–105, 128–145
6AF1.0	Students write verbal expressions and sentences as algebraic expressions and equations; they evaluate algebraic expressions, solve simple linear equations, and graph and interpret their results.	65–69, 262–263, 293–303, 393–398	330–348
🔑 **6AF1.1**	Write and solve one-step linear equations in one variable.	114–118, 149–153, 220–226	330–334, 336–340
Topic 6: Equations and Functions			
ALGEBRA AND FUNCTIONS			
🔑 **4AF1.5**	Understand that an equation such as $y = 3x + 5$ is a prescription for determining a second number when a first number is given.	393–398	388–392
🔑 **4AF2.0**	Students know how to manipulate equations.	114–118, 149–153, 220–226	65–69, 262–263, 293–303, 393–398

Standard	Text of Standard	Primary Citations	Supporting Citations
 4AF2.1	Know and understand that equals added to equals are equal.	114–118, 149–153, 220–226, 264–269	220–226, 264–269
4AF2.2	Know and understand that equals multiplied by equals are equal.	114–118, 149–153, 220–226, 264–269	330–348
Topic 7: The Coordinate Plane			
ALGEBRA AND FUNCTIONS			
5AF1.4	Identify and graph ordered pairs in the four quadrants of the coordinate plane.	366–370	371–381
MEASUREMENT AND GEOMETRY			
4MG2.0	Students use two-dimensional coordinate grids to represent points and graph lines and simple figures:	366–370	371–381, 429–438
4MG2.1	Draw the points corresponding to linear relationships on graph paper (e.g., draw 10 points on the graph of the equation $y = 3x$ and connect them by using a straight line).	371–372	393–398
4MG2.2	Understand that the length of a horizontal line segment equals the difference of the x-coordinates.	393–405	400–405, 429–433
4MG2.3	Understand that the length of a vertical line segment equals the difference of the y-coordinates.	393–405	400–405, 429–433

Get Ready!

Algebra is a way of using math to describe relationships and solve problems. In this course, you will be preparing for algebra. But, did you know you've been getting ready for algebra since you started learning math? Use the Prerequisite Skills section to refresh your memory about things you have learned in other math classes.

1 Whole Numbers

2 Place Value

3 Addition and Subtraction

4 Multiplication and Division

5 Exponents

6 Prime Factorization

7 Fractions

8 Decimals

9 Least Common Multiple

10 Estimation

11 Angles

12 Probability

Let's Go!

Prerequisite Skills Pre-Test

▷ Vocabulary and Concept Check

1. In the number sentence $27 \div 3 = 9$, identify the dividend, divisor, and quotient.

2. How are whole numbers different from counting numbers?

3. What is the least common multiple of two numbers?

4. Which of the numbers 2, 3, 4, 5, and 6 are prime? How do you know?

5. The probability of pulling a red marble out of a bag is certain. What colors of marbles are in the bag? Explain.

6. Write an example of a terminating decimal.

7. In 5^3, which number is the base? Which number is the exponent?

8. Is the marked angle acute, right, or obtuse? How do you know?

▷ Skills Check

Compare. Replace the ● with >, <, or = to make a true statement.

9. 2 ● 6
10. 10 ● 4
11. 19 ● 16
12. 17 ● 25

Order each set of numbers from least to greatest.

13. 5, 3, and 9
14. 8, 12, and 7

Write the value of each underlined digit.

15. <u>7</u>38
16. 9<u>4</u>6
17. 3<u>0</u>2
18. 4,<u>9</u>05

Write each number in expanded form.

19. 384
20. 1,629
21. 2,055
22. 7,020

Add.

23. $253 + 341$
24. $2,433 + 1,324$
25. $4,583 + 768$
26. $3,697 + 1,875$

Subtract.

27. $896 - 453$

28. $7,859 - 3,341$

29. $635 - 388$

30. $5,732 - 2,975$

Multiply.

31. $53 \cdot 72$

32. $147 \cdot 21$

33. $206 \cdot 61$

34. $328 \cdot 32$

Divide.

35. $756 \div 6$

36. $3,375 \div 3$

37. $612 \div 9$

38. $4,776 \div 4$

Simplify.

39. 7^2

40. 4^3

41. 9^1

42. 2^5

Write each product using exponents.

43. $3 \cdot 3 \cdot 3 \cdot 3$

44. $5 \cdot 5 \cdot 5 \cdot 5 \cdot 5 \cdot 5$

45. $y \cdot y$

46. $k \cdot k \cdot k \cdot k \cdot k$

Write the prime factorization of each number.

47. 18

48. 25

49. 30

50. 36

Write the fraction.

51. What fraction of the fruit is apples?

52. What fraction of the pizza has been eaten?

Identify the decimal as terminating or repeating. Write any terminating decimal as a fraction.

53. 0.91

54. $0.\overline{53}$

55. 0.003

56. $0.\overline{1}$

Find the least common multiple of each pair of numbers.

57. 2 and 4

58. 3 and 5

59. 4 and 6

Round to the nearest ten to estimate the answer.

60. $2,662 + 4,913$

61. $584 - 229$

62. $37 \cdot 21$

63. $638 \div 321$

Use compatible numbers to estimate the answer.

64. $275 \div 3$

65. $419 \div 6$

Classify each triangle as acute, obtuse, or right.

66.

67.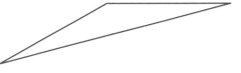

68. Classify the angles of the corners of this poster.

A fruit bowl has 7 purple plums and 1 red plum. Find the probability of selecting each of the following at random.

69. a purple plum

70. a red plum

A bag has red and blue marbles. You pull one marble from the bag without looking and return it. The tally chart shows the results of 10 pulls.

Outcome	Tally	Total
red	卌 II	7
blue	III	3

Based on the results, identify each event as certain, likely, unlikely, or impossible.

71. pulling a red marble

72. pulling a blue marble

▷ Problem-Solving Check

73. TRAVEL A pilot turns her plane halfway around to prepare to land. How many degrees does the plane rotate?

74. FINANCE Mala's bank statement shows that she has $300 + $70 + $5 in her account. How much money is in her account?

75. PACKAGING Keirnan buys plastic forks in sets of 8. He buys plastic spoons in sets of 10. He sets tables with 1 fork and 1 spoon at each place. How many sets of each should he buy so that no forks or spoons are left over?

LANDSCAPING Oscar is planting 325 tulips and 435 daisies in a public garden.

76. How many flowers will Oscar plant in all?

77. He has planted 228 flowers so far. How many more flowers does he have left to plant?

78. **SPORTS** This table shows the number of points scored by the top scorers in a basketball game. Who scored the greater number of points, Berto or Joe?

Player	Points Scored
Lon	9
Berto	16
Silvio	13
Joe	18
Rod	15

79. **BAKING** Tyrese is baking 9 cakes for a party. He has baked 5 so far. What fraction of the cakes does he have left to bake?

80. **HOBBIES** Dalila's design for a square quilt has 6 squares on each side. So, the total number of squares is 6^2. How many squares will Dalila need to make the entire quilt?

81. **STATISTICS** The chance of this spinner landing on red is 0.333. . . . Is this decimal terminating or repeating? Explain.

82. **HOBBIES** Elio built a jewelry box with a volume of 18 cubic inches. The volume equals the length times the width times the height. Each side is a prime number of inches long. What are the lengths of the sides?

FASHION A designer makes 36 dresses for each of 24 stores. She puts the dresses in four equal groups to store in a warehouse.

83. How many dresses did she make in all?

84. How many dresses are in each group?

85. **TRAVEL** Joanne drove 437 miles on Saturday. She drove 382 miles on Sunday. Round to the nearest ten to estimate the total number of miles she drove on both days.

86. **PROBABILITY** A bag holds green and yellow beads. You pull a bead from the bag and replace it. After 10 pulls, you have pulled a total of 9 green beads and 1 yellow bead. Is it certain, likely, unlikely, or impossible that you will pull a yellow bead?

Whole Numbers

Vocabulary

counting number (p. 6)

whole number (p. 6)

Standard 1NS1.1 **Count, read, and write whole numbers to 100.**

Standard 1NS1.2 **Compare and order whole numbers to 100 by using the symbols for less than, equal to, or greater than (<, =, >).**

 The What: I will define, identify, and compare whole numbers.

 The Why: Whole numbers are used every day. They are used to label lockers and show grades on a test.

Suppose someone asked you to count to 10. You would say, "1, 2, 3, 4, 5, 6, 7, 8, 9, 10." This makes sense because the set of numbers 1, 2, 3, and so on are called the **counting numbers.**

Notice that this set of numbers does not include zero. The set of **whole numbers** includes all of the counting numbers and zero.

counting numbers

0, 1, 2, 3, . . .

whole numbers

Whole Numbers	NOT Whole Numbers
5	−5
29	2.8
43	$\frac{2}{5}$
394	$6\frac{1}{3}$

Some fractions, mixed numbers, and decimals are not whole numbers. However, the whole number 1 can be written as a fraction $\left(\frac{2}{2}\right)$ or as a decimal (1.000).

EXAMPLE Identify Whole Numbers

① Identify the whole numbers in the set $\frac{3}{4}$, 12, 4.1.

$\frac{3}{4}$ $\frac{3}{4}$ is a fraction. It is not a whole number.

12   12 is a whole number.

4.1   4.1 is a decimal. It is not a whole number.

So, 12 is the whole number.

Your Turn

Identify the whole numbers in each set.

a. $\frac{5}{8}$, 2.3, 19

b. 3.6, 0, $4\frac{1}{2}$

> **Remember!**
> A mixed number "mixes" a counting number and a fraction.
> $1\frac{1}{2}$ is a mixed number.

Numbers to the right on a number line are **greater than** numbers to the left. Numbers to the left are **less than** numbers to the right.

Two symbols, > meaning "greater than" and < meaning "less than," are used to show these relationships.

EXAMPLE Compare Whole Numbers

② Compare 5 ● 8. Replace the ● with >, <, or = to make a true statement.

8 is greater than 5.   8 is to the right of 5.

8 > 5   The > symbol means greater than.

Your Turn

Compare. Replace the ● with >, <, or = to make a true statement.

c. 3 ● 9

d. 6 ● 4

> **Talk Math**
> **WORK WITH A PARTNER** Say an example of a whole number and a number that is not a whole number. Have your partner check your examples.

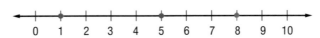
EXAMPLE **Order Whole Numbers**

3 Order the numbers 5, 1, 8 from least to greatest.

```
 ◄──┼──◆──┼──┼──┼──◆──┼──┼──◆──┼──┼──►
    0  1  2  3  4  5  6  7  8  9  10
```

1 is the least number.   1 is farthest to the left.

8 is the greatest number.   8 is farthest to the right.

The numbers in order from least to greatest are 1, 5, 8.

Your Turn

Order each set of numbers from least to greatest.

e. 9, 2, 6   f. 15, 11, 8

Skills, Concepts, and Problem Solving

Vocabulary Review
counting number
whole number

Example 1
(page 7)

VOCABULARY

1. Explain how whole numbers and counting numbers are alike. Explain how they are different.

Example 1
(page 7)

Identify the whole numbers in each set.

2. $7.7, 15, \frac{7}{10}$ 3. $3\frac{3}{4}, 0.5, 8$

4. $36, 8.3, \frac{5}{6}$ 5. $\frac{1}{12}, 5\frac{7}{8}, 9$

Example 2
(page 7)

Compare. Replace ● with >, <, or = to make a true statement.

6. 8 ● 2   7. 6 ● 3

8. 7 ● 10   9. 14 ● 9

10. 11 ● 18   11. 19 ● 25

Example 3
(page 8)

Order each set of numbers from least to greatest.

12. 1, 7, 2   13. 4, 9, 5

14. 8, 6, 3   15. 10, 12, 7

16. **FINANCE** Sally has $20 in her bank account. Lazara has $15 in his bank account. Who has the greater amount? Explain.

17. **SCHOOL** Zacaria's locker number is 195. Rhonda's locker number is 214. Li's locker number is 98. Order these locker numbers from least to greatest.

18. *Talk Math* Work with a partner to summarize the rules for recognizing a whole number.

19. *Writing in Math* Write three whole numbers less than 15. Explain how you know that they are whole numbers.

Progress Check 1
(Lesson 1)

▷ Vocabulary and Concept Check

> counting numbers (p. 6)   whole numbers (p. 6)

Choose the term that *best* completes each statement. (Example 1)

1. The _____?_____ include 1, 2, 3, 4, 5, . . .
2. The _____?_____ include 0, 1, 2, 3, 4, 5, . . .

▷ Skills Check

Identify the whole numbers in each set. (Example 1)

3. $7, 5.3, \frac{3}{10}$

4. $\frac{1}{6}, 0.4, 16$

5. $8\frac{7}{8}, 75, 0.9$

6. $0.64, 24\frac{3}{4}, 382$

Compare. Replace ● with >, <, or = to make a true statement. (Example 2)

7. 7 ● 3

8. 2 ● 4

9. 8 ● 9

10. 11 ● 8

11. 15 ● 17

12. 18 ● 20

Order each set of numbers from least to greatest. (Example 3)

13. 7, 2, 4

14. 9, 1, 0

15. 10, 5, 16

16. 20, 14, 11

▷ Problem-Solving Check

17. **FITNESS** Tyler bikes 14 miles on Saturday. He bikes 9 miles on Sunday. On which day does he bike farther? Explain. (Example 2)

18. **ASTRONOMY** Saturn has 35 named moons. Jupiter has 38 named moons. Neptune has 9 named moons. Order these planets from least number of named moons to greatest number of named moons. Explain. (Example 3)

19. **NUMBER THEORY** Identify the only number that is a whole number, but not a counting number. (Example 1)

20. **H.O.T.** Problem Fidel says that all counting numbers are whole numbers. Jewel says that all whole numbers are counting numbers. Who do you think is correct? Explain. (Example 1)

21. **REFLECT** Explain how you know when a number is not a whole number. (Example 1)

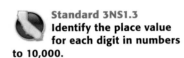

Place Value

Vocabulary

digit (p. 10)

place value (p. 10)

expanded form (p. 10)

Standard 3NS1.3
Identify the place value for each digit in numbers to 10,000.

Standard 3NS1.5 **Use expanded notation to represent numbers.**

 The What: I will identify the place value for digits in numbers. I will write numbers in expanded form.

The Why: Place value and expanded form represent the structure of our number system. They are useful for counting money and determining prices.

The number 5,037 has the **digits** 5, 0, 3, and 7. The value of each digit varies by its place in the number. **Place value** is the value given to a digit because of its place in a number. The 3 in 5,037, for example, is in the tens place. Each digit of a number can be put into a place-value chart.

Place-Value Chart for 5,037

1000	100	10	1
thousands	hundreds	tens	ones
5	0	3	7

You can use a place-value chart to write numbers in **expanded form.** Expanded form shows the value of each digit.

$$5,037 = 5 \cdot 1,000 + 0 \cdot 100 + 3 \cdot 10 + 7 \cdot 1$$

$$= 5,000 + 0 + 30 + 7$$

thousand   hundreds   tens   ones

1 Write the value of the underlined digit in 3,099.

1000	100	10	1
thousands	hundreds	tens	ones
3	0	9	9

3,099 The digit 3 is in the thousands place.

3,000 3 thousands is 3,000.

Your Turn

Write the value of each underlined digit.

a. 6,<u>4</u>81

b. 1,59<u>8</u>

Talk Math

WORK WITH A PARTNER Work together to write a 4-digit number. Take turns pointing to a digit in the number. Then name the place and value of that digit.

EXAMPLE Expanded Form

2 Write 2,657 in expanded form.

1000	100	10	1
thousands	hundreds	tens	ones
2	6	5	7

2 thousands = 2,000
6 hundreds = 600
5 tens = 50
7 ones = 7
2,657 = 2,000 + 600 + 50 + 7

Remember!

Numbers written in expanded form show the sum of the values of the places.

Your Turn

Write each number in expanded form.

c. 3,851

d. 9,402

Examples 1–2
(page 11)

VOCABULARY

1. What is the place value of a digit in a number?

2. Explain how to write a number in expanded form.

Example 1
(page 11)

Write the value of each underlined digit.

3. <u>9</u>50

4. 2<u>7</u>9

5. <u>5</u>03

6. 28<u>0</u>

7. 2,95<u>3</u>

8. 4,<u>9</u>78

9. 6,4<u>3</u>8

10. <u>6</u>,492

11. 3,1<u>6</u>7

12. 5,<u>4</u>09

Example 2
(page 11)

Write each number in expanded form.

13. 487

14. 5,925

15. 1,402

16. 65

17. 984

18. 503

19. 2,954

20. 6,075

21. 4,003

22. 8,060

23. **FINANCE** Phil saw online that he has $400 + $90 + $3 in his bank account. How much money is in his account?

24. **FINANCE** The second digit in 1,?95 on Juana's check was erased. The value of the digit was 200. What was the digit?

25. *Writing in Math* Write a four-digit number with 1, 4, 5, and 8. Write the number so that the value of the 5 is 500 and the value of the 4 is 40. Name the value of the other digits in the number.

26. *Talk Math* Discuss with a partner how to write the number 295 in expanded form.

Progress Check 2

(Lesson 2)

▷ Vocabulary and Concept Check

> digit (p. 10)   place value (p. 10)
> expanded form (p. 10)

Choose the term that *best* completes each statement. (Examples 1–2)

1. The ___?___ of 5 in 4,592 is the hundreds.

2. The expression $300 + 80 + 4$ shows a number in ___?___.

▷ Skills Check

Write the value of each underlined digit. (Example 1)

3. <u>8</u>55   4. 42<u>9</u>

5. 6,349 (6 underlined)   6. 1,7<u>3</u>5

7. 2,<u>0</u>41   8. 3,2<u>5</u>7

Write each number in expanded form. (Example 2)

9. 448   10. 305

11. 1,583   12. 4,655

13. 2,093   14. 7,030

▷ Problem-Solving Check

15. **FINANCE** Kenyon is checking his bank statement. He writes the amount $4,500 in expanded form. Show what he wrote. (Example 2)

16. **NUMBER THEORY** A three-digit number has 4 tens. There are twice as many ones as tens. There is one more hundred than the number of tens. What is the number? (Examples 1–2)

17. **H.O.T.** Problem Create a place-value chart for a number that has 5 tens, 3 hundreds, 8 ones, and 7 thousands. (Examples 1–2)

18. **REFLECT** Explain how you find the value of a digit in the thousands place. (Example 1)

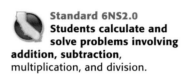

Addition and Subtraction

Vocabulary

add (p. 14)

sum (p. 14)

addend (p. 14)

subtract (p. 15)

difference (p. 15)

 Standard 6NS2.0 Students calculate and solve problems involving addition, subtraction, multiplication, and division.

 The What: I will add and subtract numbers.

 The Why: Add to find the total cost of two or more items. Subtract to find how much money you have left after making a purchase.

Here are some tips to help you **add.**

SYMBOL ➕	
Operation	**Addition**
Add to:	• combine groups • find a total or **sum**
Here Is How:	**1.** Line up the digits of the **addends** by their place values. **2.** Add the columns. Start with the ones and move right to left. **3.** Rename as necessary.
Key Words:	plus, sum of, increased by, added to, more than, total of, in all

EXAMPLE Add Whole Numbers

1 Find 538 + 294.

$$
\begin{array}{r}
\ \overset{1}{5}\ 3\ 8 \\
+\ 2\ 9\ 4 \\
\hline
2
\end{array}
\longrightarrow
\begin{array}{r}
\overset{1}{}\ \overset{1}{5}\ 3\ 8 \\
+\ 2\ 9\ 4 \\
\hline
3\ 2
\end{array}
\longrightarrow
\begin{array}{r}
\overset{1}{}\ \overset{1}{5}\ 3\ 8 \\
+\ 2\ 9\ 4 \\
\hline
8\ 3\ 2
\end{array}
$$

• Align the digits.
• Add the ones. 8 + 4 = 12.
• Rename 12 ones as 1 ten and 2 ones.

• Add the tens. 3 + 9 + 1 = 13
• Rename 13 tens as 1 hundred and 3 tens.

• Add the hundreds. 5 + 2 + 1 = 8

So, 538 + 294 = 832.

Your Turn
Add.

a. 487 + 819

b. 2,564 + 4,182

Here are some tips to help you **subtract.**

SYMBOL —

Operation	Subtraction
Subtract to:	• find a **difference** • compare two numbers • take away and find how many are left • find how many more are needed to complete a group or set
Here Is How:	1. Line up the digits of the numbers by their place values. 2. Subtract the columns. Start with the ones and move right to left. 3. Rename as necessary.
Key Words:	minus, difference, subtracted from, decreased by, less than, fewer than

EXAMPLE **Subtract Whole Numbers**

2 Find $849 - 583$.

$$
\begin{array}{r} 8\ 4\ 9 \\ -\ 5\ 8\ 3 \\ \hline 6 \end{array}
\qquad
\begin{array}{r} {}^{7}\ {}^{14} \\ 8\ \cancel{4}\ 9 \\ -\ 5\ 8\ 3 \\ \hline 6\ 6 \end{array}
\qquad
\begin{array}{r} {}^{7}\ {}^{14} \\ \cancel{8}\ \cancel{4}\ 9 \\ -\ 5\ 8\ 3 \\ \hline 2\ 6\ 6 \end{array}
$$

• Align the digits.
• Subtract the ones.
  $9 - 3 = 6$

• Notice $8 > 4$.
• Rename 8 hundreds and 4 tens as 7 hundreds and 14 tens.
• Subtract the tens. $14 - 8 = 6$

• Subtract the hundreds. $7 - 5 = 2$

So, $849 - 583 = 266$.

Your Turn
Subtract.

c. $697 - 328$

d. $3,248 - 1,793$

Talk Math

WORK WITH A PARTNER Write one number each. Discuss how you would add the numbers. Then discuss how you would subtract the lesser number from the greater one.

3 **SCHOOL** The seventh-grade class at Davis Middle School has 1,103 students. The eighth-grade class has 1,427 students. How many students are there in all?

Add 1,103 and 1,427 to find the total.

$$\begin{array}{r} {\scriptstyle 1} \\ 1\,1\,0\,3 \\ +\ 1\,4\,2\,7 \\ \hline 2\,5\,3\,0 \end{array}$$ Add the digits.
Add and rename as necessary.

There are 2,530 students altogether.

Your Turn

Remember!
Line up digits when adding or subtracting numbers.

e. **SALES** A theater sold 948 tickets on Monday and 3,395 tickets on Tuesday. How many tickets were sold in all?

f. **CONSTRUCTION** A park needs 855 feet of fencing. So far, 685 feet have been built. How many feet of fencing still need to be built?

Skills, Concepts, and Problem Solving

Vocabulary Review
add
sum
addend
subtract
difference

Examples 1–3
(pages 14–16)

VOCABULARY

1. Give an example of when you would add to find an answer.

2. Give an example of when you would subtract to find an answer.

Example 1
(page 14)

Add.

3. 325 + 231
4. 867 + 324
5. 3,485 + 2,213
6. 1,294 + 2,336
7. 2,524 + 378
8. 806 + 5,579

Example 2
(page 15)

Subtract.

9. 948 − 716
10. 481 − 253
11. 6,807 − 418
12. 2,493 − 1,252
13. 4,950 − 2,477
14. 7,654 − 7,561

Example 3
(page 16)

15. **BUSINESS** Sydney worked 1,750 hours last year. This year, she worked 1,975 hours. How many hours has she worked in the last two years?

16. **COMMUNITY SERVICE** Lana needs 1,500 volunteers for a charity walk-a-thon. She has signed up 823 people so far. How many more volunteers does she need?

17. **Talk Math** Discuss with a partner how you would add 87 and 109.

18. **Writing in Math** Write two numbers with a sum of 950. Then, use one of these numbers to write a subtraction sentence with a difference of 300.

Progress Check 3
(Lesson 3)

▷ Vocabulary and Concept Check

add (p. 14)   subtract (p. 15)
addend (p. 14)   sum (p. 14)
difference (p. 15)

Choose the term that *best* completes each statement. (Examples 1–2)

1. _____?_____ to find the difference between numbers.

2. _____?_____ to find the total of two groups.

▷ Skills Check

Add. (Example 1)

3. $613 + 285$

4. $1,214 + 4,512$

5. $3,814 + 3,349$

6. $4,769 + 858$

Subtract. (Example 2)

7. $628 - 515$

8. $4,586 - 2,245$

9. $3,583 - 1,756$

10. $8,405 - 8,283$

▷ Problem-Solving Check

11. **INTERIOR DESIGN** Jared is putting carpet in his 1,355 square foot basement. He has already covered 490 square feet. How many square feet does he have left to cover? (Example 3)

12. **POPULATION** Weston had a population of 8,143. Over the next year, the town's population increased by 795. What is the population now? (Example 3)

13. **TRAVEL** Ayana is flying 3,249 miles to another city. She has flown 1,165 miles so far. How many more miles does she have to go? (Example 3)

14. **H.O.T. Problem** Find the two numbers that have a sum of 1,000 and a difference of 310. (Example 3)

15. **REFLECT** How are addition and subtraction of multi-digit numbers alike? How are they different? (Examples 1–3)

Multiplication and Division

Vocabulary

multiply (p. 18)

factor (p. 18)

product (p. 18)

divide (p. 19)

dividend (p. 19)

divisor (p. 19)

quotient (p. 19)

 **Standard 6NS2.0
Students calculate and
solve problems involving**
addition, subtraction,
multiplication, and division.

 The What: I will multiply and divide numbers.

The Why: Multiply to find the total cost when buying several items with the same price. Divide to equally share the cost of a movie with friends.

Here are some tips to help you **multiply.**

SYMBOL	✕ • ( )
Operation	**Multiplication**
Multiply to:	combine equal-sized groups into one group
Here Is How:	Multiply the **factors.** The result is the **product.** Factors Product $25 \cdot 12 = 300$ 1. Line up the digits of the factors by their place values. 2. Find partial products. Multiply each digit, right to left, in one factor by the other factor. 3. Rename as necessary. 4. Add partial products.
Key Words:	times, product, multiplied by, at, of, factor

EXAMPLE   Multiply by a Two-Digit Number

① Find $317 \cdot 24$.

$$
\begin{array}{r} 3\ 1\ 7 \\ \times\ \ \ \ 2\ 4 \\ \hline 1\ 2\ 6\ 8 \end{array}
\quad\rightarrow\quad
\begin{array}{r} 3\ 1\ 7 \\ \times\ \ \ \ 2\ 4 \\ \hline 1\ 2\ 6\ 8 \\ +\ 6\ 3\ 4\ 0 \end{array}
\quad\rightarrow\quad
\begin{array}{r} 3\ 1\ 7 \\ \times\ \ \ \ 2\ 4 \\ \hline 1\ 2\ 6\ 8 \\ +\ 6\ 3\ 4\ 0 \\ \hline 7\ 6\ 0\ 8 \end{array}
$$

First partial product:
• Multiply 317 by 4.
• $317 \cdot 4 = 1,268$

Second partial product:
• Multiply 317 by 20.
• $317 \cdot 20 = 6,340$

Product:
• Add 1,268 and 6,340.

$317 \cdot 24 = 7,608$

Your Turn
Multiply.

a. $318 \cdot 24$

b. $298 \cdot 32$

Here are some tips to help you **divide.**

SYMBOL	÷ / ⟋
Operation	**Division**
Divide to:	separate a group into equal-sized groups
Here Is How:	Divide the **dividend** by the **divisor.** The result is the **quotient.** $$\begin{array}{r} 35 \leftarrow \text{Quotient} \\ \text{Divisor} \rightarrow 8)\overline{280} \leftarrow \text{Dividend} \end{array}$$ **1.** Divide each digit in the dividend by the largest possible number, moving from left to right. **2.** Multiply the digit in the quotient by the divisor. Keep digits aligned. **3.** Subtract. **4.** Bring down the next number.
Key Words:	divided by, the quotient of, the ratio of, per

EXAMPLE Divide by a One-Digit Number

2 Find 752 ÷ 8.

$$8)\overline{752} \quad \rightarrow \quad 8)\overline{752} \atop -72 \quad \rightarrow \quad \begin{array}{r} 9 \\ 8)\overline{752} \\ -72 \\ \hline 3 \end{array}$$

Divide by the largest possible number. Since 7 is smaller than 8, move to the right.

Multiply. 8 · 9 = 72

Subtract. 75 − 72 = 3

$$\begin{array}{r} 9\ 4 \\ 8)\overline{7\ 5\ 2} \\ -7\ 2\downarrow \\ \hline 3\ 2 \\ -3\ 2 \\ \hline 0 \end{array}$$

Bring down the next digit, 2, and repeat the steps. Divide 32 ÷ 8 = 4. Multiply 8 · 4 = 32. Subtract 32 − 32 = 0. Since there are no other digits to bring down, the answer is complete.

752 ÷ 8 = 94

Your Turn
Divide.

c. 4,615 ÷ 5

d. 5,404 ÷ 7

Talk Math

WORK WITH A PARTNER Write a number. Discuss how you multiply that number by 28 and divide it by 4.

③ GARDENING Brady has 750 flower bulbs. He has 6 flower beds. If he put an equal number of bulbs in each flower bed, how many bulbs are in each bed?

$750 \div 6 = 125$ Divide to solve.

He put 125 bulbs in each bed.

Remember!

Be careful to line up digits as you work through a division problem.

Your Turn

e. **READING** Joan has 1,064 pages left to read in a book. She has 8 days to finish the book before school starts. If she reads the same number each day, how many pages per day must she read?

Skills, Concepts, and Problem Solving

Examples 1–3
(pages 18–20)

VOCABULARY

Vocabulary Review
multiply
factor
product
divide
dividend
divisor
quotient

1. Give an example of when you would multiply to find an answer.

2. Explain the roles of the dividend and divisor in a division problem.

Example 1
(page 18)

Multiply.

3. $29 \cdot 37$

4. $89 \cdot 74$

5. $106 \cdot 82$

6. $312 \cdot 23$

7. $1,003 \cdot 15$

8. $1,194 \cdot 26$

Example 2
(page 19)

Divide.

9. $936 \div 4$

10. $672 \div 7$

11. $1,256 \div 2$

12. $5,043 \div 3$

13. $3,174 \div 6$

14. $2,565 \div 9$

Example 3
(page 20)

15. **TRAVEL** A small plane that takes people to an island carries 8 passengers. Suppose 56 people want to travel to the island on one day. What is the fewest number of trips the small plane will need to make to the island that day?

16. **BAKING** A bakery makes 624 muffins. Suppose the muffins are equally placed in 8 boxes. How many muffins are placed in each box?

17. **Talk Math** Where and why do you use 0 as a placeholder when you multiply $385 \cdot 61$? Discuss this question with a partner.

18. **Writing in Math** Explain how partial products can be used to multiply by two-digit numbers. You can use the example $273 \cdot 46$ in your explanation.

Progress Check 4

(Lesson 4)

▷ Vocabulary and Concept Check

> divide (p. 19)
> dividend (p. 19)
> divisor (p. 19)
> factor (p. 18)
>
> multiply (p. 18)
> product (p. 18)
> quotient (p. 19)

Choose the term that *best* completes each statement. (Examples 1–3)

1. When finding a quotient, you divide the ___?___ into the ___?___.

2. ___?___ to combine equal-sized groups.

▷ Skills Check

Multiply. (Example 1)

3. $263 \cdot 13$

4. $342 \cdot 21$

5. $111 \cdot 54$

6. $1,022 \cdot 36$

7. $243 \cdot 41$

8. $485 \cdot 73$

Divide. (Example 2)

9. $1,374 \div 3$

10. $6,055 \div 7$

11. $4,096 \div 8$

12. $6,894 \div 6$

13. $3,588 \div 6$

14. $9,108 \div 9$

▷ Problem-Solving Check

15. **FITNESS** Each day Frank bikes 9 miles. Suppose the odometer on Frank's bike reads 2,295 miles after one year. How many days did he ride? (Example 3)

16. **ADVERTISING** A company's monthly budget for ads is $5,940. They divide the budget equally among print, TV, and radio. How much does the company spend for each type of ad? (Example 3)

17. **BUSINESS** Four friends did a catering job together. They split their earnings of $7,436 evenly. How much did each friend earn? (Example 3)

18. **H.O.T.** Problem A number is divided by 23. Then the quotient is multiplied by 6. That product is 1,536. Explain how to find the original number. (Examples 1–3)

19. **REFLECT** How do you know when to multiply and when to divide to solve a word problem? (Examples 1–3)

Exponents

Vocabulary

exponent (p. 22)

base (p. 22)

power (p. 22)

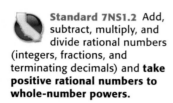 **Standard 7NS1.2** Add, subtract, multiply, and divide rational numbers (integers, fractions, and terminating decimals) and **take positive rational numbers to whole-number powers.**

Standard 7AF2.1 Interpret positive whole-number powers as repeated multiplication and negative whole-number powers as repeated division or multiplication by the multiplicative inverse. **Simplify and evaluate expressions that include exponents.**

 The What: I understand and interpret positive exponents.

 The Why: Exponents can be used when indicating square footage of a bedroom or cubic units of cereal in a box.

An **exponent** tells how many times a number, called the **base,** is used as factor. A base raised to an exponent is called a **power.**

$$\overbrace{2}^{\text{exponent}}{}^{4} = 2 \cdot 2 \cdot 2 \cdot 2 = 16$$

base

2^4 is read "two raised to the **fourth** power."

To simplify a number with an exponent, multiply the base by itself the number of times equal to the exponent. For example, $8^3 = 8 \cdot 8 \cdot 8 = 512$. The base 8 was used 3 times as a factor to find the product 512.

 EXAMPLE Simplify Exponents

1 **Simplify 3^4.**

$3^4 = 3 \cdot 3 \cdot 3 \cdot 3$ Use 3 as a factor 4 times.

$\quad = 81$ Multiply.

Your Turn
Simplify.

a. 2^3 b. 6^2 c. 4^4

To show identical factors as a number with an exponent, make the factor the base. Then, count the number of times the factor is used and write this number as an exponent. For example, $4 \cdot 4 \cdot 4 = 4^3$, because the base 4 was used as a factor 3 times, as indicated by the exponent.

EXAMPLES Write Exponents

Write the product using exponents.

2 $5 \cdot 5 \cdot 5$

$5 \cdot 5 \cdot 5$

5 is used as a factor 3 times.

5^3

5 is the base.

3 is the exponent.

3 $m \cdot m \cdot m \cdot m$

$m \cdot m \cdot m \cdot m$

m is used as a factor 4 times.

m^4

m is the base.

4 is the exponent.

Your Turn

Write each product using exponents.

d. $2 \cdot 2 \cdot 2$

e. $b \cdot b \cdot b \cdot b \cdot b$

Talk Math

WORK WITH A PARTNER Each choose a number that is 6 or less. Write a power using one number as the base and the other number as the exponent.

4 TECHNOLOGY A 6-bit computer monitor can display 2^6 colors. How many colors can the monitor display?

$2^6 = 2 \cdot 2 \cdot 2 \cdot 2 \cdot 2 \cdot 2$ Use 2 as a factor 6 times.

    $= 64$ Multiply.

The monitor can display 64 different colors.

Your Turn

f. **MUSIC** The choir is in four equal rows. The total number of singers is 4^2. How many singers are in the choir?

Skills, Concepts, and Problem Solving

Vocabulary Review
exponent
base
power

Examples 1–3
(pages 22–23)

VOCABULARY

1. What is the name for the 2 in the number 2^5?

2. 2^5 is read "two raised to the fifth ___?___."

Example 1
(page 22)

Simplify.

3. 4^2 4. 3^2 5. 4^3 6. 2^5

7. 8^2 8. 6^3 9. 8^1 10. 5^4

Examples 2–3
(page 23)

Write each product using exponents.

11. $4 \cdot 4 \cdot 4$ 12. $2 \cdot 2 \cdot 2 \cdot 2$

13. $7 \cdot 7 \cdot 7 \cdot 7 \cdot 7$ 14. $3 \cdot 3 \cdot 3 \cdot 3 \cdot 3 \cdot 3$

15. $1 \cdot 1 \cdot 1 \cdot 1 \cdot 1 \cdot 1 \cdot 1$ 16. $x \cdot x$

17. $k \cdot k \cdot k$ 18. $d \cdot d \cdot d \cdot d \cdot d$

Example 4
(page 24)

19. **HOBBIES** There are eight squares along each side of a chessboard, so there are a total of 8^2 squares. How many squares are on a chessboard?

20. **Talk Math** What are two ways to write the number 256 in exponential form? Discuss this with a partner.

21. *Writing in Math* Does $3^4 = 4^3$? Explain.

22. **H.O.T. Problem** A 4-bit computer monitor can display 2^4 colors. How many colors can an 8-bit monitor display?

Progress Check 5

(Lesson 5)

▷ Vocabulary and Concept Check

> base (p. 22)   power (p. 22)
>
> exponent (p. 22)

Choose the term that *best* completes each statement. (Examples 1–4)

1. In 4^3, 4 is the ____?____.

2. In 4^3, 3 is the ____?____.

▷ Skills Check

Simplify. (Example 1)

3. 3^3 **4.** 2^4 **5.** 4^2

6. 9^2 **7.** 3^5 **8.** 5^3

Write each product using exponents. (Examples 2–3)

9. $5 \cdot 5 \cdot 5 \cdot 5$ **10.** $8 \cdot 8 \cdot 8$

11. $6 \cdot 6 \cdot 6 \cdot 6 \cdot 6$ **12.** $2 \cdot 2 \cdot 2 \cdot 2 \cdot 2 \cdot 2 \cdot 2$

13. $b \cdot b \cdot b$ **14.** $z \cdot z \cdot z \cdot z \cdot z \cdot z$

▷ Problem-Solving Check

15. HOBBIES A bingo board has five squares along each side. It has a total of 5^2 squares. How many total squares are there on a bingo board? (Example 4)

16. LANDSCAPING The area of a square garden that is 6 feet on one side is 6^2 square feet. What is the area in square feet of a square garden that is 7 feet on one side? (Example 4)

17. H.O.T. Problem Write the missing number: $4^{\blacksquare} = 2^6$

18. REFLECT Explain what the expression 3^x means.

Prime Factorization

Vocabulary

factor (p. 26)

prime number (p. 26)

factor tree (p. 26)

prime factorization (p. 26)

exponent (p. 26)

 Standard 5NS1.4 Determine the prime factors of all numbers through 50 and write the numbers as the product of their prime factors by using exponents to show multiples of a factor.

The What: I will write the prime factors of a number.

The Why: Computer software uses very large prime numbers to keep information safe. When you secure e-mail, online purchases, or downloads, prime number "keys" make the data private.

A **factor** is a number or expression that is multiplied by another to yield a product. For example, the factors of 6 are 1, 2, 3, and 6 because $1 \cdot 6 = 6$ and $2 \cdot 3 = 6$.

A **prime number** is a whole number greater than 1 that has exactly two factors, 1 and itself. The number 5 is a prime number because its only factors are 1 and 5.

PRIME FACTORIZATION

Use a **factor tree** to find the prime factors of any number.

Keep factoring until all of the factors are prime numbers.

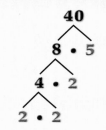

factor tree

Prime factorization shows a number as a product of only prime numbers.

$$40 = 2 \cdot 2 \cdot 2 \cdot 5 = 2^3 \cdot 5$$

Exponents indicate repeating factors.

$$2^3 = 2 \cdot 2 \cdot 2$$

exponent

base

Study Tip

Some common prime numbers are 2, 3, 5, 7, 11, 13, 17, and 19.

To find prime factors, start with any two factors. Continue finding factors until only prime factors are listed.

EXAMPLE **Prime Factorization**

1 **Write the prime factorization of 36.**

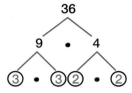

or

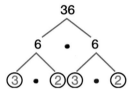

or

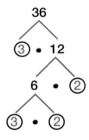

$$36 = 3 \cdot 3 \cdot 2 \cdot 2 = 3^2 \cdot 2^2$$

Your Turn

Write the prime factorization of each number.

a. 20

b. 42

Talk Math

WORK WITH A PARTNER Discuss with a partner how factor trees help you find the prime factorization of a number.

2 INDUSTRY A storage company rents a locker with a volume of 12 cubic feet. The volume equals the length times the width times the height. Each side of the locker is a prime number of feet long. What is the length of each side?

$12 = 2 \cdot 2 \cdot 3$ Find the prime factors of 12.

The sides of the locker are 2 feet, 2 feet, and 3 feet.

Your Turn

c. RETAIL A packaging store sells a box with a volume of 18 cubic feet. The volume equals the length times the width times the height. Each side of the box is a prime number of feet long. What is the length of each side?

Skills, Concepts, and Problem Solving

Vocabulary Review
factor
prime number
factor tree
prime factorization
exponent

Examples 1–2
(pages 27–28)

VOCABULARY

1. What is a prime number? Use the term *factor* in your definition.

2. List all prime numbers less than 20.

3. Explain how to make a factor tree.

4. What is the prime factorization of a number?

Example 1
(page 27)

Write the prime factorization of each number.

5. 15	**6.** 20	**7.** 21	**8.** 25
9. 27	**10.** 32	**11.** 35	**12.** 40
13. 42	**14.** 45	**15.** 50	**16.** 36

Example 2
(page 28)

17. **ART** An artist makes a photo box so that each side has a length that is a prime number. The total volume is 30 cubic inches. The volume is the product of the length times the width times the height. What is the length of each side in inches?

18. **HOBBIES** Hugo builds a dog house with a volume of 20 cubic feet. The volume equals the length times the width times the height. Each side is a prime number of feet long. What is the length of each side?

19. *Talk Math* How many different factor trees can you write for the number 48? Discuss this with a partner.

20. *Writing in Math* How is prime factorization different from writing a number's factors?

Progress Check 6

(Lesson 6)

▷ Vocabulary and Concept Check

> exponent (p. 26)
> factor (p. 26)
> factor tree (p. 26)
>
> prime factorization (p. 26)
> prime number (p. 26)

Choose the term that *best* completes each statement. (Examples 1–2)

1. In 2^8, 8 is a(n) ____?____.

2. 2, 3, 5, 7, and 11 are examples of ____?____.

3. The ____?____ of a number shows the number as a product of primes.

▷ Skills Check

Write the prime factorization of each number. (Example 1)

4. 8
5. 12
6. 18
7. 28
8. 30
9. 39
10. 44
11. 46
12. 49

▷ Problem-Solving Check

13. **ART** Len built a music box with a volume of 70 cubic inches (volume = length · width · height). Each side is a prime number of inches long. What is each side's length? (Example 2)

14. **PACKAGING** Luisa's storage unit has 45 cubic feet of space (volume = length · width · height). Each side is a prime number of feet long. How long is each side? (Example 2)

15. **H.O.T. Problem** Students use combination locks to secure school lockers. Most combinations use three numbers. Suppose the greatest possible number is 40. Create five locker combinations that use only prime numbers. (Examples 1–2)

16. **NUMBER SENSE** What is the missing number in this prime factorization of 48? (Example 1)

$$48 = 2^{\blacksquare} \cdot 3$$

17. **REFLECT** Write the numbers 1 to 50 in rows of 10. Circle the prime numbers. Describe two patterns that you see. (Example 1)

Fractions

Vocabulary

fraction (p. 30)

numerator (p. 30)

denominator (p. 30)

 Standard 2NS4.2 Recognize fractions of a whole and parts of a group.

Standard 2NS4.3 Know that when all fractional parts are included, such as four-fourths, the result is equal to the whole and to one.

 The What: I will recognize and name fractions.

The Why: Fractions identify parts of a whole, like the amount of uneaten pizza. Fractions identify parts of a group, like the part of a soccer team traveling to an away game.

Suppose a pizza is cut into eight slices. The slices are the same size. You eat three slices. What part of the pizza did you eat?

A **fraction** can show the amount of pizza you ate. Fractions represent equal parts of a whole or group.

	FRACTIONS	
Term	**Numerator**	**Denominator**
Words	The **numerator** is the number above the bar in a fraction. It tells the part of the whole or group.	The **denominator** is the number below the bar in a fraction. It tells the total number of equal parts.
Numbers	$\frac{3}{8}$ ← numerator ← denominator	You ate 3 slices. There are 8 equal slices.
Picture		

EXAMPLE **Write Fractions**

1 What fraction of the apples is green?

7 green apples

10 apples in all

$\frac{7}{10}$ of the apples are green.

Your Turn

a. What fraction of the pie
has been eaten?

b. What fraction of the
beads is blue?

When the numerator and denominator of a fraction are the
same number, the fraction represents a whole and is equal to 1.

$\frac{2}{2} = 1$

$\frac{4}{4} = 1$

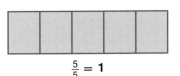

$\frac{5}{5} = 1$

Talk Math

WORK WITH A PARTNER Say and write a fraction that
is less than 1. Listen as your partner says and writes a
fraction equal to 1. Work together to write a fraction
greater than 1.

Real-World EXAMPLE

2 MUSIC A band is giving seven concerts. Four of the
concerts are sold out. What fraction of the concerts is
not sold out?

$7 - 4 = 3$

$\frac{3}{7}$ ← concerts not sold out

   ← total concerts

$\frac{3}{7}$ of the total number of concerts is not sold out.

Your Turn

c. **POLITICAL SCIENCE** Lakita is making ten posters for a
school election. She has finished seven so far. What
fraction of the posters does she still have to make?

Examples 1–2
(page 31)

VOCABULARY

1. Explain what the fraction $\frac{3}{5}$ means.

2. Write two fractions with denominators of 7.

Example 1
(page 31)

3. What fraction of the pie has been eaten?

4. What fraction of the pizza is left?

5. What fraction of the windows is open?

Example 2
(page 31)

6. **GOVERNMENT** Suppose there are nine people on the city council. Seven people voted in favor of a bill. What fraction voted in favor of the bill?

7. **ART** A jeweler makes a bracelet with a "green, green, yellow, green, green, yellow" bead pattern. What fraction of the beads is green?

8. **COOKING** Joyce uses five of the dozen eggs in a carton to make an omelet. What fraction of the eggs is left?

9. **Talk Math** Discuss with a partner how the fractions $\frac{3}{5}$ and $\frac{5}{3}$ compare to 1.

10. **Writing in Math** On a quiz, students were asked to draw a picture for $\frac{1}{3}$. One student drew the following.

Does this picture represent $\frac{1}{3}$? Why or why not?

Progress Check 7

(Lesson 7)

▷ Vocabulary and Concept Check

denominator (p. 30)   numerator (p. 30)

fraction (p. 30)

Choose the term that *best* completes each statement.

1. A _____?_____ represents part of a whole or group.

2. In $\frac{4}{9}$, 4 is the _____?_____.

3. In $\frac{3}{5}$, 5 is the _____?_____.

▷ Skills Check

Write the fraction. (Example 1)

4. What fraction of the pizza has pepperoni?

5. What fraction of the coins is pennies?

▷ Problem-Solving Check

6. **BUSINESS** At a shirt factory, 2 of 15 shirts are dress shirts. What fraction of the shirts is dress shirts? (Example 2)

7. **FASHION** A designer decorates belts with a "blue, blue, yellow, blue, yellow" bead pattern. What fraction of the beads is yellow? (Example 2)

8. **FOOD SERVICE** Cal is making 30 meals for a party. He has made 19 meals so far. What fraction of meals does he have left to make? (Example 2)

9. **H.O.T. Problem** How are the two pictures alike? How are the pictures different? (Examples 1–2)

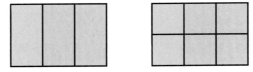

10. **REFLECT** Draw three pictures that represent $\frac{1}{4}$. Explain why each picture shows this fraction.

Decimals

Vocabulary

decimal (p. 34)

decimal point (p. 34)

terminating decimal
(p. 34)

repeating decimal (p. 34)

Standard 7NS1.5 Know that every rational number is either a terminating or repeating decimal and be able to convert terminating decimals into reduced **fractions.**

Standard 4NS1.0 Students understand the place value of whole numbers and decimals to two decimal places and how whole numbers and decimals relate to fractions. Students use the concepts of negative numbers.

The What: I will understand decimals and identify decimals as terminating or repeating.

The Why: Money amounts are written using decimals.

Place value is the value given to a digit because of its place in a number. The 3 in 537, for example, is in the tens place.

Decimals extend the place-value system to values less than 1. The **decimal point** shows where this happens.

This is the place-value chart for 23.64.

10	1	0.1	0.01
tens	ones	tenths	hundredths
2	3 .	6	4

Study Tip

Three dots at the end of a decimal indicate that the repeating digits go on forever.

These dots are called ellipses.

A horizontal bar over the repeating numbers means the same thing.

$0.333 \ldots = 0.\overline{3}$

DECIMALS

Term	Words	Numbers
terminating decimals	A **terminating decimal** has a finite (certain) number of digits to the right of the decimal point.	2.65 3.06 0.94
repeating decimals	A **repeating decimal** has a pattern of digits that repeats to the right of the decimal point.	0.333 . . . 1.414141 . . . 0.7656565 . . .

Identify the decimal as terminating or repeating.

① **0.36363636 . . .**

0.36363636 . . .   The sequence 36 repeats forever.

So, the decimal is repeating.

② **0.25**

0.25   There are a finite (certain) number of digits after the decimal point.

So, the decimal is terminating.

Your Turn

Identify the decimal as terminating or repeating.

a. 0.33

b. 0.55555 . . .

EXAMPLE Rewriting Terminating Decimals

Rewrite the terminating decimal as a fraction.

③ **0.73**

0.73 is 73 hundredths or $\frac{73}{100}$.

The terminating decimal 0.73 is rewritten as the fraction $\frac{73}{100}$.

Your Turn

Rewrite the terminating decimal as a fraction.

c. 0.33

Talk Math

WORK WITH A PARTNER How do you know if 0.77 is a terminating or repeating decimal? If it is a terminating decimal, how do you write it as a fraction? Discuss these issues with a partner.

4 SCHOOL Madison answered 0.89 of the questions on a test correctly. What fraction of the questions did she answer correctly?

Rewrite the terminating decimal 0.89 as a fraction.

0.89 is 89 hundredths or $\frac{89}{100}$.

Madison answered $\frac{89}{100}$ of the questions correctly.

Your Turn

d. **POLITICAL SCIENCE** In the last election, 0.43 of the people voted. What fraction of the people voted?

Skills, Concepts, and Problem Solving

> **Vocabulary Review**
> decimal
> decimal point
> terminating decimal
> repeating decimal

Examples 1–3
(page 35)

VOCABULARY

1. How is the number 32 different from 32.4? Use the words *decimal* and *decimal point* in your answer.

2. Make a place-value chart for 23.57.

3. Write an example of a terminating decimal.

4. Write an example of a repeating decimal.

Examples 1–2
(page 35)

Identify the decimal as terminating or repeating.

5. 0.545454 . . .   6. 0.87   7. 0.009

8. 0.231   9. 1.21212 . . .   10. 3.3 . . .

Example 3
(page 35)

Rewrite each terminating decimal as a fraction.

11. 1.59   12. 0.87   13. 0.231

Example 4
(page 36)

14. **STATISTICS** The chance of tossing a 6 on a six-sided number cube is 0.1666. . . . Is this decimal terminating or repeating? Explain.

15. **POPULATION** A recent United States census showed that about 0.51 of Americans were women. What fraction of Americans were women?

16. **Talk Math** What kind of decimals do $\frac{1}{6}$ and $\frac{1}{3}$ become? Discuss this with a partner.

17. **Writing in Math** Explain the difference between terminating and repeating decimals. How are they alike? How are they different?

Progress Check 8

(Lesson 8)

▷ Vocabulary and Concept Check

> decimal (p. 34)   repeating decimal (p. 34)
> decimal point (p. 34)   terminating decimal (p. 34)

Choose the term that *best* completes each statement.

1. 8.55 is a _____?_____.

2. 0.5676767 . . . is a _____?_____.

▷ Skills Check

Identify the decimal as terminating or repeating. (Examples 1–2)

3. 0.111 . . .   4. 0.13   5. 0.3535 . . .

6. 0.040404 . . .   7. 0.69 . . .   8. 0.69

Rewrite each terminating decimal as a fraction. (Example 3)

9. 1.33   10. 0.007   11. 0.13

▷ Problem-Solving Check

12. **STATISTICS** When you flip a coin, the chance of landing on heads three times in a row is 0.125. Is this decimal terminating or repeating? Explain. (Example 4)

13. **POPULATION** A recent United States census showed that about 0.49 of Americans were men. What fraction of Americans were men? (Example 4)

14. **H.O.T.** Problem Sort these numbers into two groups. Explain how you sorted and justify your reason for sorting the way you did. Can you then sort the numbers a different way? (Example 1)

    23.23   23.23 . . .

    0.23   23.0023

    0.232323   0.23 . . .

15. **REFLECT** What is the difference between a terminating decimal and a repeating decimal? (Examples 1–2)

Prerequisite 9

Least Common Multiple

Vocabulary

multiple (p. 38)

common multiple (p. 38)

least common multiple (LCM) (p. 38)

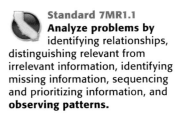

Standard 7MR1.1 **Analyze problems by** identifying relationships, distinguishing relevant from irrelevant information, identifying missing information, sequencing and prioritizing information, and **observing patterns.**

Standard 2NS3.1 **Use** repeated addition, arrays, and **counting by multiples** to do multiplication.

The What: I will identify the least common multiple of two or more numbers.

The Why: Least common multiples are used to solve problems, such as determining how many packages of hot dogs and hot dog buns to buy.

You find **multiples** of a number when you multiply by the whole numbers. For example, find the multiples of 2 by multiplying 2 by 1, 2, 3, 4 . . . .

$$1 \cdot 2 = 2 \qquad 3 \cdot 2 = 6$$
$$2 \cdot 2 = 4 \qquad 4 \cdot 2 = 8$$

Multiples of **2 = 2, 4, 6, 8, . . .**

MULTIPLES

Term	Common Multiple	Least Common Multiple
Words	A **common multiple** of two or more numbers is a number that is a multiple of each of the numbers.	The **least common multiple** of two or more numbers is the least of their common multiples, excluding zero.
Numbers	**9:** 9, 18, 27, (36), 45, 54, 63, (72), 81, 90, 99, (108), . . . **12:** 12, 24, (36), 48, 60, (72), 84, 96, (108), 120, . . . The numbers 9 and 12 have common multiples of 36, 72, and 108. If the lists continued, they would show many more common multiples.	**9:** 9, 18, 27, (36), 45, 54, 63, 72, 81, 90, . . . **12:** 12, 24, (36), 48, 60, 72, 84, 96, . . . The least common multiple of 9 and 12 is 36.

① **Find the LCM of 4 and 6.**

4: 4, 8, 12, 16, 20, 24	List multiples of each number.
6: 6, 12, 18, 24, 30, 36	

4: 4, 8, ⑫ 16, 20, 24	Identify common multiples.
6: 6, ⑫ 18, 24, 30, 36	Circle the LCM.

The LCM of 4 and 6 is 12.

Your Turn

Find the LCM of each pair of numbers.

a. 2 and 8　　　　　　**b.** 3 and 5　　　　　　**c.** 4 and 10

Study Tip

Read backward to remember the meaning of LCM.

M stands for: Find MULTIPLES.

C stands for: Identify COMMON multiples.

L stands for: Choose the LEAST (smallest) one.

Real-World EXAMPLE

② **SHOPPING** Hot dogs come in packages of eight and hot dog buns come in packages of six. You put one hot dog in each bun. What is the fewest number of packages of each that you can buy so that none will be left over?

8: 8, 16, ㉔ 32, 40, 48	Find the LCM of 8 and 6.
6: 6, 12, 18, ㉔ 30, 36, 48	

Hot dogs	Divide the LCM by each
$24 \div 8 = 3$	number to find the number
	of each item needed.
Buns	
$24 \div 6 = 4$	

You need three packages of hot dogs and four packages of buns.

Your Turn

d. **SHOPPING** A card shop sells cards in packages of eight and envelopes in packages of ten. One card will go in each envelope. What is the fewest number of packages of each item you can buy so that no cards or envelopes will be left over?

WORK WITH A PARTNER Choose two different numbers that are 12 or less. Work together to find the least common multiple.

Skills, Concepts, and Problem Solving

Examples 1–2
(page 39)

VOCABULARY

1. How do you know that 5, 10, 15, and 20 are multiples of 5?

2. Explain the difference between a common multiple and the least common multiple.

Example 1
(page 39)

Find the LCM of each set of numbers.

3. 2 and 6
4. 5 and 10
5. 3 and 4

6. 3 and 5
7. 5 and 8
8. 6 and 10

9. 8 and 12
10. 2, 3, and 4
11. 3, 5, and 6

Example 2
(page 39)

12. **SHOPPING** A store sells 4 hamburger patties in a package. It sells buns in packages of 8. Tom puts 1 hamburger in each bun. What is the fewest number of each he can buy so that none will be left over?

13. **PACKAGING** Pat buys paper plates in sets of 10. She buys paper cups in sets of 12. She sets the table with 1 cup for every plate. How many sets of plates and cups should she buy so that no cups or plates are left over?

14. **PACKAGING** Charo is making gift bags for a party. Each bag gets 1 hat and 1 noisemaker. Bags come in packages of 3. Hats come in packages of 5. Noisemakers come in packages of 6. What is the fewest number of each item Charo can buy so that no pieces are left over?

15. **Talk Math** Work with a partner to find the least common multiple of two numbers greater than 12.

16. **Writing in Math** Paloma's cousin said that the product of two numbers is always their least common multiple. Paloma disagreed. Who do you think is correct? Why?

Progress Check 9

(Lesson 9)

▷ Vocabulary and Concept Check

> common multiple (p. 38)   multiple (p. 38)
> least common multiple
>   (LCM) (p. 38)

Choose the term that *best* completes each statement.

1. The numbers 6, 12, and 18 are _____?_____ of 2 and 3.

2. The number 6 is the _____?_____ of 2 and 3.

▷ Skills Check

Find the LCM of each set of numbers. (Example 1)

3. 4 and 5   4. 3 and 8

5. 6 and 8   6. 4 and 10

7. 10 and 12   8. 2, 4, and 5

▷ Problem-Solving Check (Example 2)

9. **FASHION** Jake buys shirts in packages of four. Ties come in packages of three. He wants to match each tie with one shirt. What is the fewest number of each he must buy so that no shirts or ties are left over?

10. **ART** A gallery buys frames in sets of ten and prints in sets of four. One print goes in each frame. What is the fewest number of sets of each the gallery can buy so that no frames or prints are left over?

11. **PACKAGING** A teacher buys pens in packages of eight. He buys pencils in packages of ten. He plans to give each of his students one pen and one pencil. What is the fewest number of packages of each item he must buy so that no pens or pencils are left over?

12. **H.O.T.** Problem The LCM of two numbers is 24. One number is two more than the other. What are the numbers?

13. **REFLECT** Explain how you would find the LCM of two numbers.

Estimation

Vocabulary

estimate (p. 42)

round (p. 42)

compatible numbers (p. 43)

 Standard 6NS2.0 Students calculate and solve problems involving addition, subtraction, multiplication, and division.

Standard 4NS1.3 Round whole numbers through the millions **to the nearest ten,** hundred, thousand, ten thousand, and hundred thousand.

 The *What*: I will use different strategies to estimate answers to problems.

The *Why*: You can estimate the number of CDs you can buy or the number of chairs you need to set up for a concert.

To **estimate** means to find an answer that is close to the exact answer. You estimate when an exact answer is not needed. You estimate to check exact calculations. Two types of estimation are rounding and using compatible numbers.

To **round** means to approximate a number to a given place value. In this lesson, you will round to the nearest ten.

ROUNDING TO THE NEAREST TEN

- Look at the ones digit.
- If the digit is less than 5, round down.
- If the digit is 5 or more, round up.

EXAMPLES Estimate Using Rounding

Round to the nearest ten to estimate each answer.

1 $326 + 142$

Look at the ones digits.

$326 \rightarrow 330$ 6 > 5 Round up.

$142 \rightarrow 140$ 2 < 5 Round down.

$330 + 140 = 470$

Add rounded numbers.

$326 + 142$ is about 470.

2 $89 \div 31$

Look at the ones digits.

$89 \rightarrow 90$ 9 > 5 Round up.

$31 \rightarrow 30$ 1 < 5 Round down.

$90 \div 30 = 3$

Divide rounded numbers.

$89 \div 31$ is about 3.

Your Turn

Round to the nearest ten to estimate each answer.

a. $476 - 353$

b. $84 \cdot 29$

Compatible numbers are numbers in a problem that are easy to work with mentally. The numbers 720 and 90 are compatible numbers for division because $72 \div 9 = 8$.

EXAMPLE Estimate Using Compatible Numbers

3 Use compatible numbers to estimate $274 \div 9$.

274 is close to 270.

Think $27 \div 9 = 3$.   The numbers 270 and 9 are compatible.

$270 \div 9 = 30$ Divide using compatible numbers.

$274 \div 9$ is about 30.

Your Turn

Use compatible numbers to estimate each answer.

c. $323 \div 4$ d. $489 \div 7$

Remember!

Exact answers are not needed when you see the words about, around, approximate, and estimate.

Real-World EXAMPLES

4 COMPETITION There are 431 dogs and 375 cats at a cat and dog show. About how many animals are there in all? Find the sum of 431 and 375 to solve.

$431 \to 400$ $3 < 5$ Round down.

$375 \to 400$ $7 > 5$ Round up.

$400 + 400 = 800$ Add rounded numbers.

There are about 800 animals at the cat and dog show.

5 BUSINESS May stocks books at a bookstore. She has 206 books to put on 7 shelves. She needs to put about the same number of books on each shelf. About how many books belong on each shelf? Divide 206 by 7 to solve.

$206 \div 7$ Write the problem as if finding an exact answer.

$210 \div 7$ Replace 206 with a number compatible with 7.

$210 \div 7 = 30$ Divide using compatible numbers.

May should put about 30 books on each shelf.

Your Turn

e. **SHOPPING** Mirna spent $89 on a jacket and $43 on a backpack. About how much did she spend?

f. **MONEY MATTERS** Three friends earn a total of $275. About how much will each friend have when they evenly divide the money?

WORK WITH A PARTNER Suppose that you have an estimation problem with addition. Would you be more likely to round or use compatible numbers? Explain. What about a problem with subtraction? Multiplication? Division?

Skills, Concepts, and Problem Solving

Examples 1–5
(pages 42–43)

VOCABULARY

1. Identify two ways to estimate an answer. Explain the differences between the two strategies.

2. Give an example of compatible numbers and explain why they are compatible.

Examples 1–2
(page 42)

Round to the nearest ten to estimate each answer.

3. $243 + 395$

4. $16 \cdot 38$

5. $347 \div 69$

6. $638 + 185$

7. $437 - 196$

8. $77 \div 38$

9. $65 \cdot 51$

10. $955 - 927$

11. $895 - 434$

12. $687 + 513$

Example 3
(page 43)

Use compatible numbers to estimate each answer.

13. $184 \div 3$

14. $489 \div 5$

15. $208 \div 4$

16. $362 \div 6$

17. $349 \div 5$

18. $625 \div 9$

Examples 4–5
(page 43)

19. **SCHOOL** A school has 32 classrooms. Each room has 77 students. About how many students are in the school?

20. **HOBBIES** Keisha has 355 beads. If she makes 4 necklaces, about how many beads can she use in each necklace?

21. **BUSINESS** A car dealer plans to sell 625 cars this year. She has sold 472 so far. Round the numbers to the nearest ten to estimate the number of cars the dealer has left to sell.

22. *Talk Math* Is it necessary to round both numbers when you subtract to estimate? If you only round one of the numbers, which one would it be? Discuss these issues with a partner.

23. *Writing in Math* On a quiz, Ivana estimated $72 \cdot 81$ as $70 \cdot 80$. Will her estimated product be less than or greater than the exact answer? Explain.

Progress Check 10
(Lesson 5)

▷ Vocabulary and Concept Check

> compatible numbers (p. 43)   rounds (p. 42)
> estimate (p. 42)

Choose the term that *best* completes each statement.

1. A(n) _____?_____ is close to the exact answer.
2. To the nearest ten, 376 _____?_____ to 380.

▷ Skills Check

Round to the nearest ten to estimate. (Examples 1–2)

3. 496 − 151   4. 28 · 51
5. 482 + 935   6. 334 ÷ 27

Use compatible numbers to estimate. (Example 3)

7. 274 ÷ 3   8. 202 ÷ 7

▷ Problem-Solving Check

9. **LANDSCAPING** Alek has 325 pounds of sand to put into 4 sandboxes. He needs to put about the same amount in each sandbox. About how many pounds of sand will go in each one? (Examples 4–5)

10. **BUSINESS** Company X offers Jeb a $510 bonus if he works for them. Company Y offers a $795 bonus. Round the amounts to the nearest ten to estimate how much more money Company Y offers. (Example 4)

11. **GEOGRAPHY** Del is flying from New York to Denver through Chicago. The distance from New York to Chicago is 731 miles. The distance from Chicago to Denver is 885 miles. About how far is the entire trip? Round the distances to the nearest ten to estimate. (Example 4)

12. **FASHION** A designer sends 28 dresses to each of 39 stores. About how many dresses does she send in all? (Examples 4–5)

13. **REFLECT** How do you know when a problem is asking you to estimate?

14. **H.O.T.** Problem Create five different multiplication sentences that each round to 60 · 40.

Angles

Vocabulary

angle (p. 46)

vertex (p. 46)

right angle (p. 46)

acute angle (p. 46)

obtuse angle (p. 46)

Standard 4MG3.5 Know the definitions of a right angle, an acute angle, and an obtuse angle. Understand that 90°, 180°, 270°, and 360° are associated, respectively, with $\frac{1}{4}, \frac{1}{2}, \frac{3}{4}$, and full turns.

Standard 3MG2.4 Identify right angles in geometric figures or in appropriate objects and determine whether other angles are greater or less than a right angle.

The What: I will classify angles and identify different kinds of turns.

The Why: You can use angles when you play different sports, such as soccer, basketball, golf, or billiards. You must consider the angle of a shot, pass, or rebound to play each game well.

An **angle** is made where two rays meet at a common endpoint. The endpoint is called a **vertex**. An angle is measured in degrees. Ninety degrees, for example, is written 90°.

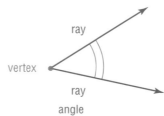

CLASSIFICATION OF ANGLES

Acute	Right	Obtuse
less than 90°	Equals 90°; the square inside the vertex of the angle indicates that this is a right angle.	greater than 90°; less than 180°

A **right angle** measures 90°. Each 90° turn takes you one-fourth of the way around a circle.

90°	$\frac{1}{4}$ turn
180°	$\frac{1}{2}$ turn
270°	$\frac{3}{4}$ turn
360°	full turn

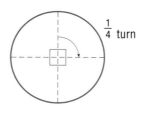

Study Tip

An index card has four right angles. Use an index card to determine whether angles are acute, right, or obtuse.

EXAMPLES Classify Angles

Classify each angle as acute, obtuse, or right.

① The angle is larger than a right angle and less than 180°.

It is an obtuse angle.

②  Each angle equals 90°.

Each of these is a right angle.

Your Turn

Classify each angle as acute, obtuse, or right.

a.

b.

Talk Math

WORK WITH A PARTNER Draw an angle and have your partner classify it as acute, right, or obtuse. Then have your partner draw an angle so you can classify it. Discuss how you know what kind of angles you and your partner have drawn.

Real-World EXAMPLE

③ ASTRONOMY Ilana rotates her telescope 270°. How far around a circle did the telescope turn?

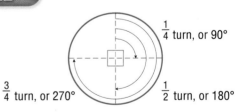

$270° = 90° \cdot 3$ Each 90° is a $\frac{1}{4}$ turn.

$\frac{3}{4}$ turn   Three 90° turns is a $\frac{3}{4}$ turn.

The telescope went $\frac{3}{4}$ around the circle.

Your Turn

c. TRANSPORTATION Kwang turns his car around 180°. How far around a circle did the car turn?

Vocabulary Review
angle
vertex
right angle
acute angle
obtuse angle

Examples 1–2
(page 47)

VOCABULARY

1. What part of an angle is the vertex?

2. Draw a right angle.

3. Compare and contrast acute and obtuse angles. How are they alike? How are they different?

Examples 1–2
(page 47)

Classify each indicated angle as acute, obtuse, or right.

4.

5.

6.

7.

8. angles in this yield sign

9. angles in this stop sign

Example 3
(page 47)

10. **TRANSPORTATION** A car makes a 90° turn. What kind of turn is this?

11. **DANCE** A dancer makes a 270° turn. What kind of turn is this?

12. **EARTH SCIENCE** Earth rotates one full turn each day. How many degrees does it rotate?

13. **Talk Math** Discuss with a partner how to classify the angle made by a wall and ceiling.

14. *Writing in Math* The expression "doing a one-eighty" can mean to change your mind or change direction. Explain how this expression relates to what you know about 180° and turns.

Progress Check 11
(Lesson 11)

▷ Vocabulary and Concept Check

> acute angle (p. 46) right angle (p. 46)
>
> angle (p. 46) vertex (p. 46)
>
> obtuse angle (p. 46)

Choose the term that *best* completes each statement. (Examples 1–2)

1. A(n) ____?____ is a 90° angle.

2. A(n) ____?____ has a measure greater than a right angle.

3. An angle is two rays that meet at a common endpoint called a ____?____.

▷ Skills Check

Classify each indicated angle as acute, obtuse, or right. (Examples 1–2)

4.   5.   6.

7. angles in notebook   8. angle 1

   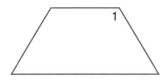

▷ Problem-Solving Check

9. **TRANSPORTATION** Lora rotates her steering wheel 90°. What kind of turn is this? (Example 3)

10. **ASTRONOMY** Larry rotates his telescope 180°. What kind of turn is this? (Example 3)

11. **H.O.T. Problem** Draw a figure that has two right angles, one acute angle, and one obtuse angle. Label each angle in the figure. (Examples 1–3)

12. **REFLECT** Explain what is meant by $\frac{1}{4}$ turn. (Examples 1–3)

Probability

Vocabulary

probability (p. 50)

certain (p. 50)

likely (p. 50)

unlikely (p. 50)

impossible (p. 50)

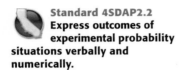

Standard 4SDAP2.2 Express outcomes of experimental probability situations verbally and numerically.

Standard 3SDAP1.1 Identify whether common events are certain, likely, unlikely, or impossible.

 The What: I will identify the probability of different events.

 The Why: Understanding probability helps people predict how likely or unlikely an event is.

Suppose you could pick an apple from this bowl of fruit without looking. Notice that 7 red apples and 3 green apples are in the bowl. What is the likelihood that you will pick a green one?

Probability is the likelihood that an event will occur. The bowl holds a total of 10 apples. P(green) represents the probability of picking a green apple.

$$\text{probability of green} = \frac{\text{number of green apples}}{\text{total number of apples}} = \frac{3}{10}$$

The probability of picking a green apple is 3 out of 10. There are four ways to describe the likelihood of an event occurring.

LIKELIHOOD OF EVENTS	
Certain events *must* occur.	You are sure (certain) to pick an apple.
Likely events have a *good chance* to occur.	You have a good chance (are likely) to pick a red apple.
Unlikely events are *possible*, but do not occur very often.	You have a lesser chance (are unlikely) to pick a green apple.
Impossible events *cannot* occur.	It is not possible (is impossible) to pick a banana.

Identify the Likelihood

A spinner has red, yellow, and blue sections. Identify each event as certain, likely, unlikely, or impossible.

1 blue

Only 1 of 12 spins will be blue.
So, a blue spin is unlikely.

2 white

The spinner has no white sections.
0 of 12 spins will be white.
So, a white spin is impossible.

Your Turn

Identify each spin as certain, likely, unlikely, or impossible.

a. red

b. green

EXAMPLES Find the Probability

Miguel has 5 white shirts, 2 blue shirts, and 3 green shirts in his closet. He picks a shirt without looking. Find the probability of each event.

3 green shirt

$\frac{3}{10}$ 3 green shirts
      10 shirts in all

The probability that Miguel will pick a green shirt is $\frac{3}{10}$.

4 blue shirt

$\frac{2}{10}$ 2 blue shirts
      10 shirts in all

$\frac{2}{10} = \frac{1}{5}$ Simplify.

The probability that Miguel will pick a blue shirt is $\frac{1}{5}$.

Your Turn

Clarissa has 2 white blouses, 3 blue blouses, and 6 green blouses in her closet. She picks a blouse without looking. Find the probability of each event.

c. green blouse

d. white blouse

Talk Math

WORK WITH A PARTNER Toss a coin 10 times and record the outcome each time. According to the results, what is the probability that the coin will land on heads?

Vocabulary Review
probability
certain
likely
unlikely
impossible

Examples 1–4
(page 51)

VOCABULARY

1. Explain the meaning of *probability* in your own words.

2. Which two vocabulary terms have opposite meanings?

Examples 1–2
(page 51)

A bag has black and white marbles. The tally chart shows the results of 15 pulls.

Outcome	Tally	Number
black	IIII	4
white	JHT JHT I	11

Based on the results, identify each event as certain, likely, unlikely, or impossible.

3. pulling a white marble

4. pulling a red marble

5. pulling either a black marble or a white marble

Examples 3–4
(page 51)

You pick one apple from this fruit bowl without looking. Find the probability of each event.

6. yellow apple

7. green apple

8. red apple

9. brown apple

Examples 3–4
(page 51)

You draw one of these marbles from the bag without looking. Find the probability of each event.

10. black marble

11. red marble

12. green marble

13. white marble

14. marble that is not white

Examples 1–2
(page 51)

15. **PROBABILITY** A spinner has 4 equal-sized sections. All 4 of the sections are yellow. What term best describes the likelihood of the pointer landing on yellow?

Examples 3–4
(page 51)

16. **PROBABILITY** A bag holds red pens and blue pens. Without looking, you pick a pen and replace the pen in the bag. You do this 10 times, picking 8 blue pens and 2 red pens. Based on these trials, what is the probability of picking a red pen?

17. **Talk Math** Discuss with a partner the probability of a coin landing heads both times when tossed twice.

H.O.T. Problems Suppose you have a bag of 7 red and 3 blue marbles.

18. You want to make the probability of picking a red marble unlikely. What could you change?

19. You want to make the probability of picking a red marble certain. How could you do this?

Progress Check 12
(Lesson 12)

▷ Vocabulary and Concept Check

> certain (p. 50)   probability (p. 50)
> impossible (p. 50)   unlikely (p. 50)
> likely (p. 50)

Choose the term that *best* completes each statement. (Examples 1–4)

1. ____?____ measures the likelihood that something will happen.

2. ____?____ events have a good chance to occur.

3. Picking one banana from a bunch of bananas is ____?____.

▷ Skills Check

A bag has red, blue, and black marbles. The tally chart shows the results of 10 pulls. The marble is returned to the bag after each pull.

Based on the results, identify each event as certain, likely, unlikely, or impossible. (Examples 1–2)

4. pulling a red marble

5. pulling a black marble

6. pulling a blue marble

Outcome	Tally	Total
red	I	1
blue	I	1
black	ЖЖ III	8

Marina has 4 blue socks, 3 yellow socks, and 3 red socks in her drawer. She pulls out a sock without looking. Find the probability of each event.
(Examples 3–4)

7. yellow sock

8. blue sock

9. white sock

10. yellow or blue sock

▷ Problem-Solving Check

11. **PROBABILITY** A bag holds green paper clips and blue paper clips. Without looking, you pick a paper clip and put it back. You do this 10 times, picking 9 green paper clips and 1 blue one. What term best describes the likelihood of picking a green paper clip? (Examples 1–2)

12. **H.O.T. Problem** What number describes a certain probability? What number describes an impossible probability? (Examples 1–4)

13. **REFLECT** How do you know when an event is likely, unlikely, certain, or impossible? (Examples 1–4)

Prerequisite Skills Post-Test

▷ Vocabulary and Concept Check

1. In the equation $6 \cdot 8 = 48$, identify the factor(s) and product(s).

2. What number is a whole number but not a counting number? Explain.

3. Explain how you could find the least common multiple of 4 and 6.

4. Which of the numbers 2, 3, 4, 5, and 6 are *not* prime? How do you know?

5. The probability of pulling a red marble out of a bag is impossible. What colors of marbles could be in the bag? Explain.

6. Write an example of a repeating decimal.

7. In 2^7, which number is the base? Which number is the exponent?

8. Is the marked angle acute, right, or obtuse? How do you know?

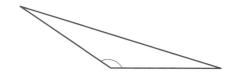

▷ Skills Check

Compare. Replace the ⬤ with >, <, or = to make a true statement.

9. 7 ⬤ 4

10. 5 ⬤ 11

11. 17 ⬤ 14

12. 24 ⬤ 19

Order each set of numbers from least to greatest.

13. 8, 4, and 7

14. 6, 15, and 10

Write the value of the underlined digit.

15. 4<u>8</u>5

16. <u>7</u>39

17. 96<u>0</u>

18. <u>2</u>,513

Write each number in expanded form.

19. 495

20. 2,456

21. 1,304

22. 5,008

Add.

23. $132 + 422$

24. $2,312 + 1,456$

25. $3,368 + 587$

26. $4,946 + 1,948$

Subtract.

27. $763 - 342$

28. $5{,}785 - 2{,}431$

29. $854 - 597$

30. $3{,}645 - 1{,}677$

Multiply.

31. $47 \cdot 68$

32. $136 \cdot 24$

33. $92 \cdot 16$

34. $341 \cdot 37$

Divide.

35. $924 \div 7$

36. $4{,}972 \div 4$

37. $744 \div 8$

38. $6{,}429 \div 3$

Simplify.

39. 5^3 **40.** 3^4 **41.** 8^1 **42.** 2^6

Write each product using exponents.

43. $4 \cdot 4 \cdot 4$

44. $7 \cdot 7 \cdot 7 \cdot 7 \cdot 7$

45. $f \cdot f \cdot f \cdot f$

46. $z \cdot z \cdot z \cdot z \cdot z \cdot z$

Write the prime factorization of each number.

47. 20   **48.** 24   **49.** 32   **50.** 45

Write the fraction.

51. What fraction of the fruit is apples?

52. What fraction of the pie has been eaten?

Identify the decimal as terminating or repeating. Rewrite any terminating decimal as a fraction.

53. $0.1717 \ldots$

54. 0.31

55. 0.007

56. $3.333 \ldots$

Prerequisite Skills Post-Test

Find the least common multiple of each pair of numbers.

57. 2 and 5   **58.** 6 and 12   **59.** 4 and 10

Round to the nearest ten to estimate the answer.

60. $789 + 634$ **61.** $684 - 235$

62. $42 \cdot 28$ **63.** $243 \div 121$

Use compatible numbers to estimate the answer.

64. $326 \div 4$ **65.** $485 \div 7$

Classify each indicated angle as acute, obtuse, or right.

66. 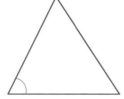   **67.**

68. the angles in this sign

A fruit bowl has 3 red apples and 5 yellow apples. Find the probability of selecting each of the following at random.

69. a yellow apple   **70.** a red apple

A bag has green and black marbles. You pull one marble from the bag without looking and return it. The tally chart shows the results of 10 pulls.

Outcome	Tally	Total
green	I	1
black	⊔⊔⊔ IIII	9

Based on the results, identify each event as certain, likely, unlikely, or impossible.

71. pulling a green marble   **72.** pulling a black marble

▷ Problem-Solving Check

73. TRAVEL A driver turns her truck one fourth of the way around to enter a parking lot. How many degrees does the truck rotate?

74. PACKAGING Sherita buys plastic cups in sets of 6. She buys plastic plates in sets of 8. She sets tables with 1 cup and 1 plate at each place. How many sets of each should she buy so that no cups or plates are left over?

75. FINANCE Sue's bank statement shows that she has $600 + $5 in her account. How much money is in her account?

COMMUNITY SERVICE Kellen is riding his bike to raise money for a charity event. He plans to ride a total of 137 miles today and 129 miles tomorrow. He has already ridden 92 miles today.

76. How many miles will Kellen ride in all?

77. How many miles does Kellen have left to ride for the charity event?

78. CHEMISTRY This table shows the boiling point of four liquids. Which has the higher boiling point, acetone or ethanol?

Liquid	Boiling Point (°C)
Acetone	50
Ethanol	78
Petrol	95
Water	100

79. SPORTS Edita made 6 of her 11 free throws in a basketball game. What fraction of the free throws did she miss?

80. ENGINEERING Neva upgrades her computer RAM by adding a chip that has 2^6 megabytes of memory. How many megabytes of memory does the chip have?

81. STATISTICS The chance of this spinner landing on red is 0.25. Is this decimal terminating or repeating? Explain.

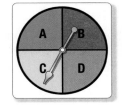

82. HOBBIES Mavis built a birdhouse with a volume of 45 cubic inches. The volume equals the length times the width times the height. Each side is a prime number of inches long. What is the length of each side?

LANDSCAPING Flora plants 32 flowers in each of 22 flower beds. She has an equal number of red, pink, yellow, and white flowers.

83. How many flowers will she plant in all?

84. How many flowers are pink?

85. ENTERTAINMENT Liam downloaded songs from the Internet. He saved 386 of the songs to CDs. He saved the remaining 243 songs to his MP3 player. Round to the nearest ten to estimate the total number of songs he downloaded in all.

86. PROBABILITY A bag holds pink and green erasers. You pull an eraser from the bag and replace it. After 10 pulls, you have pulled a total of 3 pink erasers and 7 green erasers. What is the likelihood of pulling a pink eraser?

Variables, Expressions, and Properties

Here's What I Need to Know

Standard 7AF1.1 Use variables and appropriate operations to write an expression, an equation, an inequality, or a system of equations or inequalities **that represents a verbal description.**

Standard 7AF1.3 Simplify numerical expressions by applying properties of rational numbers, and justify the process used.

Vocabulary Preview

variable A symbol or letter used to represent a number (p. 65)

$$3 + a = 6$$

variable

expression A combination of numbers, variables, and operation symbols (p. 65)

$$6x + 7$$

property A statement that is true for all numbers (p. 77)

Commutative Property of Multiplication:

The order in which you multiply two numbers does not change their product.
$4 \times 5 = 5 \times 4$

The What

I will learn to:

- change words into expressions and equations.
- evaluate expressions by using the order of operations.
- apply properties to simplify expressions.

The Why

Expressions and equations are used to represent point totals in basketball scores. They also can represent the cost of many items.

Option 1

Math Online Are you ready for Chapter 1? Take the Online Readiness
Quiz at ca.algebrareadiness.com to find out.

Option 2

Get ready for Chapter 1. Review these skills.

Vocabulary and Skills Review

Match each sentence with the correct vocabulary word.

1. In $2 + 2 = 4$, 4 is the ___?___.
2. In $10 \div 5 = 2$, 2 is the ___?___.
3. In $2 \cdot 7 = 14$, 7 is a ___?___.

   A. difference
   B. factor
   C. quotient
   D. sum

Copy each problem. Find the missing number.

4. $10 + \underline{\quad?\quad} = 25$
5. $15 - \underline{\quad?\quad} = 8$
6. $4 \cdot \underline{\quad?\quad} = 32$
7. $50 \div \underline{\quad?\quad} = 10$

Find the sum or difference.

8. $25 + 32$
9. $43 + 58$
10. $36 - 15$
11. $62 - 27$

Tips
- Make lists instead of writing sentences.
- Use symbols instead of words.

Note-taking

Chapter Vocabulary	My Notes
A **numerical expression** contains numbers and may contain operations. (p. 65)	numerical expression • numbers: 1, 2, 3, . . . • operations: $+, -, \cdot, \div$ • $6 - 2$
An **algebraic expression** contains at least one variable and may contain operations. (p. 65)	algebraic expression • variables: n, x, y . . . • $6n - 2$
An **equation** is a mathematical sentence that contains an equal sign (=). (p. 67)	equation • left side = right side • $3 + 2 = 4 + 1$

1-1 A Plan for Problem Solving

Vocabulary

understand (p. 60)

plan (p. 60)

solve (p. 60)

check (p. 60)

Standard 7AF1.1 Use variables and appropriate operations to write an expression, an equation, an inequality, or a system of equations or inequalities **that represents a verbal description.**

Standard 7MR1.2 Formulate and justify mathematical conjectures based on a general description of the mathematical question or problem posed.

Standard 7MR3.3 Develop generalizations of the results obtained and the strategies used and apply them to new problem situations.

The What: I will learn how to approach problems and solve them.

The Why: A problem-solving plan will help me find answers to real-life questions, like figuring out how many hours I need to work to buy a computer.

Atepa has an after-school job. He earns $8 an hour. He is saving to buy a computer that costs $1,760. Atepa wants to buy the computer in 5 months. How many hours does Atepa need to work each month to buy the computer?

You can use four steps to solve problems like Atepa's.

PROBLEM-SOLVING STEPS

Step 1 Understand

- Identify the problem to be solved.
- Ask what information you know.

Step 2 Plan

- Ask what you want to know.
- Think about all of the problem-solving strategies you know.
- Choose the strategy you think will work best.

Step 3 Solve

- Use the strategy you chose.

Step 4 Check

- Check your answer. Does it make sense? Is it the only possible answer?
- Could you have used a different strategy to solve the problem?

Study Tip

Each chapter has a lesson with a different problem-solving strategy. Learn these strategies. Then choose the one you think will work best.

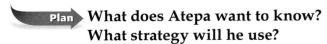

1 **Use the four-step plan to solve Atepa's problem given on page 60.**

> **Understand** **What does Atepa know?**

- He earns $8 an hour.
- The computer costs $1,760.
- He wants to buy the computer in 5 months.

> **Plan** **What does Atepa want to know?**
> **What strategy will he use?**

Atepa wants to know how many hours each month he needs to work. He can use the work backward strategy.

> **Solve** **Atepa works backward from the computer's cost.**

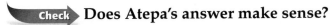

Amount needed each month:

$$\frac{\text{computer cost}}{\text{number of months}} = \frac{1{,}760}{5} = 352$$

Atepa must earn $352 per month.

Number of monthly hours:

$$\frac{\text{monthly earnings}}{\text{hourly pay rate}} = \frac{352}{8} = 44$$

Atepa must work 44 hours per month.

> **Check** **Does Atepa's answer make sense?**

Suppose Atepa works 44 hours each month. He will earn 44 · $8 or $352 each month.

$352 · 5 = $1,760   Multiply $352 by 5 months.

The answer checks.

Your Turn **Solve using the four-step problem-solving plan. Use the *work backward* strategy.**

a. **BUSINESS** Jennifer has an after-school job. She earns $7 an hour. She is saving to buy a sound system that costs $896. Jennifer wants to buy it in 2 months. How many hours a month does Jennifer need to work?

BUSINESS Jennifer works 8 weeks in 2 months.

Talk Math

WORK WITH A PARTNER Look through the Table of Contents on pages ix–xvii. Read the names of different problem-solving strategies. Talk with your partner about how each strategy could be used to solve problems.

Real-World Challenge· · · · ·

After removing the 1-point and 3-point baskets, how many of Laura's points were made from 2-point baskets?

Real-World EXAMPLE

2 **Solve using the four-step problem-solving plan.**

· · · · · · · · · · **SPORTS** Laura scored 19 points in her last basketball game. She made two free throws for 1 point each. She also made a total of seven 2-point and 3-point baskets. How many 2-point baskets did she make?

 Understand **What do you know?**

- Laura scored 19 points.
- She made two 1-point free throws.
- She made seven 2-point and 3-point baskets.

Plan **What do you want to know? What strategy will you use?**

You want to know how many 2-point baskets she made. You can use the guess-and-check strategy.

Solve **Guess three 2-point and four 3-point baskets.**

2-Point Basket Points	+	3-Point Basket Points	+	Free Throws	=	Total Points
$3 \cdot 2$	+	$4 \cdot 3$	+	$2 \cdot 1$	=	Total Points
6	+	12	+	2	=	20

The total is too great. Try a greater number of 2-point baskets and a lesser number of 3-point baskets.

2-Point Basket Points	+	3-Point Basket Points	+	Free Throws	=	Total Points
$4 \cdot 2$	+	$3 \cdot 3$	+	$2 \cdot 1$	=	Total Points
8	+	9	+	2	=	19

Check **Try other combinations.**

- five 2-point baskets + two 3-point baskets + two free throws = 18 points
- six 2-point baskets + one 3-point basket + two free throws = 17 points

So, Laura made four 2-point baskets.

Your Turn **Solve using the four-step problem-solving plan. Use the guess-and-check strategy.**

b. **SPORTS** Rafael scored 18 points in his last basketball game. He made three free throws for 1 point each. He also made a total of seven 2-point and 3-point baskets. How many 3-point baskets did he make?

Study Tip

What if your guess results in 17 total points? How should you improve your guess?

Examples 1–2
(pages 61–62)

VOCABULARY Match each action to the problem-solving step where it is done.

1. Choose a problem-solving strategy.

2. Identify what you know.

3. Make sure the answer makes sense.

4. Use the strategy to solve the problem.

Examples 1–2
(pages 61–62)

SPORTS In one class, 12 students play basketball and 9 students play soccer. Eighteen students in the class play at least one of these sports. How many students play both sports?

5. What do you know?

6. What do you want to know? What strategy will you use?

7. How many students play both sports?

8. Does your answer make sense? How do you know?

9. **RETAIL** Haley bought 14 school items at the store. She bought twice as many pencils as she did notebooks. She bought twice as many markers as she did pencils. How many of each school item did she buy?

Skills, Concepts, and Problem Solving

HOMEWORK HELP

For Exercises	See Example(s)
10–13	1
14–17	2
18–20	1
21	2

DIGITAL CAMERA Marcus is saving to buy a digital camera that costs $552. He earns $6 an hour at a part-time job. Marcus wants to buy the camera in 4 months. How many hours a month does Marcus need to work to buy the digital camera?

10. What do you know?

11. What do you want to know? What strategy will you use?

12. How many hours a month does Marcus need to work?

13. Does your answer make sense? How do you know?

BUSINESS Hannah earns $5 an hour working on the weekends. Her younger brother earns $4 an hour. Last weekend they worked a total of nine hours and earned $38. How many hours did Hannah work?

14. What do you know?

15. What do you want to know? What strategy will you use?

16. How many hours did Hannah work?

17. Does your answer make sense? How do you know?

18. **RECREATION** LaToya works part time. She wants to buy a video game system. She makes $8 an hour. The system costs $240. How many hours does LaToya need to work?

TRAVEL You have a 3:30 P.M. flight to New York. It takes $1\frac{1}{2}$ hours to check in at the airport and board the plane. It takes 45 minutes to drive to the airport and park the car.

19. At what time should you arrive at the airport?

20. At what time should you leave your home to drive to the airport?

21. **SPORTS** Moira scored 15 points in her last basketball game. She made two free throws for 1 point each. She also made a total of six 2-point baskets and 3-point baskets. How many 2-point baskets did she make?

22. *Writing in Math* Which operation(s) did you use to solve Exercise 18? Why?

23. **H.O.T.** Problem Draw a diagram that illustrates Exercise 18.

STANDARDS PRACTICE

Choose the *best* answer.

24 Two prices were rounded to the nearest dollar and then added together. The sum of this estimation was $6. Which two prices were rounded, and then added?

A $4.25 + $2.79

B $3.68 + $2.93

C $3.79 + $2.34

D $5.89 + $0.89

25 The graph shows Ciro's height for the years 2002 through 2007. During which two-year period did Ciro grow the most in height?

F 2002 and 2004

G 2003 and 2005

H 2004 and 2006

J 2005 and 2007

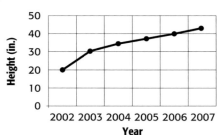

Spiral Review

Add, subtract, multiply, or divide. (Prerequisite Skills 3 and 4)

26. $248 + 928$ 27. $490 - 321$ 28. $28 \cdot 42$ 29. $876 \div 12$

30. $5,672 - 4,893$ 31. $2,192 + 4,516$ 32. $952 \div 14$ 33. $1,153 \cdot 14$

34. **HOBBIES** Miguel had 48 baseball cards. He gave 14 of them to Hallie. How many cards did he have left? (Prerequisite Skills 3)

Expressions and Equations

Vocabulary

numerical expression
(p. 65)

algebraic expression
(p. 65)

variable (p. 65)

equation (p. 67)

Standard 7AF1.1
Use variables and appropriate operations to write an expression, an equation, an inequality, or a system of equations or inequalities **that represents a verbal description.**

Standard 6AF1.0 Students write verbal expressions and sentences as algebraic expressions and equations; they evaluate algebraic expressions, solve simple linear equations, and graph and interpret their results.

 The What: I will represent verbal expressions and sentences as algebraic expressions and equations.

 The Why: The number of items in a set, such as the number of songs on a CD, can be shown using an expression.

A **numerical expression** contains numbers and may contain operations.

$$6 + 4 \cdot 3$$

An **algebraic expression** contains at least one variable and may contain operations. A **variable** is a symbol or letter used to represent a number.

variable
$$6n + 4$$

Suppose every CD has 16 songs on it. Numerical expressions can represent the number of songs.

16 · 1 is the number of songs on 1 CD.
16 · 2 is the number of songs on 2 CDs.
16 · 3 is the number of songs on 3 CDs.

The number of songs on any number of CDs is 16 times the number of CDs. An algebraic expression can represent this situation.

The number of songs **times** the number of CDs

16   · n

Talk Math

WORK WITH A PARTNER Suppose one hamburger costs $1.50. Write the numerical expressions that represent the cost of 1, 2, and 10 hamburgers. What is an algebraic expression that shows the cost of x hamburgers?

To solve word problems, you may need to write algebraic expressions for verbal expressions. When you read problems, look for these key words.

SYMBOL ✚ ➖ ✖ • () ➗				
Operation	**Addition**	**Subtraction**	**Multiplication**	**Division**
Look for these key words in word problems.	• plus • sum • increased by • more than • added to • total	• minus • difference • decreased by • less than • fewer than • subtracted from	• times • product • multiplied by • at • of • factor	• divided by • quotient • ratio • per
Example(s)	$x + 5$	$m - 5$	$5y$ $5 \cdot y$ $5(y)$ $(5)(y)$	$n \div 5$ $\dfrac{n}{5}$

Remember!

Factors are the numbers or **variables** being multiplied. The result is the **product.**

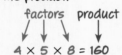

factors   product
$4 \times 5 \times 8 = 160$

EXAMPLES Write Algebraic Expressions

Write an algebraic expression for each verbal expression.

① the sum of m and 18

$m + 18$

② x divided by y

$x \div y$ or $\frac{x}{y}$

③ the product of 5 and t

$5t$ or $5 \cdot t$

④ three less than n times 4

$4n - 3$

Your Turn Write an algebraic expression for each verbal expression.

a. the quotient of x and 10

b. 4 more than 8 times k

EXAMPLES Write Verbal Expressions

Write a verbal expression for each algebraic expression.

⑤ $32 - b$

thirty-two minus b

⑥ $(y + 4) \div 9$

the sum of y and 4, divided by 9

Your Turn Write a verbal expression for each algebraic expression.

c. $15v$

d. $r - \frac{t}{d}$

An **equation** is a mathematical sentence that contains an equal sign (=). Both sides of an equation have the same value.

EXAMPLES Write Equations

Write an equation for each sentence.

7 Three times *g* equals 21.

Three times *g* equals 21.

$$3g \qquad = \qquad 21$$

$$3g = 21$$

8 Five more than two times *n* is 15.

Five more than two times *n* is 15.

$$5+ \qquad\qquad 2n \qquad = 15$$

$$5 + 2n = 15$$

Your Turn Write an equation for each sentence.

e. The sum of 5 and *h* is equal to 12.

f. 20 decreased by *x* is the same as 6.

EXAMPLES Write Sentences

Write a sentence for each equation.

9 $x - 2 = 14$

$$x - 2 \qquad = \qquad 14$$

Two less than *x* equals 14.

10 $7y + 6 = 34$

$$7y \qquad + 6 = 34$$

Seven times *y* plus 6 is 34.

Your Turn Write a sentence for each equation.

g. $\frac{x}{4} = 6$

h. $3n - 10 = 19$

Real-World EXAMPLE

11 **SHOPPING** Ramon bought 5 shirts that cost $7 each and 3 hats that cost $15 each. Write an expression for the total amount that Ramon spent.

$$(5 \cdot 7) + (3 \cdot 15)$$

Your Turn

i. **SHOPPING** Mr. Menendez bought 5 pens that cost $1 each and 3 notebooks that cost $2 each. Write an expression for the total amount that Mr. Menendez spent.

SHOPPING Ramon buys clothes for school.

Guided Practice

Examples 1–4, 7–8, and 11
(pages 66–67)

VOCABULARY Write three examples of each.

1. numerical expression

2. algebraic expression

3. equation

Examples 1–4
(page 66)

Write an algebraic expression for each verbal expression.

4. t more than s

5. the product of 7 and m

Examples 5–6
(page 66)

Write a verbal expression for each algebraic expression.

6. $\frac{d}{7}$

7. $20m + 4$

Examples 7–8
(page 67)

Write an equation for each sentence.

8. A number m added to 6 equals 17.

9. Ten is the same as four times r minus 6.

Examples 9–10
(page 67)

Write a sentence for each equation.

10. $20 - x = 3$

11. $4n = 80$

Example 11
(page 67)

12. **NATURE** Monica likes to watch birds and keep track of those she sees. One week, Monica saw 25 cardinals and 12 finches. Write an expression for the total number of cardinals and finches she saw.

13. **BIOLOGY** A smile requires 26 fewer muscles than a frown. Write an equation to represent the number of muscles a person uses to smile in terms of the number of muscles a person uses to frown.

14. *Talk Math* Have a classmate write an equation. Say at least two equivalent verbal sentences.

Skills, Concepts, and Problem Solving

HOMEWORK HELP	
For Exercises	See Example(s)
15–20	1–4
21–24	5–6
25–28	7–8
29–32	9–10
33–34	11

Write an algebraic expression for each verbal expression.

15. the sum of 35 and z

16. the difference of a number and 7

17. the product of 16 and p

18. the quotient of 25 and a number

19. 18 and three times d

20. 49 increased by twice a number

Write a verbal expression for each algebraic expression.

21. $\frac{b}{11}$

22. $6 - y$

23. $2m - 1$

24. $3r + 8$

Write an equation for each sentence.

25. Three plus w equals 15.

26. Two is equal to six divided by x.

27. Five times r equals 7.

28. 12 is the same as n minus 2.

Write a sentence for each equation.

29. $10 - k = h$

30. $g + 7 = 3$

31. $3r = 18$

32. $\dfrac{t}{4} = 1$

33. RETAIL Karim bought 2 shovels for $15 each and 3 trees for $25 each. Write an expression for the total amount that Karim spent.

34. FINANCE Eva worked 3 hours for $4 an hour and 5 hours for $6 an hour. Write an expression for the total amount that Eva earned.

35. *Writing in Math* Choose a variable. Write an equation for this sentence: *Twice a number minus seven equals the sum of 15 and the number.* Rewrite the sentence without changing the meaning.

H.O.T. Problems A phone company charges one amount for the first minute and a different amount for additional minutes. The table shows the cost of several phone calls.

Time (min)	Cost ($)
1	0.20
5	0.60
10	1.10

36. What is the cost of the first minute of a call?

37. What is the cost for each additional minute?

38. Use the table to write an algebraic expression for the cost of a phone call when you know the number of minutes for the call.

STANDARDS PRACTICE

Use the graph to find the *best* answer choice.

39 A pet store donated 50¢ to the dog shelter for every 5-pound bag of dog food it sold on Saturday and Sunday. How much money did the pet store donate?

A $5.50

B $9.00

C $45.00

D $90.00

Bags of Dog Food Sold

Saturday

Sunday

= 10 bags sold

 Review

Solve using the four-step problem-solving plan. (Lesson 1-1)

40. STATISTICS There are 80 students enrolled in the music program at King High School. There are 52 students in chorus and 49 students in band. How many students are enrolled in both band and chorus?

Add, subtract, multiply, or divide. (Prerequisite Skills 3 and 4)

41. $14 + 15$

42. $21 + 29$

43. $45 - 36$

44. $29 \cdot 17$

45. $38 \div 19$

46. $804 \div 12$

Vocabulary

order of operations
(p. 70)

base (p. 72)

exponent (p. 72)

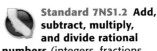

Standard 7NS1.2 Add, subtract, multiply, and divide rational numbers (integers, fractions, and terminating decimals) and **take positive rational numbers to whole-number powers.**

Standard 7AF2.1 Interpret positive whole-number powers as repeated multiplication and negative whole-number powers as repeated division or multiplication by the multiplicative inverse. **Simplify and evaluate expressions that include exponents.**

Standard ALG4.0 Students simplify expressions before solving linear equations and inequalities in one variable, such as $3(2x - 5) + 4(x - 2) = 12$.

 The What: I will learn to use the order of operations to evaluate expressions.

 The Why: Use the order of operations to find the number of goals scored by a team member during a season.

Trisha scored 3 goals in her sixth soccer game. She scored 2 goals in each of her first five games. Trisha's goal total is represented by the expression $3 + 2 \cdot 5$.

goals in sixth game ⌐   ⌐————goals in first 5 games

$$3 + 2 \cdot 5 \leftarrow$$

The expression $3 + 2 \cdot 5$ has two operations: addition and multiplication. The value of the expression depends on the order in which these operations are performed.

$$3 + 2 \cdot 5$$
$$3 + 10 = 13$$

Order of operations is the set of rules to find the correct value of an expression.

Remember!
Grouping symbols include parentheses ( ), brackets [ ], and fraction bars $\frac{8+1}{9}$.

ORDER OF OPERATIONS

❶ Find the values of expressions inside **grouping symbols: parentheses, brackets, or braces.**

Always simplify the expression inside the innermost grouping symbol first.

$$2\left[\frac{8 + 1}{9} + 4(7 + 3)\right]$$

❷ Simplify terms involving **exponents.**

❸ **Multiply** and **divide** in order from left to right.

❹ **Add** and **subtract** in order from left to right.

EXAMPLE Evaluate Expressions

1 Find the value of $30 \div 5 + 4 \cdot 5 - 2$. Show your work.

$$30 \div 5 + 4 \cdot 5 - 2 = 6 + 4 \cdot 5 - 2 \qquad \text{Divide 30 by 5.}$$
$$= 6 + 20 - 2 \qquad \text{Multiply 4 and 5.}$$
$$= 26 - 2 \qquad \text{Add 6 and 20.}$$
$$= 24 \qquad \text{Subtract 2 from 26.}$$

Your Turn Find the value of each expression. Show your work.

a. $6 \div 2 + 36$

b. $3 \cdot 4 - 32 \div 4$

c. $5 \cdot 4 - 3 + 6$

d. $8 \cdot 2 + 4 \cdot 3$

Real-World Challenge

The team scored 26 goals in the first 6 games. How many goals were scored by players other than Trisha?

Real-World EXAMPLE

2 **SPORTS** Trisha scored 2 goals in each of her first 5 soccer games. She scored 3 goals in the sixth game. What was her total number of goals?

Total goals:

goals in the first 5 games plus the goals in the sixth game

$$2 \cdot 5 \qquad + \qquad 3$$

$$2 \cdot 5 + 3 = 10 + 3 \qquad \text{Multiply 2 and 5.}$$
$$= 13 \qquad \text{Add 10 and 3.}$$

Your Turn

e. **CELL PHONES** Shane's monthly cell phone rate was $38, plus an additional per-minute charge. He talked for 45 minutes. Each minute costs $0.10. What was his total cell phone bill?

EXAMPLE Expressions with Parentheses

3 Find the value of $3(18 + 2) - 7$.

$$3(18 + 2) - 7 = 3(20) - 7 \qquad \text{Add 18 and 2.}$$
$$= 60 - 7 \qquad \text{Multiply 3 and 20.}$$
$$= 53 \qquad \text{Subtract 7 from 60.}$$

Your Turn Find the value of each expression. Show your work.

f. $3(4 + 6) - 8$

g. $4(4 + 3) - 6$

h. $9(22 - 17) - 7$

i. $6(5 - 3) - 2$

WORK WITH A PARTNER Ricardo had $50 for an MP3 player. He saved $10 per week for 20 weeks. His total was 50 + 10 · 20. Did Ricardo have $1,200 or $250? Explain.

An exponent is a shorter way to show multiplication. The expression 2 · 2 · 2 · 2 can be written as 2^4. The **base** is the number being multiplied. The **exponent** tells how many times the base is used as a factor.

exponent

base → 2^4

$3 \cdot 3 = 3^2$	$9 \cdot 9 \cdot 9 = 9^3$	$5 \cdot 5 \cdot 5 \cdot 5 = 5^4$
3 to the second power or 3 squared	9 to the third power or 9 cubed	5 to the fourth power

EXAMPLE **Expressions with Exponents**

4 Find the value of $10 + 5^2 \cdot 4$.

$$10 + 5^2 \cdot 4 = 10 + 25 \cdot 4 \quad \text{Find } 5^2. \ 5 \cdot 5 = 25$$
$$= 10 + 100 \quad \text{Multiply 25 and 4.}$$
$$= 110 \quad \text{Add 10 and 100.}$$

Your Turn Find the value of each expression. Show your work.

j. $8 + 5^2 \cdot 3$ k. $2 + 4^3 \cdot 2$

EXAMPLE **Expressions with Grouping Symbols and Exponents**

5 Find the value of $2^3 + [2(24 - 9)]$.

$$2^3 + [2(24 - 9)] = 2^3 + [2(15)] \quad \text{Start with the innermost grouping symbol. Subtract 9 from 24.}$$
$$= 2^3 + 30 \quad \text{Multiply 2 and 15.}$$
$$= 8 + 30 \quad \text{Find } 2^3. \ 2 \cdot 2 \cdot 2 = 8$$
$$= 38 \quad \text{Add 8 and 30.}$$

Remember!

Brackets enclose parentheses. Find the value inside the parentheses and then find the value inside the brackets.

$[9 - (5 + 2)]$

Your Turn Find the value of each expression. Show your work.

l. $[5(8 - 6)] + 4^2$ m. $3^2 + [2(31 - 4)]$

Examples 1–4
(pages 71–72)

1. **VOCABULARY** An expression contains division and subtraction. Which operation should you do first?

2. How are $5 \cdot 3 + 5$ and $5(3 + 5)$ different? Explain to a classmate.

3. In 7^4, what is the number 7 called?

Example 1
(page 71)

Find the value of each expression. Show your work.

4. $54 - 42 \div 7$

5. $36 \div 9 \cdot 3$

6. $8 \cdot 5 + 3 \cdot 7$

7. $6 + 21 \div 7 \cdot 2$

Example 2
(page 71)

8. **CAREER CONNECTION** Teresa rents a car for 5 days and drives 300 miles. The car company charges $25 a day and $1.10 a mile. What was the total cost for renting the car?

CAREER CONNECTION In 2005, nearly 23,000 Californians worked in the automotive rental industry.

Examples 3–5
(pages 71–72)

Find the value of each expression. Show your work.

9. $4(5 + 7)$

10. $14 - (4 + 5)$

11. $12 + 3^2 \cdot 2$

12. $2^5 + 3$

13. $7 + 5 - 3^2$

14. $2^3 + [18 \div (7 - 4)]$

15. $10^2 + [8(3 + 2)]$

16. $[2(14 + 6)] - 1^3$

Skills, Concepts, and Problem Solving

HOMEWORK HELP	
For Exercises	**See Example(s)**
17–24	1
25–29	2
30–33	3
34–39	4
40–45	5

Find the value of each expression. Show your work.

17. $10 \cdot 3 + 5$

18. $10 + 3 \cdot 5$

19. $6 \cdot 2 + 7 \cdot 3$

20. $9 \cdot 7 - 4 \cdot 3$

21. $42 + 21 \div 7$

22. $9 + 18 \div 3$

23. $21 \div 7 + 4 \cdot 11$

24. $36 \div 9 + 3 \cdot 1$

MONEY Write an expression for each situation. Then use the information in the table to find the solution. Show your work.

Item	Cost
Pack of balloons	$2
Roll of streamers	$3
Pack of plates	$6
Pack of napkins	$4
Centerpiece	$7

25. Israel buys a pack of balloons, two rolls of streamers, four packs of plates, and four packs of napkins for a school dance. What is the total cost?

26. He decides to return the balloons to the store and purchase six centerpieces instead. Now what is the total cost?

27. **TICKETS** Write an expression that shows the total cost of fifty $6 movie tickets and eight $12 play tickets. Then, find the total cost.

REAL ESTATE The Phams had originally planned to live in the Northeast region of the United States. They have now decided to live in the West. They will be making monthly house payments for a 20-year period.

28. Use the graph to estimate the cost of living in both the Northeast region and in the West.

29. Estimate how much more the Phams will be spending on house payments in the West than they would have spent living in the Northeast.

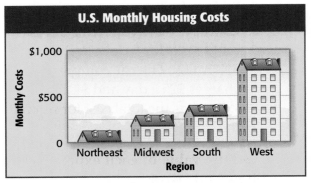

Source: U.S. Census Bureau, Census 2000 Summary

Find the value of each expression. Show your work.

30. $40 \cdot (6 - 2)$

31. $30 \cdot (8 - 5)$

32. $(9 - 6)(6 + 4)$

33. $5(6 - 1) + 10$

34. $5^3 + 4^2$

35. $5^2 + 4^3$

36. $6^2 - 2^3$

37. $8^2 - 3^3$

38. $24 \div (5 - 3)^2 + 3$

39. $25 \div (8 - 3)^2 + 5$

40. $[5(4 + 6)] - 3^2$

41. $[6(12 - 3)] - 4^2$

42. $10^2 + [3(6 - 5)]$

43. $2^4 + [5(7 - 1)]$

44. $5^3 + [8(6 + 1)]$

45. $2^4 + [10(8 - 6)]$

46. *Writing in Math*   Write a paragraph that describes the steps in the order of operations.

47. **NUMBER SENSE** Emily and Moses are evaluating $24 \div 2 \cdot 3$. Who is correct? Explain your thinking.

Emily	Moses
$24 \div 2 \cdot 3 = 12 \cdot 3$	$24 \div 2 \cdot 3 = 24 \div 6$
$= 36$	$= 4$

H.O.T. Problems Expressions with a value of 1 can be created using only four 4s.

$$4 + \frac{4}{4} - 4 = 1$$

48. Create an expression using only four 4s that has a value of 2.

49. Create two expressions using only four 4s that have values of 3, 4, 5, 6, 7, 8, and 9.

50. **MAKE UP A PROBLEM** Write an expression involving addition and multiplication in which you should add first.

STANDARDS PRACTICE

Choose the *best* answer.

51 What is the value of $3[2(5 - 4) + 3(6 - 4)]$?

A 24   **C** 58

B 32   **D** 60

52 The desks in Mr. Yoshio's room and Ms. Brown's room are arranged in 6 rows. Each row has 4 desks. How many total desks are there in both classrooms?

F 10   **H** 24

G 20   **J** 48

53 Julio recorded songs and music videos onto a CD. The CD holds 1,000 megabytes (MB) of data. How many megabytes are left on the CD?

Song or Video	Size (MB)
Song #1	35
Song #2	40
Song #3	37
Video #1	125
Video #2	140

A 377 MB   **C** 888 MB

B 623 MB   **D** 1,377 MB

Spiral Review

Write an algebraic expression for each verbal expression. (Lesson 1-2)

54. seven more than x

55. the quotient of m and n

56. the difference of g and 10

57. 5 plus the product of 8 and y

Write a verbal expression for each algebraic expression. (Lesson 1-2)

58. $20n$

59. $d - 2$

60. $5k + 2$

61. $m \div n + 10$

62. **LANGUAGE** In one class, ten students study only Spanish, and six students study only French. Eighteen students study at least one of these languages. How many students study both? (Lesson 1-1)

Progress Check 1

(Lessons 1-1, 1-2, and 1-3)

▷ Vocabulary and Concept Check

algebraic expression (p. 65)	exponent (p. 72)	solve (p. 60)
base (p. 72)	numerical expression (p. 65)	understand (p. 60)
check (p. 60)	order of operations (p. 70)	variable (p. 65)
equation (p. 67)	plan (p. 60)	

Choose the term that *best* completes each sentence. (Lessons 1-1, 1-2, 1-3)

1. A(n) _____?_____ contains variables.

2. In the expression 8^3, the number 3 is called the _____?_____.

3. A(n) _____?_____ is made up of two equal expressions and has an equal sign.

▷ Skills Check

Write an algebraic expression for each verbal expression. (Lesson 1-2)

4. 12 more than s

5. the difference between the quotient of 24 and b and 36

Write an equation for each sentence. (Lesson 1-2)

6. Eight plus d equals 29.

7. Five fewer than b is equal to c divided by 30.

Find the value of each expression. (Lesson 1-3)

8. $6 \cdot 10 + 20$

9. $5^2(3 + 1) + 30$

▷ Problem-Solving Check

Solve using the four-step problem-solving plan. (Lesson 1-1)

10. **SPORTS** Sherrill plays basketball. In one game, she made 7 baskets for a total of 19 points. If she made 2-point and 3-point baskets, how many 3-point baskets did she make?

11. **FINANCE** RaShonda purchased a portable music player for $99. Music downloads cost $1.99 each. RaShonda downloaded 400 songs. Write an expression for the total amount of money RaShonda spent. (Lesson 1-2)

12. **REFLECT** Which operations did you use in your expression for Exercise 11? Why? (Lesson 1-3)

Commutative and Associative Properties

Vocabulary

property (p. 77)

Commutative Property of Addition (p. 77)

Commutative Property of Multiplication (p. 78)

Associative Property of Addition (p. 79)

Associative Property of Multiplication (p. 79)

Standard 7AF1.3 Simplify numerical expressions by applying properties of rational numbers and justify the process used.

Standard 3AF1.5 Recognize and use the commutative and associative properties of multiplication.

The What: I will use the Commutative and Associative Properties to simplify expressions.

The Why: Rearranging numbers can make calculations easier when finding the number of hours worked.

Jamal volunteered for 10 hours in May and 16 hours in June at the food pantry. Malia volunteered for 16 hours in May and 10 hours in June. Jamal and Malia each volunteered for 26 hours.

	Hours in May		Hours in June		Total Hours
Jamal	10	+	16	=	26
Malia	16	+	10	=	26

A statement that is true for all numbers is called a **property.** The **Commutative Property of Addition** states that the order in which two numbers are added does not change their sum.

COMMUTATIVE PROPERTY OF ADDITION

Pictures:

Numbers: 2 + 3 = 3 + 2

Words: The order in which two numbers are added does not change their sum.

Algebra: For any numbers a and b, $a + b = b + a$.

EXAMPLES Commutative Property of Addition

Complete each equation.

❶ 7 + 3 = 3 + _?_

7 + 3 = 3 + _7_

❷ 2 + 9 = _?_ + _?_

2 + 9 = _9_ + _2_

Your Turn Complete each equation.

a. 5 + 4 = 4 + _?_

b. 4 + 7 = _?_ + _?_

The **Commutative Property of Multiplication** states that the order in which you multiply numbers does not change their product.

COMMUTATIVE PROPERTY OF MULTIPLICATION

Pictures: =

Numbers: $3 \cdot 2 = 2 \cdot 3$

Words: The order in which two numbers are multiplied does not change their product.

Algebra: For any numbers a and b, $a \cdot b = b \cdot a$

Remember!

Subtraction is not commutative.

$10 - 2 \neq 2 - 10$

Division is not commutative, either.

$6 \div 3 \neq 3 \div 6$

EXAMPLES Commutative Property of Multiplication

Complete each equation.

3 $7 \cdot 3 = 3 \cdot \underline{\ \ ?\ \ }$

$7 \cdot 3 = 3 \cdot \underline{\ \ 7\ \ }$

4 $2 \cdot 9 = \underline{\ \ ?\ \ } \cdot \underline{\ \ ?\ \ }$

$2 \cdot 9 = \underline{\ \ 9\ \ } \cdot \underline{\ \ 2\ \ }$

Your Turn Complete each equation.

c. $4 \cdot 8 = \underline{\ \ ?\ \ } \cdot 4$

d. $5 \cdot 6 = \underline{\ \ ?\ \ } \cdot \underline{\ \ ?\ \ }$

Talk Math

WORK WITH A PARTNER Create five examples of the Commutative Property of Addition. If you say, "3 plus 5," then your partner says, "5 plus 3." You both say, "3 plus 5 equals 5 plus 3." Next, create five examples of the Commutative Property of Multiplication.

The numbers 2, 3, and 5 appear in the same order on both sides of the equation below. Parentheses are used to group the numbers differently.

$$(2 + 3) + 5 = 2 + (3 + 5)$$
$$5 + 5 = 2 + 8$$
$$10 = 10$$

Even though the numbers are grouped differently, each side of the equation $(2 + 3) + 5 = 2 + (3 + 5)$ has a value of 10.

The **Associative Property of Addition** states that the way numbers are grouped does not change their sum.

ASSOCIATIVE PROPERTY OF ADDITION

Pictures: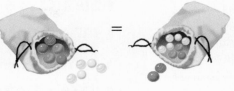

Numbers: $(2 + 3) + 5 = 2 + (3 + 5)$

Words: The way numbers are grouped when they are added does not change their sum.

Algebra: For any numbers a, b, and c,
$$(a + b) + c = a + (b + c)$$

The **Associative Property of Multiplication** states that the way numbers are grouped does not affect their product.

ASSOCIATIVE PROPERTY OF MULTIPLICATION

Pictures: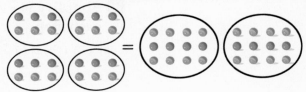

Numbers: $(2 \cdot 3) \cdot 4 = 2 \cdot (3 \cdot 4)$

Words: The way numbers are grouped when they are multiplied does not change their product.

Algebra: For any numbers a, b, and c,
$$(a \cdot b) \cdot c = a \cdot (b \cdot c)$$

EXAMPLES Associative Properties

Complete each equation.

5 $(15 + 9) + 1 = \underline{?} + (9 + 1)$
$(15 + 9) + 1 = \underline{15} + (9 + 1)$

6 $8 \cdot (10 \cdot 3) = (8 \cdot \underline{?}) \cdot \underline{?}$
$8 \cdot (10 \cdot 3) = (8 \cdot \underline{10}) \cdot \underline{3}$

Your Turn Complete each equation.

e. $(7 + 6) + 1 = 7 + (\underline{?} + 1)$

f. $(9 \cdot 2) \cdot 5 = 9 \cdot (\underline{?} \cdot \underline{?})$

Study Tip

The commutative properties involve order. The associative properties involve grouping.

Remember which is which by letting the first two letters of commutative stand for "change order."

EXAMPLES Identify Properties

Name the property shown by each statement.

7 $4 \cdot 11 = 11 \cdot 4$ Commutative Property of Multiplication

8 $(n + 12) + 5 = n + (12 + 5)$ Associative Property of Addition

9 $17 + (20 + 3) = 17 + (3 + 20)$ Commutative Property of Addition

Your Turn Name the property shown by each statement.

g. $(5 \cdot 4) \cdot 3 = 5 \cdot (4 \cdot 3)$ **h.** $6 + n = n + 6$

You can use the Commutative and Associative Properties to simplify expressions. Numbers can be reordered and regrouped for easier calculations.

$$75 + (112 + 25) = 75 + (25 + 112) \quad \text{Commutative Property of Addition}$$
$$= (75 + 25) + 112 \quad \text{Associative Property of Addition}$$
$$= \quad 100 \quad + 112$$
$$= \quad 212$$

EXAMPLE Simplify Numerical Expressions

10 Simplify $(4 \cdot 7) \cdot 25$. Name the property used in each step.

$$(4 \cdot 7) \cdot 25 = (7 \cdot 4) \cdot 25 \quad \text{Commutative Property of Multiplication}$$
$$= 7 \cdot (4 \cdot 25) \quad \text{Associative Property of Multiplication}$$
$$= 7 \cdot 100$$
$$= 700$$

Your Turn Simplify each expression. Name the property used in each step.

i. $(14 + 27) + 6$ **j.** $20 \cdot (9 \cdot 5)$

Real-World EXAMPLE

11 SHOPPING Mr. Rodriguez spent $5, $28, and $35 on three grocery shopping trips. Use the Associative Property of Addition to write two different expressions that represent the total amount he spent.

$$(5 + 28) + 35 = 33 + 35 = 68$$
$$5 + (28 + 35) = 5 + 63 = 68$$

Real-World Challenge. . . . . .
How much did Mysti spend in all?

Your Turn

k. RETAIL Mysti bought three DVDs for $19, $24, and $21. Use the Associative Property of Addition to write two different expressions that represent Mysti's total cost.

Examples 1–10
(pages 77–80)

VOCABULARY

1. Write an equation that shows the Commutative Property of Addition.

2. Write an equation that shows the Associative Property of Multiplication.

Examples 1–6
(pages 77–79)

Copy and complete each equation.

3. $9 + 2 = 2 + \underline{?}$

4. $10 + 8 = \underline{?} + \underline{?}$

5. $4 \cdot 7 = \underline{?} \cdot 4$

6. $6 \cdot 5 = \underline{?} \cdot \underline{?}$

7. $(10 + 8) + 3 = \underline{?} + (8 + 3)$

8. $4 \cdot (9 \cdot 3) = (4 \cdot \underline{?}) \cdot \underline{?}$

9. $(12 \cdot 2) \cdot 5 = \underline{?} \cdot (\underline{?} \cdot 5)$

10. $13 + (7 + 21) = (\underline{?} + \underline{?}) + 21$

Examples 7–9
(page 80)

Name the property shown by each statement.

11. $16 + 9 = 9 + 16$

12. $12 + (14 + 20) = (12 + 14) + 20$

13. $(10 \cdot 4) \cdot 3 = 10 \cdot (4 \cdot 3)$

14. $5 \cdot 8 = 8 \cdot 5$

Example 10
(page 80)

Simplify each expression. Name the property used in each step.

15. $12 + (39 + 8)$

16. $(5 \cdot 18) \cdot 2$

Example 11
(page 80)

17. **SHOPPING** Kadijah bought the items shown on the receipt. Write an expression that represents Kadijah's total purchase amount. Use the Commutative Property of Addition to write a second expression that represents her purchase.

```
      RECEIPT
Milk     $2.95
Bread    $1.29
Cheese   $6.05

  THANK YOU
  COME AGAIN
```

18. **TRAVEL** Sergio travels on his job. His routes include stops after 16 miles, 48 miles, 24 miles, and 22 miles. Use the Commutative and Associative Properties to write two expressions that represent his total distance traveled.

19. *Writing in Math* Write an expression like $(5 + 2) + 8$. Have a classmate rewrite the expression using the Associative Property. Then rewrite your classmate's expression using the Commutative Property.

Skills, Concepts, and Problem Solving

Copy and complete each equation.

For Exercises	See Example(s)
20–27	1–4
28–35	5–6
36–43	7–9
44–49	10–11

HOMEWORK HELP

20. $9 + 2 = 2 + \underline{?}$

21. $10 + 15 = \underline{?} + 10$

22. $7 + 8 = \underline{?} + \underline{?}$

23. $x + y = \underline{?} + \underline{?}$

24. $7 \cdot 6 = \underline{?} \cdot 7$

25. $4 \cdot 12 = 12 \cdot \underline{?}$

26. $8 \cdot 9 = \underline{?} \cdot \underline{?}$

27. $c \cdot d = \underline{?} \cdot \underline{?}$

28. $(10 + 8) + 3 = \underline{\ ?\ } + (8 + 3)$ 29. $4 \cdot (9 \cdot 3) = (4 \cdot 9) \cdot \underline{\ ?\ }$

30. $(28 \cdot 2) \cdot 5 = \underline{\ ?\ } \cdot (\underline{\ ?\ } \cdot 5)$ 31. $19 + (11 + 25) = (19 + \underline{\ ?\ }) + \underline{\ ?\ }$

32. $4 + (2 + 6) = (\underline{\ ?\ } + \underline{\ ?\ }) + 6$ 33. $7 \cdot (3 \cdot 10) = (7 \cdot \underline{\ ?\ }) \cdot \underline{\ ?\ }$

34. $(r + s) + t = \underline{\ ?\ } + (s + t)$ 35. $x \cdot (y \cdot z) = (x \cdot \underline{\ ?\ }) \cdot \underline{\ ?\ }$

Name the property shown by each statement.

36. $15 + 6 = 6 + 15$ 37. $19 \cdot 20 = 20 \cdot 19$

38. $20 + 35 = 35 + 20$ 39. $12 + (14 + 6) = (12 + 14) + 6$

40. $(6 \cdot 7) \cdot 5 = 6 \cdot (7 \cdot 5)$ 41. $12 + (23 + 7) = (12 + 23) + 7$

42. $(18 + 15) + 32 = (15 + 18) + 32$ 43. $(6 \cdot 7) \cdot 5 = (7 \cdot 6) \cdot 5$

Simplify each expression. Name the property used in each step.

44. $14 + (37 + 6)$ 45. $5 \cdot (39 \cdot 2)$

46. $25 + (42 + 5)$ 47. $2 \cdot (71 \cdot 5)$

48. **RETAIL** Kim bought three different kinds of flowers for a bouquet, each of which cost a different amount. He paid $12, $10, and $18 for the flowers. Use the Associative Property of Addition to write two different expressions that represent the total amount he spent.

49. **SPORTS** Carie served for 5, 6, and 4 points in her first three volleyball games. Use the Associative Property of Addition to write two different expressions that represent her number of points off serve.

50. **MENTAL MATH** Use the Commutative and Associative Properties to solve mentally.

$$75 + 187 + 25$$

51. **DECORATING** Kendra noticed that for the placemat to the right she could either multiply 15 times 12 or 12 times 15 to find its area of 180 square inches. What property allows her to do this?

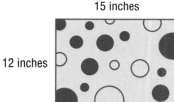

15 inches

12 inches

H.O.T. Problems

52. Hector stated that the expression $(40 \div 8) \div 2$ is equal to $40 \div (8 \div 2)$ because of the Associative Property. Marty disagrees. Who is correct? Why?

53. **MAKE UP A PROBLEM** Write an equation that illustrates the Commutative Property of Multiplication and has a value of 24 on each side.

54. *Writing in Math* People put on their socks before they put on their shoes when they get dressed. The actions of putting on socks and putting on shoes are *not* commutative. What other actions from your life are not commutative?

55. TRAVEL Edie wants to visit her aunt and her grandfather. Use the Associative and Commutative Properties of Addition. Show that Edie will travel the same distance no matter which relative she visits first.

Distances Between Houses (in miles)	
Edie and Aunt	17
Edie and Grandfather	22
Aunt and Grandfather	19

56. SPORTS The Badger football team scored the following points in their first four games: 25, 16, 35, 24. Explain how the points should be arranged to make them easier to add.

STANDARDS PRACTICE

Choose the *best* answer.

57 At a used bookstore, Rico bought 5 hardcover books and 4 paperback books. How much did Rico spend in all?

A $8

B $30

C $34

D $38

58 Use the Commutative and Associative Properties to compute the product.

$$2 \cdot 4 \cdot 2.5 \cdot 15 \cdot 5 \cdot 10$$

F 1,500   **H** 15,000

G 12,000   **J** 120,000

59 The rate of erosion of a certain river basin is b millimeters per year. Which equation best describes the total amount of erosion (T) of the basin over a 15-year period?

A $T = b + 15$ **C** $T = b \div 15$

B $T = 15b$ **D** $T = 15 \div b$

Spiral Review

Find the value of each expression. **Show your work.** (Lesson 1-3)

60. $6 + 14 \cdot 2$

61. $(15 + 5) \div 2 + 9$

62. $36 \div 3^2 + 10$

63. $9 \cdot 2 + 5 \cdot 4$

Write an algebraic expression for each verbal expression. (Lesson 1-2)

64. the sum of 3 and n

65. the product of m and n

66. five plus the quotient of 8 and x **67.** the difference of 7 and t

68. SCHOOL A student scored 35 points on a quiz. Questions were worth two and five points. The student answered 13 questions correctly. How many 2-point questions did the student answer correctly? (Lesson 1-1)

Distributive Property
(p. 84)

 Standard 7AF1.3 Simplify numerical expressions by applying properties of rational numbers and justify the process used.

Standard 5AF1.3 Know and use the distributive property in equations and **expressions** with variables.

 The *What*: I will use the Distributive Property to simplify expressions.

The *Why*: The Distributive Property can make it simpler to determine the cost of purchases.

At a local restaurant, Jonathan and each of his three friends ordered a hamburger, pretzels, and a lemonade. Each hamburger cost $3. Each bag of pretzels cost $2. Each lemonade cost $1. There are two ways to find the total cost of the order.

Method 1: Multiply the cost of one complete order by the number of people ordering.

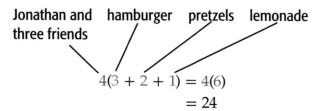

$$4(3 + 2 + 1) = 4(6)$$
$$= 24$$

Method 2: Add the costs of all the hamburgers, pretzels, and lemonades.

$$4 \text{ hamburgers} + 4 \text{ pretzels} + 4 \text{ lemonades}$$
$$(4 \cdot 3) \quad + \quad (4 \cdot 2) \quad + \quad (4 \cdot 1) \quad = 12 + 8 + 4$$
$$= \$24$$

Both methods produce a total of $24. In Method 2, the 4 is multiplied by $3, $2, and $1. This is an example of the Distributive Property of Multiplication over Addition.

Method 1		**Method 2**
$4(3 + 2 + 1)$	$=$	$(4 \cdot 3) + (4 \cdot 2) + (4 \cdot 1)$
$4(6)$	$=$	$12 + 8 + 4$
24	$=$	24

The **Distributive Property** states that multiplying the sum of two or more numbers by another number is the same as multiplying each of those numbers by the other number.

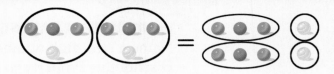

DISTRIBUTIVE PROPERTY

Numbers: $2(3 + 1) = (2 \cdot 3) + (2 \cdot 1)$ or
$(3 + 1)2 = (3 \cdot 2) + (1 \cdot 2)$

Words: Multiplying the sum of two numbers by a third number is the same as multiplying each of the two numbers by the third.

Algebra: For any numbers a, b, and c,
$a(b + c) = a \cdot b + a \cdot c$, or
$(b + c)a = b \cdot a + c \cdot a$.

EXAMPLES Distributive Property

Complete each equation.

① $5(6 + 4) = (5 \cdot 6) + (5 \cdot \underline{\ ?\ })$

$5(6 + 4) = (5 \cdot 6) + (5 \cdot \underline{\ 4\ })$

② $(9 + 5)10 = (\underline{\ ?\ } \cdot \underline{\ ?\ }) + (\underline{\ ?\ } \cdot \underline{\ ?\ })$

$(9 + 5)10 = (\underline{\ 9\ } \cdot \underline{\ 10\ }) + (\underline{\ 5\ } \cdot \underline{\ 10\ })$

Your Turn Complete each equation.

a. $2(3 + 5) = (2 \cdot 3) + (2 \cdot \underline{\ ?\ })$

b. $(7 + 1)6 = (\underline{\ ?\ } \cdot \underline{\ ?\ }) + (\underline{\ ?\ } \cdot \underline{\ ?\ })$

The Distributive Property can also be used to rewrite expressions based on common factors.

$$(5 \cdot 2) + (5 \cdot 3) = 5(2 + 3)$$

Study Tip

For mental math, think of large numbers as the sum of smaller numbers.
Example:
$5 \cdot 26 = 5(20 + 6)$
$= (5 \cdot 20) + (5 \cdot 6)$
$= 100 + 30$
$= 130$

EXAMPLES Distributive Property

Complete each equation.

③ $(5 \cdot 6) + (5 \cdot 4) = 5(6 + \underline{\ ?\ })$

$(5 \cdot 6) + (5 \cdot 4) = 5(6 + \underline{\ 4\ })$

④ $(7 \cdot 3) + (7 \cdot 2) = 7(\underline{\ ?\ } + \underline{\ ?\ })$

$(7 \cdot 3) + (7 \cdot 2) = 7(\underline{\ 3\ } + \underline{\ 2\ })$

Your Turn Complete each equation.

c. $(10 \cdot 4) + (10 \cdot 8) = 10(4 + \underline{\ ?\ })$

d. $(2 \cdot 9) + (2 \cdot 1) = 2(\underline{\ ?\ } + \underline{\ ?\ })$

Talk Math

WORK WITH A PARTNER Name a one-digit number. Name a two-digit number. Have your partner use the Distributive Property to find the product of your numbers. Then, your partner will name a one-digit number and a two-digit number for you to multiply. Talk about how the Distributive Property can make it easier to find products.

Using the Distributive Property can make calculations easier to do. Remember the two methods used to find Jonathan's total order at the beginning of the lesson. Use these two methods to find Luanda's total grocery purchase.

 Real-World EXAMPLE

RETAIL Luanda went to the store and bought 4 cans of soup and 4 cans of corn. Each can of soup cost $3. Each can of corn cost $2.

5 Write two equations representing the total amount of money T Luanda spent at the store.

Equation 1: $T = 4(3) + 4(2)$
Equation 2: $T = 4(3 + 2)$

6 Determine the total amount Luanda spent at the store.

Equation 1:

$T = 4(3) + 4(2)$	Multiply 4 and 3. Multiply 4 and 2.
$= 12 + 8$	Add 12 and 8.
$= 20$	

Equation 2:

$T = 4(3 + 2)$	Add 3 and 2.
$= 4(5)$	Multiply 4 and 5.
$= 20$	

Using either method, Luanda spent $20 at the store.

Your Turn
ENTERTAINMENT A family purchased tickets for 3 adults and 3 children to an amusement park. Each adult's ticket cost $25. Each child's ticket cost $15.

e. Write two equations representing the total cost T of the tickets.

f. Determine the family's total ticket cost.

ENTERTAINMENT Suppose one adult could not go to the amusement park. What is the new total ticket cost?

Examples 1–4
(page 85)

1. VOCABULARY Write an equation that shows the Distributive Property.

Copy and complete each equation using the Distributive Property.

2. $3(2 + 6) = (3 \cdot 2) + (3 \cdot \underline{\;?\;})$

3. $(3 + 7)10 = (\underline{\;?\;} \cdot \underline{\;?\;}) + (\underline{\;?\;} \cdot \underline{\;?\;})$

4. $(6 \cdot 11) + (6 \cdot 3) = 6(11 + \underline{\;?\;})$

5. $(5 \cdot 4) + (5 \cdot 2) = 5(\underline{\;?\;} + \underline{\;?\;})$

Examples 5–6
(page 86)

SPORTS Rich bought two baseballs and two basketballs.

Basketballs $22 each

Baseballs $4 each

6. Write two equations representing the total amount of money T Rich spent for his purchases.

7. Determine the total amount Rich spent at the store.

RETAIL Isabela bought three apples, three peaches, and three bananas. Apples cost 49¢ each, peaches are 59¢ each, and bananas are 22¢ each.

8. Write two equations representing the total amount of money T spent for the fruit Isabela bought.

9. Determine the total cost for the fruit.

HOMEWORK HELP	
For Exercises	**See Example(s)**
10–13	1–2
14–17	3–4
18–25	5–6

Copy and complete each equation using the Distributive Property.

10. $2(3 + 7) = (2 \cdot 3) + (2 \cdot \underline{\;?\;})$

11. $2(6 + 1) = (2 \cdot 6) + (2 \cdot \underline{\;?\;})$

12. $(9 + 1)8 = (\underline{\;?\;} \cdot \underline{\;?\;}) + (\underline{\;?\;} \cdot \underline{\;?\;})$

13. $(4 + 1)8 = (\underline{\;?\;} \cdot \underline{\;?\;}) + (\underline{\;?\;} \cdot \underline{\;?\;})$

14. $(12 \cdot 10) + (12 \cdot 7) = 12(10 + \underline{\;?\;})$

15. $6 \cdot 9 + 6 \cdot 3 = 6(9 + \underline{\;?\;})$

16. $8 \cdot 3 + 8 \cdot 2 = 8(\underline{\;?\;} + \underline{\;?\;})$

17. $7 \cdot 2 + 7 \cdot 5 = 7(\underline{\;?\;} + \underline{\;?\;})$

RETAIL Mr. Williams bought school clothes for his son. He purchased five shirts for $10 each and five pairs of pants for $20 each.

18. Write two equations that represent the total amount of money T Mr. Williams spent at the store.

19. Determine the total cost of the shirts and pants.

BUSINESS Marie and Mark work at a department store. Each earns $6 per hour. Maria works 20 hours per week. Mark works 30 hours per week.

20. Write two equations that represent the total amount of money T earned in one week by Marie and Mark.

21. Determine Marie and Mark's total earnings for the week.

ENTERTAINMENT Heather's family wants to attend the school play. Heather bought two adult tickets for $7.50 each and two student tickets for $5.00 each.

22. Write two equations that represent the total amount of money T Heather spent on tickets.

23. Determine the total cost of the tickets.

BUSINESS Mr. Jackson works 40 hours a week at $27.50 an hour. Mrs. Toney works the same number of hours at $30.00 an hour.

24. Write two equations that represent the total amount of money T earned in one week by Mr. Jackson and Mrs. Toney.

25. How much do Mr. Jackson and Mrs. Toney earn in a week?

26. *Writing in Math* Create a word problem that can be represented by the equation $T = 5(4 + 3)$.

H.O.T. Problems The Grant family went to the store and bought the items shown in the table.

Item	Cost per Item	Quantity Purchased
can of green beans	$0.50	4
can of soup	$1.00	4
bag of apples	$2.00	4
box of crackers	$3.50	2
jar of jelly	$2.50	2

Real-World Challenge

If green beans are buy one get one free, how can the expression for the cost of the items be rewritten to reflect the change?

27. Use the Distributive Property to write an expression that represents the cost of the items.

28. Find the change Ms. Grant received if she gave the clerk $30.

29. How much money did Ms. Grant give the clerk if she received $14 in change?

Choose the *best* answer.

30 Which property is used in the equation below?

$$12(5 + 4) = 60 + 48$$

A Associative Property of Multiplication

B Distributive Property

C Commutative Property of Addition

D Associative Property of Addition

31 Tom completed a 300-piece puzzle in 12 hours. At this rate, how long will it take him to complete a 750-piece puzzle?

F 18 hours   **H** 30 hours

G 24 hours   **J** 32 hours

32 What value of h makes the equation true?

$$168 \div h = 24$$

A 11   **C** 8

B 9   **D** 7

33 The steps Joseph took to simplify the expression $(36 - 3) \div 3$ are shown below.

$$(36 - 3) \div 3 = 36 - 1$$
$$= 35$$

What should Joseph have done differently in order to simplify the expression correctly?

F divided $(36 - 3)$ by 3

G divided (-3) by $(36 - 3)$

H subtracted $(3 \div 3)$ from 36

J subtracted 3 from $(36 - 3)$

34 Which of the following equations is equivalent to $m = 4(6) + 4(20)$?

A $m = 4(6) + 20$

B $m = 4 + 4(20)$

C $m = 4(6) + 4$

D $m = 4(26)$

Spiral Review

Name the property shown by each statement. (Lesson 1-4)

35. $5 + 4 = 4 + 5$

36. $10 + (3 + 12) = (10 + 3) + 12$

37. $(6 \cdot 7) \cdot 2 = 6 \cdot (7 \cdot 2)$

38. $a \cdot b = b \cdot a$

Find the value of each expression. (Lesson 1-3)

39. $28 - 3 \times 5$

40. $36 + 20 \div 4$

41. $10^2 \div (75 + 25)$

42. $100 - 5^2 + 2^3$

43. **MONEY MATTERS** A long-distance telephone call costs 25¢ for the first minute plus 10¢ for each additional minute. What is the total cost of a phone call that lasts 15 minutes? (Lesson 1-3)

Write a verbal sentence for each equation. (Lesson 1-2)

44. $10 + x = 20y$

45. $n - 20 = 100 \div p$

Progress Check 2

(Lessons 1-4 and 1-5)

▷ Vocabulary and Concept Check

Associative Property of Addition (p. 79)   Commutative Property of Multiplication (p. 78)
Associative Property of Multiplication (p. 79)   Distributive Property (p. 84)
Commutative Property of Addition (p. 77)   property (p. 77)

Choose the property that *best* completes each sentence. (Lessons 1-4 and 1-5)

1. The equation $(6 \cdot 3) \cdot 5 = 6 \cdot (3 \cdot 5)$ is an example of the
   _____?_____.

2. The equation $6 + 3 = 3 + 6$ is an example of the _____?_____.

3. The _____?_____ states that $5(4 + 1)$ equals $(5 \cdot 4) + (5 \cdot 1)$.

▷ Skills Check

Copy and complete each equation using the Commutative Property. (Lesson 1-4)

4. $6 + 5 = 5 + \underline{\ ?\ }$

5. $15 \cdot 7 = \underline{\ ?\ } \cdot \underline{\ ?\ }$

Copy and complete each equation using the Associative Property. (Lesson 1-4)

6. $(8 + 9) + 10 = \underline{\ ?\ } + (9 + 10)$

7. $(4 \cdot 5) \cdot 6 = \underline{\ ?\ } \cdot (\underline{\ ?\ } \cdot 6)$

Copy and complete each equation using the Distributive Property. (Lesson 1-5)

8. $(6 \cdot 3) + (6 \cdot 5) = \underline{\ ?\ }(3 + 5)$

9. $7(13 + 10) = (7 \cdot \underline{\ ?\ }) + (7 \cdot \underline{\ ?\ })$

Name the property shown by each statement. (Lessons 1-4 and 1-5)

10. $12 + 6 = 6 + 12$

11. $13 + (15 + 7) = (13 + 15) + 7$

12. $4(21 + 15) = (4 \cdot 21) + (4 \cdot 15)$

13. $(6 \cdot 3) + (6 \cdot 2) = 6(3 + 2)$

▷ Problem-Solving Check

MONEY MATTERS Joaquin went to a yard sale on Saturday. He bought 6 pairs of jeans for $2 each and 4 sweaters also for $2 each. (Lesson 1-5)

14. Write two equations representing the total amount of money T Joaquin spent at the store.

15. How much money did Joaquin spend?

16. **REFLECT** Write a summary that explains the Commutative, Associative, and Distributive Properties. Include examples of each property.
    (Lessons 1-4 and 1-5)

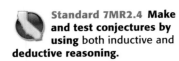

1-6 Problem-Solving Strategy: Guess and Check

Vocabulary

guess and check (p. 91)

Standard 7MR2.4 Make and test conjectures by using both inductive and **deductive reasoning.**

Standard 7AF1.1 Use variables and **appropriate operations to write an expression, an equation,** an inequality, or a system of equations or inequalities **that represents a verbal description.**

Standard 7MR2.8 Make precise calculations and check the validity of the results from the context of the problem.

Use the Strategy

Mr. Wilson is reading a newspaper article about his grandson's football game. He reads that the team made a total of 7 field goals and touchdowns, and three successful point-after-touchdown (PAT) kicks. The team scored a total of 36 points.

Mr. Wilson wants to know how many touchdowns the team scored.

Understand **What does Mr. Wilson know?**

- The team scored 3 points from PAT kicks.
- The team made at least 3 touchdowns.
- Field goals are 3 points each.
- Touchdowns are 6 points each.
- The team made a total of 7 field goals and touchdowns.
- The team scored a total of 36 points.
- PAT kicks are one point each.

Plan **What does Mr. Wilson want to know?**

Mr. Wilson wants to know the number of touchdowns. He can guess the number and calculate the total number of points. If his total is not 36 points, he needs to change his guess and try again.

Solve **Mr. Wilson guesses 3 touchdowns and 4 field goals.**

$$\text{Touchdown Points} + \text{Field-Goal Points} + \text{PAT Kicks} = \text{Total Points}$$
$$3 \cdot 6 \quad + \quad 4 \cdot 3 \quad + \quad 3 \quad = \quad 33$$

The total is too small. Mr. Wilson tries a larger number of touchdowns.

$$\text{Touchdown Points} + \text{Field-Goal Points} + \text{PAT Kicks} = \text{Total Points}$$
$$4 \cdot 6 \quad + \quad 3 \cdot 3 \quad + \quad 3 \quad = \quad 36$$

The team scored 4 touchdowns.

Check **Mr. Wilson checks other possible answers.**

- 5 touchdowns + 2 field goals + 3 extra points = 39 points
- 6 touchdowns + 1 field goal + 3 extra points = 42 points

No other possibility has a total of 36 points.

> **Study Tip**
>
> Think about your answer. What if Mr. Wilson's guess results in a total score of 39? How should he change his guess?

Use the guess-and-check strategy to solve the problem.

1. One hundred tickets were sold for the school play. Adult tickets cost $5 each. Student tickets cost $2 each. The total earned from ticket sales was $290. How many adult tickets were sold?

   Understand What do you know?

   Plan What are you trying to find? How can you figure it out?

   Solve Make a guess. Calculate the total. Change your guess if needed.

   Check How can you be sure that your answer is the only possibility?

2. *Writing in Math* The tricycles and bicycles in Henry's yard have 26 wheels and 10 seats. How many tricycles and bicycles are there? Explain why the guess-and-check strategy would be useful in solving this problem.

Problem-Solving Practice

Solve using the guess-and-check strategy.

3. **FOOD** For a party, Jaya made turkey sandwiches for the adults and peanut butter sandwiches for the children. Jaya made 10 more peanut butter sandwiches than turkey sandwiches. If she made 60 sandwiches, how many peanut butter sandwiches did she make?

4. **MONEY** Mrs. Martinez has $1.80 in dimes and quarters in her pocket. She has twice as many quarters as dimes. How many coins does she have?

SPORTS In a basketball game, the home team scored a total of 26 points. Players made 4 free throws worth 1 point each. They also made a total of nine 2-point and 3-point baskets.

5. How many 2-point baskets did the team make?

6. How many 3-point baskets did the team make?

Solve using any strategy.

7. **SAVING** Lucinda is saving money for a school trip. If the pattern continues, how much money will she have saved by the end of the 7th month?

Month	Total Saved ($)
1	15
2	30
3	45
4	60

8. **RECREATION** Pablo is going to a movie. He has to be home by 7:00. The movie runs for 120 minutes. It takes 30 minutes for Pablo to get home. What is the latest starting time Pablo can choose for the movie?

9. **FOOD** At a party, 15 teens ate cake for dessert, and 20 teens ate ice cream. Of these, 5 teens ate both cake and ice cream. How many teens ate dessert?

1-7 Other Properties

Vocabulary

Identity Property of Addition (p. 93)

Identity Property of Multiplication (p. 94)

Multiplicative Property of Zero (p. 94)

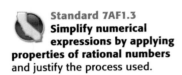 **Standard 7AF1.3** **Simplify numerical expressions by applying properties of rational numbers** and justify the process used.

 The What: I will learn how adding zero, multiplying by zero, and multiplying by one affects the result.

The Why: It is important to understand that a number does not change when multiplied by one or added to zero.

Gaspar grew three bean plants for a science project. For two weeks he measured and recorded each plant's growth. The following table shows his measurements.

Bean Plant Measurements						
Week	1			2		
Plant	1	2	3	1	2	3
Height (cm)	14	12	13	15	12	15

He noticed that Plants 1 and 3 grew. Plant 2 did not grow at all.

Plant 1: $14 + 1 = 15$ Plant 2: $12 + 0 = 12$ Plant 3: $13 + 2 = 15$

The **Identity Property of Addition** states that when you add zero to a number, the sum is that number.

IDENTITY PROPERTY OF ADDITION

Words: The sum of any number and zero is the original number.

Numbers: $45 + 0 = 45$

Algebra: For any number a, $a + 0 = 0 + a = a$.

EXAMPLES Add Zero

Complete each equation.

① $29 + \underline{\ ?\ } = 29$

$29 + \underline{\ 0\ } = 29$

② $\underline{\ ?\ } + 0 = 15$

$\underline{\ 15\ } + 0 = 15$

Your Turn Complete each equation.

a. $\underline{\ ?\ } + 43 = 43$

b. $78 = 0 + \underline{\ ?\ }$

Remember!

The Commutative Properties state that order does not matter when adding or multiplying.

$14 + 0 = 0 + 14 = 14$

$25 \times 1 = 1 \times 25 = 25$

The **Identity Property of Multiplication** states that the product of any number and one is the original number.

IDENTITY PROPERTY OF MULTIPLICATION

Pictures:

Words: When any number is multiplied by one, the product is the original number.

Numbers: $3 \cdot 1 = 3$

Algebra: For any number a, $a \cdot 1 = 1 \cdot a = a$.

EXAMPLES Multiply by One

Complete each equation.

3 $7 \cdot \underline{\ ?\ } = 7$

$7 \cdot \underline{\ 1\ } = 7$

4 $\underline{\ ?\ } \cdot 1 = 42$

$\underline{\ 42\ } \cdot 1 = 42$

Your Turn

Complete each equation.

c. $\underline{\ ?\ } \cdot 699 = 699$

d. $2 = 1 \cdot \underline{\ ?\ }$

Talk Math

WORK WITH A PARTNER Look at Examples 1–4. Talk with your partner and explain why the two properties are called Identity Properties.

Another property is the **Multiplicative Property of Zero**. The Multiplicative Property of Zero states that when you multiply a number by zero, the product is zero.

MULTIPLICATIVE PROPERTY OF ZERO

Words: When any number is multiplied by zero, the product is zero.

Numbers: $45 \cdot 0 = 0$

Algebra: For any number a, $a \cdot 0 = 0 \cdot a = 0$.

EXAMPLES Multiply by Zero

Complete each equation.

5 $0 \cdot 61 = $ _?_

$0 \cdot 61 = $ _0_

6 $45 \cdot $ _?_ $= 0$

$45 \cdot $ _0_ $= 0$

Your Turn Complete each equation.

e. $14 \cdot 0 = $ _?_

f. _?_ $\cdot 15 = 0$

These three properties apply to variables as well as numbers.

EXAMPLES Use Properties with Variables

Complete each equation.

7 $0 \cdot r = $ _?_

$0 \cdot r = $ _0_

9 _?_ $\cdot ps = ps$

_1_ $\cdot ps = ps$

8 $d + $ _?_ $= d$

$d + $ _0_ $= d$

10 _?_ $\cdot 5m = 0$

_0_ $\cdot 5m = 0$

Your Turn Complete each equation.

g. $7n \cdot $ _?_ $= 0$

h. $ur \cdot $ _?_ $= ur$

i. $s + $ _?_ $= s$

j. $pt + $ _?_ $= pt$

Real-World EXAMPLES

ANIMALS Juan takes his dog to the veterinarian. At each visit, she weighs the dog. In March, the dog weighed p pounds. In July, the dog weighed the same.

11 Write an equation that compares the weight in March to the weight in July.

$$p \quad + \quad 0 \quad = \quad p$$

original weight   change in weight   final weight

12 Name the property represented by the equation.

Identity Property of Addition

Your Turn

RETAIL Mr. O'Malley sells cars. He receives $550 for each car he sells. The first week in July, he sold 1 car. The second week, he did not sell any cars.

k. Write an equation that shows how much money he earned in the first two weeks of July. Let T represent the total amount of money he earned in the first two weeks of July.

l. Name the property represented by the equation.

ANIMALS Suppose the dog weighed 27 pounds in March. How much did it weigh in July?

Examples 1–12
(pages 93–95)

VOCABULARY Write two equations that show each property.

1. Identity Property of Addition

2. Identity Property of Multiplication

3. Multiplicative Property of Zero

Copy and complete each equation.

Examples 1–10
(pages 93–95)

4. $26 + \underline{\ ?\ } = 26$

5. $523 = 0 + \underline{\ ?\ }$

6. $353 = \underline{\ ?\ } \cdot 353$

7. $\underline{\ ?\ } \cdot 1 = 409$

8. $0 \cdot 6 = \underline{\ ?\ }$

9. $0 = 37 \cdot \underline{\ ?\ }$

10. $0 \cdot f = \underline{\ ?\ }$

11. $\underline{\ ?\ } \cdot 4j = 0$

Examples 11–12
(page 95)

RETAIL Mr. Rollins sets the weekly price of oil in an automotive store. One week, oil was $1.29 a quart. The next week it was $1.29 a quart.

12. Write an equation that compares the price for the oil each week.

13. Name the property represented by the equation.

14. **Talk Math** Which is easier for you to calculate? Explain to a partner.

$$99 + 0 \qquad 99 \cdot 0$$

For Exercises	See Example(s)
15–18	1–2
19–22	3–4
23–26	5–6
27–32	7–10
33–36	11–12

HOMEWORK HELP

Copy and complete each equation.

15. $3 + \underline{\ ?\ } = 3$

16. $\underline{\ ?\ } + 32 = 32$

17. $\underline{\ ?\ } + 0 = 57$

18. $98 = \underline{\ ?\ } + 0$

19. $38 \cdot \underline{\ ?\ } = 38$

20. $\underline{\ ?\ } \cdot 19 = 19$

21. $\underline{\ ?\ } \cdot 1 = 88$

22. $67 = 1 \cdot \underline{\ ?\ }$

23. $0 \cdot 12 = \underline{\ ?\ }$

24. $29 \cdot 0 = \underline{\ ?\ }$

25. $31 \cdot \underline{\ ?\ } = 0$

26. $0 = \underline{\ ?\ } \cdot 159$

27. $0 \cdot h = \underline{\ ?\ }$

28. $\underline{\ ?\ } = 0 \cdot 4v$

29. $c + \underline{\ ?\ } = c$

30. $he + \underline{\ ?\ } = he$

31. $xy = \underline{\ ?\ } + xy$

32. $5e \cdot \underline{\ ?\ } = 0$

RETAIL Julian works in a clothing store. He is paid $2 for each shirt he sells, $3 for each pair of pants he sells, and $1 for each smaller item he sells. One day he sells 5 pairs of pants, 4 ties, and no shirts.

33. Write an equation that shows how much he earns for his sales.

34. Name the property used to find out how much his shirt sales affected his total sales.

HEALTH Erin was not feeling well. She knows a normal temperature is 98.6°F. Her temperature is 98.6°F.

35. Write an equation that compares her temperature and normal temperature.

36. Name the property represented by the equation.

37. *Writing in Math* Ginnie works in a music store. She is paid $125 a week plus $1 for each CD she sells. Explain how to find Ginnie's weekly earnings using the properties in this lesson.

38. **H.O.T.** Problem Jason pays Cami $35 a day to care for his child. One week, the child attended day care for 4 days. Use the Multiplicative Property of Zero and the Identity Property of Addition to find how much Jason paid for child care that week.

STANDARDS PRACTICE

Choose the *best* answer.

39 The table shows how much a student spent when shopping for ingredients for a fruit salad. Which property is used to find the amount the student spent on oranges?

Cost of Fruit					
Fruit	bananas	apples	oranges	berries	pears
Price/lb	0.49	1.29	1.19	4.99	0.99
Cost ($)	1.75	2.98	0	4.49	1.98

   A Multiplicative Property of Zero

   B Identity Property of Multiplication

   C Identity Property of Addition

   D Identity Property of Multiplication, Identity Property of Addition, and Multiplicative Property of Zero

Spiral Review

Solve using the guess-and-check strategy. (Lesson 1-6)

40. **RECREATION** Marty has a total of 28 CDs and DVDs. He has three times more DVDs than he has CDs. How many of each does he have?

Name the property shown by each equation. (Lessons 1-4 and 1-5)

41. $25 + n = n + 25$

42. $5(a + b) = 5a + 5b$

43. $(2 + 1) + 3 = 2 + (1 + 3)$

44. $(r \cdot p) \cdot rp = r \cdot (p \cdot rp)$

Simplify each expression. (Lesson 1-3)

45. $3[4(5 - 3) \div 2^2]$

46. $14 - 12 \div 4 \cdot 2 + 8$

Math Lab
Simplify Expressions Using Tiles

Materials

algebra tiles

 Standard 7AF1.3 Simplify numerical expressions by applying properties of rational numbers and justify the process used.

Standard 7MR2.5 Use a variety of methods, such as words, numbers, symbols, charts, graphs, tables, diagrams, and **models,** to **explain mathematical reasoning.**

 The *What:* I will use algebra tiles to model and regroup expressions.

 The *Why:* Tiles with like shapes show algebraic expressions with like terms.

Algebra tiles can be used to model and simplify algebraic expressions.

REPRESENTING EXPRESSIONS WITH ALGEBRA TILES

- Use tiles to represent numbers and variables.

| 1 | x |
| numbers | variables |

- Group all like tiles together. Like tiles are the same size, shape, and color.

- Count the number of each type of tile.

- Write the simplified expression.

EXAMPLES   Model Expressions

Use algebra tiles to model each expression.

1 $2x$

$2x$

2 $2x + 1$

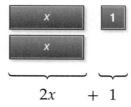

$2x \quad + 1$

Your Turn Use algebra tiles to model each expression.

a. $4x$

b. 6

c. $x + 2$

d. $5 + 2x$

3 Regroup algebra tiles to simplify $2 + x + 6$.

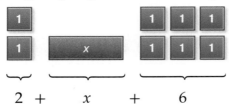

$$2 \;+\; x \;+\; 6$$

$$= \;x\; +\; 8$$

Your Turn
Regroup algebra tiles to simplify each expression.

e. $3x + 2 + x$ f. $3 + 2x + 4$

g. $2x + 3 + 4x + 6$ h. $4x + 1 + x + 5$

Analyze the Results

Use or draw algebra tiles to model each expression.

1. 4 2. 11

3. $3x$ 4. $8b$

5. $n + 5$ 6. $2x + 3$

7. $7 + 4d$ 8. $6 + k$

Regroup algebra tiles to simplify each expression.

9. $x + 4 + 3x$ 10. $4x + 10 + 2x$

11. $7 + 2x + 6$ 12. $3 + 3x + 8$

13. $x + 8 + 4x + 5$ 14. $5x + 2 + x + 6$

15. $2 + x + 3 + 4x$ 16. $7 + 6x + 3 + 2x$

17. **Talk Math** Talk with a partner about how you know how to group like tiles.

18. **WHO IS CORRECT?** Steve says that $x + 3 + 2x - 1 = 3x + 3$. Linh disagrees. She thinks that $x + 3 + 2x - 1 = 3x + 2$. Use algebra tiles or drawings to explain who is correct.

Vocabulary

term (p. 100)

like term (p. 100)

unlike term (p. 100)

coefficient (p. 101)

simplest form (p. 102)

evaluate (p. 103)

Standard 7AF1.3
Simplify numerical expressions by applying properties of rational numbers and justify the process used.

The *What*: I will apply the Distributive, Associative, and Commutative Properties to combine like terms.

The *Why*: Combining like terms results in simplified expressions. Simplified expressions are easier to use when calculating costs and prices.

Students in the Math Club charged d dollars to wash a car. They washed 15 cars on Saturday and 20 cars on Sunday.

cars washed on Saturday ⟶   ⟵ cars washed on Sunday

$15d + 20d$ ⟵ total amount

In the expression $15d + 20d$, $15d$ and $20d$ are called terms. A **term** is a number, a variable, or a product of numbers and variables. Terms are separated by $+$ and $-$ signs.

$$15d + 20d \qquad\qquad 5b + c - 3$$

two terms   three terms

The terms $15d$ and $20d$ have the same variable. They are called **like terms**. The terms $5b$ and c in the expression $5b + c - 3$ are **unlike terms** because the variables are not alike.

Terms with the same variable but different exponents are called unlike terms, such as $8m$ and $6m^3$.

Remember!
Remember that r is the same as $1 \cdot r$.

Like Terms	Unlike Terms
$7r + r$	$7r + s$
$8m - 5m$	$8m^2 - 5m$
$4xy + 9xy - 2xy$	$4xy + 9yz - 2xz$
$3y^2 + 6y^2 - 2y^2$	$3y + 6y^2 - 2y^5$

Determine if each expression has like terms. If so, state them.

1 $5x + 8x$

$5x + 8x$

like terms: $5x$, $8x$

2 $4d - 3c$

$4d - 3c$

unlike terms

3 $6ab + 2bc + 7bc$

$6ab + 2bc + 7bc$

like terms: $2bc$, $7bc$

4 $4b^2 - b + 8b^2$

$4b^2 - b + 8b^2$

like terms: $4b^2$, $8b^2$

Your Turn Determine if each expression has like terms. If so, state them.

a. $4p + 7t$

b. $8n - 2n$

c. $14rs + 10rs + 7$

d. $4d^2 - 5d + 2d^2$

Talk Math

WORK WITH A PARTNER Say four like terms using the variable x. Listen as your partner says four like terms using the variable y.

Coefficient

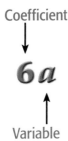

Variable

A **coefficient** is a number that is multiplied by a variable in an algebraic expression. In the expression $6a$, the coefficient is 6. The variable is a.

The order of operations, along with the Commutative, Associative, and Distributive Properties, can be used to simplify expressions that contain like terms.

EXAMPLE Simplify Expressions

5 Simplify $3t + t$. Provide a reason for each step.

$3t + t = 3 \cdot t + 1 \cdot t$ Identity Property of Multiplication

$= (3 + 1)t$ Distributive Property

$= 4t$ Add 3 and 1.

Your Turn Simplify each expression. Provide a reason for each step.

e. $4y + 7y$

f. $x + 6x$

g. $b + 2b$

h. $3(x + 4)$

EXAMPLE Simplify Expressions

6 Simplify $5p + 7 + 3p$. Provide a reason for each step.

$$5p + 7 + 3p = 5p + 3p + 7 \quad \text{Commutative Property}$$
$$= (5 + 3)p + 7 \quad \text{Distributive Property}$$
$$= 8p + 7 \quad \text{Add 5 and 3.}$$

Your Turn Simplify the expression. Provide a reason for each step.

i. $4y + 10 + 7y$ j. $x + 9 + 6x$

k. $12 + 2n + 8n$ l. $6 + b + 10b$

In Example 6, the terms $8p$ and 7 are unlike terms. The expression $8p + 7$ cannot be simplified any further. Expressions whose terms cannot be combined are in **simplest form.**

EXAMPLE Simplify Expressions

7 Simplify $7(3h + 4)$. Provide a reason for each step.

$$7(3h + 4) = (7 \cdot 3h) + (7 \cdot 4) \quad \text{Distributive Property}$$
$$= 21h + 28 \quad \text{Multiply 7 and 3. Multiply 7 and 4.}$$

The terms $21h$ and 28 are unlike terms.

The expression $21h + 28$ is in simplest form.

Your Turn Simplify the expression. Provide a reason for each step.

m. $6(5y + 4)$ n. $8(4k + 3)$

o. $2(x + y)$ p. $5(7c + 4)$

EXAMPLE Simplify Expressions

8 Simplify $6r + 5(2r + 3s)$. Provide a reason for each step.

$$6r + 5(2r + 3s) = 6r + 5 \cdot 2r + 5 \cdot 3s \quad \text{Distributive Property}$$
$$= 6r + 10r + 15s \quad \text{Multiply } 5 \cdot 2 \text{ and } 5 \cdot 3.$$
$$= (6 + 10)r + 15s \quad \text{Distributive Property}$$
$$= 16r + 15s \quad \text{Add 6 and 10.}$$

The terms $16r$ and $15s$ are unlike terms.

The expression $16r + 15s$ is in simplest form.

Your Turn Simplify each expression. Provide a reason for each step.

q. $3(2c + 4d)$ r. $10(4a + 5b) - 20a$

s. $2x + 3(x + xy)$ t. $7b + 4(b + bc)$

WORK WITH A PARTNER Explain the meaning of the phrases *like terms* and *simplest form*. Then, work with your partner to give an example of each.

When you know the value of a variable, substitute the value in the expression and simplify. This is called **evaluating** the expression.

EXAMPLES Write and Evaluate Expressions

FINANCE Students in the Math Club charged d dollars to wash a car. The students washed 15 cars on Saturday and 20 cars on Sunday. They also received $40 in donations.

FINANCE The Math Club received $40 in donations.

9 Write an expression in simplest form to represent the total amount of money raised from the car wash.

15 cars washed Saturday	·	d cost per car	+	20 cars washed Sunday	·	d cost per car	+	40 donations	=	total money raised

$$15d + 20d + 40 = (15 + 20)d + 40 \qquad \text{Distributive Property}$$
$$= 35d + 40 \qquad \text{Order of Operations}$$

10 If the students charged $5 per car, how much money did they raise?

$$35d + 40 = 35(5) + 40 \qquad \text{Substitute 5 for } d.$$
$$= 175 + 40 \qquad \text{Order of Operations}$$
$$= 215$$

The students raised $215.

Your Turn

FINANCE Charlie worked 4 hours on Friday and 5 hours on Saturday. He earned $20 in tips. Let x represent Charlie's hourly rate of pay.

u. Write an expression in simplest form to represent Charlie's total earnings.

v. If Charlie's hourly rate of pay is $10, how much money did he earn including tips?

Vocabulary Review
term
like term
unlike term
coefficient
simplest form
evaluate

Examples 1–4
(page 101)

VOCABULARY Write three examples of each.

1. like terms

2. unlike terms each having a coefficient of 5

Examples 1–4
(page 101)

Determine if each expression has like terms. If so, state them.

3. $6y - 3y$

4. $7a^3 + 8a$

5. $5b + 9b^2$

6. $8rs + rs$

Examples 5–8
(pages 101–102)

Simplify each expression. Provide a reason for each step.

7. $13f + 3f$

8. $8m + m$

9. $6y + 4 + 3y$

10. $8(6m + 5)$

11. $8(4g + 7n)$

12. $10q + 3(2q + 8qv)$

Examples 9–10
(page 103)

BUSINESS At a bake sale, the freshmen sold 100 cookies and 50 brownies. The same price was charged for cookies and brownies. Let n represent the amount charged for each cookie and brownie.

13. Write an expression in simplest form to represent the total amount of money raised by the freshmen.

14. If each cookie and brownie cost $0.50, how much money did the freshmen raise?

15. Write an expression in simplest form to represent the total amount of money raised if the cookies and brownies have different prices.

Skills, Concepts, and Problem Solving

For Exercises	See Example(s)
16–23	1–4
24–27	5
28–29	6
30–31	7
32–35	8
36–39	9–10

HOMEWORK HELP

Determine if each expression has like terms. If so, state them.

16. $4b + b$

17. $9a - 8a$

18. $6m - 4m^2$

19. $3n^2 + 4n$

20. $6a^2 + 4a$

21. $3y^3 + 3y$

22. $6mn + mn$

23. $8t + 3qt$

Simplify each expression. Provide a reason for each step.

24. $9c + 3c$

25. $15n + 3n$

26. $b + 5b$

27. $3n + n$

28. $7g + 30 + 10g$

29. $2n + 5 + 4n$

30. $6(9y + 3)$

31. $4(8j + 5)$

32. $4(5r + 6m)$

33. $8(2v + 9w)$

34. $4g + 6(5g + 3gh)$

35. $10c + 8(2cd + 3c)$

RETAIL Temecka bought 5 CDs on Saturday and 7 CDs on Tuesday. She paid a total of $4.80 in sales tax. Let y represent the cost of each CD.

36. Write an expression to represent Temecka's total cost.

37. If each CD costs $10, what was Temecka's total cost?

LITERATURE Emily started on page 14 of her book. She read for 2 hours on Thursday and 3 hours on Friday. Let r represent the number of pages she read per hour.

38. Write an expression in simplest form to represent the total number of pages that Emily has read.

39. If Emily read 25 pages an hour, how many pages total has she read?

40. *Writing in Math* Tomas bought cartons of milk on Monday, Tuesday, and Wednesday. The cost of a carton of milk is m. Write a word problem in which you combine like terms to find the total cost of the cartons of milk.

41. **H.O.T. Problem** Gwen joined a video club in June. She paid a membership fee of $12 and pays $2 to rent a movie. Rose rents movies without a membership. She pays $5 to rent a movie. Both girls rented seven movies in June. Who paid less money?

STANDARDS PRACTICE

42 A class is recycling paper. The graph shows how many pounds of paper they recycled last month. Which choice shows the weight of the recycled paper in simplest terms?

A $(225 + 150 + 350 + 400)$

B $(350 + 225 + 400 + 150)$

C $(350 + 225) + (400 + 150)$

D $1,125$

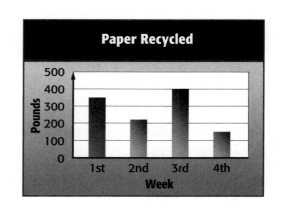

Paper Recycled

Spiral Review

Copy and complete each equation. (Lesson 1-7)

43. $256 + \underline{\ ?\ } = 256$

44. $18 \cdot 1 = \underline{\ ?\ }$

45. $5 \cdot \underline{\ ?\ } = 0$

46. $6y = \underline{\ ?\ } + 6y$

47. $1 \cdot 4k = \underline{\ ?\ }$

48. $0 = 9t \cdot \underline{\ ?\ }$

49. **FITNESS** Raoul paid a membership fee of $50 to join a health club. He is also charged a monthly fee of $30. Write an expression for the amount Raoul pays after 5 months. (Lesson 1-2)

Study Guide

Study Tips
• Make lists instead of writing sentences.
• Use symbols instead of words.

Understanding and Using the Vocabulary

After completing the chapter, you should be able to define each term, property, or phrase and give an example of each.

algebraic expression (p. 65)	Identity Property of Addition (p. 93)
Associative Property of Addition (p. 79)	Identity Property of Multiplication (p. 94)
Associative Property of Multiplication (p. 79)	like terms (p. 100)
base (p. 72)	Multiplicative Property of Zero (p. 94)
check (p. 60)	numerical expression (p. 65)
coefficient (p. 101)	order of operations (p. 70)
Commutative Property of Addition (p. 77)	plan (p. 60)
Commutative Property of Multiplication (p. 78)	property (p. 77)
Distributive Property (p. 84)	simplest form (p. 102)
equation (p. 67)	solve (p. 60)
evaluate (p. 103)	term (p. 100)
exponent (p. 72)	understand (p. 60)
guess and check (p. 91)	unlike terms (p. 100)
	variable (p. 65)

Complete each sentence with the correct mathematical term or phrase.

1. $8m$ and $4n$ cannot be combined because they are _____?_____.

2. In the expression $5x + 3$, 5 is a(n) _____?_____.

3. A mathematical sentence that contains an equal sign is a(n) _____?_____.

4. A(n) _____?_____ contains numbers, variables, and may contain operations.

5. A statement that is true for all numbers is a(n) _____?_____.

6. According to the _____?_____, the order in which two numbers are added does not change the sum.

7. A letter that stands for a number is a(n) _____?_____.

8. Terms that contain exactly the same variables are called _____?_____.

9. In the number 4^5, 5 is a(n) _____?_____.

10. According to the _____?_____, zero can be added to any number without changing the value of the number.

Skills and Concepts

Objectives and Examples

Review Exercises

LESSON 1-1 pages 60–64

Approach problems and how to solve them.

Step 1 **Understand** What do you know?

Step 2 **Plan** What do you want to know? What strategy will you use?

Step 3 **Solve** Use the strategy.

Step 4 **Check** Does your answer make sense? Are there other possible answers?

Solve using the four-step problem-solving plan.

11. **FINANCE** Lupe earns $4 an hour, and Phillip earns $5 an hour. They work a total of 6 hours and earn $27. How many hours does Lupe work?

12. **SPORTS** Miguel owns four types of sports socks and two pairs of running shoes. How many combinations of socks and shoes does he have?

LESSON 1-2 pages 65–69

Represent verbal expressions and sentences as algebraic expressions and equations.

Write an algebraic expression for 7 decreased by the quotient of x and 2.

7	decreased by	quotient of x and 2
7	−	$x \div 2$

Write an algebraic expression for each verbal expression.

13. the product of 5 and n

14. 8 less than m

15. the sum of 2 and three times x

16. the quotient of a and b

Write a sentence for each equation.

17. $30 - x = 4$

18. $2y + 6 = 14$

LESSON 1-3 pages 70–75

Use the order of operations to evaluate expressions.

$$2 \cdot 7 + 2 \cdot 3 = 14 + 2 \cdot 3 \quad \text{Multiply 2 and 7.}$$
$$= 14 + 6 \quad \text{Multiply 2 and 3.}$$
$$= 20 \quad \text{Add 14 and 6.}$$

Find the value of each expression.

19. $3 + 8 \div 2$

20. $12 \div 4 + 15 \cdot 3$

21. $29 - 3(6 - 4)$

22. $4(11 - 1) + 2 \cdot 8$

23. $8 + 12 \div 4 \cdot 3$

24. $10^2(3 + 1)$

Objectives and Examples

Review Exercises

LESSON 1-4 pages 77–83

Use the Commutative and Associative Properties.

$3 + x + 2 = 3 + 2 + x$

Commutative Property of Addition

$5 \cdot (8 \cdot y) = (5 \cdot 8) \cdot y$

Associative Property of Multiplication

Name the property shown by each statement.

25. $6 + (7 + b) = (6 + 7) + b$

26. $2 \cdot c \cdot 10 = 2 \cdot 10 \cdot c$

27. $a + b + c = a + c + b$

28. $(g \cdot 4) \cdot 6 = g \cdot (4 \cdot 6)$

LESSON 1-5 pages 84–89

Use the Distributive Property.

Complete the equation using the Distributive Property.

$8(3 + 7) = 8 \cdot \underline{?} + 8 \cdot \underline{?}$

$8(3 + 7) = 8 \cdot \underline{3} + 8 \cdot \underline{7}$

Copy and complete each equation using the Distributive Property.

29. $3(8 + 4) = (3 \cdot 8) + (3 \cdot \underline{?})$

30. $(6 \cdot 4) + (6 \cdot 7) = 6(4 + \underline{?})$

31. $12(4 + 6) = (\underline{?} \cdot \underline{?}) + (\underline{?} \cdot \underline{?})$

LESSON 1-7 pages 93–97

Use the Identity Properties and Multiplicative Property of Zero.

$10 + 0 = 10$

Identity Property of Addition

$12 \cdot 1 = 12$

Identity Property of Multiplication

$7 \cdot 0 = 0$

Multiplicative Property of Zero

Copy and complete each equation.

32. $5 + \underline{?} = 5$

33. $0 = 4 \cdot \underline{?}$

34. $6 = \underline{?} \cdot 6$

35. $\underline{?} \cdot x = x$

LESSON 1-8 pages 100–105

Apply the Distributive, Associative, and Commutative Properties to combine like terms.

$$10x + 2x + 7 = 12x + 7$$

Simplify each expression. Provide a reason for each step.

36. $9m + 3m$

37. $12x + 3y + 4x$

38. $4u + 12z + 20 + 8u + 6z + 4$

Chapter Test

▷ Vocabulary and Concept Check

1. Write an algebraic expression that contains two different operations and at least two unlike terms.

▷ Skills Check

Write an algebraic expression for each verbal expression.

2. x increased by 12

3. the product of 5 and y

Find the value of each expression. Show your work.

4. $13 + 4 \cdot 5$

5. $12 + 6 \div 3 - 4$

6. $3(8 + 2) - 7$

7. $15 + 4^2 \cdot 3$

Name the property shown by each statement.

8. $8 + 10 = 10 + 8$

9. $6(m + 2) = 6 \cdot m + 6 \cdot 2$

10. $14 + 0 = 14$

11. $xy = yx$

12. $0 \cdot 24x = 0$

13. $(a + b) + c = a + (b + c)$

Simplify each expression.

14. $5v + 3v$

15. $3t + 6u - 2t + u$

16. $6x + 9y + 5 + 4x + 3y + 1$

17. $12n + m + 9 - 3n + 7m + 5$

▷ Problem-Solving Check

ENTERTAINMENT Joshua and three friends went to the movies on Friday night. Tickets cost $8.00 each. Each of them bought a soda for $4.50, and two of them bought popcorn for $5.50.

18. Let T represent total amount spent. Write an equation that represents the total amount spent by all four friends.

19. Find the total amount spent.

20. **NUMBER SENSE** Regina has a total of 8 quarters and dimes. She has a total of $1.55. How many dimes does she have?

PART 1 Multiple Choice

Choose the *best* answer.

1 The table shows how much Miranda spent on her lunch during one week.

Day	Amount
Monday	$3.05
Tuesday	$3.15
Wednesday	$3.65
Thursday	$2.55
Friday	$4.05

Which is the *best* estimate for the total amount Miranda spent on lunches that week?

A $13

B $17

C $19

D $20

2 What is the order of operations for the expression $27 \div (4 + 5) \cdot 2$?

F divide, add, multiply

G divide, multiply, add

H add, divide, multiply

J add, multiply, divide

3 Which property is illustrated by the equation below?

$$6 + (2 + 5) = (2 + 5) + 6$$

A Associative Property of Addition

B Commutative Property of Addition

C Distributive Property

D Identity Property of Addition

4 Which is an algebraic expression?

F 12

G $3 \cdot 4 + 8$

H $7w + 5 = 12$

J $2x + 4$

5 Which expresses the equation below in words?

$$3(x - 4) = 7x + 5$$

A Three times a number minus four is seven times that number plus five.

B Three times a number minus four is seven times the sum of that number and five.

C Three times the difference of a number and four is seven times that number plus five.

D Three times the difference of a number and four is seven times the sum of that number and five.

6 Amy had 20 grapes. Diego gave her 18 more. Amy ate some grapes and had 13 left. Which equation shows how many grapes Amy ate?

F $(20 + 18) + x = 13$

G $(20 + 18) - x = 13$

H $(20 - 18) + x = 13$

J $(20 - 18) - x = 13$

7 Which expression is *not* equivalent to $5(2x + 4y) + 6y$?

A $5 \cdot 2x + 4y + 6y$

B $5 \cdot 2x + 5 \cdot 4y + 6y$

C $10x + 20y + 6y$

D $10x + 26y$

8 The expression $5x + x$ is the same as $5x + 1x$. Which property states that this is true?

F Identity Property of Addition

G Identity Property of Multiplication

H Distributive Property

J Multiplicative Property of Zero

Record your answers on the answer sheet provided by your teacher or on a separate sheet of paper.

9 The graph shows the number of flat-screen computer monitors sold during the last 6 months at Marvel Computers. Estimate the monthly average number of monitors sold to the nearest 10.

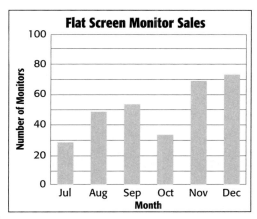

10 One pound of coffee makes 100 cups. If 300 cups of coffee are served at each football game, how many pounds are needed for 7 games?

11 The eighth grade ordered 216 hot dogs for their end-of-the-year party. There are 8 hot dogs in a single package. How many packages did they buy?

12 What is the value of the expression?
$$\frac{5 - 3}{8 + 6 \div 2 \cdot 4}$$

13 Mrs. Miller joined a book club. The table shows how much money Mrs. Miller has spent on books each of the first five months.

Suppose x represents the cost of 1 book. Write an expression that represents the total amount that Mrs. Miller has spent.

Month	1st	2nd	3rd	4th	5th
Amount Spent ($)	x	$3x$	$5x$	x	$4x$

Record your answers on the answer sheet provided by your teacher or on a separate piece of paper.

14 To raise money for a charity, the ninth-grade science class asked a student group to perform a benefit concert in the school's 400-seat auditorium.

Tickets for the 180 seats near the stage sold for $30 each. Tickets for the other seats were sold at a lower price. The benefit concert sold out, raising a total of $9,360.

a. Write an equation for the price p of each ticket in the section *not* near the stage.

b. Find the price of each ticket in the section *not* near the stage.

NEED EXTRA HELP?														
If You Missed Question...	1	2	3	4	5	6	7	8	9	10	11	12	13	14
Go to Lesson...	1-1	1-3	1-4	1-2	1-2	1-2	1-8	1-7	1-1	1-6	1-2	1-3	1-1	1-2
For Help with Algebra Readiness Standard...	MR2.0	7NS1.2	7AF1.3	7AF1.1	7AF1.1	7AF1.1	7AF1.3	7AF1.3	MR2.0	MR2.0	7AF1.1	7NS1.2	MR2.0	7AF1.1

CHAPTER 2

Integers and Equations

 Here's What I Need to Know

◀— **Standard ALG5.0 Students solve multi-step problems, including word problems, involving linear equations in one variable and provide justification for each step.**

◀— **Standard 7NS1.2 Add, subtract, multiply, and divide rational numbers (integers, fractions, and terminating decimals) and take positive rational numbers to whole-number powers.**

◀— **Standard 7AF1.3 Simplify numerical expressions by applying properties of rational numbers and justify the process used.**

Vocabulary Preview

equation A mathematical sentence stating that two expressions are equal (p. 114)

$$6 + 4 = 3 + 7$$

integers The set of counting numbers, their additive inverses, and zero (p. 119)

$$\{ \ldots, -3, -2, -1, 0, 1, 2, 3, \ldots \}$$

 The What

I will learn to:

- solve equations.
- add, subtract, multiply, and divide integers.

🌐 **The Why**

Integers are used in bank accounts, temperature, or in sports. You can solve equations to determine the cost of renting a movie at the local video store.

Option 1

Math Online Are you ready for Chapter 2? Take the Online Readiness Quiz at ca.algebrareadiness.com to find out.

Option 2

Get ready for Chapter 2. Review these skills.

Vocabulary and Skills Review

Match each sentence with the correct vocabulary word.

1. In $9 \cdot 8 = 72$, 72 is the __?__.

2. A(n) __?__ is a mathematical sentence stating that two expressions are equal.

3. In $x - 4$, x is the __?__.

A. equation

B. product

C. sum

D. variable

Add, subtract, multiply, or divide.

4. $43 + 28$

5. $65 - 49$

6. $12 \cdot 6$

7. $63 \div 7$

Simplify each expression.

8. $5y + y$

9. $7m - 2m$

10. $10c - 2 + 2c$

11. $8r + 3 - 4r$

Note-taking

Tips
Summarize the most important ideas.

Chapter Instruction	My Notes
You can **solve** an equation by finding the value (or values) of the variable that make the equation true. (p. 114)	To solve an equation, find numbers for variables to make it true.
You will often have to write an equation for a word problem. First **define** the variable by identifying the unknown and choosing a variable to represent it. Then use numbers and operations to write the equation. (p. 116)	For a word problem, choose a variable then write an equation for the unknown.

2-1 Equations

Vocabulary

equation (p. 114)

solve (p. 114)

solution (p. 114)

define (p. 116)

Standard 6AF1.1 Write and solve one-step linear equations in one variable.

 The What: I will write and solve equations using substitution and mental math.

The Why: Equations can represent costs based on a per-use charge, such as online movie rental clubs and cell phones.

An **equation** is a mathematical sentence that contains an equal sign (=). Each side of an equation is an expression.

EXPRESSIONS	EQUATIONS
$3 + 5$ and $5 + 3$	$3 + 5 = 5 + 3$
$3x - 2$ and 7	$3x - 2 = 7$
$a + b$ and 10	$a + b = 10$
A and $\ell \cdot w$	$A = \ell \cdot w$

An online movie rental company charges $5 for each movie rented. The equation $5m = 20$ shows the number of movies that can be rented for $20.

$$5m \qquad = \qquad 20$$

movie cost ($5) times number of movies (m) — is equal to — total

You can **solve** an equation by finding the value or values of the variable that make the equation true.

Any number that makes an equation true is called a **solution.** The solution to $5m = 20$ is 4 because $5(4) = 20$. So, 4 movies can be rented for $20.

WORK WITH A PARTNER A store charges $7 for each movie rented. Say an equation that represents how many movies you could rent for $21. What is the solution?

EXAMPLE Solve Equations by Substitution

1 Which of the numbers 1, 2, 3, or 4 is the solution for $x + 5 = 8$?

$x + 5 = 8$	Replace x with each number to find the solution.
$1 + 5 \neq 8$	Try $x = 1$.
$2 + 5 \neq 8$	Try $x = 2$.
$3 + 5 = 8$ ✓	Try $x = 3$. This is a true statement.
$4 + 5 \neq 8$	Try $x = 4$.

The solution is 3.

Remember!

When you solve by substitution, try the given numbers in order until you find the answer.

Your Turn

Which of the numbers 6, 7, 8, or 9 is the solution for each equation?

a. $g + 2 = 11$

b. $2 = m \div 3$

EXAMPLES Solve Equations Using Mental Math

Solve each equation using mental math.

2 $8 + y = 15$

$8 + y = 15$	What number added to 8 equals 15?
$y = 7$	$8 + 7 = 15$

3 $2x = 8$

$2x = 8$	What number multiplied by 2 equals 8?
$x = 4$	$2 \times 4 = 8$

4 $\frac{n}{3} = 5$

$\frac{n}{3} = 5$	What number divided by 3 equals 5?
$n = 15$	$15 \div 3 = 5$

Your Turn

Solve each equation using mental math.

c. $9 + x = 17$

d. $12 - r = 9$

e. $9n = 36$

f. $\frac{h}{2} = 4$

You will often have to write an equation in order to solve a word problem. First **define** the variable by identifying the unknown and choosing a variable to represent it. Then use numbers and operations to write the equation.

EXAMPLES Write and Solve Equations

Write an equation. Solve your equation.

5 AGES Ed is twice as old as his sister. Ed is 14. How old is Ed's sister?

Let s represent the sister's age.   Define the variable.

$$\underbrace{14}_{\text{Ed's age}} \overset{=}{\underset{\text{is}}{\wedge}} \underbrace{2s}_{\substack{\text{twice his} \\ \text{sister's age}}}$$

Write an equation for Ed's age. The word *twice* means two times. What number multiplied by 2 equals 14?

$$2s = 14$$
$$s = 7$$

Ed's sister is 7 years old.

6 NUMBER SENSE A number is 5 more than another number. If the greater number is 17, what is the lesser number?

Let n represent the lesser number.   Define the variable.

$$\underbrace{n}_{\text{number}} \underbrace{+\;5}_{\text{5 more than}} = \underbrace{17}_{\substack{\text{greater} \\ \text{number}}}$$

Write an equation for the greater number. What number increased by 5 equals 17?

$$n + 5 - 5 = 17 - 5$$
$$n = 12$$

The lesser number is 12.

Your Turn

Write an equation. Solve your equation.

g. **NUMBER SENSE** Three times a number is 18. Find the number.

h. **AGES** Ana is 16. She is 4 years older than her brother. How old is Ana's brother?

Talk Math

WORK WITH A PARTNER Say three equations that have a solution of 5. Listen as your partner says three other equations with a solution of 5.

Vocabulary Review
equation
solve
solution
define

Examples 1–6
(pages 115–116)

VOCABULARY

1. Write an equation that involves addition.

2. Write an equation with a solution of 6.

Example 1
(page 115)

Which of the numbers 2, 4, 8, or 12 is the solution for each equation?

3. $x + 3 = 7$

4. $9 = n - 3$

Which of the numbers 12, 15, 18, or 20 is the solution for each equation?

5. $p \div 2 = 10$

6. $2m = 24$

Examples 2–4
(page 115)

Solve each equation using mental math.

7. $x + 7 = 14$

8. $9d = 27$

9. $g - 7 = 15$

10. $\frac{n}{4} = 6$

Examples 5–6
(page 116)

Write an equation. Solve your equation.

11. **PHOTOGRAPHY** Four photographs can be made from a sheet of printing paper. How many sheets of printing paper are needed to make 24 photographs?

12. **MARKETING** Shoe Palace sells basketball shoes for $50. This is $10 more than the price of the same shoes at Foot World. How much do the shoes cost at Foot World?

Skills, Concepts, and Problem Solving

HOMEWORK HELP	
For Exercises	**See Example(s)**
13–20	1
21–28	2–4
29–32	5–6

Which of the numbers 3, 4, 5, or 6 is the solution for each equation?

13. $4 + y = 7$

14. $11 = k + 6$

15. $2 = g - 4$

16. $b - 2 = 2$

Which of the numbers 9, 12, 15, or 18 is the solution for each equation?

17. $5q = 45$

18. $60 = 4n$

19. $4 = x \div 3$

20. $2 = h \div 9$

Solve each equation using mental math.

21. $y + 2 = 13$

22. $r + 11 = 17$

23. $19 - q = 5$

24. $17 = k - 3$

25. $5x = 25$

26. $9 = 3n$

27. $7 = \frac{y}{3}$

28. $\frac{p}{2} = 8$

Write an equation. Solve your equation.

SPORTS Surf lessons cost $20 per hour.

29. Write and solve an equation to determine how many hours of lessons you can buy for $60.

30. Write and solve an equation to determine how many hours of lessons you can buy for $100.

31. **EMPLOYMENT** Toya worked 3 more days this month than she worked last month. She worked 15 days this month. How many days did Toya work last month?

32. **SCIENCE** The chirping rate of crickets increases as the temperature increases. In the equation $T = c + 40$, T is the temperature in °F and c is the number of chirps in 15 seconds. About how many chirps will a cricket make in 15 seconds when the temperature is 60°F?

33. *Writing in Math* Explain what it means to "solve an equation."

34. **H.O.T.** Problem Write five different equations that have a solution of 10.

🌐 **Real-World Challenge**
There are about 1,000 miles of surfing beaches in California.

STANDARDS PRACTICE

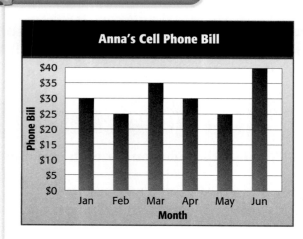

Use the graph to solve.

A cell phone company charges Anna $0.10 for each minute of usage.

35. How many minutes did Anna use in February?

   A 250   **C** 150

   B 200   **D** 100

36. How many minutes did Anna use in March and April together?

   F 300   **H** 650

   G 350   **J** 700

Spiral Review

Simplify each expression. Provide a reason for each step. (Lesson 1-8)

37. $2x + 3x$

38. $y + 5 + y$

39. $6p + 6 + 9 + 4p$

40. $2(d + 3) + 5d$

41. **EMPLOYMENT** Pat worked for 5 hours for $8 an hour. He received $16 in tips. Write an expression that shows the total amount Pat earned. (Lesson 1-2)

Vocabulary

integer (p. 119)

opposite (p. 119)

absolute value (p. 119)

 Standard ALG2.0
Students understand and use such operations as taking the opposite, finding the reciprocal, taking a root, and raising to a fractional power. They understand and use the rules of exponents.

Remember!

Zero is neither positive nor negative.

 The What: I will compare and graph integers.

 The Why: Integers are used to represent a decrease in temperature.

A newscast says the temperature is 4 degrees above zero. Later, the temperature may drop to 4 degrees below zero. The reading −4° appears on the screen. What does −4 mean?

INTEGERS

Negative and positive numbers can be used to show opposite situations. These numbers are often shown on a number line.

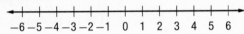

The numbers −1, −2, −3, . . . are called negative integers. Negative four is written as −4.

The numbers 1, 2, 3, . . . are called positive integers. Positive four is written as +4 or 4.

The set of **integers** can also be written as the set { . . . , −3, −2, −1, 0, 1, 2, 3, . . . }, where . . . means the pattern continues and never ends.

The numbers 4 and −4 are **opposites**. Opposites are numbers that are the same distance from 0 on a number line, but have different signs.

The **absolute value** of a number is its distance from zero on a number line. The absolute value of both 4 and −4 is 4, since each is 4 units from zero.

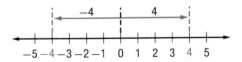

Talk Math

WORK WITH A PARTNER Say an integer between −10 and 10. Listen as your partner says its opposite. Together create an equation using both numbers, such as $4 + (−4) = 0$.

-25°

EXAMPLES **Find Opposites**

Answer each question.

1 **What is the opposite of 7?**

The opposite of 7 is −7.
Both 7 and −7 are 7 units
from zero on a number line.

2 **What is the opposite of −3?**

The opposite of −3 is 3.
Both 3 and −3 are
3 units from zero on
a number line.

3 **If 25° above zero is +25, then 25° below zero is __?__.**

25° below zero is −25°.

25° above zero and 25°
below zero are opposites.

4 **What is the absolute value of −9?**

The absolute value of
−9 is 9.
−9 is 9 units from zero
on a number line.

Your Turn

Answer each question.

a. What is the opposite of 6?

b. What is the opposite of −8?

c. If a 10-pound gain is +10, then a 10-pound loss is __?__.

d. What is the absolute value of −7?

Letters can be used to label points on number lines. On the
following number line, *A* represents −4.

EXAMPLE **Identify Integers on a Number Line**

5 **Name the integers represented by *A*, *C*, and *F*.**

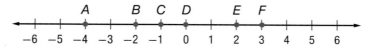

Point *A* is −4.   Point *C* is −1.   Point *F* is 3.

Your Turn

e. Name the integers represented by *B*, *D*, and *E*.

EXAMPLE **Graph Integers on a Number Line**

6 **Graph the integers −6, −3, 2, and 4 on a number line.**

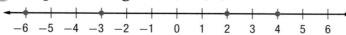

Your Turn

f. Graph the integers −5, −3, 0, and 1 on a number line.

On a number line, the greater integer is always to the right. For example, −2 is to the right of −4. So, −2 is greater than −4. −4 is to the left of −2. So, −4 is less than −2.

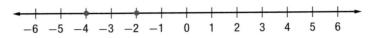

Remember!
The wider part of the inequality symbol faces the greater number.
−3 > −8 and −10 < −6

−2 is greater than **−4**   **−4** is less than **−2**

−2 > −4   **−4 < −2**

EXAMPLES   Compare Integers

Use the number line to complete each statement with one of the two symbols < or >.

7 3 ⬤ −4   **8** −5 ⬤ −3

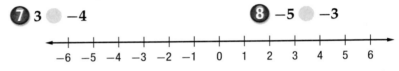

3 is to the right of −4.   −5 is to the left of −3.
3 is greater than −4.   −5 is less than -3.
3 **>** −4   −5 **<** −3

Your Turn Use a number line to complete each statement with one of the two symbols < or >.

g. 1 ⬤ −6   h. −6 ⬤ −2

Real-World EXAMPLE

9 **Lamar is playing miniature golf at Mini Golf World. At the first hole, his score was one under par. His score at the next hole was two over par. Describe these terms by using integers. Then use one of the two symbols < or > to compare them.**

One under par can be represented by –1.
Two over par can be represented by +2.
Compare: –1 < 2

Your Turn

i. **SPACE EXPLORATION** About 6 seconds before liftoff of the space shuttle, the three main shuttle engines start. About 120 seconds after liftoff, the solid rocket boosters burn out. Describe these times by using integers. Then use one of the two symbols < or > to compare them.

Vocabulary Review
integer
opposite
absolute value

Examples 1–4
(page 120)

VOCABULARY

1. Write an integer and its opposite. Explain how you know that the numbers are opposites.

2. Write the absolute value of −16. Explain how you know the absolute value.

Examples 1–4
(page 120)

Answer each question.

3. What is the opposite of 5?

4. What is the opposite of −6?

5. If 32 feet above sea level is 32, then 32 feet below sea level is ____?____.

6. What is the absolute value of −5?

Example 5
(page 120)

7. Name the integers represented by A, B, and C.

8. Graph the integers 4, −1, and −7 on a number line.

Example 6
(page 120)

Examples 7–8
(page 121)

Use a number line to complete each statement with one of the two symbols < or >.

9. 4 ⬤ −9

10. −5 ⬤ −1

Example 9
(page 121)

11. **TEMPERATURE** The high temperature one day was 5° above zero. The low temperature was 2° below zero. Describe each temperature using an integer. Then use one of the two symbols < or > to compare them.

12. **CHEMISTRY** The melting point of radon is −71°C. Its boiling point is −62°C. Use one of the two symbols < or > to compare these two integers.

13. **Talk Math** Use one of the two symbols < or > to compare two negative integers. Then use the comparison to write a word problem.

Skills, Concepts, and Problem Solving

Answer each question.

HOMEWORK HELP	
For Exercises	See Example(s)
14–19	1–4
20–21	5
22–23	6
24–27	7–8
28–31	9

14. What is the opposite of 9?

15. What is the opposite of −2?

16. What is the absolute value of −2?

17. What is the absolute value of 11?

18. If 10 meters below sea level is −10, then 10 meters above sea level is ____?____.

19. If an 8-yard loss in football is −8, then an 8-yard gain is _____?_____.

20. Name the integers represented by *A*, *B*, and *C*.

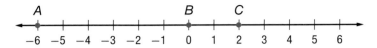

21. Name the integers represented by *G*, *H*, and *K*.

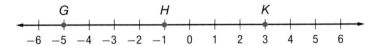

22. Graph the integers −5, −2, 0, and 7 on a number line.

23. Graph the integers −7, −3, 1, and 5 on a number line.

Use a number line to complete each statement with one of the two symbols < or >.

24. 3 ⬤ −3

25. −1 ⬤ 0

26. −2 ⬤ −1

27. −3 ⬤ −6

Solve.

28. **TEMPERATURE** The low temperature one day was 6° below zero and the high temperature was 2° below zero. Describe each temperature using an integer. Then use one of the two symbols < or > to compare them.

29. **GEOGRAPHY** Death Valley, California, has the lowest altitude in the United States. Its elevation is 282 feet below sea level. Describe this elevation using an integer and explain your choice.

30. **SPACE EXPLORATION** Five minutes before liftoff, the space shuttle's auxiliary power units start up. Two minutes before liftoff, the astronauts close and lock their visors. Describe these times using integers. Then use one of the two symbols < or > to compare them.

Space Exploration In 1981, the first space shuttle mission safely ended when astronauts, scientists, and engineers worked together to land Columbia at Edwards Air Force Base.

31. *Writing in Math* Write a word problem that compares 4 and −6. Use a sports situation, such as football.

H.O.T. Problems Cesar threw a 3 on the number cube and moved forward three spaces. He landed on a space that said: "Take a card." The card said: "Go back 5 spaces."

32. Where is Cesar in relation to his starting position before he rolled the number cube?

33. Describe each move using an integer.

34. Compare the integers of his starting and ending positions using one of the two symbols $<$ or $>$.

STANDARDS PRACTICE

Choose the *best* answer.

35 Which number is greater than −6 and less than −3?

A −8	**C** −4
B −6	**D** −2

36 The prices of three stocks go down during the week by $4.50, $2.35, and $3.10. Which choice correctly orders the losses?

F −$2.35 < −$3.10 < −$4.50

G −$4.50 < −$3.10 < −$2.35

H −$2.35 > −$4.50 > −$3.10

J −$3.10 > −$2.35 > −$4.50

Spiral Review

Solve each equation using mental math. (Lesson 2-1)

37. $a + 4 = 7$ **38.** $b - 5 = 6$ **39.** $4c = 20$ **40.** $\frac{d}{2} = 8$

41. SPORTS There are 53 players on an NFL football team roster. This is 30 more than the number on an NHL hockey team. Write and solve an equation for the number of players on an NHL roster. (Lesson 2-1)

Complete each equation. (Lesson 1-5)

42. $2(3 + 4) = (2 \cdot 3) + (2 \cdot \underline{\ ?\ })$

43. $5(6 + 3) = (5 \cdot \underline{\ ?\ }) + (5 \cdot 3)$

44. $(4 \cdot 7) + (4 \cdot 5) = 4(7 + \underline{\ ?\ })$

45. $(6 \cdot 4) + (6 \cdot 3) = 6(\underline{\ ?\ } + \underline{\ ?\ })$

Progress Check 1

(Lessons 2-1 and 2-2)

▷ Vocabulary and Concept Check

> absolute value (p. 119) opposite (p. 119)
> define (p. 116) solution (p. 114)
> equation (p. 114) solve (p. 114)
> integers (p. 119)

Choose the term that *best* completes each statement.

1. A(n) _____?_____ is any number that makes an equation true.

2. The _____?_____ of 9 is −9.

3. Two numbers that are the same distance from zero have the same _____?_____.

4. _____?_____ include counting numbers, their opposites, and zero.

▷ Skills Check

Which of the numbers 7, 8, 9, or 10 is the solution for each equation? (Lesson 2-1)

5. $x + 5 = 12$

6. $a - 3 = 7$

7. $3d = 27$

8. $\frac{t}{4} = 2$

9. Graph the integers −4, −1, and 0 on a number line. (Lesson 2-2)

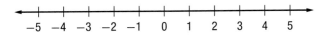

Use a number line to complete each statement with one of the two symbols < or >. (Lesson 2-2)

10. −1 ⬤ 1

11. −3 ⬤ −5

▷ Problem-Solving Check

12. **CIVICS** The House and Senate make up the two legislative branches of the United States government. The House has 435 members. This number is 335 more than the number of senators. How many members of the Senate are there? (Lesson 2-1)

13. **CHEMISTRY** Sulfur dioxide is a gas often produced by volcanoes. Its boiling point is −10°C. Its melting point is about −72°C. Use one of the two symbols < or > to compare these temperatures. (Lesson 2-2)

14. **REFLECT** Explain how you would solve the equation $3x = 18$. (Lesson 2-1)

Math Lab
Model Integer Operations Using Cubes

Materials

blue and red centimeter cubes

 Standard 7NS1.2 Add, subtract, multiply, and divide **rational numbers (integers,** fractions, and terminating decimals) and take positive rational numbers to whole-number powers.

Standard 6NS2.0 Students calculate and solve problems involving addition, subtraction, multiplication, and division.

Standard 7MR3.3 Develop generalizations of the results obtained and the strategies used and apply them to new problem situations.

 The What: I will use cubes to model adding and subtracting integers.

 The Why: Cubes can model integers. Integers can represent temperature readings and sports statistics.

You can use cubes to model adding integers.

 Let red centimeter cubes represent hot coals with a value of +1. These are positive numbers.

 Let blue centimeter cubes represent ice cubes with a value of −1. These are negative numbers.

 When a red centimeter cube (+1) is paired with a blue centimeter cube (−1), the value is 0. This is called a "zero pair."

EXAMPLE   Add Integers Using Models

1 Use cubes to find $-4 + 2$.

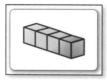

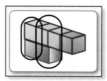

Start with 4 ice cubes.

Add 2 hot coals. Pair each ice cube with a hot coal.

Remove the zero pairs. There are 2 ice cubes left.

$$-4 + 2 = -2$$

Your Turn
Use cubes to find each sum.

a. $5 + 2$ **b.** $-5 + 2$ **c.** $5 + (-2)$ **d.** $-5 + (-2)$

You can also use cubes to model subtracting integers.

2 Use cubes to find −5 − (−2).

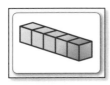

Start with 5 ice cubes.

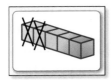

Remove 2 ice cubes. There are 3 ice cubes left.

$$-5 - (-2) = -3$$

3 Use cubes to find 5 − (−3).

Start with 5 hot coals.

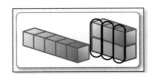

To subtract 3 ice cubes, first add 3 zero pairs. The value does not change.

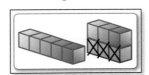

Now remove the 3 ice cubes. There are 8 hot coals left.

$$5 - (-3) = 8$$

Subtracting a negative is the same as adding a positive.

$$5 - (-3) = 5 + 3 = 8$$

Your Turn

Use cubes to find each difference.

e. −2 − (−1)

f. −4 − (−2)

g. −3 − (−2)

h. 3 − (−2)

i. 4 − (−1)

j. 3 − (−3)

Adding and Subtracting Integers

Vocabulary

additive inverse (p. 129)

Distributive Property
(p. 131)

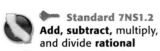 **Standard 7NS1.2**
Add, subtract, multiply, and divide **rational numbers** (integers, fractions, and terminating decimals) and take positive rational numbers to whole-number powers.

Standard 7AF1.3
Simplify numerical **expressions by applying properties of rational numbers** and justify the process used.

Standard 7MR3.2 Note the method of deriving the solution and demonstrate a conceptual understanding of the derivation by solving similar problems.

The What: I will understand and apply the rules for adding and subtracting integers.

The Why: Integers are used in banking and in different sports.

One week ago, Sharece had saved $30. Yesterday, she withdrew $15 for the movies. How much money is left?

$$30 + (-15) = 15$$

beginning amount   money for movies   current amount

Positive integers represent deposits. Negative integers represent withdrawals. Sharece's total is $15.

EXAMPLES   Use a Number Line to Add Integers

Use the number line to find each sum.

You can use a number line to add integers. Always start at zero. For positive numbers, move right. For negative numbers, move left.

1 $-5 + 2$

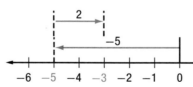

Step 1: Start at zero.
Step 2: Move left five.
Step 3: Move right two. Where do you end?

$-5 + 2 = -3$

2 $7 + (-3)$

Step 1: Start at zero.
Step 2: Move right seven.
Step 3: Move left three. Where do you end?

$7 + (-3) = 4$

Your Turn Draw a number line to find each sum.

a. $-7 + 3$ b. $-6 + 2$ c. $4 + (-1)$

Think about depositing and withdrawing money when you add integers.

EXAMPLES **Add Integers**

Find each sum.

③ 2 + (−5)

$2 + (−5) = −3$ Think about depositing $2 and withdrawing $5.

④ −4 + 2

$−4 + 2 = −2$ Think about withdrawing $4 and depositing $2.

⑤ −3 + (−4)

$−3 + (−4) = −7$ Think about withdrawing $3 and then withdrawing $4 more.

Your Turn
Find the sum.

d. $4 + (−8)$ e. $−6 + 2$

f. $−9 + 10$ g. $−3 + (−2)$

In Lesson 2-2, you learned that 6 and −6 are opposites. Because $6 + (−6) = 0$, 6 and −6 are also called **additive inverses.**

EXAMPLES **Additive Inverse**

Draw a number line to find each sum.

⑥ 6 + (−6)

Step 1: Start at zero.
Step 2: Move right six.
Step 3: Move left six.
Where do you end?

The sum is zero, so 6 and −6 are additive inverses.

⑦ −4 + 4

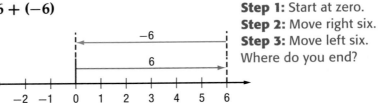

Step 1: Start at zero.
Step 2: Move left four.
Step 3: Move right four.
Where do you end?

The sum is zero, so −4 and 4 are additive inverses.

Your Turn
Draw a number line to find each sum.

h. $10 + (−10)$ i. $−8 + 8$

WORK WITH A PARTNER Create an equation with negative integers to represent this situation: Sharece started with $60. She withdrew $20. Find her total.

You can subtract integers by using additive inverses. Any subtraction can be rewritten as an addition by using the additive inverse.

$$8 - (-5) = 8 + 5$$

subtract   add the
                  additive inverse

$$5 - 8 = 5 + (-8)$$

subtract   add the
                  additive inverse

EXAMPLES   Subtract Integers

Find each difference by using additive inverses.

8 $4 - 7$

$4 - 7 = 4 + (-7)$ Rewrite as a sum using the additive inverse of 7.

$ = -3$ Add 4 and −7.

9 $-6 - 2$

$-6 - 2 = -6 + (-2)$ Rewrite as a sum using the additive inverse of 2.

$ = -8$ Add −6 and −2.

10 $-3 - (-4)$

$-3 - (-4) = -3 + 4$ Rewrite as a sum using the additive inverse of −4.

$ = 1$ Add −3 and 4.

Your Turn
Find each difference by using additive inverses.

j. $7 - 8$ k. $-6 - (-4)$

In Chapter 1, you used the **Distributive Property** to simplify expressions with like terms. When like terms have integer coefficients, use the same rules for adding and subtracting integers.

Remember!

The Distributive Property states that for any numbers a, b, and c, $(a \cdot b) + (a \cdot c) = a(b + c)$.

EXAMPLES **Simplify Expressions**

Simplify each expression.

⑪ $-2d + 5d$

$\begin{aligned} -2d + 5d &= (-2 + 5)d & \text{Rewrite using the Distributive Property.} \\ &= 3d & \text{Add } -2 \text{ and } 5. \end{aligned}$

⑫ $7y - (-3y)$

$\begin{aligned} 7y - (-3y) &= [7 - (-3)]y & \text{Rewrite using the Distributive Property.} \\ &= (7 + 3)y & \text{Rewrite as a sum using the additive inverse of } -3. \\ &= 10y & \text{Add } 7 \text{ and } 3. \end{aligned}$

⑬ $3y - 8y$

$\begin{aligned} 3y - 8y &= (3 - 8)y & \text{Rewrite using the Distributive Property.} \\ &= [3 + (-8)]y & \text{Rewrite as a sum using the additive inverse of } 8. \\ &= -5y & \text{Add } 3 \text{ and } -8. \end{aligned}$

Your Turn

Simplify each expression.

l. $-3x + 7x$
m. $2r - 5r$
n. $-7y - (-2y)$

Real-World EXAMPLE

⑭ **BANKING** Bill spent $10 more than he has in his bank account. His balance is −$10. He makes a $15 deposit. How much is in his bank account now?

−10 represents his balance. 15 represents the deposit.

beginning balance   deposit

$-10 \qquad + \qquad 15 \qquad = 5$

Bill has $5 in the account now.

Your Turn

o. **BANKING** Tito has $12 in his bank account. He writes a check for $15. What is the balance of his bank account now?

p. **WEATHER** The temperature is −6°C. It rises 3°C. What is the temperature now?

BANKING Many bank accounts will not allow customers to carry a negative balance.

Vocabulary Review
additive inverse
Distributive Property

Examples 6–7
(page 129)

VOCABULARY

1. Write a number and its additive inverse.
   How do you know that the numbers are additive inverses?

Examples 1–2
(page 128)

Draw a number line to find each sum.

2. $-6 + 7$
3. $8 + (-4)$

Examples 3–5
(page 129)

Find each sum.

4. $7 + (-9)$
5. $-6 + 2$
6. $-4 + 11$
7. $-5 + (-6)$

Examples 6–7
(page 129)

Draw a number line to find each sum.

8. $-8 + 8$
9. $2 + (-2)$

Examples 8–10
(page 130)

Find each difference by using additive inverses.

10. $5 - 9$
11. $-5 - 2$
12. $-4 - (-3)$
13. $-7 - (-9)$

Examples 11–13
(page 131)

Simplify each expression.

14. $4b + (-5b)$
15. $2n - 7n$

Example 14
(page 131)

16. **SPORTS** A football team loses 3 yards on its first play and 5 yards on its second play. Write the team's total yardage on the two plays as an integer.

17. **WEATHER** The high temperature one day was $-1°C$, and the low temperature was $-6°C$. What was the difference between the high and low temperatures?

18. **Talk Math** Explain how to use a number line to find $5 + (-2)$.

Skills, Concepts, and Problem Solving

HOMEWORK HELP	
For Exercises	**See Example(s)**
19–22	1–2
23–34	3–5
35–38	6–7
39–46	8–10
47–50	11–13
51–52	14

Draw a number line to find each sum.

19. $-3 + 2$
20. $-5 + 4$
21. $8 + (-2)$
22. $8 + (-7)$

Find each sum.

23. $5 + (-8)$
24. $6 + (-10)$
25. $3 + (-7)$
26. $5 + (-11)$
27. $-9 + 3$
28. $-7 + 4$
29. $-7 + 1$
30. $-6 + 8$
31. $-6 + (-10)$
32. $-6 + (-3)$
33. $-4 + (-8)$
34. $-6 + (-9)$

Draw a number line to find each sum.

35. $5 + (-5)$
36. $7 + (-7)$
37. $-14 + 14$
38. $-12 + 12$

Find each difference by using additive inverses.

39. $8 - 11$
40. $7 - 4$
41. $-5 - (-9)$
42. $-2 - (-3)$
43. $-5 - 3$
44. $-4 - 4$
45. $-9 - (-11)$
46. $-9 - (-12)$

Simplify each expression.

47. $8k + (-4k)$ **48.** $9a + (-6a)$ **49.** $6r - 11r$ **50.** $7y - 10y$

51. RECREATION A scuba diver descends from -5 feet to -62 feet. How many feet did she descend?

52. NATURE Two California landmarks, Mt. Whitney and Death Valley, are the highest and lowest points in the lower 48 states. The difference in elevation is 4,504 meters. The elevation of Death Valley is -86 meters. What is the elevation of Mt. Whitney?

NATURE Although Mt. Whitney and Death Valley differ dramatically in elevation, they are less than 80 miles apart.

53. *Writing in Math* Write two algebraic expressions with a sum of $-8m$.

54. H.O.T. Problem A scientist at a weather station records the temperature each hour. She records $-11°C$, $-8°C$, $-5°C$, and $-2°C$. Describe the pattern. If the pattern continues, identify the temperature the scientist will record in the next hour.

STANDARDS PRACTICE

Choose the *best* answer.

55 Which expression is equivalent to $3x + 12$?

A $3(x + 12)$ **C** $3(x + 4)$

B $15x$ **D** $x + 15$

56 Simplify $-4g + (3g) - 5g$.

F $-6g$ **H** $2g$

G $-4g$ **J** $6g$

Spiral Review

Use a number line to complete each statement with one of the two symbols < or >. (Lesson 2-2)

57. -4 ⬤ 1 **58.** -2 ⬤ -5 **59.** 6 ⬤ 0

60. WEATHER The low temperatures on Saturday and Sunday were $-2°C$ and $-4°C$. Describe each temperature using an integer. Then use one of the two symbols < or > to compare them. (Lesson 2-2)

Write an algebraic expression for each verbal expression. (Lesson 1-2)

61. twice a certain number n **62.** s increased by t

Progress Check 2
(Lesson 2-3)

▷ Vocabulary and Concept Check

> additive inverse (p. 129)   **Distributive Property** (p. 131)

Choose the term that *best* completes each statement.

1. The ____?____ states that $-2k + 4k = (-2 + 4)k$.

2. The sum of a number and its ____?____ is zero.

▷ Skills Check

Find each sum. (Lesson 2-3)

3. $6 + (-1)$ 4. $-4 + 7$

5. $1 + (-3)$ 6. $-9 + 5$

7. $-8 + (-4)$ 8. $-11 + 8$

Find each difference by using additive inverses. (Lesson 2-3)

9. $3 - 8$ 10. $0 - 10$

11. $2 - (-5)$ 12. $-7 - 1$

13. $-6 - (-11)$ 14. $4 - 8$

Simplify each expression. (Lesson 2-3)

15. $-12m + 9m$ 16. $4n - 9n$

▷ Problem-Solving Check

17. **ARTS** A school theater group earned $85 from ticket sales of its last play. The group spent $93 in production expenses. Represent the group's overall profit or loss as an integer. (Lesson 2-3)

18. **BANKING** Malina's bank account has a balance of −$28 on Monday and −$16 on Tuesday. Represent the change in Malina's account as an integer. (Lesson 2-3)

19. **SPORTS** In Saturday's game, the football team had these yardages on the first three plays: −15, −7, and 18. Find the total net yardage for the first three plays. (Lesson 2-3)

20. **REFLECT** Explain how you use the additive inverse to find $3 - (-2)$. (Lesson 2-3)

2-4 Multiplying Integers

Vocabulary

factor (p. 135)

product (p. 135)

 Standard 7NS1.2 Add, subtract, **multiply**, and divide **rational numbers (integers**, fractions, and terminating decimals) **and take** positive **rational numbers to whole-number powers.**

 **Standard 7AF1.3 Simplify** numerical **expressions** by applying properties of rational numbers and justify the process used.

> **The What:** I will understand and apply the rules for multiplying integers.
>
> **The Why:** Integers can be multiplied when finding money amounts or determining a golfer's score.

On Tuesday, Carlos took $4 out of his wallet to buy a binder for math. On Wednesday and on Thursday, he took $4 out of his wallet to buy binders for science and social studies. How much did the money in his wallet change?

One way to find the total amount Carlos's money changed is to find $-4 + (-4) + (-4)$. Since multiplication is repeated addition, another way is to multiply -4 and 3.

$$-4 + (-4) + (-4) = -4 \cdot 3 = -12$$

$\underbrace{\qquad\qquad}_{-4 \text{ added } 3 \text{ times}}$ $\underbrace{\quad}_{\text{factors}}$ $\underbrace{\quad}_{\text{product}}$

Factors are numbers or expressions that are multiplied together. The result is called a **product.** The product -12 represents how much Carlos's money changed in all.

> **Remember!**
>
> Multiplication can be shown in different ways, such as:
>
> -4×3
>
> $(-4)(3)$
>
> $-4 \cdot 3$.

INTEGER MULTIPLICATION

Positive Number × Positive Number	Negative Number × Positive Number	Negative Number × Negative Number
$1 \cdot 3 = 3$	$-1 \cdot 3 = -3$	$-1 \cdot (-3) = 3$
$2 \cdot 3 = 6$	$-2 \cdot 3 = -6$	$-2 \cdot (-3) = 6$
$3 \cdot 3 = 9$	$-3 \cdot 3 = -9$	$-3 \cdot (-3) = 9$
$4 \cdot 3 = 12$	$-4 \cdot 3 = -12$	$-4 \cdot (-3) = 12$
The product is positive.	The product is negative.	The product is positive.

Multiply integers the same way you multiply whole numbers. Multiply the numbers. Then check the signs to find the sign of the product.

RULES FOR MULTIPLYING INTEGERS

Positive Number · Positive Number = Positive Number
Negative Number · Negative Number = Positive Number
Positive Number · Negative Number = Negative Number
Negative Number · Positive Number = Negative Number

EXAMPLES Multiply Integers

Find each product.

1 $-4 \cdot 6$

$-4 \cdot 6 = -24$

Multiply 4 and 6.
Then check the signs.
The signs are different.
The product is negative.

2 $5 \cdot (-3)$

$5 \cdot (-3) = -15$

Multiply 5 and 3.
Then check the signs.
The signs are different.
The product is negative.

3 $-2 \cdot (-8)$

$-2 \cdot (-8) = 16$

Multiply 2 and 8.
Then check the signs.
The signs are the same.
The product is positive.

4 $6 \cdot 7$

$6 \cdot 7 = 42$

Multiply 6 and 7.
Then check the signs.
The signs are the same.
The product is positive.

Your Turn Find each product.

a. $-6 \cdot 5$ b. $8 \cdot (-3)$ c. $-9 \cdot (-4)$

Study Tip

$2^5 = 2 \cdot 2 \cdot 2 \cdot 2 \cdot 2$
or
$(-2)^5 = (-2)(-2)(-2)(-2)(-2)$
Parentheses can indicate multiplication.

The Associative Property of Multiplication can be used to group three or more factors.

If there is an odd number of negative signs, the product is negative.

$$(-2)(-2)(-2)(-2)(-2) = \underbrace{(-2)(-2)} \cdot \underbrace{(-2)(-2)} \cdot (-2)$$
$$= \underbrace{4 \quad \cdot \quad 4} \quad \cdot (-2)$$
$$= \underbrace{16} \quad \cdot (-2)$$
$$= -32$$

Using -2 as a factor five times results in a negative product.

If there is an even number of negative signs, the product is positive.

$$(-2)(-2)(-2)(-2) = \underbrace{(-2)(-2)} \cdot \underbrace{(-2)(-2)}$$
$$= \underbrace{4 \quad \cdot \quad 4}$$
$$= 16$$

Using -2 as a factor four times results in a positive product.

Talk Math

WORK WITH A PARTNER You and your partner each write an integer on different index cards. Look at the numbers together and decide if the products will be positive, negative, or equal to zero. Discuss how the integer factors could be changed in order to find products that are negative, positive, or equal to zero.

EXAMPLES Multiply Three or More Integers

Find each product.

5 $(-2)(4)(-3)$

$$(-2)(4)(-3) = \underbrace{(-8)(-3)}$$ Multiply −2 and 4.

$$= 24$$ Multiply −8 and −3.

6 $(-2)^2(3)(-4)(-5)$

$$(-2)^2(3)(-4)(-5) = (-2)(-2)(3)(-4)(-5)$$ $(-2)^2$ equals $-2 \cdot -2$.

$$= \underbrace{(4)(3)}(-4)(-5)$$ Simplify $(-2)^2$.

$$= \underbrace{(12)(-4)}(-5)$$ Multiply 4 and 3.

$$= \underbrace{(-48)(-5)}$$ Multiply 12 and −4.

$$= 240$$ Multiply −48 and −5.

Your Turn

Find each product.

d. $(-3)(3)(-4)$

e. $(-2)(-5)(-3)$

f. $(-1)(2)(-3)(4)$

g. $(-3)^2(4)(-2)(-1)$

In Chapter 1, you learned to substitute the value for each variable when simplifying algebraic expressions. The next step is to follow the order of operations. When expressions contain integers, follow the rules for multiplying integers.

Evaluate Expressions

Find the value of each expression.

7 $5x$, if $x = -4$

$$5x = (5)(-4) \qquad \text{Substitute } -4 \text{ for } x.$$

$$= -20 \qquad \text{Multiply 5 and } -4.$$

8 $-5gh$, if $g = -3$ and $h = 7$

$$-5gh = (-5)(-3)(7) \qquad \text{Substitute } -3 \text{ for } g \text{ and } 7 \text{ for } h.$$

$$= (15)(7) \qquad \text{Multiply } -5 \text{ and } -3.$$

$$= 105 \qquad \text{Multiply 15 and 7.}$$

Your Turn Find the value of each expression.

h. $-3m$, if $m = 6$ **i.** $2xy$, if $x = -4$ and $y = 3$

Real-World EXAMPLES

9 WEATHER Each hour, the temperature changes $-2°C$. How much will the temperature change after 5 hours?

$$(-2)(5) = -10 \qquad \text{Multiply } -2 \text{ and 5 to solve.}$$

The temperature changes -10 degrees.

10 BUSINESS A stock has a change of -2 dollars each day for 4 days. What is the overall change of the stock?

$$(-2)(4) = -8 \qquad \text{Multiply } -2 \text{ and 4 to solve.}$$

The stock changed -8 dollars.

Your Turn

j. FINANCE Five people each have a balance of $-\$12$. How much is the total balance for all five people?

Guided Practice

Vocabulary Review
factor
product

Examples 1–8
(pages 136–138)

VOCABULARY

1. The product of 20 and -1 is -20. Name three other multiplication problems that have a product of -20.

Examples 1–4
(page 136)

Find each product.

2. $-5 \cdot 2$ 3. $4 \cdot (-3)$

4. $-6 \cdot (-5)$ 5. $4 \cdot 7$

Examples 5–6
(page 137)

6. $(-2)(3)(-4)$ 7. $(-2)(-1)(-3)$

8. $(-3)(2)(-3)(-1)$ 9. $(-2)(3)(-2)(-2)$

10. $(-2)^2(3)(-2)(-2)$ 11. $(-3)^2(-1)(-1)(2)$

Examples 7–8
(page 138)

Find the value of each expression.

12. $-5n$, if $n = -7$

13. $-8b$, if $b = 6$

14. $-3ab$, if $a = 5$ and $b = -2$

15. $-2xy$, if $x = -2$ and $y = -5$

Examples 9–10
(page 138)

16. GOLF In golf, scores below par are posted as negative integers. If a golfer shoots -3 each day of a 4-day tournament, what is the golfer's overall score?

17. EARTH SCIENCE Suppose a glacier melts at a rate of 9 centimeters a day from its original size. What amount does the glacier change after 10 days?

18. 𝒯𝒶𝓁𝓀 𝑀𝒶𝓉𝒽 Have a classmate say an integer. Decide if the product of that number and -1 will be positive or negative. Find a rule for multiplying any integer by -1.

Skills, Concepts, and Problem Solving

HOMEWORK HELP

For Exercises	See Example(s)
19–26	1–4
27–38	5–6
39–44	7–8
45–47	9–10

Find each product.

19. $-3 \cdot 6$

20. $-5 \cdot 7$

21. $5 \cdot (-8)$

22. $6 \cdot (-9)$

23. $-7 \cdot (-8)$

24. $-9 \cdot (-7)$

25. $10 \cdot 2$

26. $12 \cdot 5$

27. $(-2)(4)(-2)$

28. $(-4)(4)(-2)$

29. $(-5)(-2)(-3)$

30. $(-6)(-2)(-2)$

31. $(-4)(3)(-2)(-2)$

32. $(-2)(1)(-3)(-5)$

33. $(-3)(-3)(-1)(-2)$

34. $(2)(6)(-3)(-1)$

35. $(2)(-2)^2(-3)(-2)$

36. $(-1)(-3)^2(-2)(3)$

37. $(-1)^2(5)(-2)(-1)$

38. $(-4)^2(-2)(-2)(1)$

Find the value of each expression.

39. $-7a$, if $a = -4$

40. $-k$, if $k = -9$

41. $-4y$, if $y = 5$

42. $-5p$, if $p = 1$

43. $-3cd$, if $c = 3$ and $d = -6$

44. $-4gh$, if $g = 5$ and $h = -2$

45. FINANCE Accounting programs use positive integers to show credits and negative integers to show payments. If Mrs. Carmona pays a monthly phone bill of $10, what total number will an accountant enter in the financial books for the year?

46. WEATHER Suppose the outside temperature has changed at a rate of $-2°F$ each hour. The temperature is now $0°F$. Predict the temperature 5 hours from now.

47. **CAREER CONNECTION** Scientists use under-ocean crafts to explore, map, and study the world below the sea. The air pressure at sea level is 1 atmosphere (atm). Every 10 meters the craft descends into the ocean, the pressure increases by about 1 atm. If the total pressure is 5 atm, what integer represents the craft's depth in meters? (Hint: Positive integers represent elevations above sea level.)

Under-ocean crafts are designed to withstand pressures deep in the ocean.

48. *Writing in Math* Write three different multiplication expressions that have a product equal to −30.

49. **H.O.T.** Problem Find the values of $(-1)^2$, $(-1)^3$, $(-1)^4$, and $(-1)^5$. Describe the pattern and explain how you could find the value of $(-1)^{15}$ without multiplying.

STANDARDS PRACTICE

Choose the *best* answer.

50 Stock ABC opened trading on May 1 at $65. This chart shows the price change of Stock ABC for the next three weeks. Use multiplication to find the price on May 19.

Daily Price Change for Stock ABC					
May 1	−$4	May 8	+$2	May 15	−$1
May 2	−$4	May 9	+$2	May 16	no change
May 3	−$4	May 10	no change	May 17	−$1
May 4	−$4	May 11	+$2	May 18	−$1
May 5	−$4	May 12	no change	May 19	−$1

A $47   C $59

B $51   D $65

51 Solve for x.
$$3x - 8 = 22$$
F −10   H 20

G 10   J 30

52 A scientist finds that the temperature of a region of the ocean is modeled by the expression $2xy$. What is the region's temperature when $x = -4$ and $y = 3$?

A 24°   C 12°

B −12°   D −24°

Spiral Review

Add or subtract. (Lesson 2-3)

53. $-2 + 5$ 54. $-3 + (-4)$ 55. $3 - 8$ 56. $-1 - (-3)$

57. **NUMBER SENSE** Marco is 17 years old. He is three years older than his sister. How old is his sister? (Lesson 2-1)

Dividing Integers

Vocabulary

quotient (p. 141)

 Standard 7NS1.2 Add, subtract, multiply, and **divide rational numbers** (**integers,** fractions, and terminating decimals) **and take** positive **rational numbers to whole-number powers.**

Standard 7AF1.3 Simplify numerical **expressions** by applying properties of rational numbers and justify the process used.

 The What: I will understand and apply the rules for dividing integers.

 The Why: Dividing integers can be used to equally share expenses.

Marissa and two friends went to the movies and out to dinner. They spent a total of $39. Suppose they decide to equally share the expense. How much did each person owe?

The girls are dividing the $39 in expenses into three equal parts.

$$-39 \div 3 = -13$$

total expenses   number of people   quotient

The result from dividing two numbers or expressions is called a **quotient.** The quotient −13 represents each girl's expense.

INTEGER DIVISION		
Negative Number ÷ **Negative Number**	**Negative Number** ÷ **Positive Number**	**Positive Number** ÷ **Negative Number**
−12 ÷ (−3) = 4	−12 ÷ 3 = −4	12 ÷ (−3) = −4
−9 ÷ (−3) = 3	−9 ÷ 3 = −3	9 ÷ (−3) = −3
−6 ÷ (−3) = 2	−6 ÷ 3 = −2	6 ÷ (−3) = −2
−3 ÷ (−3) = 1	−3 ÷ 3 = −1	3 ÷ (−3) = −1
The quotient is positive.	The quotient is negative.	The quotient is negative.

Divide integers the same way you divide whole numbers. Divide the numbers. Then check the signs to find the sign of the quotient.

RULES FOR DIVIDING INTEGERS

Positive Number ÷ Positive Number = Positive Number

Negative Number ÷ Negative Number = Positive Number

Positive Number ÷ Negative Number = Negative Number

Negative Number ÷ Positive Number = Negative Number

Talk Math

WORK WITH A PARTNER You and your partner each write a nonzero integer on different index cards. Look at the numbers together. If you divide the greater number by the lesser one, will the quotient be positive or negative? Explain. Repeat a total of 10 times.

Remember!

Division can be shown in different ways, such as:

−15 ÷ 3

$\frac{-15}{3}$

EXAMPLES Divide Integers

Find each quotient.

1 −18 ÷ 6 — Divide 18 and 6. Then check the signs.

−18 ÷ 6 = −3 — The signs are different. The quotient is negative.

2 35 ÷ (−5) — Divide 35 and 5. Then check the signs.

35 ÷ (−5) = −7 — The signs are different. The quotient is negative.

3 −36 ÷ (−9) — Divide 36 and 9. Then check the signs.

−36 ÷ (−9) = 4 — The signs are the same. The quotient is positive.

4 45 ÷ 5 — Divide the 45 and 5. Then check the signs.

45 ÷ 5 = 9 — The signs are the same. The quotient is positive.

Your Turn Find each quotient.

a. 15 ÷ (−3)　　　　b. −24 ÷ (−8)

c. −30 ÷ 5　　　　d. 36 ÷ 9

You may need to simplify expressions with integers and exponents. Remember the order of operations.

1. Perform all operations inside parentheses first.

2. Simplify exponents.

3. Multiply and divide from left to right.

4. Add and subtract from left to right.

EXAMPLE   **Exponents in Expressions**

5 Find the value of $\left(\frac{-6}{2}\right)^3$.

$$\left(\frac{-6}{2}\right)^3 = \underbrace{(-3)^3} \qquad \text{Simplify inside the parentheses first.}$$

$$= \underbrace{(-3)(-3)(-3)} \qquad \text{Simplify exponents.}$$

$$= -27$$

Another method is to multiply the factors in pairs in order to decide if your product is positive or negative.

$$\left(\frac{-6}{2}\right)^3 = \underbrace{(-3)^3} \qquad \text{Simplify inside the parentheses first.}$$

$$= \underbrace{(-3)(-3)} \times (-3) \qquad \text{Multiply } (-3)(-3).$$

$$= \underbrace{(9)(-3)} \qquad \text{Multiply } (9)(-3).$$

$$= -27 \qquad \text{The solution is negative.}$$

Your Turn

Find the value of each expression.

e. $\left(\frac{-12}{4}\right)^4$ 　　　　　　　　　　f. $\left(\frac{6}{-3}\right)^3$

EXAMPLE   **Evaluate Expressions**

6 Find the value of $\frac{-21}{a}$, if $a = -3$.

$$\frac{-21}{a} = \frac{-21}{-3} \qquad \text{Substitute } -3 \text{ for } a.$$

$$= 7 \qquad \text{Divide } -21 \text{ by } -3.$$

Your Turn

Find the value of each expression.

g. $\frac{w}{-5}$, if $w = -45$ 　　　　　　　　h. $\frac{16}{z}$, if $z = -4$

Real-World EXAMPLE

7 **FINANCE** The price of a stock changed −$10 in 5 days. The price changed by the same amount each day. What was the change on the last day?

$-10 \div 5 = -2$ 　Divide −10 and 5 to solve.

The stock changed −$2 on the last day.

Your Turn

i. **WEATHER** The elevation of a weather balloon changes −60 feet in 5 hours. If the elevation changes the same amount each hour, how many feet does it change in the first hour?

Examples 1–6
(pages 142–143)

VOCABULARY

1. The quotient of 18 and −6 is −3.
   Name three other division problems that have a quotient of −3.

Examples 1–4
(page 142)

Find each quotient.

2. −24 ÷ 8

3. 30 ÷ (−5)

4. −54 ÷ (−9)

5. 35 ÷ 7

Find the value of each expression.

Example 5
(page 143)

6. $\left(\dfrac{6}{-3}\right)^5$

7. $\left(\dfrac{-10}{2}\right)^4$

Example 6
(page 143)

8. $\dfrac{h}{-6}$, if $h = 42$

9. $\dfrac{x}{y}$, if $x = -24$ and $y = 8$

Example 7
(page 143)

10. **SPORTS** In golf, scores below par are written as negative integers. A golfer has a score of −8 after a 4-day tournament. She had the same score each day. What was her score on the last day?

11. **BUSINESS** Stock ABC lost $15 in a 5-day trading period. It lost the same amount each day. What was the loss on the first day?

12. **Talk Math** You and a classmate each say an integer. Decide if the greater number divided by the lesser one will be positive or negative. Give examples to support your answer.

Skills, Concepts, and Problem Solving

HOMEWORK HELP	
For Exercises	**See Example(s)**
13–24	1–4
25–28	5
29–34	6
35–38	7

Find each quotient.

13. −21 ÷ 7

14. −28 ÷ 4

15. 33 ÷ (−3)

16. 42 ÷ (−7)

17. −56 ÷ (−8)

18. −45 ÷ (−9)

19. 72 ÷ 8

20. 95 ÷ 5

21. −36 ÷ 6

22. −72 ÷ 9

23. 96 ÷ (−12)

24. 68 ÷ (−4)

Find the value of each expression.

25. $\left(\dfrac{2}{-2}\right)^7$

26. $\left(\dfrac{12}{-4}\right)^3$

27. $\left(\dfrac{-9}{3}\right)^4$

28. $\left(\dfrac{-6}{3}\right)^6$

29. $\dfrac{k}{-5}$, if $k = 35$

30. $\dfrac{q}{-4}$, if $q = -32$

31. $\dfrac{-81}{m}$, if $m = -9$

32. $\dfrac{32}{c}$, if $c = -8$

33. $\dfrac{a}{b}$, if $a = 42$ and $b = -7$

34. $\dfrac{x}{y}$, if $x = -44$ and $y = -11$

35. **SPORTS** A football team had a total loss of 60 yards in penalties on 6 plays. Each penalty cost the team the same number of yards. How much did each penalty cost the team?

36. **WEATHER** The temperature changed −9°C in 3 hours. It changed the same amount each hour. How much did the temperature change each hour?

37. **RECREATION** Bob went scuba diving. He recorded his original and final depths as well as his start and stop times. His depth changed the same number of meters each minute. How many meters did the depth change each minute?

	Time	Depth (meters)
Start	3:10	0
Stop	3:30	−60

38. **FINANCE** The Spanish Club bank account has a balance of −$36. Each of the 9 members pays an equal share to bring the balance to zero. How much will each member pay?

39. *Writing in Math* Write three different division expressions that have a quotient of −4.

40. **H.O.T.** Problem Find two integers such that $a + b = -8$ and $\frac{a}{b} = 3$.

Choose the *best* answer.

41 This graph shows that the temperature changed −8°C in 4 hours. What was the rate of change?

 A −8°C each hour

 B −4°C each hour

 C −2°C each hour

 D 4°C each hour

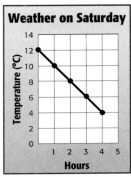

Weather on Saturday

42 Your class puts on a play. The expenses include $45 for the script, $189 for sets and costumes, and $36 for programs and other expenses. You sell 30 tickets at $3 each. How much money does your class make on each ticket?

 F −$6   H −$4

 G −$5   J −$3

Spiral Review

Copy and complete each equation. (Lesson 1-7)

43. $150 + \underline{\ ?\ } = 150$ 44. $37 \times 1 = \underline{\ ?\ }$ 45. $6r \cdot \underline{\ ?\ } = 0$

46. **FOOTBALL** A football player was charged with three 15-yard penalties during a game. What was the total change in yardage because of the penalties? (Lesson 2-4)

Problem-Solving Strategy:
Look for a Pattern

Vocabulary

look for a pattern
(p. 146)

 Standard 7NS1.2
Add, subtract, multiply,
and divide rational
numbers (integers, fractions, and
terminating decimals) and take
positive rational numbers to
whole-number powers.

Standard 7MR1.1 Analyze
problems by identifying
relationships, distinguishing
relevant from irrelevant
information, identifying missing
information, sequencing and
prioritizing information, **and**
observing patterns.

Standard 7MR2.6 Express the
solution clearly and logically
by using the appropriate
mathematical notation and
terms and clear language;
support solutions with evidence
in both verbal and symbolic
work.

$-96 + 8 = -88$
$-88 + 8 = -80$
$-80 + 8 = -72$
$-72 + 8 = -64$
$-64 + 8 = -56$
$-56 + 8 = -48$

Use the Strategy

Kevin went scuba diving. This chart shows his depth at
different times. If this pattern continued, at what time did
Kevin arrive at the surface?

Time	10:00 A.M.	10:01 A.M.	10:02 A.M.	10:03 A.M.	10:04 A.M.	10:05 A.M.	10:06 A.M.
Depth (feet)	−96	−88	−80	−72	−64	−56	−48

 What do you know?

• the time each measurement was taken

• the depth at each time

Plan **What do you want to know?**

You want to know what time Kevin arrived at the surface, or
0 feet. **Look for a pattern** to see how many feet he rose each
minute. Use the pattern to find how many more minutes it
took for Kevin to reach the surface.

Solve **Add 8 to each measurement until the sum is 0 feet.**

• Kevin rose 8 feet per minute.
• At 10:06 A.M., he was still 48 feet below the surface.

$$48 \text{ feet} \div 8 \text{ feet/minute} = 6 \text{ minutes}$$

feet remaining   increase per minute

It will take Kevin another 6 minutes to reach the surface. He
will arrive at 10:12 A.M.

 Continue the pattern in the chart to confirm
your answer.

Time	10:00 A.M.	10:01 A.M.	10:02 A.M.	10:03 A.M.	10:04 A.M.	10:05 A.M.	10:06 A.M.	10:07 A.M.	10:08 A.M.	10:09 A.M.	10:10 A.M.	10:11 A.M.	10:12 A.M.
Depth (feet)	−96	−88	−80	−72	−64	−56	−48	−40	−32	−24	−16	−8	0

Look for a pattern to solve the problems.

1. Each minute, a scuba diver's depth changes. In the first 3 minutes, the depth changes from −85 to −80 to −75 to −70 feet. How many more minutes until the diver reaches the surface?

Understand **What do you know?**

Plan **What are you trying to find? How can you figure it out?**

Solve **Look for a pattern. Use the pattern to find the answer.**

Check **How can you be sure that your answer is correct?**

2. *Writing in Math* Meryl borrowed $104 from her dad to buy an MP3 player. Each week she gives her dad her allowance to pay back the loan. Meryl wants to know how many weeks it will be until she has paid her debt. Explain why looking for a pattern will help find the answer.

Date	Balance
Week 1	-$104
Week 2	-$96
Week 3	-$88
Week 4	-$80

Problem-Solving Practice

Solve by looking for a pattern.

3. **WEATHER** Willa records the outside temperature each hour, starting at 6 A.M. She records −10°C, −8°C, −6°C, and −4°C. If the pattern continues, at what time will the temperature be 6°C?

4. **BANKING** Mr. and Mrs. Blake make regular weekly deposits to their bank account. This chart shows the balance during the first four weeks. If the pattern continues, how many more weeks until the account reaches $0?

Date	Description	Balance
Week 1	Deposit	-$135
Week 2	Deposit	-$120
Week 3	Deposit	-$105
Week 4	Deposit	-$90

Solve using any strategy.

5. **AUTOMOBILE** There are 10 total cars and motorcycles in a parking lot. Camila counts a total of 34 wheels. How many cars are there?

6. **DOGS** A dog breeder had a certain number of dogs. After a litter was born, he had triple the original number. He sold 5 dogs and had 4 left. What was the original number of dogs?

7. **BUSINESS** There are 5 people at a meeting. Each person will shake hands one time with each other person in the room. How many handshakes will there be?

8. **COOKING** Amal's recipe for potato salad needs twice as many potatoes as eggs. If he uses a total of 15 potatoes and eggs, how many eggs does he use?

Progress Check 3

(Lessons 2-4 and 2-5)

▷ Vocabulary and Concept Check

factors (p. 135)   quotient (p. 141)
product (p. 135)

Choose the term that *best* completes each statement.

1. The result from dividing two numbers or expressions is called a _____?_____.

2. _____?_____ are numbers or expressions that are multiplied together.

▷ Skills Check

Find each product. (Lesson 2-4)

3. $-7 \cdot 6$

4. $-4 \cdot (-8)$

5. $9 \cdot (-8)$

6. $5 \cdot (-2) \cdot 5$

Find each quotient. (Lesson 2-5)

7. $48 \div (-8)$

8. $-54 \div (-6)$

9. $-35 \div 7$

10. $81 \div (-9)$

Find the value of each expression. (Lessons 2-4 and 2-5)

11. $2n + 5$, if $n = -4$

12. $8 - 3m$, if m is 7

13. $\frac{n}{5} - 6$, if $n = -20$

14. $\frac{-24}{d} + 7$, if $d = 3$

▷ Problem-Solving Check

15. **GOLF** Suppose a golfer has a score of -5 on each day of a 3-day tournament. What is her total score for the tournament? (Lesson 2-4)

16. **FINANCE** A stock changed $-\$28$ in 4 days. The price changed the same amount each day. How much did the stock change on the first day? (Lesson 2-5)

17. **REFLECT** Explain how you can determine the sign of a product by looking at the signs of the two factors. (Lesson 2-4)

2-7 Solving Equations

Vocabulary

inverse operation   (p. 149)

one-step equation
   (p. 150)

two-step equation
   (p. 150)

Standard ALG5.0
Students solve multi-step problems, including word problems, involving linear equations in one variable and provide justification for each step.

Standard 7AF4.1 Solve two-step linear equations and inequalities in one variable over the rational numbers, interpret the solution or solutions in the context from which they arose, and verify the reasonableness of the results.

Standard 6AF1.1 Write and solve one-step linear equations in one variable.

 The *What*: I will solve equations using inverse operations.

The *Why*: Equations can be used to determine the cost of international phone calls and store purchases.

Amateya made an international phone call to her grandparents. The total cost of the call was \$13. Suppose the call costs \$2 for each minute plus a \$3 fee. How long did Amateya talk?

The equation $2m + 3 = 13$ can be used to find the number of minutes m that Amateya talked.

$$\underbrace{2m}_{\text{cost for } m \text{ minutes}} + \underbrace{3}_{\text{fee}} = \underbrace{13}_{\text{total cost}}$$

In Lesson 2-1, you solved equations using the guess-and-check strategy and mental math.

Another method for solving equations is to isolate the variable. Isolating a variable means the variable is separated from the rest of the equation and appears alone on one side.

To isolate a variable, use the **inverse operation.** Addition and subtraction are inverse operations. Multiplication and division are also inverse operations.

Talk Math

WORK WITH A PARTNER Name an operation and a number. Listen as your partner provides the inverse. For example, you say "add 6" and your partner replies "subtract 6." Repeat five times.

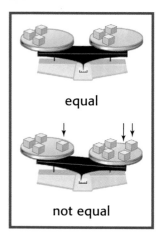

equal

not equal

A **one-step equation** has only one operation in the equation. When you perform an inverse operation on one side of an equation, you must perform the same operation on the other side. For example, if you add 2 to one side, you must add 2 to the other side. This keeps both sides of the equation equal.

EXAMPLES Solve One-Step Equations

Solve each equation.

❶ $t + 3 = 13$

$t + 3 = 13$	Use the inverse of addition.
$t + 3 - 3 = 13 - 3$	Subtract 3 from each side.
$t = 10$	
$10 + 3 = 13$	Substitute for t to check your answer.
$13 = 13$ ✓	The answer checks.

❷ $-10 = 2x$

$2x = -10$	Use the inverse of multiplication.
$(2x) \div 2 = -10 \div 2$	Divide each side by 2.
$x = -5$	
$2 \cdot (-5) = -10$	Substitute for x to check your answer.
$-10 = -10$ ✓	The answer checks.

Your Turn
Solve each equation.

a. $9 = w + 6$ b. $r - 7 = -2$

c. $5p = 20$ d. $-9 = \dfrac{d}{4}$

The equation $2m + 3 = 13$ is a **two-step equation** because it has two operations. You will need to use two steps to solve for m.

Think about wrapping a present. You put on the paper first and the bow second. You reverse this order when you open a present. You remove the bow first and then remove the paper.

Work backward when solving two-step equations. To solve for m in $2m + 3 = 13$, start by "undoing" the addition. Then "undo" the multiplication.

EXAMPLES Solve Two-Step Equations

Solve each equation.

③ $2m + 3 = 13$

$2m + 3 = 13$	Multiplication and addition are used.
$2m + 3 - 3 = 13 - 3$	First, use the inverse of addition.
$2m = 10$	Subtract 3 from each side.
$(2m) \div 2 = 10 \div 2$	Then, use the inverse of multiplication.
$m = 5$	Divide each side by 2.
$2(5) + 3 = 13$	Substitute for m to check your answer.
$10 + 3 = 13$	
$13 = 13 \checkmark$	The answer checks.

④ $16 = 3f - 2$

$16 = 3f - 2$	Multiplication and subtraction are used.
$16 + 2 = 3f - 2 + 2$	First, use the inverse of subtraction.
$18 = 3f$	Add 2 to each side.
$18 \div 3 = (3f) \div 3$	Then, use the inverse of multiplication.
$6 = f$	Divide each side by 3.
$16 = 3(6) - 2$	Substitute for f to check your answer.
$16 = 18 - 2$	
$16 = 16 \checkmark$	The answer checks.

Your Turn

Solve each equation.

e. $4c - 6 = 14$ f. $7n + 5 = 19$

g. $27 = 8y + 3$ h. $-16 = 6w + 14$

Real-World EXAMPLE

⑤ TRANSPORTATION A taxi company charges $2 plus $3 for each mile. If a trip cost $11, how many miles was the trip?

$3m + 2 = 11$	Assign a variable. Write an equation.
$3m + 2 - 2 = 11 - 2$	Use the inverse of addition.
$3m = 9$	Subtract 2 from each side.
$3m \div 3 = 9 \div 3$	Use the inverse of multiplication.
$m = 3$	Divide each side by 3.

The trip is 3 miles.

Your Turn

i. **COMMUNICATION** A phone company charges $5 each month for international service and $2 for each call. If this month's bill is $29, how many calls were made?

Vocabulary Review
inverse operation
one-step equation
two-step equation

Examples 1–5 **VOCABULARY**
(pages 150–151)

1. Write a one-step equation using one variable and subtraction.

2. Write a two-step equation using one variable, multiplication, and addition.

Solve each equation.

Examples 1–2   **3.** $b + 8 = 37$ **4.** $9 = n - 5$
(page 150)   **5.** $-27 = 3t$ **6.** $\frac{n}{6} = 12$

Examples 3–4   **7.** $2 + 3n = 11$ **8.** $18 = 2y + 12$
(page 151)   **9.** $36 = 5y - 4$ **10.** $3n - 8 = 25$

Example 5   **11. GEOMETRY** The sum of the measures of a triangle's angles is 180°.
(page 151)   Angles A and B have the same measure. Angle C has a measure of
                40°. What is the measure of Angle A?

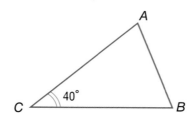

12. MUSIC A music shop charges $10 for each CD and then $2 to ship any purchase. Suppose Larisa spent $52 to send a gift to her cousin. How many CDs did she buy?

Skills, Concepts, and Problem Solving

Solve each equation.

HOMEWORK HELP	
For Exercises	**See Example(s)**
13–20	1–2
21–28	3–4
29–32	5

13. $h + 6 = 6$ **14.** $12 = k + 8$

15. $2 = d - 14$ **16.** $j - 6 = 6$

17. $7t = -56$ **18.** $-42 = 6n$

19. $7 = \frac{m}{2}$ **20.** $\frac{r}{3} = 9$

21. $3n + 5 = 14$ **22.** $5 + 7n = 33$

23. $12 = 3z + 9$ **24.** $24 = 5m + 9$

25. $4m - 16 = -4$ **26.** $-12 = 5k - 17$

27. $10 = 3t - 2$ **28.** $5a - 8 = 32$

29. LANDSCAPING The perimeter of a square P is four times the length of a side s: $P = 4s$. What is the length of a side of a square garden if the perimeter is 48 feet?

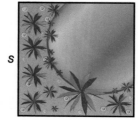

s

30. BUSINESS A taxi driver charged $2 a mile, plus a flat fee of $4. If the driver charged $18, how many miles was the trip?

31. **PHYSICS** Speed is equal to distance divided by time. The formula $s = \frac{d}{t}$ represents this relationship. A race car travels at an average speed of 150 miles per hour. How far can it travel in 3 hours?

32. **TOURISM** The Ling Family deposited $200 to start saving for a vacation. They plan to deposit $50 per month. How many months has the account been open if it contains $450?

33. *Writing in Math* Write a two-step equation for which the solution is 3.

34. **H.O.T. Problem** Contrast solving $2m - 9 = 5$ and $2m + 9 = 5$. Discuss differences in the steps taken to solve the equations.

Real-World Connection · · ·

TOURISM In 2005, the tourist industry in San Francisco generated more than 910,000 jobs.

STANDARDS PRACTICE

Choose the *best* answer.

The relationship between Fahrenheit and Celsius temperatures is given by the equation

$C = \dfrac{5(F - 32)}{9}$. This graph shows the normal high temperatures in San Francisco from January to June.

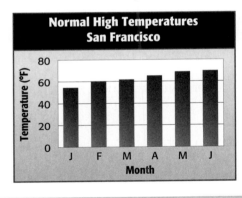

Normal High Temperatures San Francisco

35 What was the approximate normal high temperature in April in degrees Celsius?

A 17°C   C 34°C

B 32°C   D 47°C

36 A taxicab charges $3 for the first mile, then $2 for each mile after that. After a cab ride, which includes a $1 tip, Ike gets $4 change from a $20 bill. How many miles was the cab ride?

F 4   H 6

G 5   J 7

Spiral Review

Evaluate each expression. (Lesson 2-5)

37. $\frac{a}{4}$, if $a = -20$ 38. $\frac{b}{-2}$, if $b = -16$ 39. $\frac{c}{-3}$, if $c = 12$

40. **STATISTICS** Five years ago, East High School had 250 more students than today. The number of students decreased by the same amount each year. What was the change in population each year? (Lesson 2-5)

Find each product. (Lesson 2-4)

41. $-7 \cdot 4$ 42. $5 \cdot (-5)$ 43. $-3 \cdot (-9)$

Study Guide

Study Tips
Summarize the most important ideas.

Understanding and Using the Vocabulary

After completing the chapter, you should be able to define each term, property, or phrase and give an example of each.

absolute value (p. 119)	look for a pattern (p. 146)
additive inverse (p. 129)	one-step equation (p. 150)
define (p. 116)	opposite (p. 119)
Distributive Property (p. 131)	product (p. 135)
equation (p. 114)	quotient (p. 141)
factor (p. 135)	solution (p. 114)
integers (p. 119)	solve (p. 114)
inverse operation (p. 149)	two-step equation (p. 150)

Complete each sentence with the correct mathematical term or phrase.

1. The distance a number is from zero is its _____?_____.

2. The _____?_____, or additive inverse, of 8 is −8.

3. A(n) _____?_____ is the value of a variable that makes an equation true.

4. A(n) _____?_____ has only one operation in the equation.

5. _____?_____ include the counting numbers, their opposites, and zero.

6. Examples of a(n) _____?_____ are addition and subtraction or multiplication and division.

7. The _____?_____ is the answer in a multiplication example.

8. You _____?_____ an equation when you find the value (or values) that makes the equation true.

9. You _____?_____ a variable by identifying an unknown and choosing a variable to represent it.

10. The sum of a number and its _____?_____ is zero.

Skills and Concepts

Objectives and Examples

Review Exercises

LESSON 2-1 pages 114–118

Solve equations using substitution and mental math.

Solve $x - 2 = 5$.

$5 - 2 \neq 5$	Try $x = 5$.
$6 - 2 \neq 5$	Try $x = 6$.
$7 - 2 = 5$ ✓	Try $x = 7$.
$x = 7$	

Which of the numbers 1, 2, 3, 4, or 5 is the solution for each equation?

11. $x + 2 = 7$ **12.** $n - 1 = 2$

Which of the numbers 7, 8, 9, or 10 is the solution for each equation?

13. $6 + d = 13$ **14.** $5g = 45$

15. $b - 4 = 4$ **16.** $\frac{z}{2} = 5$

LESSON 2-2 pages 119–124

Graph and compare positive and negative integers.

Name the integers represented by A, B, and C.

Point A is -4.   Point B is -1.   Point C is 2.

Compare -4 and -1.

-4 is left of -1 on the number line.
-4 is less than -1 on the number line.
$-4 < -1$

Draw a number line for each problem and graph the integers.

17. -3, -2, and 0

18. -1, 2, and -3

19. -5, 1, and -4

Use a number line to complete each statement with one of the two symbols < or >.

20. 2 ⬤ -4 **21.** -1 ⬤ -2

22. 0 ⬤ -5 **23.** -6 ⬤ -3

LESSON 2-3 pages 128–133

Add and subtract integers.

$1 + (-4) = -3$	$-5 + 3 = -2$
$-1 + 4 = 3$	$-2 + (-4) = -6$

$2 - 5 = 2 + (-5) = -3$
$-1 - (-3) = -1 + 3 = 2$

$5k + (-6k) = -k$

Find each sum.

24. $-1 + (-4)$ **25.** $-3 + 7$

26. $2 + (-5)$ **27.** $-8 + 1$

Find each difference by using the additive inverse.

28. $5 - 8$ **29.** $-4 - (-5)$

Simplify.

30. $3b + (-6b)$ **31.** $-2m + 4m$

Study Guide

Objectives and Examples

Review Exercises

LESSON 2-4 pages 135–140

Multiply integers.

If the two factors have the same sign, the product is positive.

$2 \cdot 3 = 6$ $\qquad$ $-2 \cdot (-3) = 6$

If the two factors have different signs, the product is negative.

$-2 \cdot 3 = -6$ $\qquad$ $2 \cdot (-3) = -6$

Find each product.

32. $-5 \cdot 6$

33. $-4 \cdot (-7)$

34. $8 \cdot (-9)$

35. $-3 \cdot (-12)$

36. $9 \cdot (-6)$

37. $-7 \cdot (-8)$

LESSON 2-5 pages 141–145

Divide integers.

If the two numbers have the same sign, the quotient is positive.

$6 \div 2 = 3$ $\qquad$ $-6 \div (-2) = 3$

If the two numbers have different signs, the quotient is negative.

$6 \div (-2) = -3$ $\qquad$ $-6 \div 2 = -3$

Find each quotient.

38. $-24 \div 6$

39. $-32 \div (-8)$

40. $-35 \div (-5)$

41. $63 \div (-7)$

42. $-40 \div (-8)$

43. $-81 \div 9$

LESSON 2-7 pages 149–153

Solve equations using inverse operations.

$3n - 1 = 11$

$3n - 1 + 1 = 11 + 1$

$3n = 12$

$\dfrac{3n}{3} = \dfrac{12}{3}$

$n = 4$

$3(4) - 1 = 11\checkmark$

Solve each equation.

44. $5x - 2 = 28$

45. $27 = \dfrac{m}{4} + 3$

46. $2 + 3w = -19$

47. $8d - 1 = 55$

Chapter Test

▷ Vocabulary and Concept Check

1. Name an integer and its additive inverse.

2. Write an equation with one variable and a subtraction operation. Explain what it means to find the solution.

▷ Skills Check

Use a number line to complete each statement with < or >.

3. $-3 \bigcirc 0$

4. $-6 \bigcirc -7$

5. $-2 \bigcirc -5$

Find each sum, difference, product, or quotient.

6. $-9 + 4$

7. $5 + (-2)$

8. $-6 + (-3)$

9. $4 - 9$

10. $-2 - 1$

11. $-1 - (-4)$

12. $5 \cdot (-5)$

13. $-8 \cdot (-4)$

14. $-9 \cdot (-5)$

15. $-3 \cdot 6$

16. $\frac{42}{-6}$

17. $\frac{-54}{9}$

Find the value of each expression.

18. $4m$, if $m = -3$

19. $-k$, if $k = -2$

20. $j + k$, if $j = -5$ and $k = 3$

21. $-3ab$, if $a = -4$ and $b = -5$

22. $\frac{n}{-6}$, if $n = -36$

23. $\frac{x}{y}$, if $x = 36$ and $y = 4$

Solve each equation.

24. $d - 7 = 11$

25. $2 + p = -10$

26. $-30 = 6m$

27. $12 = \frac{y}{3}$

28. $3n + 2 = 17$

29. $1 + 4h = 17$

30. $20 = 2y - 10$

31. $6m - 48 = 0$

▷ Problem-Solving Check

32. **RECREATION** A scuba diver changes elevation at a rate of -3 feet each minute. If she starts at sea level (0 feet), what is her elevation after 12 minutes?

33. **ARTS** The conductor of an orchestra decides to hire 3 more violin players than flute players. There are 17 violinists. Identify a variable for the number of flute players. Write an equation to find the number of flute players.

Standards Practice

Choose the *best* answer.

1 Solve for x.

$2x - 3 = 9$

   A -6 **C** 3

   B -3 **D** 6

2 Which expression has the same value as 2^6?

   F 6^2 **H** 4^3

   G $(-2)^6$ **J** $(-6)^2$

3 Which integer is less than -1 and greater than -4?

   A -6 **C** 0

   B -2 **D** 3

4 If a is a positive number and b is a negative number, which expression is always positive?

   F $a \div b$ **H** $a + b$

   G $a \cdot b$ **J** $a - b$

5 On a winter afternoon, the temperature was 30°F at 3:00 P.M. By 6:00 P.M., the temperature dropped to 15°F. If the temperature continued to drop at the same rate, what would the temperature be at 10:00 P.M.?

   A 5°F   **C** -5°F

   B 0°F   **D** -10°F

6 In a room, the number of chairs c is equal to five times the number of tables t. Which equation matches the information?

   F $5t = c$ **H** $5c = t$

   G $t + c = 5$ **J** $c \cdot t = 5$

7 Which steps are used to solve the equation $5x + 2 = 12$?

   A Add 2 to each side of the equation. Then multiply each side by 5.

   B Add 2 to each side of the equation. Then divide each side by 5.

   C Subtract 2 from each side of the equation. Then multiply each side by 5.

   D Subtract 2 from each side of the equation. Then divide each side by 5.

8 This table shows the opening and closing prices of a stock during one trading week (Monday to Friday). The stock changed the same amount each day. How much did it change on Tuesday?

Day	Price (dollars)
Opening	15
Closing	5

   F $-\$10$

   G $-\$2$

   H $-\$5$

   J 0

9 The variable x is a negative number. Which expression is always negative?

   A $x \cdot x \cdot x$

   B $x \cdot x$

   C $x \div x$

   D $x \cdot x \cdot x \cdot x$

Record your answers on the answer sheet provided by your teacher or on a separate sheet of paper.

10 The temperature at 9 A.M. was −5°C. It rose 8°C during the next 6 hours. It then fell 6°C during the 6 hours after that. What was the temperature at 9 P.M.?

11 The school football team had two false starts, two holding penalties, and one clipping penalty. What was the total change in yards from these penalties?

Penalty	Yards
Clipping	−15
False Start	−5
Holding	−10
Roughing the Kicker	−15
Tripping	−10

12 The formula for finding the volume of a cube is $V = s^3$, where s represents the length of a side and V represents the volume.

Suppose a packing store sells boxes in the sizes shown below. The store charges $2 per cubic foot to pack a box. How much does it cost to pack a medium box?

Box	Side Length (feet)
Small	2
Medium	3
Large	4

13 On a school camping trip, two parent volunteers are needed for each student. Five teachers will accompany the group. There are 47 adults going on the trip.

  a. Write an equation to find the number of students.

  b. Solve your equation.

14 What is the solution of $\frac{k}{-4} = 5$?

Record your answers on the answer sheet provided by your teacher or on a separate sheet of paper.

15 The tenth grade student government sells muffins to raise money for charity. They plan to make 60 muffins.

Their expenses include $35 for ingredients and advertising. The group wants to make a profit of $85.

  a. Write an equation to determine the price p of a muffin.

  b. How much will each muffin cost?

  c. Suppose the group decides to make 40 larger muffins instead. How much will each muffin need to cost in order to make the same profit?

NEED EXTRA HELP?															
If You Missed Question...	1	2	3	4	5	6	7	8	9	10	11	12	13	14	15
Go to Lesson...	2-1 2-7	2-2	2-2	2-3	2-1 2-4	2-3	2-7	2-3 2-5	2-4 2-5	2-3	2-3 2-5	2-2	2-1 2-7	2-1 2-5 2-7	2-1 2-7
For Help with Algebra Readiness Standard...	6AF1.1 ALG5.0	ALG2.0	ALG2.0	7NS1.2	6AF1.1 7NS1.2	7NS1.2	ALG5.0	7NS1.2	7AF1.3	7NS1.2	7NS1.2	ALG2.0	6AF1.1 ALG5.0	6AF1.1 7NS1.2 ALG5.0	ALG5.0

Fractions

Here's What I Need to Know

Standard ALG5.0 Students solve multistep problems, including word problems, involving linear equations in one variable and provide justification for each step (excluding inequalities).

Standard 7NS1.2 Add, subtract, multiply, and divide rational numbers (integers, fractions, and terminating decimals) and take positive rational numbers to whole-number powers.

Standard 7AF4.1 Solve two-step linear equations and inequalities in one-variable over the rational numbers, interpret the solution or solutions in the context from which they arose, and verify the reasonableness of the results.

Vocabulary Preview

fraction A number that represents part of a whole or group (pp. 162, 164)

$$\frac{1}{3} \qquad \frac{1}{3}$$

rational number A number that can be written in the form $\frac{a}{b}$ $(b \neq 0)$ (p. 164)

$$\frac{5}{6}$$

$$3.5 \rightarrow 3\frac{1}{2} \text{ or } \frac{7}{2}$$

$$-3 \rightarrow \frac{-3}{1}$$

simplest form A fraction that does not have any common factor other than one in its numerator and denominator (p. 179)

The What

I will learn to:

- add, subtract, multiply, and divide fractions and mixed numbers.
- solve equations and problems with rational solutions.

The Why

Fractional pieces are used to form the rhythm of music. The length of sound can be represented by whole, half, quarter, eighth, and sixteenth notes.

Option 1

Math Online Are you ready for Chapter 3? Take the Online Readiness Quiz at ca.algebrareadiness.com to find out.

Option 2

Complete the Quick Check below.

Vocabulary and Skills Review

Match each sentence with the correct vocabulary word.

1. __?__ means to find the value of a variable in an equation.

2. $x + 2 = 4$ is an example of a(n) __?__.

3. Addition is the __?__ of subtraction.

4. The number $\frac{4}{5}$ is an example of a(n) __?__.

A. equation

B. expression

C. fraction

D. inverse operation

E. solve

Find the missing number.

5. $1 \cdot$ __?__ $= 24$

6. $2 \cdot$ __?__ $= 24$

7. $3 \cdot$ __?__ $= 24$

8. $6 \cdot$ __?__ $= 24$

Solve.

9. $x + 5 = 8$

10. $\frac{y}{5} = 2$

Note-taking

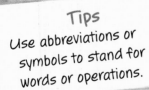

Tips

Use abbreviations or symbols to stand for words or operations.

Chapter Instruction	My Notes
A **fraction,** such as $\frac{1}{3}$, represents part of a whole or group. The fraction $-\frac{1}{3}$ is a negative fraction. (p. 162)	fraction $\frac{1}{3}$ = part of whole $-\frac{1}{3}$ = negative fraction
A new group of numbers is needed to define $\frac{1}{3}$ and $-\frac{1}{3}$. This group is called the **rational numbers.** (p. 164)	Rational Numbers $\frac{1}{3}$ and $-\frac{1}{3}$ = numbers you can write as fractions.

Math Lab
Represent Fractions

Vocabulary

fraction (p. 162)

Materials

colored pencils or markers (optional)

Standard 7MR2.5 Use a variety of methods, such as words, numbers, symbols, charts, graphs, tables, **diagrams, and models to explain mathematical reasoning.**

Standard 6NS1.2 Interpret and use ratios in different contexts to show the relative size of two quantities using appropriate notations.

 The What**:** I will color shapes to represent different fractions.

The Why**:** Fractions are used to represent the number of friends wearing jeans. Pictures are useful tools for illustrating fractions.

Fractions are numbers that represent parts of a set or parts of a whole. Four teenagers are standing in a cafeteria line. Three teenagers are wearing jeans. What fraction represents the number wearing jeans?

You can use set models, rectangular models, circle models, and triangular models to represent fractions.

ACTIVITY   **Use a Set Model**

① Create a set model that represents the fraction of teenagers wearing jeans. Then, write the fraction.

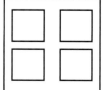

Each small square represents 1 classmate. There are 4 classmates, so draw 4 blank squares.

Shade 3 squares to show the 3 classmates wearing jeans.

3 out of 4 classmates, or $\frac{3}{4}$, are wearing jeans.

2 Represent the fraction $\frac{3}{4}$. Use a rectangular model, a circle model, and a triangular model.

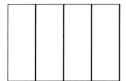

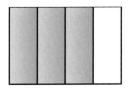

Draw a rectangle. Divide it into 4 equal parts

Shade 3 sections to show 3 out of 4 parts.

Draw a circle. Divide it into 4 equal parts.

Shade 3 sections to show 3 out of 4.

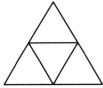

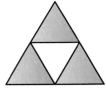

Draw a triangle with 3 equal sides. Divide it into 4 equal parts.

Shade 3 sections to show 3 out of 4.

Analyze the Results

Pick a random group of 8 classmates. Create set models, rectangular models, circle models, and triangular models to represent each of the following.

1. What fraction of the group is wearing sneakers?

2. What fraction of the group is wearing jeans?

3. What fraction of the group has brown hair?

4. **REFLECT** Describe a situation for which it would be difficult to model a fraction using drawings.

Vocabulary

fraction (p. 164)

rational number (p. 164)

additive inverse (p. 165)

 Standard 6NS1.1
Compare and order positive and negative fractions, decimals, and mixed numbers **and place them on a number line.**

 The What: I will compare and order fractions using a number line.

The Why: Fractions can be used to show parts of a whole, like the blue tiles in a floor.

Mr. Chung tiled his kitchen floor. One out of every three tiles was blue. The fraction $\frac{1}{3}$ represents the blue part of the floor.

A **fraction,** such as $\frac{1}{3}$, represents part of a whole or group.

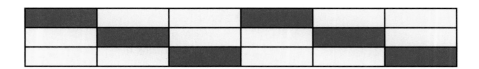

Study Tip

The fraction $-\frac{1}{3}$ can also be written as $\frac{-1}{3}$ or $\frac{1}{-3}$.

Recall that integers like 5 and −5 are opposites, or additive inverses. The opposite of $\frac{1}{3}$ is the fraction $-\frac{1}{3}$.

A new group of numbers is needed to define $\frac{1}{3}$ and $-\frac{1}{3}$. This group is called the **rational numbers.** A rational number can be written in the form $\frac{a}{b}$ where $b \neq 0$.

Talk Math

WORK WITH A PARTNER Name three integers and three fractions. Be sure that at least one of the fractions is negative. Listen as your partner names three integers and three fractions.

You can use a number line to show $\frac{1}{3}$ and $-\frac{1}{3}$.

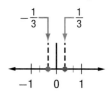

EXAMPLE   Graph Fractions

1 Graph the fraction $-\frac{2}{3}$ on a number line.

Divide the distance from 0 to −1 into thirds.

Count back 2 sections from 0.

Label the point.

Your Turn

Graph each fraction on a number line.

a. $-\frac{1}{2}$

b. $-\frac{3}{5}$

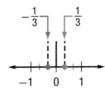

The fractions $-\frac{1}{3}$ and $\frac{1}{3}$ are opposites or **additive inverses.**
The sum of additive inverses is zero.

EXAMPLES   Find Additive Inverses

Write the additive inverse of each number.

2 $\frac{1}{2}$

The opposite of $\frac{1}{2}$ is $-\frac{1}{2}$.
So, the additive inverse of $\frac{1}{2}$ is $-\frac{1}{2}$.

3 $-\frac{4}{3}$

The opposite of $-\frac{4}{3}$ is $\frac{4}{3}$.
So, the additive inverse of $-\frac{4}{3}$ is $\frac{4}{3}$.

Your Turn

Write the additive inverse of each number.

c. $\frac{4}{5}$

d. $-\frac{5}{6}$

The additive inverse of any number is the same distance from 0 on the number line as the number. The sign is different to indicate the opposite direction from 0.

EXAMPLE   Graph Additive Inverses

4 Write the additive inverse of $\frac{2}{3}$. Then graph the number and its additive inverse.

The additive inverse, $-\frac{2}{3}$, has the opposite sign.

Your Turn

Write the additive inverse of each number. Then graph the number and its additive inverse.

e. $\frac{3}{4}$

f. $-\frac{1}{2}$

EXAMPLE Compare Fractions

❺ Use a number line to determine which number is greater, $-\frac{1}{4}$ or $-\frac{3}{4}$.

Graph the fractions on the same number line.

$-\frac{1}{4}$ is greater than $-\frac{3}{4}$. $-\frac{1}{4}$ is to the right of $-\frac{3}{4}$.

$-\frac{1}{4} > -\frac{3}{4}$

Your Turn

Use a number line to determine which number is greater.

g. $\frac{3}{8}$ or $\frac{5}{8}$

h. $-\frac{1}{6}$ or $-\frac{5}{6}$

EXAMPLE Order Fractions

❻ Order the fractions $-\frac{5}{8}, -\frac{3}{8}, -\frac{7}{8}$ from least to greatest.

Graph the fractions on the same number line.

$-\frac{7}{8}$ is the least fraction. $-\frac{7}{8}$ is farthest to the left.

$-\frac{3}{8}$ is the greatest fraction. $-\frac{3}{8}$ is farthest to the right.

$-\frac{7}{8} < -\frac{5}{8} < -\frac{3}{8}$

Your Turn

Order each set of fractions from least to greatest.

i. $\frac{2}{6}, \frac{5}{6}, \frac{3}{6}$

j. $-\frac{4}{5}, -\frac{2}{5}, -\frac{3}{5}$

Real-World EXAMPLE

❼ **FASHION** Marguerite changed the hem on her jeans by $\frac{1}{8}$ inch. She changed the hem on her slacks by $\frac{3}{8}$ inch. Which is the greater fraction?

$\frac{3}{8}$ is greater than $\frac{1}{8}$. $\frac{3}{8}$ is to the right of $\frac{1}{8}$.

$\frac{3}{8} > \frac{1}{8}$

Your Turn

k. **COOKING** A chef changes the amount of tomatoes in a recipe by $\frac{1}{4}$ cup and the amount of olives by $\frac{3}{4}$ cup. Which is the greater fraction?

Examples 1–7
(pages 165–166)

VOCABULARY

1. Explain how a positive fraction and its opposite are alike and different.

Example 1
(page 165)

Graph each fraction on a number line.

2. $-\dfrac{1}{8}$

3. $\dfrac{4}{5}$

Examples 2–3
(page 165)

Write the additive inverse of each number.

4. $\dfrac{1}{6}$

5. $-\dfrac{5}{9}$

Example 4
(page 165)

Write the additive inverse. Then graph both fractions.

6. $\dfrac{1}{8}$

7. $-\dfrac{4}{5}$

Example 5
(page 166)

Use a number line to determine which number is greater.

8. $\dfrac{1}{3}$ or $\dfrac{2}{3}$

9. $-\dfrac{3}{5}$ or $-\dfrac{2}{5}$

Example 6
(page 166)

Order each set of fractions from least to greatest.

10. $\dfrac{3}{7}, \dfrac{5}{7}, \dfrac{2}{7}$

11. $-\dfrac{4}{9}, -\dfrac{7}{9}, -\dfrac{5}{9}$

Example 7
(page 166)

12. **TRANSPORTATION** Winona has $\dfrac{3}{4}$ tank of gas in her car. Fatima has $\dfrac{1}{4}$ tank of gas. Graph the numbers on a number line. Who has more gas?

13. **PLUMBING** To install a new sink, Edna changed the length of one pipe by $\dfrac{1}{5}$ inch, and the second pipe by $\dfrac{3}{5}$ inch. Which is the greater fraction?

14. **Talk Math** Have a partner tell you a number. Decide if the number is an integer or a rational number.

Skills, Concepts, and Problem Solving

Graph each fraction on a number line.

For Exercises	See Example(s)
15–18	1
19–22	2–3
23–26	4
27–29	5
30–32	6
33–34	7

15. $-\dfrac{1}{5}$

16. $-\dfrac{2}{9}$

17. $\dfrac{4}{7}$

18. $\dfrac{3}{8}$

Write the additive inverse of each number.

19. $\dfrac{5}{6}$

20. $\dfrac{3}{4}$

21. $-\dfrac{2}{5}$

22. $-\dfrac{6}{7}$

Write the additive inverse of each number. Then graph the number and its additive inverse.

23. $\dfrac{4}{5}$

24. $\dfrac{2}{3}$

25. $-\dfrac{2}{7}$

26. $-\dfrac{3}{4}$

Use a number line to determine which number is greater.

27. $\frac{2}{5}$ or $\frac{4}{5}$

28. $\frac{1}{4}$ or $\frac{3}{4}$

29. $-\frac{1}{8}$ or $-\frac{3}{8}$

Order each set of fractions from least to greatest.

30. $\frac{1}{8}, \frac{5}{8}, \frac{3}{8}$

31. $-\frac{1}{7}, -\frac{6}{7}, -\frac{5}{7}$

32. $-\frac{8}{9}, -\frac{5}{9}, -\frac{4}{9}$

33. Printed photographs are sensitive to changes in temperature. A museum keeps the storage room temperature within a range of $+\frac{3}{4}$°C to $-\frac{3}{4}$°C. Graph these fractions on a number line and compare them.

34. **TEMPERATURE** A scientist records changes of $-\frac{7}{8}$°C and $-\frac{3}{8}$°C during a science experiment. Which is the greater fraction?

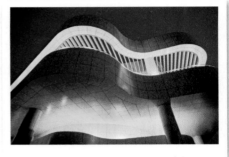

PHOTOGRAPHY Many California museums, including the Getty Museum in Los Angeles, have exhibits of photography collections.

35. *Writing in Math* Write an inequality comparing two negative fractions, each with a 5 in the denominator.

36. **H.O.T.** Problem Identify any two fractions less than $-\frac{1}{8}$ and greater than $-\frac{1}{2}$.

Choose the *best* answer.

37 Which rational number is greater than $-\frac{3}{4}$ and less than $-\frac{1}{3}$?

A $-\frac{5}{6}$

B $-\frac{1}{2}$

C $-\frac{1}{4}$

D 0

38 A scientist records these temperature changes while cooling a liquid. Which choice correctly orders the numbers?

F $-\frac{7}{9} > -\frac{4}{9} > -\frac{2}{9}$

G $-\frac{4}{9} > -\frac{2}{9} > -\frac{7}{9}$

H $-\frac{2}{9} > -\frac{7}{9} > -\frac{4}{9}$

J $-\frac{2}{9} > -\frac{4}{9} > -\frac{7}{9}$

Hour	Temperature Change (°C)
1	$-\frac{4}{9}$
2	$-\frac{7}{9}$
3	$-\frac{2}{9}$

Spiral Review

Solve. (Lesson 2-7)

39. $2x + 3 = 11$

40. $5p - 1 = 19$

41. $2 + \frac{n}{3} = 8$

42. $1 = \frac{k}{6} - 4$

43. **SPORTS** The Hawks scored 5 fewer than 3 times as many points as the Tornadoes. The Hawks scored 37 points. How many points did the Tornadoes score? (Lesson 2-7)

3-2 Fractions and Mixed Numbers

Vocabulary

proper fraction (p. 169)

improper fraction (p. 169)

mixed number (p. 169)

equivalent fraction (p. 172)

 Standard 6NS1.1 Compare and order positive and negative fractions, decimals, and mixed numbers and place them on a number line.

The *What:* I will compare and order improper fractions and mixed numbers.

The *Why:* You may use mixed numbers and improper fractions when you measure baking ingredients.

In a **proper fraction,** the absolute value of the numerator is less than the absolute value of the denominator. In an **improper fraction,** the absolute value of the numerator is greater than or equal to the absolute value of the denominator.

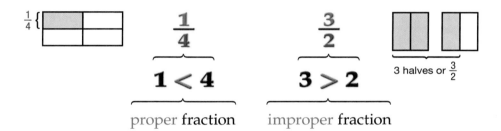

PROPER FRACTIONS
$\frac{5}{6}$
$\frac{42}{95}$
$-\frac{1}{3}$
$-\frac{7}{8}$

IMPROPER FRACTIONS
$\frac{11}{6}$
$\frac{112}{101}$
$-\frac{4}{3}$
$-\frac{10}{10}$

You can rewrite an improper fraction as a **mixed number.** Mixed numbers combine or "mix" an integer and a fraction. Examples include $2\frac{1}{2}$ and $-6\frac{2}{5}$.

Remember that any number that can be written in the form $\frac{a}{b}$, where $b \neq 0$, is a rational number. So, improper fractions and mixed numbers are rational numbers.

Lesson 3-2 Fractions and Mixed Numbers **169**

WORK WITH A PARTNER Name five improper fractions.
Listen as your partner names five mixed numbers.

You can look at the numerator and denominator to decide if
a fraction is proper or improper.

 **EXAMPLES** Identify Proper and Improper Fractions

Identify each fraction as proper or improper.

❶ $\frac{3}{5}$

$3 < 5$

The numerator is less than
the denominator.

proper

❷ $\frac{7}{7}$

$7 = 7$

The numerator is equal
to the denominator.

improper

Your Turn

Identify each fraction as proper or improper.

a. $\frac{9}{4}$

b. $\frac{3}{8}$

You can rewrite an improper fraction as a mixed number
using division.

EXAMPLE Change Improper Fractions to
Mixed Numbers

❸ Write $\frac{7}{2}$ as a mixed number.

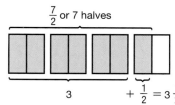

$\frac{7}{2}$ or 7 halves

3 $+ \frac{1}{2} = 3\frac{1}{2}$

$$\begin{array}{r} 3R1 \\ 2\overline{)7} \end{array}$$ Divide 7 by 2. numerator ÷ denominator

$3\frac{1}{2}$ Write the answer as a mixed number. Place the
remainder of 1 over the divisor 2.

Your Turn

Write each improper fraction as a mixed number.

c. $\frac{8}{3}$

d. $-\frac{5}{2}$

Study Tip

You can also rewrite a
mixed number as an
improper fraction.

$3\frac{1}{2}$ is seven halves, or $\frac{7}{2}$.

$4\frac{2}{3}$ is fourteen thirds, or $\frac{14}{3}$.

You can use a number line to graph improper fractions and mixed numbers.

EXAMPLE **Improper Fractions and Mixed Numbers**

4️⃣ **Graph** $-2\frac{3}{4}$ **and** $\frac{5}{4}$ **on a number line.**

Rewrite $\frac{5}{4}$ as $1\frac{1}{4}$.

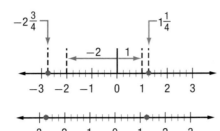

Divide each unit into fourths.

Starting from zero, count to the integer. Then count to the fractional increment.

Your Turn
Graph each pair of numbers on a number line.

e. $4\frac{1}{3}, -\frac{5}{3}$

f. $-2\frac{2}{5}, \frac{8}{5}$

To compare improper fractions and mixed numbers, graph them on a number line. The number farthest right is the greater number.

Real-World EXAMPLE **Compare Numbers**

5️⃣ **CONSTRUCTION** Walt uses $1\frac{3}{8}$-inch nails to attach drywall and $\frac{17}{8}$-inch nails to attach paneling. Which nails are longer?

Write $\frac{17}{8}$ as $2\frac{1}{8}$.

Graph $2\frac{1}{8}$ and $1\frac{3}{8}$ on a number line.

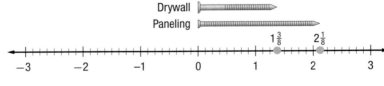

$2\frac{1}{8} > 1\frac{3}{8}$

The paneling nails are longer.

Your Turn

g. **BAKING** A recipe calls for $1\frac{3}{4}$ cups of raisins and $2\frac{1}{4}$ cups of walnuts. Does the recipe use more raisins or walnuts?

Suppose a recipe calls for $\frac{1}{2}$ teaspoon salt. Your only measuring spoon is $\frac{1}{4}$ teaspoon. How many $\frac{1}{4}$ teaspoons of salt do you need to equal $\frac{1}{2}$ teaspoon?

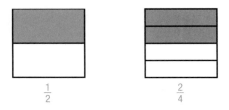

$\frac{1}{2}$ $\frac{2}{4}$

Notice the green and purple shaded portions are equal.

The fractions $\frac{1}{2}$ and $\frac{2}{4}$ are equivalent.

Equivalent fractions, such as $\frac{1}{2}$ and $\frac{2}{4}$, have the same value.

EXAMPLES   **Show Equivalent Fractions**

Use a model to show each pair of fractions is equivalent.

6 $\frac{2}{3}, \frac{4}{6}$

The fractions $\frac{2}{3}$ and $\frac{4}{6}$ are equivalent.

7 $\frac{10}{8}, \frac{5}{4}$

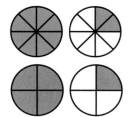

The fractions $\frac{10}{8}$ and $\frac{5}{4}$ are equivalent.

Your Turn

Use a model to show each pair of fractions is equivalent.

h. $\frac{3}{4}, \frac{6}{8}$ i. $\frac{2}{5}, \frac{4}{10}$

Examples 1–7
(pages 170–172)

VOCABULARY

1. Write an example of a proper fraction and an improper fraction.

2. What is a mixed number? How are mixed numbers related to improper fractions?

Examples 1–2
(page 170)

Identify each fraction as proper or improper.

3. $-\dfrac{7}{2}$

4. $\dfrac{3}{8}$

Example 3
(page 170)

Write each improper fraction as a mixed number.

5. $\dfrac{8}{5}$

6. $\dfrac{9}{2}$

7. $\dfrac{5}{4}$

8. $\dfrac{12}{5}$

Example 4
(page 171)

Graph each pair of numbers on a number line.

9. $1\dfrac{2}{3}, -\dfrac{5}{3}$

10. $-2\dfrac{1}{5}, \dfrac{7}{5}$

11. $-2\dfrac{3}{4}, \dfrac{5}{4}$

12. $-3\dfrac{1}{2}, \dfrac{3}{2}$

Example 5
(page 171)

13. **FITNESS** On Saturday, Benito walked $2\dfrac{3}{10}$ miles. On Sunday, he walked $1\dfrac{9}{10}$ miles. On which day did he walk farther?

14. **BAKING** A recipe calls for $\dfrac{5}{2}$ cups of peanut butter. What is this amount written as a mixed number?

Examples 6–7
(page 172)

Use a model to show each pair of fractions is equivalent.

15. $\dfrac{1}{3}, \dfrac{2}{6}$

16. $\dfrac{3}{4}, \dfrac{6}{8}$

17. $\dfrac{18}{4}, \dfrac{9}{2}$

18. $\dfrac{5}{3}, \dfrac{10}{6}$

19. **Talk Math** Listen as your partner tells you a mixed number. Tell your partner a mixed number that is greater than that number.

Skills, Concepts, and Problem Solving

HOMEWORK HELP

For Exercises	See Example(s)
20–23	1–2
24–31	3
32–39	4
40–41	5
42–45	6–7

Identify each fraction as proper or improper.

20. $\dfrac{11}{5}$

21. $\dfrac{4}{7}$

22. $\dfrac{8}{3}$

23. $\dfrac{27}{19}$

Write each improper fraction as a mixed number.

24. $\dfrac{5}{3}$

25. $\dfrac{11}{2}$

26. $\dfrac{9}{4}$

27. $\dfrac{12}{5}$

28. $\frac{7}{4}$

29. $\frac{9}{7}$

30. $\frac{15}{7}$

31. $\frac{13}{6}$

Graph each pair of numbers on a number line.

32. $2\frac{1}{2}, -\frac{3}{2}$

33. $1\frac{3}{4}, -\frac{7}{4}$

34. $3\frac{1}{3}, -\frac{7}{3}$

35. $-1\frac{1}{5}, \frac{8}{5}$

36. $1\frac{7}{8}, -\frac{9}{8}$

37. $-2\frac{5}{6}, \frac{8}{6}$

38. $-2\frac{4}{5}, \frac{10}{5}$

39. $-3\frac{3}{8}, -\frac{12}{8}$

40. **COOKING** A soup recipe uses $3\frac{1}{4}$ cups of carrots and $\frac{11}{4}$ cups of celery. Does the recipe use more carrots or celery?

41. **CAREER CONNECTION** A small business owner is responsible for hiring and managing employees, collecting bills and taxes, advertising, maintaining records, and much more.

An owner of a bakery has $4\frac{1}{8}$ pounds of coffee and $\frac{37}{8}$ pounds of tea in stock. Does he have more coffee or tea?

BUSINESS In California, a small business can be certified as small or micro, depending on income and number of employees.

Use a model to show each pair of fractions is equivalent.

42. $\frac{1}{5}, \frac{2}{10}$

43. $\frac{3}{7}, \frac{6}{14}$

44. $\frac{3}{2}, \frac{9}{6}$

45. $\frac{5}{5}, \frac{10}{10}$

46. **GEOGRAPHY** Parts of the United States have different time zones. California is in the Pacific time zone. Parts of Newfoundland, Canada, use Newfoundland time, which is $-3\frac{1}{2}$ hours from standard Greenwich time. Graph this number. Then compare it to -8 hours, the reference for the Pacific time zone.

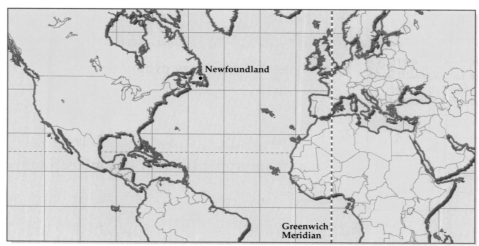

47. **TEMPERATURE** Lois finds that she can save \$500 by lowering her store's thermostat by $\frac{3}{2}$°F. Write this temperature change as a mixed number.

48. **ARCHITECTURE** An architect widens a doorway by $\frac{21}{8}$ inches. Write this change as a mixed number.

49. *Writing in Math* Write a paragraph explaining how to show two fractions are equivalent.

50. **H.O.T.** Problem Write a mixed number less than −2 and greater than −3.

STANDARDS PRACTICE

Choose the *best* answer.

51 Which mixed number is less than $-2\frac{3}{4}$ and greater than $-3\frac{1}{2}$?

A $-2\frac{1}{2}$

B $-2\frac{7}{8}$

C $-3\frac{5}{8}$

D $-3\frac{3}{4}$

52 The table shows the number of miles Emilio, Aaron, and Lois run every day.

Person	Miles Ran
Emilio	$1\frac{1}{2}$
Aaron	$2\frac{1}{2}$
Lois	$1\frac{3}{4}$

Which choice correctly orders the mixed numbers from least to greatest?

F $1\frac{3}{4},\ 1\frac{1}{2},\ 2\frac{1}{2}$

G $1\frac{1}{2},\ 1\frac{3}{4},\ 2\frac{1}{2}$

H $2\frac{1}{2},\ 1\frac{3}{4},\ 1\frac{1}{2}$

J $2\frac{1}{2},\ 1\frac{1}{2},\ 1\frac{3}{4}$

Spiral Review

Use a number line to determine which number is greater. (Lesson 3-1)

53. $-\frac{5}{6}$ or $-\frac{3}{6}$

54. $-\frac{6}{8}$ or $-\frac{7}{8}$

55. $\frac{5}{7}$ or $\frac{6}{7}$

Find each quotient. (Lesson 2-5)

56. $54 \div (-9)$

57. $-56 \div (-8)$

58. $63 \div 7$

59. **WEATHER** Suppose it rained 7 inches in April last year. The *change* for the same month this year was −3 inches. How many inches did it rain in April this year? (Lesson 2-3)

Progress Check 1

(Lessons 3-1 and 3-2)

▷ Vocabulary and Concept Check

> additive inverse (p. 165)   mixed number (p. 169)
> equivalent fraction (p. 172)   proper fraction (p. 169)
> fraction (p. 164)   rational number (p. 164)
> improper fraction (p. 169)

Choose the term that *best* completes each statement. (Lessons 3-1 and 3-2)

1. A(n) _____?_____ has a denominator that is greater than the numerator.

2. The fractions $\frac{1}{3}$ and $-\frac{1}{3}$ are _____?_____.

3. $3\frac{1}{2}$ is an example of a(n) _____?_____.

▷ Skills Check

Write the additive inverse. Then, graph the number and its additive inverse. (Lesson 3-1)

4. $-\frac{1}{4}$

5. $\frac{4}{5}$

Use a number line to determine which number is greater. (Lesson 3-1)

6. $-\frac{2}{3}$ and $-\frac{1}{3}$

7. $-\frac{3}{8}$ and $-\frac{5}{8}$

Graph each pair of numbers on a number line. (Lesson 3-2)

8. $2\frac{2}{3}, -\frac{5}{3}$

9. $-1\frac{3}{4}, \frac{6}{4}$

10. $-1\frac{1}{3}, \frac{8}{3}$

11. $3\frac{3}{4}, -\frac{9}{4}$

Use a model to show each pair of fractions is equivalent. (Lesson 3-2)

12. $\frac{3}{5}, \frac{6}{10}$

13. $\frac{7}{2}, \frac{14}{4}$

▷ Problem-Solving Check

14. **TEMPERATURE** During the first hour of a storm, the temperature changes $-\frac{1}{4}°$F. During the second hour, the temperature changes $-\frac{3}{4}°$F. Graph these fractions on a number line and compare them. (Lesson 3-1)

15. **COOKING** A chef uses $1\frac{5}{8}$ cups of onions and $\frac{11}{8}$ cups of peppers in a sauce. Which of these amounts is greater? (Lesson 3-2)

16. **REFLECT** Explain how you would graph the fraction $-\frac{3}{4}$. (Lesson 3-1)

Factors and Simplifying Fractions

Vocabulary

prime number (p. 177)

composite number (p. 177)

factor (p. 177)

common factor (p. 179)

greatest common factor (GCF) (p. 179)

simplest form (p. 179)

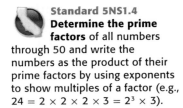

Standard 5NS1.4 Determine the prime factors of all numbers through 50 and write the numbers as the product of their prime factors by using exponents to show multiples of a factor (e.g., $24 = 2 \times 2 \times 2 \times 3 = 2^3 \times 3$).

The What: I will identify prime and composite numbers, find greatest common factors, and simplify fractions.

The Why: Simplifying fractions can make it easier to compare statistics, such as the number of students in a group who belong to one class.

Any whole number greater than 1 is either a **prime number** or a **composite number.** The numbers 0 and 1 are neither prime nor composite.

A prime number has exactly two factors, 1 and itself.

A composite number has more than two factors.

PRIME NUMBERS	
Numbers	Factors
2	1, 2
5	1, 5
7	1, 7
11	1, 11
23	1, 23
29	1, 29

COMPOSITE NUMBERS	
Number	Factors
6	1, 2, 3, 6
9	1, 3, 9
10	1, 2, 5, 10
16	1, 2, 4, 8, 16
21	1, 3, 7, 21
28	1, 2, 4, 7, 14, 28

When you **factor** a number, you list all of the numbers that can be multiplied together, two at a time, to get that number. For example, $6 = 1 \cdot 6$ and $6 = 2 \cdot 3$, so the factors of 6 are 1, 2, 3, and 6.

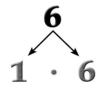

← factors

Identify each number as prime or composite.

① 3

factors: 1, 3

The only factors are
1 and 3.

So, 3 is prime.

② 12

factors: 1, 2, 3, 4, 6, 12

There are more than
2 factors.

So, 12 is composite.

Your Turn

Identify each number as prime or composite.

a. 6   **b.** 5   **c.** 11   **d.** 9

One way to find the factors for a number is to make an organized list. List all of the numbers from 1 to half (or just over half for odd numbers) of the number. If a number is a factor of the number, then write the other factor. If it is not a factor of the number, or if that factor is already listed, cross it off.

EXAMPLES List Factors

List the factors of each number.

③ 8

List the numbers from
1 to half of 8.
Decide if each number
is a factor.

1 ⟶ 8
2 ⟶ 4
3̶ Not a factor
4̶ Already listed

The factors of 8 are
1, 2, 4, and 8.

④ 12

List the numbers from
1 to half of 12.
Decide if each number
is a factor.

1 ⟶ 12
2 ⟶ 6
3 ⟶ 4
4̶ Already listed
5̶ Not a factor
6̶ Already listed

The factors of 12 are
1, 2, 3, 4, 6, and 12.

Your Turn

List the factors of each number.

e. 10   **f.** 9   **g.** 11   **h.** 14

Talk Math

WORK WITH A PARTNER List the factors of 18. Listen
as your partner lists the factors of 24.

If two or more numbers have the same factors, they have common factors. To find the **common factors,** list the factors of each number. Then circle the factors common to each list.

> **EXAMPLE** Find Common Factors
>
> **⑤ Find the common factors of 10 and 15.**
>
> 10: 1, 2, 5, 10   List the factors of 10.
> 15: 1, 3, 5, 15   List the factors of 15.
>
> 10: ①, 2, ⑤, 10
> 15: ①, 3, ⑤, 15   Circle the common factors.
> The common factors are 1 and 5.
>
> **Your Turn**
> **Find the common factors of each set of numbers.**
>
> i. 4 and 10   j. 6 and 18

The greatest of the common factors is called the **greatest common factor (GCF).** To find the greatest common factor, list the common factors for each number. Then circle the greatest one.

> **EXAMPLE** Find the Greatest Common Factor
>
> **⑥ Find the greatest common factor (GCF) of 8 and 20.**
>
> 8:   1, 2, 4, 8
> 20: 1, 2, 4, 5, 10, 20   List the factors.
> 8:   <u>1, 2, 4,</u> 8
> 20: <u>1, 2, 4,</u> 5, 10, 20   Underline the common factors.
> 8:   1, 2, ④, 8
> 20: 1, 2, ④, 5, 10, 20   Circle the greatest common factor.
> The GCF of 8 and 20 is 4.
>
> **Your Turn**
> **Find the greatest common factor (GCF) of the numbers.**
>
> k. 12 and 18   l. 7 and 14

Study Tip

Two or more numbers may have many common factors, but only one greatest common factor.

Equivalent fractions, such as $\frac{1}{2}$ and $\frac{2}{4}$, have the same value. A fraction is in **simplest form** when the greatest common factor of the numerator and denominator is 1.

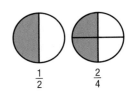

$\frac{1}{2}$ $\frac{2}{4}$

The fraction $\frac{21}{28}$ is not in simplest form. The greatest common factor of 21 and 28 can be used to simplify $\frac{21}{28}$.

To write fractions in simplest form, start by finding the greatest common factor of the numerator and denominator. Then, divide the numerator and denominator by the GCF.

Factors of 21:   Factors of 28:

$$1, 3, \textcircled{7}, 21 \qquad 1, 2, 4, \textcircled{7}, 14, 28$$

greatest common factor

$$\frac{21 \div 7}{28 \div 7} = \frac{3}{4}$$

EXAMPLE Write Fractions in Simplest Form

7 Write $\frac{8}{12}$ in simplest form.

8: 1, 2, $\textcircled{4}$, 8

12: 1, 2, 3, $\textcircled{4}$, 6, 12   List the factors. Find the GCF.

$\frac{8 \div 4}{12 \div 4} = \frac{2}{3}$ Divide both numerator and denominator by the GCF.

So, $\frac{8}{12} = \frac{2}{3}$ in simplest form.

Your Turn

Write each fraction in simplest form.

m. $\frac{6}{9}$ n. $\frac{12}{20}$

Real-World EXAMPLE

8 SPORTS A football team made 18 of the last 24 field goals it attempted. What was their success rate as a fraction in simplest form?

18: 1, 2, 3, $\textcircled{6}$, 9, 18   List the factors. Find the GCF.

24: 1, 2, 3, 4, $\textcircled{6}$, 8, 12, 24

$\frac{18 \div 6}{24 \div 6} = \frac{3}{4}$ Divide both numerator and denominator by the GCF.

The football team's success rate in simplest form is $\frac{3}{4}$ or 3 out of 4.

Your Turn

o. SCHOOL At Star Middle School, 60 out of 150 students are 8th graders. Write the number of 8th graders as a fraction in simplest form.

Guided Practice

Examples 1–7
(pages 178–180)

VOCABULARY

1. Write an example of a prime number and explain how you know it is a prime number. Repeat for a composite number.

2. What is the greatest common factor of two numbers?

3. How do you know when a fraction is in simplest form?

Examples 1–2
(page 178)

Identify each number as prime or composite.

4. 16   5. 19   6. 21   7. 27

Examples 3–4
(page 178)

List the factors of each number.

8. 2   9. 10   10. 16   11. 27

Example 5
(page 179)

Find the common factors of each set of numbers.

12. 6 and 12   13. 5 and 8   14. 12 and 14

Example 6
(page 179)

Find the greatest common factor (GCF) of each set of numbers.

15. 4 and 16   16. 7 and 10   17. 3 and 18

Example 7
(page 180)

Write each fraction in simplest form.

18. $\frac{6}{10}$ 19. $\frac{5}{8}$ 20. $\frac{12}{16}$ 21. $\frac{8}{24}$

Example 8
(page 180)

22. **SPORTS** Della shot 16 free throws in her last basketball game. She made 10 of them. What is her success rate as a fraction in simplest form?

23. **PACKAGING** A company finds that 3 of 18 boxes it ships, or $\frac{3}{18}$, arrive damaged. What is this fraction in simplest form?

24. **Talk Math** Have a partner say a number. Decide if it is prime or composite. Name another number that is in the same category. Then challenge your partner to do the same.

Skills, Concepts, and Problem Solving

HOMEWORK HELP	
For Exercises	**See Example(s)**
25–32	1–2
33–40	3–4
41–44	5
45–48	6
49–52	7
53–55	8

Identify each number as prime or composite.

25. 2   26. 10   27. 15   28. 19

29. 24   30. 25   31. 29   32. 33

List the factors of each number.

33. 8   34. 14   35. 18   36. 21

37. 23   38. 25   39. 30   40. 32

Find the common factors of each set of numbers.

41. 3 and 9   **42.** 12 and 16   **43.** 15 and 20   **44.** 18 and 33

Find the greatest common factor (GCF) of the numbers.

45. 6 and 12   **46.** 8 and 15   **47.** 15 and 30   **48.** 16 and 24

Write each fraction in simplest form.

49. $\frac{6}{15}$ **50.** $\frac{12}{18}$ **51.** $\frac{12}{36}$ **52.** $\frac{8}{32}$

53. POPULATION In one city, there are 4 middle schools and 18 elementary schools or $\frac{4}{18}$. What is this fraction in simplest form?

54. FLAGS Zack is making an American flag. The white cloth is 9 inches wide. The red cloth is 12 inches wide. The stripes must all be the same width. If he does not want to waste cloth, what is the widest the stripes can be?

55. ART Tara has 18 red and 12 blue beads with which she wants to make bracelets. She wants to use all of her beads. She decides to put an equal number of beads and only one color on each bracelet. What is the greatest number of beads she can use on each bracelet?

During the 2005 school year, there were over 6 million students studying in the 9,372 California public schools.

56. *Writing in Math*   Write three numbers that have a GCF of 4.

57. H.O.T. Problem Find the GCF of any two prime numbers. Demonstrate why this pattern is true.

STANDARDS PRACTICE

Choose the *best* answer.

58 The chart shows the results of a school district survey. What fraction, in simplest form, of the total number of students in the district return to the same school as the year before?

A $\frac{2}{3}$ **B** $\frac{3}{4}$ **C** $\frac{13}{15}$ **D** $\frac{24}{32}$

Status	Number
New to district	600
New school in district	200
Same school in district	2400
Total	**3200**

Spiral Review

Use a number line to determine which number is greater. (Lesson 3-2)

59. $-2\frac{2}{3}$ or $-3\frac{1}{8}$

60. $-1\frac{3}{4}$ or $-1\frac{1}{3}$

Solve each equation. (Lesson 2-7)

61. $2x - 5 = 3$

62. $\frac{n}{3} + 1 = 2$

63. FINANCE A stock changed $-\$8$ in the last 4 months. The price changed the same amount each month. How much did it change the first month? (Lesson 2-5)

Progress Check 2
(Lesson 3-3)

▷ Vocabulary and Concept Check

> common factor (p. 179)
> composite number (p. 177)
> factor (p. 177)
> greatest common factor (GCF) (p. 179)
> prime number (p. 177)
> simplest form (p. 179)

Choose the term that *best* completes each statement.

1. The _____?_____ of 4 and 8 is 4.

2. When you _____?_____ a number, you list all the numbers that can be multiplied together, two at a time, to get that number.

3. The fraction $\frac{1}{2}$ is $\frac{4}{8}$ written in _____?_____.

4. The integers 2, 3, and 5 are examples of _____?_____.

▷ Skills Check

Identify each number as prime or composite. (Lesson 3-3)

5. 7   6. 11   7. 9   8. 10

Find the greatest common factor (GCF) of the numbers. (Lesson 3-3)

9. 2 and 9   10. 2 and 12

11. 8 and 12   12. 6 and 9

Write each fraction in simplest form. (Lesson 3-3)

13. $\frac{5}{10}$ 14. $\frac{3}{9}$ 15. $\frac{9}{12}$ 16. $\frac{8}{20}$

▷ Problem-Solving Check

17. **LANDSCAPING** In Alma's garden, 9 of the 12 trees are oak trees. What fraction of the trees, in simplest form, are oaks? (Lesson 3-3)

18. **ART** Keira makes bracelets. She has 20 square beads and 15 round beads. She puts one kind of bead on each bracelet. She puts the same number of beads on each one. If she makes 7 bracelets, what is the greatest number of beads she can use on each bracelet? (Lesson 3-3)

19. **REFLECT** In your own words, explain how to write $\frac{8}{12}$ in simplest form. (Lesson 3-3)

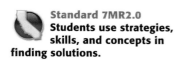
Vocabulary

draw a diagram (p. 184)

Standard 7MR2.0
Students use strategies,
skills, and concepts in
finding solutions.

Standard 7MR2.5 Use a variety
of methods, such as words,
numbers, symbols, charts, graphs,
tables, **diagrams,** and models, **to**
explain mathematical reasoning.

Standard 6NS1.2 Interpret and
use ratios in different contexts
to show the relative size of two
quantities using appropriate
notations.

Use the Strategy

Seth is playing in a school chess tournament. The tournament starts with 8 players. The winner of each match goes to the next round. What fraction of the players (in simplest form) will still be in the tournament in the third round?

CHESS At the start of a chess game, there are 32 pieces on the board. Each player has $\frac{1}{2}$ the pieces, or 16.

Understand **What does Seth know?**

- The number of chess players in the tournament.
- Only winners advance.

Plan **What does Seth want to know?**

Seth wants to compare the number of players in the third round to the number that start the tournament. He wants to write this comparison as a fraction in simplest form. Seth can **draw a diagram.**

Solve **Seth creates a diagram for 8 players.**

Round 1	**Round 2**	**Round 3**	**Round 4**
Player A			
Player B	Winner		
Player C		Winner	
Player D	Winner		
Player E			Champion
Player F	Winner		
Player G		Winner	
Player H	Winner		

So, $\frac{2}{8}$ of the players will be left. The fraction $\frac{2}{8}$ simplifies to $\frac{1}{4}$.

Check **Solve another way.**

Each round, $\frac{1}{2}$ of the players are eliminated. In Round 1, there are 8 players. So in Round 2, there are 4 players. In Round 3, there are 2 players.

$$2 \text{ of } 8 \text{ is } \frac{2}{8}, \text{ or } \frac{1}{4}.$$

The answer is correct.

Draw a diagram as you solve the problem.

1. **GAMES** How many total games will be played in a tournament of 16 players if players are eliminated after losing one game?

   Understand ▶ **What do you know?**

   Plan ▶ **What are you trying to find out? How can you figure it out?**

   Solve ▶ **Draw a diagram to solve the problem.**

   Check ▶ **How can you be sure that your answer is correct?**

2. *Writing in Math* Gabriel is organizing a chess tournament for 64 players. Players will be eliminated after losing one game. Each player plays only one game per day. Gabriel wants to know how many days the tournament will last. Why is drawing a diagram a good strategy for solving this problem?

Problem-Solving Practice

Solve using the *draw a diagram* strategy.

3. **PHYSICS** A ball is dropped from 32 meters above the ground. Each time it hits the ground, it bounces up $\frac{1}{2}$ as high as it fell. What fraction (in simplest form) of the original height is the last bounce up when it hits the ground the third time?

4. **ENTERTAINMENT** Abe, Curtis, and Chet stand in line to buy movie tickets. In how many different ways can they stand in line?

5. **GENEALOGY** Shawna made a list of her parents, grandparents, great grandparents, and great-great grandparents. If none of these people are stepparents, how many people are listed? What fraction of the total, in simplest form, are great grandparents?

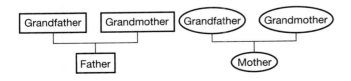

Solve using any strategy.

6. **SCHOOL** Jaime scored 89 on his math test. Questions are worth 5 points or 2 points. There is no partial credit. Suppose Jaime answered 37 questions correctly. How many 5-point questions are correct?

7. **GEOMETRY** The numbers shown here are called rectangular numbers. How many dots make up the eighth rectangular number?

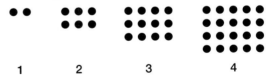

| 1 | 2 | 3 | 4 |

8. **NATURE** A snail at the bottom of a 10-foot well climbs up 3 feet each day, but slips back 2 feet at night. How many days will it take the snail to reach the top of the well and escape?

9. **BAKING** A cake that is 9 inches by 9 inches will serve 9 people. How many cakes that measure 12 inches by 12 inches will serve 48 people? (*Hint*: the answer is not 4.)

Multiplying Fractions

Vocabulary

product (p. 186)

 Standard 7NS1.2 Add, subtract, **multiply**, and divide **rational numbers** (integers, fractions, and terminating decimals) and take positive rational numbers to whole-number powers.

Standard 6NS2.1 Solve problems involving addition, subtraction, **multiplication, and division of positive fractions and explain why a particular operation was used for a given situation.**

Standard 6NS2.2 Explain the meaning of multiplication, and division of positive fractions and perform the calculations.

 The What: I will multiply fractions and raise fractions to positive powers.

The Why: You multiply fractions when you work with measurements, such as when cooking or designing interior spaces.

About $\frac{1}{3}$ of Earth's land can be used for farming. About $\frac{2}{5}$ of this farmland is used to grow grain crops. What part of Earth's land is used to grow grain?

Solve this problem by multiplying $\frac{1}{3}$ and $\frac{2}{5}$.

RULES FOR MULTIPLYING FRACTIONS

Words	To multiply fractions, multiply the numerators. Then multiply the denominators.
Numbers	$\frac{1}{3} \cdot \frac{2}{5} = \frac{1 \cdot 2}{3 \cdot 5} = \frac{2}{15}$
Pictures	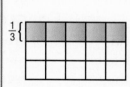
Algebra	For fractions $\frac{a}{b}$ and $\frac{c}{d}$ ($b \neq 0$, $d \neq 0$), $\frac{a}{b} \cdot \frac{c}{d} = \frac{ab}{cd}$.

Since the **product** of the fractions $\frac{1}{3}$ and $\frac{2}{5}$ is $\frac{2}{15}$, about $\frac{2}{15}$ of Earth's land is used to grow grain.

$$\frac{2}{5} \text{ of } \frac{1}{3} \text{ is } \frac{2}{15}.$$

 **Talk Math**

WORK WITH A PARTNER You and your partner each name a proper fraction. Then, work together to find the product of the two fractions.

Study Tip

Teaching another person is one of the best ways to learn the skill yourself. Explain the rule for multiplying fractions to a classmate.

EXAMPLES Multiply Fractions

Find each product.

1 $\dfrac{3}{4} \cdot \dfrac{3}{8}$

$$\dfrac{3}{4} \cdot \dfrac{3}{8} = \dfrac{3 \cdot 3}{4 \cdot 8} = \dfrac{9}{32}$$ Multiply numerators.
Multiply denominators.

2 $\dfrac{1}{5} \cdot \dfrac{3}{4} \cdot \dfrac{1}{2}$

$$\dfrac{1}{5} \cdot \dfrac{3}{4} \cdot \dfrac{1}{2} = \dfrac{1 \cdot 3 \cdot 1}{5 \cdot 4 \cdot 2} = \dfrac{3}{40}$$ Multiply numerators.
Multiply denominators.

Your Turn

Find each product.

a. $\dfrac{1}{2} \cdot \dfrac{1}{6}$

b. $\dfrac{3}{5} \cdot \dfrac{1}{2} \cdot \dfrac{3}{4}$

In Examples 1 and 2, both products were in simplest form. If the products are not in simplest form, you have two options.

Option 1: Multiply, then simplify.
Option 2: Simplify, then multiply.

EXAMPLE Write Products in Simplest Form

3 Multiply $\dfrac{5}{9} \cdot \dfrac{3}{10}$. Write the product in simplest form.

Option 1: Multiply first.

$$\dfrac{5}{9} \cdot \dfrac{3}{10} = \dfrac{5 \cdot 3}{9 \cdot 10}$$

$$= \dfrac{15}{90}$$ Multiply numerators.
Multiply denominators.

$$= \dfrac{1}{6}$$ Simplify.

Option 2: Simplify first.

$$\dfrac{5}{9} \cdot \dfrac{3}{10} = \dfrac{{}^{1}5}{9} \cdot \dfrac{3}{10_{\,2}}$$ The GCF of 5 and 10 is 5.
Divide by 5.

$$= \dfrac{1}{{}_{3}9} \cdot \dfrac{3^{\,1}}{2}$$ The GCF of 3 and 9 is 3.
Divide by 3.

$$= \dfrac{1}{3} \cdot \dfrac{1}{2}$$ Multiply numerators.
Multiply denominators.

$$= \dfrac{1 \cdot 1}{3 \cdot 2} \text{ or } \dfrac{1}{6}$$ Simplify.

Your Turn

Multiply. Write each product in simplest form.

c. $\dfrac{2}{3} \cdot \dfrac{5}{6}$

d. $\dfrac{3}{4} \cdot \dfrac{2}{5}$

e. $\dfrac{2}{9} \cdot \dfrac{3}{4}$

The rule you use for multiplying positive fractions also applies to negative fractions.

Remember!

Use the rules for multiplying integers.

positive number (+)
·
positive number (+)
=
positive number (+)

negative number (−)
·
negative number (−)
=
positive number (+)

positive number (+)
·
negative number (−)
=
negative number (−)

negative number (−)
·
positive number (+)
=
negative number (−)

EXAMPLES Multiply Negative Fractions

Multiply. Write each product in simplest from.

④ $-\frac{4}{5} \cdot \frac{2}{3}$

$-\frac{4}{5} \cdot \frac{2}{3} = -\frac{4 \cdot 2}{5 \cdot 3} = -\frac{8}{15}$

Negative · Positive = Negative
Multiply numerators.
Multiply denominators.

⑤ $-\frac{1}{2} \cdot -\frac{3}{4}$

$-\frac{1}{2} \cdot -\frac{3}{4} = \frac{1 \cdot 3}{2 \cdot 4} = \frac{3}{8}$

Negative · Negative = Positive
Multiply numerators.
Multiply denominators.

Your Turn

Multiply. Write each product in simplest form.

f. $\frac{4}{5} \cdot -\frac{1}{3}$

g. $-\frac{3}{4} \cdot -\frac{5}{8}$

EXAMPLES Multiply Improper Fractions

Multiply. Write each product in simplest form.

⑥ $\frac{1}{2} \cdot \frac{5}{3}$

$\frac{1}{2} \cdot \frac{5}{3} = \frac{1 \cdot 5}{2 \cdot 3} = \frac{5}{6}$

Multiply numerators.
Multiply denominators.

⑦ $\frac{9}{4} \cdot \frac{5}{3}$

$= \frac{9 \cdot 5}{4 \cdot 3}$

Multiply numerators.
Multiply denominators.

$= \frac{45}{12}$

$= \frac{45 \div 3}{12 \div 3}$

Simplify.

$= \frac{15}{4}$ or $3\frac{3}{4}$

Write the answer as a mixed number.

Your Turn

Multiply. Write each product in simplest form.

h. $\frac{3}{4} \cdot \frac{5}{4}$

i. $\frac{3}{2} \cdot -\frac{7}{4}$

j. $-\frac{7}{3} \cdot -\frac{9}{4}$

To multiply positive or negative mixed numbers, change them to improper fractions first. Then multiply.

8 NUTRITION Berta eats $2\frac{1}{4}$ energy bars. Each bar has $\frac{1}{2}$ of its Calories from fat. How many bars represent the Calories that come from fat? Draw a diagram to solve the problem.

1 bar   1 bar $\frac{1}{4}$ bar

The diagram shows $2\frac{1}{4}$ bars. The green and purple sections each show $\frac{1}{2}$ of the total bars. The mixed number $1\frac{1}{8}$ represents half of the bars.

$1\frac{1}{8}$ bars represent Calories that come from fat.

9 FITNESS Leon runs $\frac{1}{2}$ lap around the track on Saturday. How many laps does he run on Sunday if he runs $\frac{1}{4}$ the distance? Explain how you solved the problem.

$\frac{1}{2} \cdot \frac{1}{4} = \frac{1 \cdot 1}{2 \cdot 4}$ Multiply numerators. Multiply denominators.

$= \frac{1}{8}$ Simplify.

Leon runs $\frac{1}{8}$ lap on Sunday.

Your Turn

k. NUTRITION Cala eats 3 energy bars. Each bar has $\frac{1}{4}$ of its Calories from protein. How many bars represent the Calories that come from protein? Draw a diagram to solve the problem.

Remember!

Mixed numbers can be written as improper fractions.

$2\frac{1}{4}$ is nine fourths, or $\frac{9}{4}$.

You have learned $x^3 = x \cdot x \cdot x$. Likewise, the expression $\left(\frac{3}{4}\right)^3$ equals $\frac{3}{4} \cdot \frac{3}{4} \cdot \frac{3}{4}$.

$$\left(\frac{3}{4}\right)^3 = \frac{3}{4} \cdot \frac{3}{4} \cdot \frac{3}{4}$$

EXAMPLE   **Evaluate Fractions Raised to Powers**

10 Evaluate $\left(\frac{2}{3}\right)^3$.

$\left(\frac{2}{3}\right)^3 = \frac{2}{3} \cdot \frac{2}{3} \cdot \frac{2}{3}$ Use the base as a factor 3 times.

$= \frac{2 \cdot 2 \cdot 2}{3 \cdot 3 \cdot 3}$ Multiply numerators. Multiply denominators.

$= \frac{8}{27}$ Simplify.

Your Turn
Evaluate.

l. $\left(\frac{1}{2}\right)^4$ m. $\left(\frac{3}{4}\right)^3$

Examples 1–10
(pages 187–189)

VOCABULARY

1. Explain how to find each product of two proper fractions.

Multiply. Write each product in simplest form.

Examples 1–3
(page 187)

2. $\frac{7}{10} \cdot \frac{3}{4}$

3. $\frac{2}{5} \cdot \frac{1}{3} \cdot \frac{2}{3}$

4. $\frac{2}{3} \cdot \frac{5}{8}$

Examples 3–5
(pages 187–188)

5. $\frac{3}{4} \cdot \frac{4}{9}$

6. $\frac{7}{8} \cdot -\frac{3}{4}$

7. $-\frac{2}{3} \cdot -\frac{1}{5}$

Examples 6–7
(page 188)

8. $\frac{7}{4} \cdot -\frac{2}{3}$

9. $\frac{3}{2} \cdot \frac{5}{3}$

10. $\frac{3}{2} \cdot \frac{3}{2}$

Examples 8–9
(page 189)

11. **INTERIOR DESIGN** The basement of a house is going to be $\frac{3}{4}$ finished as a recreation room. Carpeting will cover $\frac{1}{2}$ of that space. What fraction of the whole basement will be carpeted? Draw a diagram to solve the problem.

12. **COOKING** A banana bread recipe uses $\frac{3}{4}$ cup of oil. How many cups of oil are needed to make half the recipe? Explain how you solved this problem.

Evaluate.

Example 10
(page 189)

13. $\left(\frac{2}{3}\right)^4$

14. $\left(\frac{1}{2}\right)^7$

15. *Talk Math* Discuss with a classmate how you would find the product of the fractions $\frac{a}{b}$ and $\frac{c}{d}$.

Skills, Concepts, and Problem Solving

Multiply. Write each product in simplest form.

HOMEWORK HELP

For Exercises	See Example(s)
16–19	1–2
20–23	3
24–27	4–5
28–30	6–7
31–33	8–9
34–37	10

16. $\frac{3}{4} \cdot \frac{1}{8}$

17. $\frac{2}{3} \cdot \frac{4}{5}$

18. $\frac{4}{5} \cdot \frac{1}{3} \cdot \frac{2}{3}$

19. $\frac{3}{4} \cdot \frac{1}{2} \cdot \frac{3}{4}$

20. $\frac{2}{3} \cdot \frac{9}{10}$

21. $\frac{3}{5} \cdot \frac{5}{9}$

22. $\frac{9}{10} \cdot \frac{2}{3}$

23. $\frac{3}{4} \cdot \frac{11}{12}$

24. $\frac{1}{2} \cdot -\frac{3}{4}$

25. $-\frac{3}{4} \cdot \frac{5}{8}$

26. $-\frac{4}{5} \cdot -\frac{2}{3}$

27. $-\frac{7}{8} \cdot -\frac{3}{5}$

28. $\frac{7}{4} \cdot \frac{1}{2}$

29. $\frac{1}{2} \cdot \frac{5}{3}$

30. $\frac{3}{2} \cdot \frac{22}{6}$

31. **NUTRITION** A new energy bar has $\frac{1}{4}$ of its Calories from fat. $\frac{2}{3}$ of those Calories are unsaturated fat. What fraction of the total number of Calories comes from unsaturated fat? Draw a diagram to solve the problem.

32. CAREER CONNECTION Crop farmers grow, store, package, and market crops. A farmer plants $\frac{3}{4}$ of his farm with apple trees. Granny Smith apple trees make up $\frac{1}{3}$ of the apple trees. What fraction of the entire farm is planted with Granny Smith apple trees? Explain how you solved this problem.

33. BAKING A farmer uses $\frac{2}{3}$ cup of her apples to bake an apple cake. How many cups of apples are needed to make half the recipe? Explain how you solved this problem.

AGRICULTURE
California farmers and ranchers produce $73 million in food, fiber, and flowers every day.

Evaluate.

34. $\left(\frac{1}{2}\right)^5$ **35.** $\left(\frac{2}{3}\right)^4$ **36.** $\left(\frac{4}{5}\right)^4$ **37.** $\left(\frac{3}{10}\right)^3$

38. *Writing in Math* Write a multiplication sentence with two fractions that have a product of $\frac{1}{2}$.

39. H.O.T. Problem One factor is $\frac{2}{3}$. The product is $\frac{1}{3}$. Explain how to find the other factor.

STANDARDS PRACTICE

Choose the *best* answer.

40 Half of a garden will be planted with flowers. Roses will make up $\frac{2}{3}$ of the flowers, and $\frac{1}{2}$ of the roses will be yellow roses. What fraction of the entire garden will be yellow roses?

A $\frac{1}{3}$ **C** $\frac{1}{8}$

B $\frac{1}{6}$ **D** $\frac{1}{12}$

41 How many cups of blueberries are needed to make 18 muffins?

Blueberry Muffins   Yield 24 muffins

Ingredients:

$1\frac{3}{4}$ cups flour	1 cup sugar
$2\frac{1}{2}$ tsp baking powder	$\frac{3}{4}$ cup milk
$\frac{1}{2}$ tsp salt	1 egg
$\frac{3}{4}$ cup butter	$1\frac{2}{3}$ cups blueberries

F $\frac{3}{4}$ **H** $1\frac{1}{2}$

G $1\frac{1}{4}$ **J** 2

Spiral Review

Find the greatest common factor of the numbers. (Lesson 3-3)

42. 4 and 10   **43.** 8 and 12   **44.** 2 and 5   **45.** 6 and 18

46. SPORTS This season, Jack scored 15 of 20 penalty shots in soccer. Write his success rate as a fraction in simplest form. (Lesson 3-3)

3-6 # Dividing Fractions

Vocabulary

reciprocal (p. 192)

multiplicative inverse (p. 192)

Inverse Property of Multiplication (p. 192)

 Standard 7NS1.2 Add, subtract, multiply, and **divide rational numbers** (integers, fractions, and terminating decimals) and take positive rational numbers to whole-number powers.

Standard 6NS2.1 Solve problems involving addition, subtraction, multiplication, and **division of positive fractions and explain why a particular operation was used for a given situation.**

Standard 6NS2.2 Explain the meaning of multiplication and division of positive fractions and perform the calculations.

 The What: I will divide fractions.

The Why: Division of fractions is used for tasks that involve measurements, like building bookcases.

Selena has $\frac{3}{4}$ yard of wood to make a bookcase. Suppose she divides the wood into $\frac{1}{8}$-yard pieces. How many $\frac{1}{8}$-yard pieces are there in $\frac{3}{4}$-yard of wood?

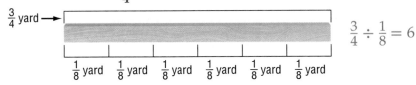

$$\frac{3}{4} \div \frac{1}{8} = 6$$

In this lesson you will use multiplication to divide fractions.

You will use the **reciprocal,** or **multiplicative inverse.**

RECIPROCALS

fraction	reciprocal	Definition
$\frac{1}{2}$	$\frac{2}{1}$	For every number $\frac{a}{b}$ ($a, b \neq 0$),
$-\frac{3}{4}$	$-\frac{4}{3}$	the reciprocal is $\frac{b}{a}$.

The **Inverse Property of Multiplication** states that the product of a number and its multiplicative inverse is 1.

$$\underbrace{\frac{5}{7}}_{\text{fraction}} \cdot \underbrace{\frac{7}{5}}_{\text{reciprocal}} = \frac{35}{35} = \underbrace{1}_{\text{product}}$$

INVERSE PROPERTY OF MULTIPLICATION

For every number $\frac{a}{b}$ ($a, b \neq 0$), there is exactly one number $\frac{b}{a}$ such that $\frac{a}{b} \cdot \frac{b}{a} = 1$.

EXAMPLES Find Reciprocals

Write the reciprocal of each fraction.

① $-\frac{5}{6}$

Find the number that gives a product of 1.

$$-\frac{5}{6} \cdot -\frac{6}{5} = 1$$

The reciprocal of $-\frac{5}{6}$ is $-\frac{6}{5}$.

② 4

Write the integer as a fraction. $4 = \frac{4}{1}$

$$\frac{4}{1} \cdot \frac{1}{4} = 1$$

The reciprocal of 4 is $\frac{1}{4}$.

Your Turn

Write the reciprocal of each fraction.

a. $\frac{7}{8}$

b. $-\frac{1}{6}$

Talk Math

WORK WITH A PARTNER Name a fraction. Have your partner name its reciprocal. Repeat four more times.

Reciprocals and the Inverse Property of Multiplication are used to divide fractions. To solve the example at the beginning of the lesson, multiply $\frac{3}{4}$ by the reciprocal of $\frac{1}{8}$.

	RULES FOR DIVIDING FRACTIONS
Words	To divide fractions, multiply the dividend by the reciprocal of the divisor. In other words, multiply by the reciprocal of the second fraction.
Numbers	$\frac{3}{4} \div \frac{1}{8} = \frac{3}{4} \cdot \frac{8}{1} = \frac{24}{4} = 6$
Algebra	For fractions $\frac{a}{b}$ and $\frac{c}{d}$ ($b, c, d \neq 0$), $\frac{a}{b} \div \frac{c}{d} = \frac{a}{b} \cdot \frac{d}{c} = \frac{a \cdot d}{b \cdot c}$

So, Selena will have six $\frac{1}{8}$-yard pieces of wood for her bookcase.

Divide Fractions

❸ Find $\frac{7}{8} \div \frac{5}{12}$. Write the quotient in simplest form.

$$\frac{7}{8} \div \frac{5}{12} = \frac{7}{8} \cdot \frac{12}{5}$$ Multiply by the reciprocal of the second fraction.

$$= \frac{7}{\overset{}{\underset{2}{8}}} \cdot \frac{\overset{3}{12}}{5}$$ Divide 8 and 12 by the GCF 4.

$$= \frac{21}{10} \text{ or } 2\frac{1}{10}$$ Multiply numerators and denominators.

Your Turn

Divide. Write each quotient in simplest form.

c. $\frac{1}{3} \div \frac{1}{2}$ d. $\frac{2}{3} \div \frac{5}{6}$

EXAMPLE   **Divide Negative Fractions**

❹ Find $\frac{2}{5} \div -\frac{3}{4}$. Write the quotient in simplest form.

$$\frac{2}{5} \div -\frac{3}{4} = \frac{2}{5} \cdot -\frac{4}{3}$$ Multiply by the reciprocal of the second fraction.

$$= -\frac{8}{15}$$ Multiply numerators. Multiply denominators.

Your Turn

Divide. Write each quotient in simplest form.

e. $\frac{2}{7} \div -\frac{1}{3}$ f. $-\frac{4}{5} \div -\frac{7}{8}$

EXAMPLE   **Divide Improper Fractions**

❺ Find $\frac{5}{2} \div \frac{3}{4}$. Write the quotient in simplest form.

$$\frac{5}{2} \div \frac{3}{4} = \frac{5}{2} \cdot \frac{4}{3}$$ Multiply by the reciprocal of the second fraction.

$$= \frac{20}{6}$$ Multiply numerators. Multiply denominators.

$$= \frac{10}{3} \text{ or } 3\frac{1}{3}$$ Simplify. Write the answer as a mixed number.

Your Turn

Divide. Write each quotient in simplest form.

g. $\frac{4}{5} \div \frac{9}{4}$ h. $\frac{11}{4} \div \frac{3}{4}$

Talk Math

WORK WITH A PARTNER In your own words, explain the process for dividing fractions. Listen as your partner explains the process in his/her own words.

6 SEWING Clara has $\frac{5}{4}$ yard of fabric to make teddy bears. Each bear requires $\frac{1}{8}$ yard of fabric. How many bears can Clara make? Draw a diagram that illustrates the problem. Then, explain how your diagram can be used to solve the problem.

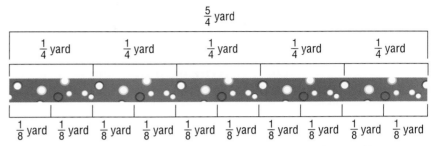

$\frac{5}{4}$ yard

| $\frac{1}{4}$ yard | $\frac{1}{4}$ yard | $\frac{1}{4}$ yard | $\frac{1}{4}$ yard | $\frac{1}{4}$ yard |

$\frac{1}{8}$ yard $\frac{1}{8}$ yard $\frac{1}{8}$ yard $\frac{1}{8}$ yard $\frac{1}{8}$ yard $\frac{1}{8}$ yard $\frac{1}{8}$ yard $\frac{1}{8}$ yard $\frac{1}{8}$ yard $\frac{1}{8}$ yard

The diagram shows the total length of fabric is $\frac{5}{4}$ yard. It shows the $\frac{1}{8}$-yard pieces needed for each bear. Count the number of $\frac{1}{8}$-yard pieces to solve the problem. Clara can make 10 bears.

7 MASONRY A mason installs a 122-inch-long brick curb. Each brick is $7\frac{5}{8}$ or $\frac{61}{8}$ inches long. Suppose the bricks are laid end-to-end. How many bricks will the mason use? Explain how you solved the problem.

$122 \div \frac{61}{8} = \frac{122}{1} \div \frac{61}{8}$ — Divide the curb length by the length of one brick. Write the integer as a fraction. $122 = \frac{122}{1}$

$= \frac{122}{1} \cdot \frac{8}{61}$ — Multiply by the reciprocal of the second fraction.

$= \frac{^2\cancel{122}}{1} \cdot \frac{8}{\cancel{61}_1}$ — The GCF of 122 and 61 is 61. Divide 122 and 61 by 61.

$= \frac{2 \cdot 8}{1 \cdot 1} = \frac{16}{1}$ — Multiply numerators. Multiply denominators.

$= 16$ — Write the answer as an integer.

Divide the total 122 inches of the curb by the $\frac{61}{8}$-inch brick.

To divide, multiply $\frac{122}{1}$ by the reciprocal $\frac{61}{8}$. $\frac{122}{1} \cdot \frac{8}{61} = 16$

The mason will use 16 bricks.

Your Turn

i. FITNESS A bike path is 2 kilometers long. There are distance markers every $\frac{1}{4}$ kilometer. How many distance markers are there? Draw a diagram that illustrates the problem. Then, explain how your diagram can be used to solve the problem.

RECREATION Over one-fourth of California's land area is set aside for state and national recreational areas.

Vocabulary Review
reciprocal
multiplicative inverse
Inverse Property of
Multiplication

Examples 1–7
(pages 193–195)

1. What is a multiplicative inverse? Explain using words.

2. Explain the Inverse Property of Multiplication in your own words.

Examples 1–2
(page 193)

Write the reciprocal of each fraction.

3. $\frac{7}{10}$

4. $-\frac{5}{4}$

Divide. Write each quotient in simplest form.

Example 3
(page 194)

5. $\frac{1}{2} \div \frac{7}{9}$

6. $\frac{3}{10} \div \frac{3}{5}$

Example 4
(page 194)

7. $-\frac{3}{7} \div -\frac{2}{5}$

8. $-\frac{5}{12} \div \frac{3}{4}$

Example 5
(page 194)

9. $\frac{11}{4} \div \frac{4}{3}$

10. $-\frac{5}{3} \div \frac{5}{6}$

Examples 6–7
(page 195)

11. **PLUMBING** A plumber has a $4\frac{1}{2}$-meter or $\frac{9}{2}$-meter pipe. A new sink requires a $\frac{1}{2}$-meter pipe. How many $\frac{1}{2}$-meter pipes can be cut from the $\frac{9}{2}$-meter pipe? Draw a diagram that illustrates the problem. Then, explain how your diagram can be used to solve the problem.

12. **INDUSTRY** A sheet of paper is $6\frac{3}{4}$ inches wide. The sheet is divided into 3 columns. How many inches wide is each column? Explain how you solved the problem.

13. **Talk Math** Write a division problem involving two fractions. Have your partner change your problem into a multiplication problem. Listen as your partner gives you a division problem to change.

Skills, Concepts, and Problem Solving

HOMEWORK HELP	
For Exercises	**See Example(s)**
14–17	1–2
18–21	3
22–25	4
26–29	5
30–33	6–7

Write the reciprocal of each fraction.

14. $\frac{5}{12}$

15. $-\frac{4}{2}$

16. $-\frac{7}{3}$

17. 19

Divide. Write each quotient in simplest form.

18. $\frac{2}{7} \div \frac{4}{7}$

19. $\frac{4}{9} \div \frac{2}{3}$

20. $\frac{3}{4} \div \frac{5}{12}$

21. $\frac{3}{5} \div \frac{9}{10}$

22. $-\frac{3}{4} \div -\frac{1}{2}$

23. $\frac{5}{6} \div -\frac{5}{8}$

24. $-\frac{7}{8} \div \frac{1}{3}$

25. $-\frac{9}{20} \div \frac{3}{4}$

26. $\frac{15}{8} \div \frac{5}{4}$

27. $\frac{3}{2} \div \frac{2}{3}$

28. $\frac{7}{3} \div \frac{7}{2}$

29. $\frac{29}{6} \div \frac{7}{3}$

30. **CAREER CONNECTION** Carpenters, plumbers, and masons make and install materials for new buildings.

A mason cuts cement slabs into fourths to make blocks. The original slab is 6 feet long. How long is each block? Draw a diagram that illustrates the problem. Then, explain how your diagram can be used to solve the problem.

31. **PLUMBING** Suppose a copper pipe is $2\frac{2}{5}$ or $\frac{12}{5}$ meters long. This pipe is cut to make 4 equal pieces. How long is each piece? Draw a diagram that illustrates the problem. Then, explain how your diagram can be used to solve the problem.

32. **COOKING** Mary needs $\frac{1}{2}$ cup of hot sauce for a stew. Suppose each bottle holds $\frac{1}{6}$ cup. How many bottles does she need? Explain how you solved the problem.

33. **HORTICULTURE** Kayla waters a tree with $\frac{25}{4}$ gallons of water once a week. Her watering can holds $\frac{5}{2}$ gallons of water. How many times does she have to fill the watering can in a week? Explain how you solved the problem.

34. *Writing in Math* Write an equation involving two fractions that have a quotient of $\frac{2}{3}$.

35. **H.O.T.** Problem Let n represent a rational number between 0 and 1. A positive number x is multiplied by n. The number x is then divided by n. Which is greater: the product or the quotient? Explain.

STANDARDS PRACTICE

Choose the *best* answer.

36 Felipe needs 12 cups of pretzels to make a snack mix. Each bag contains $2\frac{3}{4}$ cups of pretzels. How much will the pretzels cost?

$2.05

PRETZELS

$2\frac{3}{4}$ cups

A $6.15   C $10.25

B $8.20   D $12.30

37 A newspaper is $7\frac{1}{2}$ or $\frac{15}{2}$ inches wide. It is divided into two columns. The left column is divided in half again to list daily stock prices. How wide is that column?

F $\frac{15}{16}$ inch   H $\frac{15}{8}$ inches

G $\frac{15}{4}$ inches   J $\frac{15}{2}$ inches

Spiral Review

Multiply. Write each product in simplest form. (Lesson 3-5)

38. $\frac{1}{2} \cdot \frac{5}{6}$ 39. $\frac{3}{4} \cdot \frac{2}{9}$ 40. $-\frac{3}{8} \cdot \frac{4}{5}$ 41. $\frac{3}{2} \cdot \frac{7}{3}$

42. **TRANSPORTATION** Half of Greg's class is going on a field trip. The bus will hold $\frac{2}{3}$ of those students. What fraction of the entire class will ride the bus? (Lesson 3-5)

Progress Check 3

(Lessons 3-5 and 3-6)

▷ Vocabulary and Concept Check

> Inverse Property of
> Multiplication (p. 192)
> multiplicative inverse (p. 192)
>
> product (p. 186)
> reciprocal (p. 192)

Choose the term that *best* completes each statement.

1. Numbers that have a product of 1 are _____?_____
   or _____?_____.

2. The _____?_____ states that the product of a number and its
   multiplicative inverse is 1.

▷ Skills Check

Multiply. Write each product in simplest form. (Lesson 3-5)

3. $\frac{2}{3} \cdot \frac{3}{8}$

4. $-\frac{3}{4} \cdot -\frac{8}{9}$

5. $-\frac{1}{2} \cdot \frac{4}{5}$

6. $\frac{4}{3} \cdot \frac{5}{2}$

Evaluate. (Lesson 3-5)

7. $\left(\frac{1}{2}\right)^5$

8. $\left(\frac{3}{4}\right)^3$

Divide. Write each quotient in simplest form. (Lesson 3-6)

9. $\frac{5}{6} \div \frac{2}{3}$

10. $\frac{3}{10} \div \frac{3}{5}$

11. $\frac{7}{12} \div -\frac{3}{4}$

12. $\frac{7}{4} \div -\frac{3}{8}$

▷ Problem-Solving Check

13. **LANDSCAPING** A gardener is planting $\frac{2}{3}$ of a garden with daisies. Of the
    daisies, $\frac{1}{2}$ are yellow. What fraction of the whole garden is yellow daisies?
    Explain how you solved the problem. (Lesson 3-5)

14. **CONSTRUCTION** A carpenter is using $\frac{5}{8}$-inch-long pieces of plastic
    tubing. How many pieces of plastic tubing can be cut from a 5-inch-long
    tube of plastic? Draw a diagram that illustrates this situation. Then,
    explain how your diagram can be used to solve the problem. (Lesson 3-6)

15. **REFLECT** Explain how to divide a fraction by another fraction. (Lesson 3-6)

3-7

Adding and Subtracting Fractions with Like Denominators

Vocabulary

like denominators
(p. 199)

 Standard 7NS1.2 Add, subtract, multiply, and divide **rational numbers** (integers, fractions, and terminating decimals) and take positive rational numbers to whole-number powers.

Standard 6NS2.1 Solve problems involving addition, subtraction, multiplication, and division **of positive fractions and explain why a particular operation was used for a given situation.**

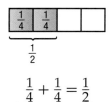

The What: I will add and subtract fractions with like denominators.

The Why: Musicians use fractions when reading music and playing instruments.

Ike plays two quarter notes on his trumpet. What is the total value of the notes?

You can solve this problem by adding $\frac{1}{4} + \frac{1}{4}$.

$$\frac{1}{4} + \frac{1}{4} = \frac{1}{2}$$

To add fractions easily, the denominators should be the same. In other words, the fractions should have **like denominators**.

$$\frac{3}{8}, \frac{5}{8} \qquad \frac{3}{4}, \frac{5}{8}$$

like denominators   unlike denominators

ADDING FRACTIONS WITH LIKE DENOMINATORS

Words	To add fractions with like denominators, add the numerators. Write the sum over the common denominator. Then simplify.
Numbers	$\frac{1}{4} + \frac{1}{4} = \frac{1+1}{4} = \frac{2}{4} = \frac{1}{2}$
Pictures	$\frac{1}{4} + \frac{1}{4}$
Algebra	For fractions $\frac{a}{c}$ and $\frac{b}{c}$ ($c \neq 0$), $\frac{a}{c} + \frac{b}{c} = \frac{a+b}{c}$.

WORK WITH A PARTNER Name two fractions with like denominators of 6 and add them. Listen as your partner names two fractions with like denominators of 7 and adds them.

EXAMPLES   Add Fractions

Add. Write each sum in simplest form.

1 $\frac{5}{8} + \frac{2}{8}$

$\frac{5}{8} + \frac{2}{8} = \frac{5+2}{8}$ The fractions have like denominators. Add the numerators.

$= \frac{7}{8}$ Write the sum over the denominator.

2 $\frac{7}{10} + \frac{9}{10}$

$\frac{7}{10} + \frac{9}{10} = \frac{7+9}{10}$ The fractions have like denominators. Add the numerators.

$= \frac{16}{10}$ Write the sum over the denominator.

$= \frac{8}{5}$ or $1\frac{3}{5}$ Simplify.

Your Turn

Add. Write each sum in simplest form.

a. $\frac{4}{9} + \frac{2}{9}$ b. $\frac{3}{4} + \frac{1}{4}$ c. $\frac{5}{6} + \frac{5}{6}$

SUBTRACTING FRACTIONS WITH LIKE DENOMINATORS

Words	To subtract fractions with like denominators, subtract the numerators. Write the difference over the common denominator. Then simplify.
Numbers	$\frac{3}{4} - \frac{1}{4} = \frac{3-1}{4} = \frac{2}{4} = \frac{1}{2}$
Pictures	$\boxed{\frac{1}{4}}\ \boxed{\frac{1}{4}}\ \boxed{\cancel{\frac{1}{4}}}\ \square$
Algebra	For fractions $\frac{a}{c}$ and $\frac{b}{c}$ ($c \neq 0$), $\frac{a}{c} - \frac{b}{c} = \frac{a-b}{c}$.

3 Find $\frac{7}{8} - \frac{3}{8}$. Write the difference in simplest form.

$$\frac{7}{8} - \frac{3}{8} = \frac{7-3}{8}$$ The fractions have like denominators. Subtract the numerators.

$$= \frac{4}{8}$$ Write the difference over the denominator.

$$= \frac{1}{2}$$ Simplify.

4 Find $\frac{13}{15} - \frac{8}{15}$. Write the difference in simplest form.

$$\frac{13}{15} - \frac{8}{15} = \frac{13-8}{15}$$ The fractions have like denominators. Subtract the numerators.

$$= \frac{5}{15}$$ Write the difference over the denominator.

$$= \frac{1}{3}$$ Simplify.

Your Turn

Subtract. Write each difference in simplest form.

d. $\frac{7}{9} - \frac{2}{9}$ **e.** $\frac{9}{10} - \frac{3}{10}$

You can also add and subtract improper fractions with like denominators. Write the answer as a mixed number.

EXAMPLE Add Improper Fractions

5 Find $\frac{19}{8} + \frac{15}{8}$. Write the sum in simplest form.

$$\frac{19}{8} + \frac{15}{8} = \frac{19 + 15}{8}$$ Add the numerators.

$$= \frac{34}{8}$$ Write the sum over the denominator.

$$= \frac{17}{4} \text{ or } 4\frac{1}{4}$$ Simplify. Write the answer as a mixed number.

Your Turn

Add. Write each sum in simplest form.

f. $\frac{7}{4} + \frac{11}{4}$ **g.** $\frac{13}{5} + \frac{24}{5}$

EXAMPLE   Subtract Improper Fractions

6 Find $\frac{31}{10} - \frac{17}{10}$. Write the difference in simplest form.

$$\frac{31}{10} - \frac{17}{10} = \frac{31 - 17}{10}$$ Subtract the numerators.

$$= \frac{14}{10}$$ Write the difference over the denominator.

$$= \frac{7}{5} \text{ or } 1\frac{2}{5}$$ Simplify. Write the answer as a mixed number.

Your Turn

Subtract. Write each difference in simplest form.

h. $\frac{8}{3} - \frac{4}{3}$ i. $\frac{29}{8} - \frac{15}{8}$

Remember!

Mixed numbers can be written as improper fractions.

$3\frac{3}{4} = 3 + \frac{3}{4}$

$\quad\quad = \frac{12}{4} + \frac{3}{4} = \frac{15}{4}$

$1\frac{3}{4} = 1 + \frac{3}{4}$

$\quad\quad = \frac{4}{4} + \frac{3}{4} = \frac{7}{4}$

Real-World EXAMPLES

MUSIC Dominic played a $3\frac{3}{4}$-minute song on the piano. Then, he played a $1\frac{3}{4}$-minute encore.

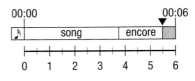

7 Which operation will you use to find how long he played altogether? Explain.

You can find how long he played by adding $3\frac{3}{4} + 1\frac{3}{4}$.

8 How long did Dominic play altogether? Write your answer in simplest form.

$$3\frac{3}{4} + 1\frac{3}{4} = \frac{15}{4} + \frac{7}{4}$$ Change the mixed numbers to improper fractions.

$$= \frac{15 + 7}{4}$$ Add the numerators.

$$= \frac{22}{4}$$ Write the sum over the like denominator.

$$= \frac{11}{2} \text{ or } 5\frac{1}{2}$$ Simplify. Write the answer as a mixed number.

Your Turn

FASHION Gloria cut $\frac{1}{4}$ foot of fabric from a $\frac{9}{4}$-foot roll.

j. Which operation will you use to find how many feet of fabric are left? Explain.

k. How many feet of fabric does Gloria have left? Write your answer in simplest form.

Examples 1–6
(pages 200–202)

VOCABULARY

1. Write two fractions with like denominators.

2. Use the fractions in Exercise 1 to explain how to add fractions with like denominators.

Examples 1–2
(page 200)

Add. Write each sum in simplest form.

3. $\frac{5}{12} + \frac{1}{12}$

4. $\frac{3}{8} + \frac{5}{8}$

Examples 3–4
(page 201)

Subtract. Write each difference in simplest form.

5. $\frac{8}{9} - \frac{7}{9}$

6. $\frac{9}{10} - \frac{1}{10}$

Example 5
(page 201)

Add. Write each sum in simplest form.

7. $\frac{7}{4} + \frac{9}{4}$

8. $\frac{15}{8} + \frac{21}{8}$

Example 6
(page 202)

Subtract. Write each difference in simplest form.

9. $\frac{15}{4} - \frac{5}{4}$

10. $\frac{13}{6} - \frac{11}{6}$

Examples 7–8
(page 202)

FASHION Katie cut $\frac{9}{4}$ feet of fabric from a $\frac{23}{4}$-foot roll.

11. Which operation will you use to find how many feet of fabric are left? Explain.

12. How many feet of fabric does Katie have left? Write your answer in simplest form.

13. **MUSIC** Jade played two $\frac{1}{2}$ notes on her bassoon. How long were the notes altogether?

14. **Talk Math** You and a classmate write a pair of fractions with like denominators. Without solving, discuss how you would find both the sum and difference of the fractions.

HOMEWORK HELP	
For Exercises	See Example(s)
15–20	1–2
21–26	3–4
27–30	5
31–34	6
35–40	7–8

Add. Write each sum in simplest form.

15. $\frac{7}{9} + \frac{1}{9}$

16. $\frac{5}{16} + \frac{7}{16}$

17. $\frac{5}{6} + \frac{1}{6}$

18. $\frac{7}{12} + \frac{1}{12}$

19. $\frac{3}{8} + \frac{3}{8}$

20. $\frac{7}{8} + \frac{5}{8}$

Subtract. Write each difference in simplest form.

21. $\frac{6}{7} - \frac{4}{7}$

22. $\frac{7}{8} - \frac{3}{8}$

23. $\frac{11}{12} - \frac{1}{12}$

24. $\frac{9}{16} - \frac{7}{16}$

25. $\frac{5}{6} - \frac{1}{6}$

26. $\frac{9}{10} - \frac{3}{10}$

Add. Write each sum in simplest form.

27. $\frac{7}{4} + \frac{11}{4}$

28. $\frac{10}{3} + \frac{5}{3}$

29. $\frac{12}{5} + \frac{8}{5}$

30. $\frac{23}{8} + \frac{27}{8}$

Subtract. Write each difference in simplest form.

31. $\frac{23}{6} - \frac{13}{6}$

32. $\frac{13}{4} - \frac{11}{4}$

33. $\frac{33}{8} - \frac{13}{8}$

34. $\frac{17}{8} - \frac{15}{8}$

FASHION Mia knits a scarf with $\frac{3}{4}$ inch of trim. She decides to add another $\frac{3}{4}$ inch of trim.

35. Which operation will you use to find how many inches of trim there are altogether? Explain.

36. How many inches of trim does Mia have altogether? Write your answer in simplest form.

37. **CONSTRUCTION** Vincent uses a $\frac{7}{8}$-inch wrench to tighten a bolt. How much wider is the opening of a $\frac{7}{8}$-inch wrench than a $\frac{3}{8}$-inch wrench?

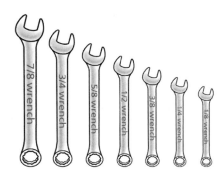

38. **NATURE** A local park is $4\frac{7}{8}$ acres. If $2\frac{1}{8}$ of the acres are closed to the public, how many acres are open to the public?

39. **MUSIC** The longest song on a new CD lasts $4\frac{7}{8}$ minutes. The shortest song lasts $3\frac{3}{8}$ minutes. How much longer is the longest song than the shortest one?

Musicians write, teach, and perform musical compositions. "I Love You, California" was written by F.B. Silverwood.

40. **MUSIC** A musician plays three eighth notes in a row. What fraction of a whole note do the notes make altogether?

41. *Writing in Math* Write an equation that involves addition of fractions and has a sum of $\frac{1}{2}$. Write an equation that involves subtraction of fractions and has a difference of $\frac{1}{2}$.

42. **H.O.T.** Problems Each of these fraction addition problems has a pattern.

$$\frac{1}{4} + \frac{1}{4} = \frac{1}{2}$$

$$\frac{3}{4} + \frac{3}{4} = \frac{3}{2}$$

$$\frac{3}{8} + \frac{3}{8} = \frac{3}{4}$$

$$\frac{5}{6} + \frac{5}{6} = \frac{5}{3}$$

The addends in each equation are the same. Notice that the denominator of the sum is one-half the denominator of one of the addends. Explain why this pattern is true.

STANDARDS PRACTICE

Choose the *best* answer.

43 To make a jacket, a designer uses $\frac{3}{8}$ bolt of red fabric, $\frac{1}{8}$ of blue fabric, and $\frac{1}{8}$ of black fabric. What fraction of a full bolt of fabric does she use?

A $\frac{5}{24}$ C $\frac{5}{8}$

B $\frac{1}{2}$ D $\frac{3}{4}$

44 This chart records the number of miles Otto runs each day. How many more miles did he run on Saturday and Sunday combined than on Friday?

Sun	Mon	Tues	Wed	Thur	Fri	Sat
$2\frac{3}{4}$	$3\frac{1}{4}$	3	$2\frac{3}{4}$	$2\frac{1}{2}$	$3\frac{1}{4}$	2

F $1\frac{1}{4}$ H $1\frac{3}{4}$

G $1\frac{1}{2}$ J 2

Spiral Review

Divide. Write each quotient in simplest form. (Lesson 3-6)

45. $\frac{5}{9} \div \frac{3}{4}$ 46. $\frac{2}{3} \div \frac{5}{6}$ 47. $\frac{3}{8} \div \frac{3}{4}$ 48. $\frac{3}{4} \div \frac{1}{2}$

49. **CONSTRUCTION** Suppose you have a piece of metal 6 inches long. How many $\frac{3}{4}$-inch bars can be cut from the piece of metal? (Lesson 3-6)

3-8 Adding Fractions with Unlike Denominators

Vocabulary

unlike denominators
(p. 206)

least common multiple (LCM) (p. 206)

least common denominator (LCD)
(p. 207)

Standard 7NS1.2 Add, subtract, multiply, and divide **rational numbers** (integers, fractions, and terminating decimals) and take positive rational numbers to whole-number powers.

Standard 6NS2.1 Solve problems involving addition, subtraction, multiplication, and division **of positive fractions and explain why a particular operation was used for a given situation.**

Standard 7MR2.1 Use estimation to verify the reasonableness of calculated resulted.

 The What: I will add fractions with unlike denominators.

 The Why: You add fractions with unlike denominators to find how much material is needed to build a wall.

Mateo and Tamika volunteer to build houses for a charitable group. They build walls with $\frac{3}{8}$ inch of drywall and $\frac{1}{2}$ inch of insulation. How thick are the walls?

You can solve this problem by adding $\frac{3}{8} + \frac{1}{2}$.

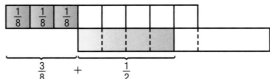

To add fractions with **unlike denominators,** replace each fraction with equivalent fractions that have like denominators. To do this, find a common multiple among the multiples of the denominators. The **least common multiple (LCM)** of two or more numbers is the smallest multiple common to the numbers.

EXAMPLES Find the Least Common Multiple

1 Find the LCM of 8 and 12.

8: 8, 16, 24   List multiples of each number until you
12: 12, 24   find a common multiple. Identify the LCM.
The LCM is 24.

2 Find the LCM of 4 and 10.

4: 4, 8, 12, 16, 20   List multiples of each number until you
10: 10, 20   find a common multiple. Identify the LCM.
The LCM is 20.

Your Turn
Find the LCM of each pair of numbers.

a. 9 and 12   b. 5 and 6

To add fractions with unlike denominators, rename one or both of the fractions so all fractions have the same denominator.

EXAMPLE Rename Fractions

❸ Rename $\frac{1}{2}$ with a denominator of 6.

$\frac{1}{2} = \frac{?}{6}$ The factor is 3 because $2 \times 3 = 6$.

$\frac{1}{2} = \frac{3}{6}$ Multiply the numerator by the same factor.

Your Turn
Rename each fraction with the given denominator.

c. $\frac{1}{2} = \frac{?}{8}$ d. $\frac{2}{3} = \frac{?}{12}$

When adding fractions, the LCM of the denominators can be used as the **least common denominator (LCD).**

EXAMPLES Find the Least Common Denominator

❹ Find the LCD of $\frac{1}{2}$ and $\frac{5}{9}$.

2: 2, 4, 6, 8, 10, 12, 14, 16, 18   List multiples of each
9: 9, 18   denominator until you find
                              a common multiple. Identify
                              the LCM.

The LCD of $\frac{1}{2}$ and $\frac{5}{9}$ is 18.

❺ Find the LCD of $\frac{5}{6}$ and $\frac{3}{8}$.

6: 6, 12, 18, 24   List multiples of each
8: 8, 16, 24   denominator until you find
                            a common multiple. Identify
                            the LCM.

The LCD of $\frac{5}{6}$ and $\frac{3}{8}$ is 24.

Your Turn
Find the LCD of each pair of fractions.

e. $\frac{2}{3}$ and $\frac{1}{6}$ f. $\frac{7}{8}$ and $\frac{3}{10}$

ADDING FRACTIONS WITH UNLIKE DENOMINATORS

To add fractions with unlike denominators, rename the fractions with a common denominator. Then add and simplify.

Refer to the beginning of the lesson. Use this process to find the thickness of Mateo and Tamika's walls.

$$\underbrace{\frac{3}{8}}_{\text{drywall}} + \underbrace{\frac{1}{2}}_{\text{insulation}} = \frac{3}{8} + \frac{4}{8} = \frac{3+4}{8} = \underbrace{\frac{7}{8}}_{\text{wall}}$$

The wall will be $\frac{7}{8}$ inch thick.

EXAMPLES   Add Fractions with Unlike Denominators

6 Find $\frac{1}{3} + \frac{3}{5}$. Write the sum in simplest form.

3: 3, 6, 9, 12, 15	List multiples of each
5: 5, 10, 15	denominator. The LCD is 15.

$\frac{1}{3} = \frac{?}{15} = \frac{5}{15} \quad \frac{3}{5} = \frac{?}{15} = \frac{9}{15}$ Rename the fractions using the LCD. Identify the factor multiplying the denominator. Multiply the numerator by the same factor.

$\frac{5}{15} + \frac{9}{15} = \frac{5 + 9}{15}$ Add the numerators.

$\qquad = \frac{14}{15}$ Write the sum over the denominator.

7 Find $\frac{3}{4} + \frac{5}{7}$. Write the sum in simplest form.

4: 4, 8, 12, 16, 20, 24, 28	List multiples of each
7: 7, 14, 21, 28	denominator. The LCD is 28.

$\frac{3}{4} = \frac{?}{28} = \frac{21}{28} \quad \frac{5}{7} = \frac{?}{28} = \frac{20}{28}$ Rename the fractions using the LCD. Identify the factor multiplying the denominator. Multiply the numerator by the same factor.

$\frac{21}{28} + \frac{20}{28} = \frac{21 + 20}{28}$ Add the numerators.

$\qquad = \frac{41}{28} \text{ or } 1\frac{13}{28}$ Simplify.

Your Turn
Add. Write each sum in simplest form.

g. $\frac{1}{8} + \frac{7}{12}$ h. $\frac{2}{5} + \frac{3}{10}$

Study Tip

Check your sums by estimating. Round fractions to 0, $\frac{1}{2}$, or 1. Then add to get an estimate.

$\frac{1}{3}$ rounds to 0.

$\frac{3}{5}$ is a little more than $\frac{1}{2}$, but not quite 1.

Compare this to your answer of $\frac{14}{15}$, which is almost 1.

WORK WITH A PARTNER Discuss with your partner how to estimate the sums of $\frac{1}{9} + \frac{7}{8}$ and $\frac{3}{4} + \frac{5}{6}$.

To add mixed numbers with unlike denominators, write the mixed numbers as improper fractions. Then rename the improper fractions with a common denominator.

Real-World EXAMPLES Add Improper Fractions

ADVERTISING Sara sold ads for her school's yearbook. She sold $3\frac{1}{4}$ or $\frac{13}{4}$ pages of ads to restaurants and $2\frac{1}{2}$ or $\frac{5}{2}$ pages to grocery stores.

8 **What operation will you use to find how much advertising she sold in all? Explain.**

You can find how much advertising she sold by adding $3\frac{1}{4}$ and $2\frac{1}{2}$. Add to find the total pages of ads.

9 **How much advertising did Sara sell in all? Write your answer as a mixed number in simplest form.**

$3\frac{1}{4} + 2\frac{1}{2} = \frac{13}{4} + \frac{5}{2}$ Write the mixed numbers as improper fractions.

4: 4
2: 2, 4 List multiples of each denominator. The LCD is 2.

$\frac{13}{4} = \frac{?}{4} = \frac{13}{4}$ $\frac{5}{2} = \frac{?}{4} = \frac{10}{4}$ Rename the fractions using the LCD.

$\frac{13}{4} + \frac{10}{4} = \frac{13 + 10}{4}$ Add the numerators.

$= \frac{23}{4}$ or $5\frac{3}{4}$ Write as a mixed number.

Your Turn

BAKING A baker makes a cake and puts icing on top. The cake is $2\frac{1}{4}$ or $\frac{9}{4}$ inches high. The icing is $\frac{1}{2}$-inch thick. How tall is the iced cake at its center?

i. What operation will you use to find the total height of the cake? Explain.

j. What is the total height of the cake? Write your answer as a mixed number in simplest form.

Over 50% of Californians work in service-related fields.

Examples 1–8
(pages 206–209)

VOCABULARY

1. Give an example of fractions with unlike denominators.

2. Explain how you use the LCM to find the LCD of two fractions.

Examples 1–2
(page 206)

Find the LCM of each pair of numbers.

3. 3 and 12

4. 4 and 10

Example 3
(page 207)

Rename the fraction with the given denominator.

5. $\frac{3}{4} = \frac{?}{8}$

6. $\frac{1}{2} = \frac{?}{10}$

Examples 4–5
(page 207)

Find the LCD of each pair of fractions.

7. $\frac{1}{2}$ and $\frac{5}{8}$

8. $\frac{3}{8}$ and $\frac{7}{12}$

Examples 6–7
(page 208)

Add. Write each sum in simplest form.

9. $\frac{5}{8} + \frac{1}{4}$

10. $\frac{1}{2} + \frac{1}{3}$

11. $\frac{3}{10} + \frac{2}{5}$

12. $\frac{1}{6} + \frac{3}{8}$

13. $\frac{1}{2} + \frac{7}{9}$

14. $\frac{11}{12} + \frac{5}{8}$

Examples 8–9
(page 209)

ENGINEERING An architect designs a floor that has a $1\frac{1}{4}$- or $\frac{5}{4}$-inch base and $\frac{7}{8}$-inch tile on top.

15. What operation will you use to find how thick the floor is altogether? Explain.

16. How thick is the floor altogether? Write your answer as a mixed number in simplest form.

California architectural landmarks include Louis Kahn's Salk Institute in La Jolla and Frank Lloyd Wright's Hollyhock House in Hollywood.

Examples 6–7
(page 208)

17. **ADVERTISING** Mr. Roth buys a $\frac{1}{4}$-page ad on the back cover of a magazine and a $\frac{1}{2}$-page ad in the middle. How much advertising does he buy in all?

18. **Talk Math** Say or write multiples of 6. Have your partner say or write multiples of 8. Work together to find the LCM of 6 and 8.

HOMEWORK HELP

For Exercises	See Example(s)
19–22	1–2
23–25	3
26–28	4–5
29–37	6–7
38–41	8–9

Find the LCM of each pair of numbers.

19. 2 and 6

20. 5 and 15

21. 4 and 5

22. 6 and 8

Rename the fraction with the given denominator.

23. $\frac{1}{2} = \frac{?}{6}$

24. $\frac{2}{3} = \frac{?}{6}$

25. $\frac{2}{3} = \frac{?}{24}$

Find the LCD of each pair of fractions.

26. $\frac{2}{3}$ and $\frac{5}{6}$

27. $\frac{3}{4}$ and $\frac{1}{3}$

28. $\frac{5}{6}$ and $\frac{5}{8}$

Add. Write each sum in simplest form.

29. $\frac{1}{2} + \frac{1}{8}$

30. $\frac{1}{4} + \frac{1}{3}$

31. $\frac{2}{3} + \frac{1}{6}$

32. $\frac{4}{5} + \frac{1}{10}$

33. $\frac{7}{10} + \frac{1}{4}$

34. $\frac{3}{8} + \frac{1}{12}$

35. $\frac{3}{4} + \frac{5}{6}$

36. $\frac{2}{3} + \frac{3}{8}$

37. $\frac{2}{5} + \frac{5}{7}$

FITNESS Leah jogs for $2\frac{3}{10}$ or $\frac{23}{10}$ miles and walks for another $1\frac{1}{2}$ or $\frac{3}{2}$ miles. How far does she jog and walk?

38. What operation will you use to find how far she exercises in all? Explain.

39. How far does Leah jog and walk? Write your answer as a mixed number in simplest form.

40. **HORTICULTURE** A gardener mixes $\frac{1}{3}$ cup of liquid fertilizer with $2\frac{1}{2}$ or $\frac{5}{2}$ cups of water. How much liquid is there in all?

41. **ROOFING** A roofer attaches roofing felt and shingles to plywood when laying down a new roof. How thick are the felt and shingles altogether?

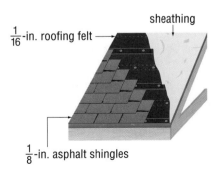

$\frac{1}{16}$-in. roofing felt

sheathing

$\frac{1}{8}$-in. asphalt shingles

Add. Write the sum in simplest form.

42. $\frac{11}{8} + \frac{3}{2}$

43. $\frac{4}{3} + \frac{17}{12}$

44. $\frac{7}{3} + \frac{7}{4}$

45. $\frac{11}{4} + \frac{17}{6}$

ENGINEERING An architect drew the following plan.

46. How thick is the sheathing and siding altogether?

47. How thick is the drywall and insulation altogether?

48. *Writing in Math* Write a fraction addition sentence with an approximate sum of 2. Use fractions with unlike denominators.

49. **H.O.T. Problem** Theo says the LCM of two numbers is always the product of the numbers. Do you agree? Explain your thinking.

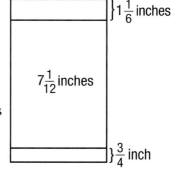

$\frac{5}{8}$-inch drywall

$3\frac{1}{2}$-inch insulation

$\frac{3}{4}$-inch wall sheathing

$\frac{7}{8}$-inch siding

STANDARDS PRACTICE

Choose the *best* answer.

50 Refer to the diagram in Exercise 47. How thick is the wall, including the drywall, insulation, sheathing, and siding?

A $4\frac{1}{8}$ inches

B $4\frac{1}{2}$ inches

C $5\frac{1}{8}$ inches

D $5\frac{3}{4}$ inches

51 A newspaper is printed with the margins and other dimensions shown here. What is the total height of the page?

F $8\frac{2}{3}$ inches

G $8\frac{3}{4}$ inches

H $8\frac{11}{12}$ inches

J 9 inches

$\}1\frac{1}{6}$ inches

$7\frac{1}{12}$ inches

$\}\frac{3}{4}$ inch

Spiral Review

Add. Write each sum in simplest form. (Lesson 3-7)

52. $\frac{5}{9} + \frac{2}{9}$

53. $\frac{1}{8} + \frac{3}{8}$

54. $\frac{5}{6} + \frac{1}{6}$

55. $\frac{7}{10} + \frac{9}{10}$

56. **FARMING** Pia plants $\frac{1}{3}$ of her farm with corn and $\frac{1}{3}$ with soybeans. What fraction of her farm is planted with corn or soybeans? (Lesson 3-7)

Subtracting Fractions with Unlike Denominators

Vocabulary

like denominators
(p. 213)

unlike denominators
(p. 213)

 Standard 7NS1.2 Add, **subtract**, multiply, and divide **rational numbers** (integers, fractions, and terminating decimals) and take positive rational numbers to whole-number powers.

Standard 6NS2.1 Solve problems involving addition, **subtraction**, multiplication, and division **of positive fractions and explain why a particular operation was used for a given situation.**

 The What: I will subtract fractions with unlike denominators.

 The Why: Cooking involves using measurements with unlike denominators.

A chef needs $\frac{5}{8}$ cup of tomatoes to make a pasta sauce. He has $\frac{1}{2}$ cup of tomatoes. How much more does the chef need?

You can solve this problem by subtracting $\frac{1}{2}$ from $\frac{5}{8}$.

$$\frac{5}{8} - \frac{1}{2}$$

$\frac{1}{8}$	$\frac{1}{8}$	$\frac{1}{8}$	$\frac{1}{8}$	$\frac{1}{8}$			
$\frac{1}{2}$							

To add fractions, the fractions should have **like denominators.** The same is true when you subtract fractions.

SUBTRACTING FRACTIONS WITH UNLIKE DENOMINATORS

To subtract fractions with **unlike denominators,** rename the fractions with a common denominator. Then subtract and simplify.

Use this process to find how many more tomatoes the chef needs.

chef's tomatoes

$$\underbrace{\frac{5}{8} - \frac{1}{2}}_{\text{total tomatoes needed}} = \frac{5}{8} - \frac{4}{8} = \underbrace{\frac{5-4}{8} = \frac{1}{8}}_{\text{amount still needed}}$$

The chef needs another $\frac{1}{8}$ cup of tomatoes.

Talk Math

WORK WITH A PARTNER You and a partner each name a fraction with a denominator 8 or less. Choose fractions with unlike denominators. Work together to subtract the fractions.

Subtract Fractions with Unlike Denominators

1 Find $\frac{5}{6} - \frac{1}{3}$. Write the difference in simplest form.

6: 6
3: 3, 6

List multiples of each denominator. The LCD is 6.

$$\overset{\times\,1}{\frac{5}{6} = \frac{?}{6} = \frac{5}{6}} \qquad \overset{\times\,2}{\frac{1}{3} = \frac{?}{6} = \frac{2}{6}}$$
$$\underset{\times\,1}{} \qquad \underset{\times\,2}{}$$

Rename the fractions using the LCD. Subtract the numerators.

$$\frac{5}{6} - \frac{2}{6} = \frac{5 - 2}{6}$$

Subtract the numerators.

$$= \frac{3}{6} \text{ or } \frac{1}{2}$$

Write the difference over the like denominator. Simplify.

Your Turn
Subtract. Write each difference in simplest form.

a. $\frac{5}{12} - \frac{1}{8}$ b. $\frac{9}{10} - \frac{2}{5}$ c. $\frac{5}{6} - \frac{3}{4}$

To subtract mixed numbers with unlike denominators, rename the fractions using a common denominator.

Subtract Improper Fractions

2 Find $\frac{17}{6} - \frac{4}{3}$. Write the difference in simplest form.

6: 6
3: 3, 6

List multiples of each denominator. The LCD is 6.

$$\overset{\times\,1}{\frac{17}{6} = \frac{?}{6} = \frac{17}{6}} \qquad \overset{\times\,2}{\frac{4}{3} = \frac{?}{6} = \frac{8}{6}}$$
$$\underset{\times\,1}{} \qquad \underset{\times\,2}{}$$

Rename the fractions using the LCD. Subtract the numerators.

$$\frac{17}{6} - \frac{8}{6} = \frac{17 - 8}{6}$$

Subtract the numerators.

$$= \frac{9}{6}$$

Write the difference over the like denominator.

$$= \frac{3}{2} \text{ or } 1\frac{1}{2}$$

Simplify. Write as a mixed number.

Your Turn
Subtract. Write each difference in simplest form.

d. $\frac{23}{6} - \frac{5}{4}$ e. $\frac{17}{4} - \frac{23}{8}$ f. $3 - \frac{5}{3}$

Remember!

To change an improper fraction to a mixed number:

Divide the numerator by the denominator.

Write the remainder over the denominator.

Use the whole number and fraction to make the mixed number.

ADVERTISING A television station has 6 hours of advertising each day. Mara has sold $3\frac{1}{4}$ or $\frac{13}{4}$ hours of the advertising time for next Wednesday.

```
1   2   3   4   5   6
        Hours
```

③ **Which operation will you use to find how many more hours she needs to sell? Explain.**

Subtract to find how much more time is needed.

④ **How many more hours of advertising does Mara need to sell? Write your answer in simplest form.**

6 hours $- 3\frac{1}{4}$ hours sold $= \underline{\quad?\quad}$ hours to sell

$6 = \frac{6}{1}$ Change the whole number to a fraction.

$3\frac{1}{4} = \frac{13}{4}$ Change the mixed number to an improper fraction.

$$\overset{\times 4}{\underset{\times 4}{\frac{6}{1} = \frac{?}{4} = \frac{24}{4}}}$$ Rename $\frac{6}{1}$ with the same denominator.

$\frac{24}{4} - \frac{13}{4} = \frac{24 - 13}{4}$ Subtract the numerators.

$= \frac{11}{4}$ or $2\frac{3}{4}$ Write the difference over the like denominator. Write as a mixed number.

Mara needs to sell $2\frac{3}{4}$ more hours of advertising.

Your Turn

BIOLOGY A biologist collects and studies salamanders. The average length of salamanders from the east coast is $2\frac{1}{4}$ or $\frac{9}{4}$ inches. The average length of salamanders from the west coast is $3\frac{1}{8}$ or $\frac{25}{8}$ inches.

g. Which operation will you use to find the difference in average length? Explain.

h. What is the difference in average length of the salamanders? Write the answer in simplest form.

BIOLOGY There are over forty species of salamanders occuring naturally in California.

Examples 1–4
(pages 214–215)

VOCABULARY

1. How do you recognize fractions with unlike denominators?

2. How do you rewrite fractions with unlike denominators so that they have like denominators?

Subtract. Write each difference in simplest form.

Example 1
(page 214)

3. $\frac{7}{8} - \frac{1}{4}$

4. $\frac{5}{6} - \frac{1}{2}$

5. $\frac{7}{12} - \frac{1}{4}$

6. $\frac{1}{6} - \frac{1}{10}$

Example 2
(page 214)

7. $\frac{13}{4} - \frac{13}{8}$

8. $\frac{23}{10} - \frac{3}{2}$

9. $3 - 1\frac{1}{2}$

10. $4 - \frac{29}{12}$

Examples 3–4
(page 215)

BUSINESS Rick needs to work $4\frac{1}{2}$ or $\frac{9}{2}$ hours on Saturday. He has worked $1\frac{3}{4}$ or $\frac{7}{4}$ hours so far.

11. Which operation will you use to find how many more hours he needs to work? Explain.

12. How many more hours does Rick need to work?

13. **BIOLOGY** A biologist collects snails from two different islands. The average length of the snails from one island is $\frac{3}{4}$ inch. The average length of the other snails is $\frac{2}{3}$ inch. What is the difference in average length?

14. **Talk Math** Talk about how to find the sum of $\frac{5}{8}$ and $\frac{1}{4}$ and the difference of $\frac{5}{8}$ and $\frac{1}{4}$. Discuss the similarities and differences of adding and subtracting fractions.

Skills, Concepts, and Problem Solving

Subtract. Write each difference in simplest form.

For Exercises	See Example(s)
15–24	1
25–32	2
33–38	3–4

HOMEWORK HELP

15. $\frac{7}{8} - \frac{3}{4}$

16. $\frac{2}{3} - \frac{1}{6}$

17. $\frac{5}{12} - \frac{1}{4}$

18. $\frac{3}{8} - \frac{1}{3}$

19. $\frac{11}{12} - \frac{7}{8}$

20. $\frac{5}{6} - \frac{5}{8}$

21. $\frac{5}{7} - \frac{1}{2}$

22. $\frac{8}{9} - \frac{5}{6}$

23. $\frac{7}{10} - \frac{2}{5}$

24. $\frac{3}{4} - \frac{2}{3}$

25. $\frac{11}{4} - \frac{3}{2}$

26. $\frac{8}{3} - \frac{13}{6}$

Subtract. Write each difference in simplest form.

27. $\dfrac{29}{6} - \dfrac{7}{3}$ **28.** $\dfrac{31}{8} - \dfrac{7}{4}$ **29.** $\dfrac{53}{12} - \dfrac{11}{4}$

30. $\dfrac{11}{6} - \dfrac{13}{8}$ **31.** $4 - \dfrac{11}{8}$ **32.** $3 - \dfrac{31}{12}$

HOBBIES Opal uses $\dfrac{3}{4}$-inch beads to make a necklace and $\dfrac{5}{8}$-inch beads for a bracelet.

33. Which operation will you use to find how much longer the necklace beads are than the bracelet beads? Explain.

34. How much longer are the necklace beads than the bracelet beads? Write your answer in simplest form.

35. **FITNESS** Omar plans to run once around the $2\dfrac{3}{4}$- or $\dfrac{11}{4}$-mile track. He has run $1\dfrac{1}{2}$ miles so far. How much farther does he have to go?

36. **FASHION** Esteban wants to use $\dfrac{5}{8}$-inch buttons. His buttonholes measure $\dfrac{11}{12}$-inch across. How much wider is the buttonhole than the button?

37. **NATURE** This map shows the area of different parts of the local nature preserve. How many more acres of flower gardens are there than woods?

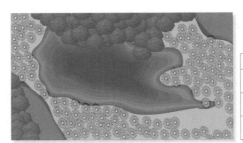

Area	Acres
Ponds and Lakes	$2\dfrac{1}{2}$
Flower Gardens	$2\dfrac{3}{8}$
Woods	$1\dfrac{3}{4}$

38. **ENGINEERING** A $6\dfrac{1}{2}$- or $\dfrac{13}{2}$-mile road is being repaved. So far $4\dfrac{7}{10}$ of the miles have been repaired. How many miles are left to be repaved?

39. *Writing in Math*   Write a fraction subtraction sentence with like denominators so that the difference is $\dfrac{1}{3}$. Then write a fraction subtraction sentence with unlike denominators so that the difference is also $\dfrac{1}{3}$.

40. **H.O.T. Problem** Seth says that the following equation is correct. Is he correct? Explain why or why not.

$$5\dfrac{1}{2} - 1\dfrac{3}{4} = 3\dfrac{3}{4}$$

CAREER CONNECTION A chef runs a kitchen in a restaurant, hotel, resort, ship, or other location. His or her responsibilities include planning the menu, cooking, purchasing ingredients, and managing the staff and budget.

CHEFS Chefs plan menus, buy ingredients, and create dishes. The restaurant industry is the largest private employer in California, with over 1.4 million employees.

41. **COOKING** A chef has chopped $\frac{3}{4}$ cup of vegetables for a recipe. He needs $2\frac{1}{2}$ cups in all. How many more cups does he need?

42. **COOKING** A fruit salad recipe includes $\frac{2}{3}$ cup of pineapple and $\frac{3}{4}$ cup of grapes. How many more cups of grapes are there than pineapple?

STANDARDS PRACTICE

Choose the *best* answer.

43 Solana plans to run twice around the $2\frac{1}{4}$- or $\frac{9}{4}$-mile track. She has run $1\frac{1}{10}$ or $\frac{11}{10}$ mile so far. How much farther does she have to run?

A $\frac{1}{20}$ mile

B $1\frac{3}{20}$ miles

C $3\frac{2}{5}$ miles

D $3\frac{9}{20}$ miles

44 This chart shows the cost for different sized ads in a newspaper. Suppose you want to place a $\frac{1}{4}$-page ad *and* a $\frac{1}{3}$-page ad. How much more space will you have than if you placed just a $\frac{1}{2}$-page ad?

Ad Size	$\frac{1}{4}$ page	$\frac{1}{3}$ page	$\frac{1}{2}$ page	Full page
Price	$40	$65	$80	$140

F $\frac{1}{12}$ page   H $\frac{1}{6}$ page

G $\frac{1}{4}$ page   J $\frac{1}{3}$ page

Spiral Review

Add. (Lesson 3-8)

45. $\frac{1}{3} + \frac{1}{2}$

46. $\frac{3}{4} + \frac{1}{6}$

47. $\frac{3}{8} + \frac{1}{2}$

48. $\frac{3}{4} + \frac{2}{3}$

49. **SPORTS** Hank spends $\frac{1}{2}$ of his practice time shooting free throws and $\frac{1}{4}$ of the time running drills. What fraction of his practice time does he spend on free throws and drills? (Lesson 3-8)

Progress Check 4

(Lessons 3-7, 3-8, and 3-9)

▷ Vocabulary and Concept Check

least common denominator (LCD) (p. 207)   like denominators (pp. 199, 213)
least common multiple (LCM) (p. 206)   unlike denominators (pp. 206, 213)

Choose the term that *best* completes each statement.

1. The fractions $\frac{3}{4}$ and $\frac{3}{8}$ have _____?_____.

2. The _____?_____ of the fractions $\frac{1}{4}$ and $\frac{1}{3}$ is 12.

▷ Skills Check

Add. Write each sum in simplest form. (Lesson 3-7)

3. $\frac{3}{10} + \frac{1}{10}$ 4. $\frac{5}{7} + \frac{3}{7}$ 5. $1\frac{1}{4} + 2\frac{3}{4}$

Subtract. Write each difference in simplest form. (Lesson 3-7)

6. $\frac{7}{8} - \frac{5}{8}$ 7. $\frac{4}{9} - \frac{1}{9}$ 8. $\frac{11}{12} - \frac{5}{12}$

Add. Write each sum in simplest form. (Lesson 3-8)

9. $\frac{1}{4} + \frac{3}{8}$ 10. $\frac{1}{2} + \frac{2}{3}$ 11. $1\frac{2}{3} + 2\frac{1}{4}$

Subtract. Write each difference in simplest form. (Lesson 3-9)

12. $\frac{7}{8} - \frac{3}{4}$ 13. $\frac{7}{10} - \frac{1}{4}$ 14. $2\frac{1}{3} - 1\frac{5}{6}$

▷ Problem-Solving Check

15. **COOKING** Ray puts $\frac{1}{3}$ cup of olive oil in a pan. Then he decides to add another $\frac{1}{3}$ cup to the recipe. How much olive oil is in there in all? (Lesson 3-7)

16. **FITNESS** Ella is biking $4\frac{1}{4}$ miles today. She has biked $2\frac{1}{2}$ miles so far. How much farther does she have to go? (Lesson 3-9)

17. **REFLECT** Explain how you decide to add or subtract when solving a word problem with fractions or mixed numbers. (Lesson 3-8)

Vocabulary

expression (p. 220)

equation (p. 221)

solve (p. 221)

inverse operations (p. 221)

Standard ALG5.0 Students solve multistep problems, including word problems, involving linear equations in one variable and provide justification for each step [excluding inequalities].

Standard 7AF4.1 Solve two-step linear equations and inequalities in one-variable over the rational numbers, interpret the solution or solutions in the context from which they arose, and verify the reasonableness of the results.

Standard 6AF1.1 Write and solve one-step linear equations in one-variable.

The What: I will simplify expressions and solve equations involving fractions.

The Why: You can solve equations with fractions to find the target heart rate when you exercise.

Fitness experts want adults to reach a target heart rate when exercising. Measuring the target heart rate helps determine if a person is exercising at the proper pace.

The **expression** below is used to find the rate. The variable a represents the person's age. What is the target rate for a 20-year-old adult?

$$\frac{3}{4}(220 - a)$$

You can find the answer by substituting 20 for the variable a.

$\frac{3}{4}(220 - a) = \frac{3}{4}(220 - 20)$ Substitute 20 for the variable a.

$\qquad = \frac{3}{4}(200)$ Simplify inside the parentheses first. Subtract 20 from 220.

$\qquad = 150$ Multiply $\frac{3}{4}$ and 200.

The target heart rate for a 20-year-old adult is 150 beats per minute.

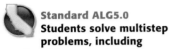

Talk Math

WORK WITH A PARTNER Use the heart rate expression to find the target heart rate for a 40-year-old adult and a 60-year-old adult.

You can use the Associative, Commutative, and Distributive Properties to simplify expressions. These properties allow you to change the order and grouping of numbers when you combine like terms.

① Simplify $\frac{2}{3} + \frac{3}{8}w + \frac{1}{6} - \frac{1}{8}w$.

$= \left(\frac{2}{3} + \frac{1}{6}\right) + \left(\frac{3}{8}w - \frac{1}{8}w\right)$	Use the Commutative Property to group like terms.
$= \left(\frac{4}{6} + \frac{1}{6}\right) + \left(\frac{3}{8}w - \frac{1}{8}w\right)$	Use the LCD to rename fractions with unlike denominators.
$= \frac{5}{6} + \frac{2}{8}w$	Add and subtract the numerators of fractions with like denominators.
$= \frac{5}{6} + \frac{1}{4}w$	Simplify.

Your Turn

Simplify each expression.

a. $\frac{1}{2} + \frac{7}{9}a - \frac{4}{9}a + \frac{3}{10}$

b. $\frac{3}{4} + \frac{4}{5}x + \frac{2}{4}$

An **equation** is a mathematical statement consisting of two expressions that are separated by an equal sign. You can **solve** an equation by finding the value or values of the variable that makes the equation true.

When solving equations with fractions, use **inverse operations** as you did with integers. Addition undoes subtraction. Multiplication undoes division.

EXAMPLE Solve One-Step Equations with Fractions

② Solve $k + \frac{2}{3} = \frac{3}{4}$.

$k + \frac{2}{3} - \frac{2}{3} = \frac{3}{4} - \frac{2}{3}$	Subtract $\frac{2}{3}$ from each side to "undo" the addition of $\frac{2}{3}$.
$k = \frac{9}{12} - \frac{8}{12}$	The LCD of 3 and 4 is 12.
$k = \frac{1}{12}$	Subtract the numerators.

Check: $k + \frac{2}{3} \stackrel{?}{=} \frac{3}{4}$

$\frac{1}{12} + \frac{2}{3} \stackrel{?}{=} \frac{3}{4}$ Substitute the answer in the original equation.

$\frac{1}{12} + \frac{8}{12} \stackrel{?}{=} \frac{9}{12}$

$\frac{9}{12} = \frac{9}{12}$ ✓   The answer checks.

Remember!

An equation is a math sentence with an = sign. Whatever you do to one side of an equation, you must do to the other side.

Your Turn

Solve each equation.

c. $\frac{1}{3} + x = \frac{2}{3}$

d. $\frac{3}{8} = \frac{1}{2} - f$

EXAMPLE   Solve One-Step Equations
with Fractions

3 Solve $\frac{3}{4}t = \frac{2}{3}$.

$\frac{3}{4}t \div \frac{3}{4} = \frac{2}{3} \div \frac{3}{4}$ Divide each side by $\frac{3}{4}$ to "undo" the multiplication.

$t = \frac{2}{3} \times \frac{4}{3}$ To divide by $\frac{3}{4}$, multiply by its reciprocal of $\frac{4}{3}$.

$t = \frac{8}{9}$ Multiply numerators. Multiply denominators.

Check: $\frac{3}{4}t \overset{?}{=} \frac{2}{3}$

$\frac{3}{4} \times \frac{8}{9} \overset{?}{=} \frac{2}{3}$ Substitute the answer in the original equation.

$\frac{24}{36} \overset{?}{=} \frac{2}{3}$

$\frac{2}{3} = \frac{2}{3}$ ✓   The answer checks.

Your Turn
Solve the equation.

e. $\frac{2}{3}n = \frac{1}{6}$ f. $\frac{3}{5}n = \frac{5}{7}$

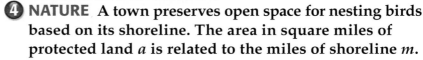

Real-World EXAMPLE

4 **NATURE** A town preserves open space for nesting birds based on its shoreline. The area in square miles of protected land a is related to the miles of shoreline m.

$$\frac{1}{3}a = m$$

The town has 4 miles of shoreline. How much land is in the protected area?

$\frac{1}{3}a = 4$ Substitute 4 for m.

$\frac{1}{3}a \div \frac{1}{3} = 4 \div \frac{1}{3}$ Divide each side by $\frac{1}{3}$ to "undo" the multiplication.

$a = 4 \times 3$ To divide by $\frac{1}{3}$, multiply by its reciprocal of $\frac{3}{1}$ or 3.

$a = 12$ Simplify.

There are 12 square miles of protected land.

Your Turn

g. **BUSINESS** The equation $\frac{2}{3}p = d$ shows the price of clothing after it goes on sale. The variable p is related to the discount price of d. Find the original price of a belt if the discount price is $12.

NATURE California has over 1,000 miles of beaches.

Two-step equations have two operations in the equation. Solve using inverse operations.

SOLVING TWO-STEP EQUATIONS

Use inverse operations:

1 Undo addition or subtraction first.

2 Then undo multiplication or division.

EXAMPLE **Solve Two-Step Equations with Fractions**

5 Solve $2x + \frac{1}{3} = \frac{3}{4}$.

$$2x + \frac{1}{3} = \frac{3}{4}$$

$$2x + \frac{1}{3} - \frac{1}{3} = \frac{3}{4} - \frac{1}{3}$$ Subtract $\frac{1}{3}$ from each side to "undo" addition.

$$2x = \frac{9}{12} - \frac{4}{12}$$ The LCD of 4 and 3 is 12.

$$2x = \frac{5}{12}$$ Subtract $\frac{4}{12}$ from $\frac{9}{12}$.

$$2x \div 2 = \frac{5}{12} \div 2$$ Divide each side by 2 to "undo" the multiplication.

$$x = \frac{5}{12} \times \frac{1}{2}$$ To divide by 2, multiply by its reciprocal, $\frac{1}{2}$.

$$x = \frac{5}{24}$$ Multiply numerators. Multiply denominators.

Check: $2x + \frac{1}{3} \stackrel{?}{=} \frac{3}{4}$

$$2 \times \frac{5}{24} + \frac{1}{3} \stackrel{?}{=} \frac{3}{4}$$ Substitute the answer in the original equation.

$$\frac{10}{24} + \frac{1}{3} \stackrel{?}{=} \frac{3}{4}$$

$$\frac{10}{24} + \frac{8}{24} \stackrel{?}{=} \frac{3}{4}$$

$$\frac{18}{24} \stackrel{?}{=} \frac{3}{4}$$

$$\frac{3}{4} = \frac{3}{4} \checkmark$$ The answer checks.

Your Turn

Solve each equation.

h. $\frac{1}{2} + 3c = \frac{3}{4}$ i. $2 - \frac{1}{2}s = \frac{1}{3}$ j. $\frac{3}{8} = \frac{3}{4}k - \frac{1}{2}$

Examples 1–5
(pages 221–223)

VOCABULARY

1. Use the term *expression* to describe an equation.

2. Explain how you can use inverse operations to solve an equation.

Example 1
(page 221)

Simplify each expression.

3. $\frac{2}{3} + \frac{3}{4}f + \frac{1}{4}$

4. $\frac{1}{2}k + \frac{3}{4} - \frac{1}{6}k$

5. $\frac{1}{3} + \frac{3}{8}b + \frac{1}{2} + \frac{1}{4}b$

Solve each equation.

Example 2
(page 221)

6. $k + \frac{5}{8} = \frac{7}{8}$

7. $z + \frac{7}{8} = 1\frac{1}{4}$

8. $a - \frac{1}{3} = \frac{5}{6}$

Solve each equation.

Example 3
(page 222)

9. $\frac{3}{4}x = \frac{1}{3}$

10. $\frac{2}{3}n = \frac{1}{2}$

Example 4
(page 222)

11. **LANDSCAPING** A landscaper decides to plant red and yellow roses according to the equation $y = \frac{2}{3}r$. In this equation, r represents red roses and y represents yellow roses. How many red roses should she plant if she has 24 yellow roses?

12. **COMMUNITY SERVICE** The sophomores and freshmen at Adams High perform service hours each week. The number of hours performed is related by the equation $\frac{2}{3}s + 1 = f$. In this equation, the number of sophomore hours is represented by s and the number of freshman hours is represented by f. If the freshmen perform 9 hours a week, how many hours will the sophomores perform?

Example 5
(page 223)

Solve each equation.

13. $2w + \frac{2}{3} = 1$

14. $\frac{2}{3}c + \frac{1}{2} = 2$

15. $\frac{3}{8}t - \frac{5}{8} = \frac{3}{4}$

16. **Talk Math** What is the difference between an expression and an equation? How are the two alike?

Simplify each expression.

HOMEWORK HELP	
For Exercises	**See Example(s)**
17–22	1
23–34	2–3
35–38	4
39–44	5

17. $\frac{3}{8}d + \frac{1}{4} + \frac{1}{8}d$

18. $\frac{1}{6} + \frac{1}{4} + \frac{1}{2}a$

19. $\frac{1}{6}t + \frac{3}{4}t - \frac{2}{3}$

20. $\frac{3}{4}k - \frac{5}{8}k + \frac{1}{2}$

21. $\frac{1}{2}m + \frac{2}{3} - \frac{1}{4}m - \frac{1}{6}$

22. $\frac{7}{8}f + \frac{11}{12} - \frac{3}{8}f - \frac{3}{4}$

Solve each equation.

23. $x + \frac{1}{4} = \frac{6}{8}$

24. $\frac{7}{8} + b = 1\frac{3}{4}$

25. $\frac{5}{12} = y + \frac{2}{6}$

26. $\frac{9}{12} = r + \frac{1}{4}$

27. $d - \frac{5}{6} = 1$

28. $r - \frac{3}{4} = \frac{1}{8}$

29. $\frac{7}{10} = b - \frac{1}{5}$

30. $\frac{1}{16} = q - \frac{5}{8}$

31. $\frac{3}{8}m = \frac{1}{4}$

32. $\frac{2}{3}z = \frac{6}{7}$

33. $\frac{1}{3}n = \frac{2}{5}$

34. $\frac{1}{4} = \frac{1}{2}t$

35. **NATURE** A town votes to preserve open space for parks. It uses the equation $p = \frac{1}{3}a$, where p is each new acre of park and a is each new acre of land that can be developed. If the town creates a $4\frac{1}{4}$ or $\frac{17}{4}$-acre park, how many acres can it develop?

36. **HEALTH** Fitness experts usually suggest that adults exercise to meet a target heart rate. This rate is about $\frac{3}{4}(220 - a)$, where a is the person's age, depending on fitness level and other heath considerations. What age matches a target heart rate of 120?

37. **CAREER CONNECTION** A company uses the equation $s = \frac{2}{5}p + 2$ to determine the sale price of its products. In this equation, s represents the sale price and p represents the original price. Suppose the sale price is $4. What was the original price?

38. **MARKET RESEARCH** A salesperson discovers that the relationship of shoppers to purchasers in her store follows the equation $p = \frac{3}{4}s$. In this equation, p is purchasers and s is shoppers. If the store had 60 purchasers on Saturday, about how many shoppers were in the store?

RETAIL In 2002, over $1\frac{1}{2}$ million Californians worked as retail salespersons.

Solve each equation.

39. $2k + \frac{1}{2} = 2$

40. $7p + \frac{3}{4} = 1$

41. $\frac{5}{6} + \frac{2}{3}m = 1$

42. $\frac{5}{8} = \frac{1}{4}d + \frac{3}{8}$

43. $\frac{1}{2}b - \frac{2}{3} = \frac{1}{6}$

44. $\frac{2}{3}q - \frac{1}{4} = \frac{7}{8}$

45. **EARTH SCIENCE** Below Earth's crust, there are three layers—the mantle, inner core, and outer core. The interior of the Earth is one-half mantle, one-third outer core, and one-sixth inner core. Simplify the expression $\frac{1}{2}m + \frac{1}{3}c + \frac{1}{6}c$. Find the fraction that represents the sum of the 2 cores.

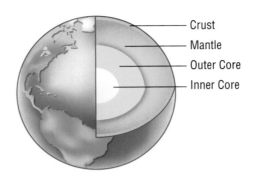

— Crust
— Mantle
— Outer Core
— Inner Core

46. *Writing in Math* Write a two-step equation that includes at least two fractions. Exchange problems with a classmate and solve.

47. **H.O.T.** Problems Kanita solves $\frac{3}{4}k + \frac{1}{2} = \frac{5}{6}$. She gets an answer of $\frac{2}{3}$. Do you agree or disagree with her solution? Why or why not?

STANDARDS PRACTICE

Choose the *best* answer.

48 Which equation has the same solution as $\frac{1}{2}x = \frac{3}{8}$?

  A $2x = \frac{8}{3}$

  B $\frac{1}{3}x = \frac{1}{4}$

  C $\frac{3}{8}x = \frac{1}{2}$

  D $\frac{2}{3}x = \frac{3}{8}$

49 How many solutions are there to the two-step equation $\frac{2}{3}x - \frac{1}{6} = \frac{3}{4}$?

  F 0

  G 1

  H 2

  J 4

Spiral Review

Subtract. (Lesson 3-9)

50. $\frac{1}{2} - \frac{1}{3}$ 51. $\frac{7}{8} - \frac{1}{2}$ 52. $\frac{3}{4} - \frac{2}{3}$ 53. $\frac{5}{6} - \frac{3}{8}$

54. **PACKAGING** There was $\frac{5}{8}$ of a pizza left in a box. Doug ate $\frac{1}{4}$ of the remaining pizza for lunch. How much pizza was there after Doug ate lunch? (Lesson 3-9)

Study Guide

Understanding and Using the Vocabulary

After completing the chapter, you should be able to define each term, property, or phrase and give an example of each.

additive inverse (p. 165)	least common multiple (LCM) (p. 206)
common factor (p. 179)	like denominators (pp. 199, 213)
composite number (p. 177)	mixed number (p. 169)
draw a diagram (p. 184)	multiplicative inverse (p. 192)
equation (p. 221)	prime number (p. 177)
equivalent fraction (p. 172)	product (p. 186)
expression (p. 220)	proper fraction (p. 169)
factor (p. 177)	rational number (p. 164)
fraction (pp. 162, 164)	reciprocal (p. 192)
greatest common factor (GCF) (p. 179)	simplest form (p. 179)
improper fraction (p. 169)	solve (p. 221)
inverse operations (p. 221)	unlike denominators (pp. 206, 213)
Inverse Property of Multiplication (p. 192)	
least common denominator (LCD) (p. 207)	

Complete each sentence with the correct mathematical term or phrase.

1. $2\frac{3}{4}$ is a(n) _____?_____.

2. The _____?_____ of $\frac{1}{3}$ and $\frac{3}{4}$ is 12.

3. A(n) _____?_____ can be written in the form $\frac{a}{b}$ ($b \neq 0$).

4. The fractions $\frac{3}{8}$ and $\frac{7}{8}$ have _____?_____.

5. The fractions $\frac{3}{4}$ and $\frac{5}{6}$ have _____?_____.

6. $\frac{5}{3}$ is a(n) _____?_____.

7. The _____?_____ of 6 and 10 is 2.

8. 2, 3, 5, and 11 are examples of a(n) _____?_____.

9. 4, 8, 14, and 20 are examples of a(n) _____?_____.

10. The _____?_____ of $\frac{5}{8}$ is $\frac{8}{5}$.

Skills and Concepts

Objectives and Examples   **Review Exercises**

LESSON 3-1   pages 164–168

Compare and order fractions using a number line.

Graph $-\frac{1}{3}$ and $-\frac{2}{3}$ on a number line.

Since $-\frac{1}{3}$ is to the right of $-\frac{2}{3}$, $-\frac{1}{3} > -\frac{2}{3}$.

Graph each fraction on a number line.

11. $\frac{3}{4}$ 12. $-\frac{1}{6}$

Order each set of fractions from least to greatest.

13. $\frac{1}{5}, \frac{3}{5}, \frac{2}{5}$ 14. $\frac{5}{8}, \frac{7}{8}, \frac{3}{8}$

15. $-\frac{6}{7}, -\frac{1}{7}, -\frac{3}{7}$ 16. $-\frac{8}{9}, -\frac{2}{9}, -\frac{5}{9}$

LESSON 3-2   pages 169–175

Compare and order fractions and mixed numbers.

Write $\frac{5}{3}$ as a mixed number.

$$\begin{array}{r} 1R2 \\ 3\overline{)5} \\ \underline{-3} \\ 2 \end{array}$$

$$\frac{5}{3} = 1\frac{2}{3}$$

Write each mixed number as an improper fraction.

17. $2\frac{2}{3}$ 18. $-3\frac{1}{4}$ 19. $-2\frac{3}{5}$

Write each improper fraction as a mixed number.

20. $\frac{7}{2}$ 21. $-\frac{8}{5}$ 22. $\frac{8}{3}$

Graph each pair of numbers on a number line.

23. $-1\frac{3}{8}, 1\frac{7}{8}$ 24. $-2\frac{4}{5}, -1\frac{1}{5}$

LESSON 3-3   pages 177–182

Simplify fractions using the greatest common factor (GCF).

Find the greatest common factor of 6 and 15.

6: 1, 2, ③, 6   List the factors of each
15: 1, ③, 5, 15   number. Circle the GCF.

The GCF of 6 and 15 is 3.

Find the greatest common factor (GCF) of the numbers.

25. 4 and 10   26. 6 and 18   27. 5 and 9

Write each fraction in simplest form.

28. $\frac{4}{8}$ 29. $\frac{6}{10}$ 30. $\frac{8}{12}$

Skills and Concepts

Objectives and Examples

Review Exercises

LESSON 3-5 pages 186–191

Multiply fractions.

$$\frac{2}{3} \cdot \frac{4}{5} = \frac{2 \cdot 4}{3 \cdot 5} = \frac{8}{15}$$
Multiply numerators.
Multiply denominators.

Multiply. Write each product in simplest form.

31. $\frac{2}{3} \cdot \frac{3}{4}$ 32. $\frac{3}{10} \cdot \frac{5}{6}$

33. $\frac{4}{5} \cdot \left(-\frac{5}{8}\right)$ 34. $\frac{5}{2} \cdot \frac{2}{3}$

Raise fractions to positive powers.

$$\left(\frac{2}{3}\right)^3 = \frac{2}{3} \cdot \frac{2}{3} \cdot \frac{2}{3} = \frac{8}{27}$$

Evaluate.

35. $\left(\frac{1}{2}\right)^5$ 36. $\left(\frac{3}{4}\right)^3$

LESSON 3-6 pages 192–197

Divide fractions.

$$\frac{7}{10} \div \frac{3}{5} = \frac{7}{10} \cdot \frac{5}{3}$$
Multiply by the reciprocal of the second fraction.

$$= \frac{^7\cancel{35}}{\cancel{30}_6} = \frac{7}{6} = 1\frac{1}{6}$$
Simplify. Write as a mixed number.

Divide. Write each quotient in simplest form.

37. $\frac{3}{5} \div \frac{2}{3}$ 38. $\frac{1}{8} \div \frac{3}{4}$

39. $-\frac{7}{12} \div \frac{5}{6}$ 40. $\frac{14}{3} \div \frac{4}{3}$

LESSON 3-7 pages 199–205

Add and subtract fractions with like denominators.

Add $\frac{7}{10} + \frac{1}{10}$.

$$\frac{7}{10} + \frac{1}{10} = \frac{7+1}{10}$$
Add the numerators. Write the sum over the denominator.

$$= \frac{8}{10} = \frac{4}{5}$$
Simplify.

Subtract $\frac{13}{15} - \frac{4}{15}$.

$$\frac{13}{15} - \frac{4}{15} = \frac{13-4}{15}$$
Subtract the numerators. Write the difference over the denominator.

$$= \frac{9}{15} = \frac{3}{5}$$
Simplify.

Add. Write each sum in simplest form.

41. $\frac{1}{3} + \frac{1}{3}$ 42. $\frac{3}{8} + \frac{1}{8}$

43. $\frac{11}{12} + \frac{7}{12}$ 44. $\frac{7}{4} + \frac{11}{4}$

Subtract. Write each difference in simplest form.

45. $\frac{3}{4} - \frac{1}{4}$ 46. $\frac{7}{8} - \frac{5}{8}$

47. $\frac{5}{12} - \frac{1}{12}$ 48. $\frac{10}{3} - \frac{5}{3}$

Study Guide

Objectives and Examples

Review Exercises

LESSON 3-8 pages 206–212

Add fractions with unlike denominators.

Add $\frac{1}{3} + \frac{7}{12}$.

$$\overset{\times 4}{\frac{1}{3}} = \frac{4}{12} \quad \overset{\times 1}{\frac{7}{12}} = \frac{7}{12} \underset{\times 1}{}$$

Rename the fractions using the LCD.

$$\frac{4}{12} + \frac{7}{12} = \frac{4+7}{12} = \frac{11}{12}$$

Add the numerators. Simplify.

Add. Write each sum in simplest form.

49. $\frac{1}{5} + \frac{7}{10}$

50. $\frac{1}{2} + \frac{5}{6}$

51. $\frac{2}{3} + \frac{3}{4}$

52. $\frac{11}{8} + \frac{9}{4}$

LESSON 3-9 pages 213–218

Subtract fractions with unlike denominators.

Subtract $\frac{5}{6} - \frac{1}{3}$.

$$\overset{\times 1}{\frac{5}{6}} = \frac{5}{6} \quad \overset{\times 2}{\frac{1}{3}} = \frac{2}{6} \underset{\times 2}{}$$

Rename the fractions using the LCD.

$$\frac{5}{6} - \frac{2}{6} = \frac{5-2}{6} = \frac{3}{6} = \frac{1}{2}$$

Subtract the numerators. Simplify.

Subtract. Write each difference in simplest form.

53. $\frac{5}{8} - \frac{1}{4}$

54. $\frac{1}{2} - \frac{1}{3}$

55. $\frac{5}{6} - \frac{1}{2}$

56. $\frac{11}{3} - \frac{7}{4}$

LESSON 3-10 pages 220–226

Solve equations with fractions.

Solve $\frac{2}{3}x - \frac{1}{2} = \frac{5}{6}$.

$$\frac{2}{3}x - \frac{1}{2} + \frac{1}{2} = \frac{5}{6} + \frac{1}{2}$$

Add $\frac{1}{2}$ to each side to "undo" subtraction.

$$\frac{2}{3}x \div \frac{2}{3} = \frac{4}{3} \div \frac{2}{3}$$

Divide each side by $\frac{2}{3}$ to "undo" the multiplication.

$$x = 2$$

Solve each equation.

57. $k - \frac{4}{5} = \frac{1}{10}$

58. $z \div \frac{3}{4} = \frac{2}{3}$

59. $\frac{2}{3}c - \frac{1}{2} = 1$

60. $\frac{3}{4}y + \frac{1}{4} = \frac{7}{12}$

Chapter Test

▷ Vocabulary and Concept Check

1. Write two examples each of improper fractions and mixed numbers.

2. Write a pair of fractions with like denominators. Write a pair of fractions with unlike denominators.

▷ Skills Check

Use a number line to determine which number is greater.

3. $\frac{7}{12}$ and $\frac{5}{12}$

4. $1\frac{5}{6}$ and $2\frac{1}{6}$

Identify each number as prime or composite.

5. 15

6. 9

Identify the greatest common factor (GCF) of the numbers.

7. 8 and 16

8. 8 and 10

Write each fraction in simplest form.

9. $\frac{5}{20}$

10. $-\frac{12}{18}$

11. $\frac{3}{15}$

12. $-\frac{14}{35}$

Multiply, divide, add, or subtract. Write each answer in simplest form.

13. $\frac{1}{4} \times \frac{4}{5}$

14. $\frac{3}{8} \times \frac{2}{3}$

15. $\frac{7}{8} \div \frac{1}{4}$

16. $\frac{5}{3} \div \frac{1}{2}$

17. $\frac{3}{10} + \frac{5}{10}$

18. $\frac{5}{8} + \frac{1}{4}$

19. $\frac{11}{12} - \frac{7}{12}$

20. $\frac{7}{8} - \frac{1}{2}$

Solve each equation.

21. $n + \frac{1}{2} = \frac{3}{4}$

22. $\frac{2}{3}g = \frac{5}{6}$

23. $2x - \frac{5}{6} = \frac{1}{3}$

24. $\frac{5}{8}m - \frac{1}{2} = \frac{3}{4}$

▷ Problem-Solving Check

AGRICULTURE A farmer plans to keep $\frac{1}{8}$ of her crop and sell $\frac{7}{8}$ to customers. She will sell $\frac{2}{3}$ of the total at local markets.

25. Which operation will you use to find the total that will be sold at local markets? Explain.

26. What fraction of the total crop will be sold at local markets? Write your answer in simplest form.

27. **MUSIC** Carmen practices the flute $2\frac{1}{2}$ or $\frac{5}{2}$ hours each day. On Saturday, she practiced $1\frac{3}{4}$ or $\frac{7}{4}$ hours. How much time did she have left to practice?

PART 1   Multiple Choice

Choose the *best* answer.

1 What number is the multiplicative inverse of $\frac{2}{3}$?

A $\frac{2}{1}$ **C** 1

B $\frac{3}{2}$ **D** $\frac{1}{3}$

2 Which mixed number is less than $-1\frac{2}{3}$?

F $-1\frac{3}{4}$ **H** -1

G $-1\frac{1}{3}$ **J** 0

3 How can you tell that a fraction is in its simplest form?

A The numerator is less than the denominator.

B The denominator is less than the numerator.

C The GCF of the numerator and denominator is 1.

D The GCF of the numerator and denominator is greater than 1.

4 Which equation has the same value for x as $\frac{3}{4}x = \frac{1}{2}$?

F $\frac{1}{2}x = \frac{3}{4}$ **H** $x + \frac{1}{2} = \frac{3}{4}$

G $\frac{2}{3}x = \frac{1}{6}$ **J** $\frac{1}{8}x = \frac{1}{12}$

5 Which of the following numbers is prime?

A 15   **C** 17

B 16   **D** 18

6 How many cups of orange juice and apple juice are needed for the punch?

Punch Recipe

Juice	Cups
Apple	$3\frac{2}{3}$
Grape	$1\frac{1}{4}$
Mango	$\frac{1}{3}$
Orange	$4\frac{1}{2}$

F $5\frac{3}{4}$ cups   **H** $7\frac{5}{6}$ cups

G $7\frac{1}{2}$ cups   **J** $8\frac{1}{6}$ cups

7 How do you divide $\frac{3}{8} \div \frac{1}{4}$?

A Multiply $\frac{3}{8}$ and $\frac{1}{4}$.

B Multiply $\frac{3}{8}$ and $\frac{4}{1}$.

C Multiply $\frac{8}{3}$ and $\frac{1}{4}$.

D Multiply $\frac{8}{3}$ and $\frac{4}{1}$.

8 What is the GCF of 6 and 12?

F 1   **H** 6

G 3   **J** 12

PART 2   Short Answer

Record your answers on the answer sheet provided by your teacher or on a separate sheet of paper.

9 **ENGINEERING** A new road will run 15 miles through a park and then 10 miles to a town. Five stop signs are placed at equal distances along the route. What is the distance between each sign?

10 This chart shows the number of minutes Elias played in each quarter of the last basketball game. What fraction of the whole game did Elias play?

Quarter	Total Minutes	Minutes Elias Played
1	15	9
2	15	6
3	15	3
4	15	12

11 In Ms. Alvarez's class, $\frac{3}{8}$ of the students are running for a position in the student government. Only $\frac{1}{3}$ of them will win their elections. What fraction of the class will hold a position in the government?

12 This chart shows the number of miles Salil hiked each day on his trip. How many more miles did he hike on Saturday and Sunday than on Monday and Tuesday?

Day	Miles Hiked
Saturday	$3\frac{3}{5}$
Sunday	$3\frac{1}{10}$
Monday	$2\frac{1}{2}$
Tuesday	3

13 Solve for z.

$$\frac{1}{3}z - \frac{2}{3} = \frac{1}{6}$$

PART 3 Extended Response

Record your answers on the answer sheet provided by your teacher or on a separate sheet of paper.

14 A chef is making stock to use for recipes in her kitchen. She combines $4\frac{1}{2}$ or $\frac{9}{2}$ cups of vegetable broth with $2\frac{1}{4}$ or $\frac{9}{4}$ cups of water.

  a. How much liquid is there in all?

  b. The chef then boils the liquid mixture until it reduces by half. How much liquid is left?

  c. The chef uses $1\frac{1}{2}$ or $\frac{3}{2}$ cups of the reduced liquid for a sauce and freezes the rest. How much stock does she freeze?

NEED EXTRA HELP?														
If You Missed Question...	1	2	3	4	5	6	7	8	9	10	11	12	13	14
Go to Lesson...	3-6	3-2	3-3	3-10	3-3	3-8	3-6	3-3	3-3	3-3	3-5	3-8 3-9	3-10	3-5 3-8 3-9
For Help with Algebra Readiness Standard...	7NS 1.2	6NS 1.1	5NS 1.4	7AF 4.1	5NS 1.4	7NS 1.2	7NS 1.2	5NS 1.4	5NS 1.4	5NS 1.4	7NS 1.2	7NS 1.2	7AF 4.1	7NS 1.2

Decimals

Here's What I Need to Know

Standard 7NS1.2 Add, subtract, multiply, and divide rational numbers (integers, fractions, and **terminating decimals) and take positive rational numbers to whole-number powers.**

Standard 7NS1.3 Convert fractions to decimals and percents **and use these representations in estimations, computations, and applications.**

Standard 7AF4.1 Solve two-step linear equations and inequalities in one variable over the rational numbers, interpret the solution or solutions in the context from which they arose, and verify the reasonableness of the results.

Vocabulary Preview

decimal A number with one or more digits to the right of the decimal point (p. 236)

1.8

terminating decimal A decimal with a finite number of digits (p. 237)

| **1.8** | **23.46** |
| tenths | hundredths |

repeating decimal A decimal with an infinite number of digits (p. 237)

0.0909 . . .   0.0̄9̄

The What

I will learn to:

- change fractions to decimals and decimals to fractions.
- perform operations using decimals.
- use decimals in expressions and equations.

The Why

Decimals are used to represent the cost of restaurant meals and bank account balances.

Option 1

Math Online Are you ready for Chapter 4? Take the Online Readiness Quiz at ca.algebrareadiness.com to find out.

Option 2

Complete the Quick Check below.

Vocabulary and Skills Review

Identify the value of the named digit in the number 254.183.
Complete each sentence with the correct term.

1. The digit 5 is in the ___?___ place.   A. hundreds
2. The digit 8 is in the ___?___ place.   B. hundredths
3. The digit 1 is in the ___?___ place.   C. tens
                                           D. tenths

Write each number in expanded form; for example, $75 = 70 + 5$.

4. 24   5. 92   6. 345   7. 802
8. 4,592   9. 1,234   10. 1,093   11. 3,040

Review of My Notes	Note-taking
	My Notes
Example: $0.5 = \frac{1}{2}$	A decimal is like a fraction—part of a whole.
<u>decimal point</u>	A decimal point separates whole numbers from fractional numbers.
Does this work for ALL decimals?	Read the decimal aloud to hear the fraction. Then, convert the fraction to lowest form.
	Most Important Ideas
	Decimals are a form of showing part of a whole. They can be converted to fractions.

Tips
Review your notes to:
- note vocabulary words.
- ask questions.
- call out the most important ideas.

Vocabulary

fraction (p. 236)

decimal (p. 236)

decimal point (p. 236)

terminating decimal (p. 237)

repeating decimal (p. 237)

 Standard 7NS1.3 Convert fractions to decimals and percents and **use these representations** in estimations, computations, and applications.

Standard 7NS1.5 Know that every rational number is either a terminating or repeating decimal and be able to convert terminating decimals into reduced fractions.

 The What: I will change fractions to decimals and decimals to fractions.

The Why: Writing parts of a whole in different ways helps compare numbers such as measurements.

In a grocery store, scales are used to weigh meats, cheeses, and produce. Spring scales show weights as fractions, such as $\frac{3}{4}$ pound. Digital scales show weights as decimals, such as 0.75 pound.

Just like a **fraction,** a **decimal** is used to represent part of a whole or part of a group. In decimals, a **decimal point** separates whole units and partial units.

$$0.75 = \frac{75}{100}$$

seventy-five hundredths

EQUIVALENT REPRESENTATIONS

Decimal	Words	Fraction
0.4	four tenths	$\frac{4}{10} = \frac{2}{5}$
0.25	twenty-five hundredths	$\frac{25}{100} = \frac{1}{4}$
0.003	three thousandths	$\frac{3}{1000}$

A decimal can be used to express a mixed number. A mixed number, such as $1\frac{15}{100}$, has both a whole number and a fraction.

$$1.15 = 1\frac{15}{100}$$

one and $\frac{15}{100}$

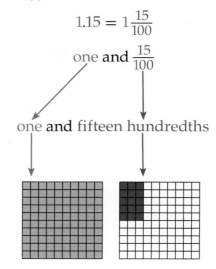

one and fifteen hundredths

EQUIVALENT REPRESENTATIONS		
Decimal	**Words**	**Mixed Number**
2.4	two and four tenths	$2\frac{4}{10} = 2\frac{2}{5}$
5.25	five and twenty-five hundredths	$5\frac{25}{100} = 5\frac{1}{4}$
10.003	ten and three thousandths	$10\frac{3}{1000}$

You will be working with two types of decimals: terminating and repeating. A **terminating decimal** can be written as a fraction with a denominator of 10, 100, 1000, and so on.

$$0.5 = \frac{5}{10} \qquad 0.274 = \frac{274}{1000}$$

A **repeating decimal** goes on forever. There are two ways to write a repeating decimal. Both ways show the pattern that repeats in the decimal. A line over digits shows that those numbers keep repeating in that pattern.

$$\frac{1}{11} = 0.0909\ldots \quad \text{or} \quad \frac{1}{11} = 0.\overline{09}$$

Talk Math

WORK WITH A PARTNER Name five terminating decimals. Listen as your partner names five repeating decimals.

Any terminating or repeating decimal can be written as a fraction. To change a terminating decimal to a fraction, look at the digits to the right of the decimal point. The place value of the digit farthest to the right tells you the fraction's denominator.

EXAMPLES Change Decimals to Fractions

Change each terminating decimal to a fraction in simplest form.

1 0.5

$0.5 = \dfrac{5}{10}$ — Write 5 in the numerator. Look at the place value of the last digit. Write 10 in the denominator.

$= \dfrac{5 \div 5}{10 \div 5}$ — Divide both numerator and denominator by the GCF, 5, to simplify.

$= \dfrac{1}{2}$

2 −0.24

$-0.24 = -\dfrac{24}{100}$ — Write 24 in the numerator. Look at the place value of the last digit. Write 100 in the denominator. The fraction will be negative.

$= -\dfrac{24 \div 4}{100 \div 4}$ — Divide both numerator and denominator by the GCF, 4, to simplify.

$= -\dfrac{6}{25}$

Your Turn
Change each terminating decimal to a fraction in simplest form.

a. 0.9   b. −0.34   c. −0.345

EXAMPLE Change Decimals to Mixed Numbers

3 Change 1.45 to a mixed number in simplest form.

$1.45 = 1\dfrac{45}{100}$ — Look at the digits to the right of the decimal. Write 45 in the numerator. Look at the place value of the last digit. Write 100 in the denominator.

$= 1\dfrac{45 \div 5}{100 \div 5}$ — Simplify the fractional part of the mixed number. Divide both numerator and denominator by the GCF, 5.

$= 1\dfrac{9}{20}$

Your Turn
Change each terminating decimal to a mixed number in simplest form.

d. 5.25   e. −7.4   f. 2.009

To divide a lesser number by a greater number:

Set up a division problem.

Put decimal points after the lesser number and in the quotient.

$$0.4$$
$$5\overline{)2.0}$$ Add zeros after the decimal point.
$$\underline{-20}$$
$$0$$ Divide until there is no remainder or until a pattern appears.

Any fraction can be changed to a decimal. Remember fractions represent divisions. For example, $\frac{2}{5}$ can be written as $2 \div 5$. To find a decimal equivalent, divide the numerator by the denominator.

$$\frac{2}{5} = 2 \div 5 = 0.4$$

EXAMPLES **Change Fractions to Decimals**

Change each fraction to a decimal.

4 $\frac{3}{8}$

$\frac{3}{8} = 3 \div 8$ Rewrite the fraction as a division problem.

$$\begin{array}{r} 0.375 \\ 8\overline{)3.000} \\ -0 \\ \hline 30 \\ -24 \\ \hline 60 \\ -56 \\ \hline 40 \\ -40 \\ \hline 0 \end{array}$$

Put a decimal point after 3 and in the quotient. Add zeros after the decimal point.

Divide until there is no remainder or until you see a pattern.

So, $\frac{3}{8} = 0.375$. There is no remainder, so this decimal is a terminating decimal.

5 $-\frac{6}{22}$

$-\frac{6}{22} = -(6 \div 22)$ Rewrite the fraction as a division problem.

$$\begin{array}{r} 0.2727 \\ 22\overline{)6.0000} \\ -0 \\ \hline 60 \\ -44 \\ \hline 160 \\ -154 \\ \hline 60 \\ -44 \\ \hline 160 \end{array}$$

Put a decimal point after 6 and in the quotient. Add zeros after the decimal point.

Divide until there is no remainder or until you see a pattern.

So, $-\frac{6}{22} = -0.2727\ldots$ Since the fraction is negative, the decimal is also negative.

There is a pattern. So, this decimal is a repeating decimal. It can be written as $-0.2727\ldots$ or $-0.\overline{27}$.

Your Turn

Change each fraction to a decimal.

g. $\frac{1}{10}$ **h.** $-\frac{2}{3}$

A mixed number can also be changed to a decimal. The whole number goes to the left of the decimal point. The quotient of the numerator divided by the denominator goes to the right.

$$4\frac{3}{4} = 4 + (3 \div 4) = 4.75$$

EXAMPLE Change Mixed Numbers to Decimals

⑥ Change $12\frac{19}{20}$ to a decimal.

$12\frac{19}{20} = 12 + \left(\frac{19}{20}\right)$ Rewrite the mixed number as the sum of a whole number and a fraction.

$= 12 + (19 \div 20)$ Rewrite the fraction as a division problem.

$$\begin{array}{r} 0.\,9\,5 \\ 2\,0\,\overline{)1\,9.\,0\,0} \\ -\,0 \\ \hline 1\,9\,0 \\ -\,1\,8\,0 \\ \hline 1\,0\,0 \\ -\,1\,0\,0 \\ \hline 0 \end{array}$$

Put a decimal point after 19 and in the quotient.

Add zeros after the decimal point.

So, $\frac{19}{20} = 0.95$.

$= 12 + 0.95$ Add the quotient to the whole number.

$= 12.95$

Your Turn

Change each mixed number to a decimal.

i. $7\frac{3}{4}$ j. $12\frac{47}{50}$

Changing fractions or mixed numbers to decimals can help you compare numbers.

EXAMPLE Compare Rational Numbers

⑦ **TRAVEL** It took Ezekiel $3\frac{4}{5}$ hours to drive to a friend's house. Candace drove 3.65 hours to visit her friend. Who had the longer trip?

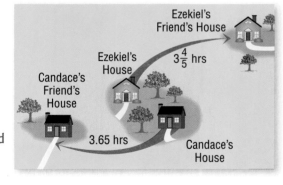

$3\frac{4}{5} = 3.8$ Rewrite the mixed number as a decimal.

$3.8 > 3.65$ Compare the two decimal numbers.

Ezekiel had a longer trip.

Your Turn

k. **BAKING** Bianca's cake used $2\frac{1}{2}$ cups of flour. Noe's cookies used 2.55 cups of flour. Who used more flour?

Examples 1–6
(pages 238–240)

VOCABULARY

Vocabulary Review
fraction
decimal
decimal point
terminating decimal
repeating decimal

1. How are fractions and decimals alike? How are they different?

Examples 1–2
(page 238)

Change each terminating decimal to a fraction in simplest form.

2. 0.2

3. -0.505

Example 3
(page 238)

Change each terminating decimal to a mixed number in simplest form.

4. 4.26

5. -3.4

Examples 4–6
(pages 239–240)

Change each fraction or mixed number to a decimal.

6. $\frac{1}{8}$

7. $-\frac{7}{15}$

8. $2\frac{3}{5}$

9. $5\frac{31}{40}$

Example 7
(page 240)

10. **PHOTOGRAPHY** Gregorio and Akiko took pictures at a wedding. They used film with the same number of exposures on each roll. Gregorio took $1\frac{3}{4}$ rolls of film, and Akiko took 1.375 rolls of film. Who took more pictures?

11. **FOOD** Frederico bought $2\frac{2}{3}$ pounds of apples. Fabio bought 2.69 pounds. Who bought more apples?

Spring scales and digital scales show the same measurement in different ways.

Skills, Concepts, and Problem Solving

HOMEWORK HELP

For Exercises	See Example(s)
12–15	1–2
16–19	3
20–27	4–6
28–30	7

Change each terminating decimal to a fraction in simplest form.

12. 0.6

13. 0.56

14. -0.9

15. -0.88

Change each terminating decimal to a mixed number in simplest form.

16. 1.40

17. -2.48

18. 11.64

19. -17.75

Change each fraction or mixed number to a decimal.

20. $\frac{5}{8}$

21. $-\frac{5}{11}$

22. $-\frac{21}{50}$

23. $\frac{11}{15}$

24. $5\frac{9}{16}$

25. $16\frac{7}{22}$

26. $11\frac{1}{9}$

27. $28\frac{1}{5}$

28. **MUSIC** Tony and Gretchen played the piano in a recital. Tony's song took 3.74 minutes to play. Gretchen's song took $3\frac{7}{16}$ minutes to play. Whose song took longer to play?

29. **CONSTRUCTION** Heather needs a board that is $7\frac{5}{8}$ feet long. She finds one that is labeled 7.5 feet. Is the board long enough? Explain.

30. **NATURE** Kaya raises bean plants inside in small pots. Each plant needs to be at least 7.4 centimeters tall before she plants it in her garden. Is a plant that has a height of $7\frac{5}{11}$ centimeters tall enough to plant in the garden? Explain.

31. **H.O.T.** Problem Mattie measured the height of the rose bush in her garden four different times. She recorded the heights in the following table.

Measurement	1	2	3	4
Height (cm)	$50\frac{5}{8}$	$50\frac{3}{4}$	$50\frac{7}{8}$	51

Describe the pattern shown in the table. If the pattern continues, what will the rose bush's height be on the seventh measurement?

32. *Writing in Math* Write a word problem in which a decimal is compared with a mixed number. Trade problems with a classmate. Solve your classmate's problem.

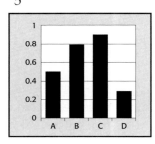
Spiral Review

Solve. (Lesson 3-10)

35. $y - \frac{3}{4} = \frac{1}{2}$ 36. $x + \frac{2}{3} = \frac{5}{6}$ 37. $2t - \frac{3}{5} = \frac{7}{10}$ 38. $3p + \frac{1}{6} = \frac{2}{3}$

39. **RETAIL** The price of clothing after it goes on sale is shown by the equation $\frac{3}{4}p = d$, where p is the original price and d is the discount price. Find the original price of a shirt if the discount price is $15.
(Lesson 3-10)

Identify the property used in each equation. (Lessons 1-4, 1-5, and 1-7)

40. $2(3 + 4) = 2 \times 3 + 2 \times 4$ 41. $3 \times 1 = 3$

4-2 Adding and Subtracting Decimals

Vocabulary

Commutative Property of Addition (p. 245)

Associative Property of Addition (p. 245)

Identity Property of Addition (p. 245)

 Standard 7NS1.2 Add, subtract, multiply, and divide **rational numbers (integers, fractions, and terminating decimals)** and take positive rational numbers to whole-number powers.

Standard 7AF1.3 Simplify numerical expressions by applying the properties of rational numbers (e.g., **identity,** inverse, **distributive, associative, commutative)** and justify the **process used.**

 The What: I will add and subtract decimals.

The Why: Decimals are added and subtracted to calculate the amount of money in a savings account or piggy bank.

Luis gets an allowance of $10.00 a week. This week, he also earned $26.47 working in a neighbor's yard. He then bought a DVD for $19.99. Below is an expression that shows how much money he has left at the end of the week.

allowance		yard earnings		DVD purchase
10.00	**+**	**26.47**	**–**	**19.99**

To find the amount of money he has left, Luis must add and subtract decimals.

ADD DECIMALS

$0.4 + 0.2 = 0.6$

$0.4 = \frac{4}{10}$ $0.2 = \frac{2}{10}$

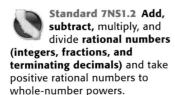

Study Tip

$7 = 7.0 = 7.00 = 7.000$

EXAMPLES Add Decimals

Add.

1 $10.00 + 26.47$

$$\begin{array}{r} 1\,0.0\,0 \\ +2\,6.4\,7 \\ \hline 3\,6.4\,7 \end{array}$$ Align decimal points.

2 $4.791 + 26.23 + 7$

$$\begin{array}{r} {\scriptstyle 1\ 1} \\ 1\,4.7\,9\,1 \\ 2\,6.2\,3\,0 \\ +\ \ 7.0\,0\,0 \\ \hline 3\,8.0\,2\,1 \end{array}$$ Align decimal points.

Add zeros as placeholders.

Your Turn

Add.

a. $79.4 + 6.7$

b. $689.3 + 0.1498 + 4$

Talk Math

WORK WITH A PARTNER Measure the length of two classroom objects to the nearest hundredth of a meter. Write the decimals that represent both lengths.

length = 0.15 m

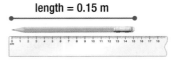

EXAMPLES Subtract Decimals

Subtract.

3 36.47 − 19.99

$$\begin{array}{r} {\scriptstyle 2\ \ 15\ 13\ 17} \\ \cancel{3}\,\cancel{6}.\cancel{4}\,\cancel{7} \quad \text{Align decimal} \\ -1\,9.9\,9 \quad \text{points.} \\ \hline 1\,6.4\,8 \end{array}$$

4 12.9 − 4.687

$$\begin{array}{r} {\scriptstyle 12\ 8\ \ 9\ 10}\quad \text{Add zeros as} \\ \cancel{1}\,2.\cancel{9}\,\cancel{0}\,\cancel{0} \quad \text{placeholders.} \\ -\ \ 4.6\,8\,7 \\ \hline 8.2\,1\,3 \quad \text{Align decimal} \\ \text{points.} \end{array}$$

Your Turn
Subtract.

c. 24.8 − 10.6   d. 48.92 − 10.34   e. 28.7 − 9.245

The amount of money Luis has left is given by the expression 10.00 + 26.47 − 19.99.

$$\begin{array}{r} 10.00 \\ +26.47 \\ \hline 36.47 \end{array} \nearrow \quad \begin{array}{r} 36.47 \\ -19.99 \\ \hline 16.48 \end{array} \quad \text{Luis has \$16.48 left.}$$

Real-World EXAMPLE

5 **FINANCE** Maggie had $142.12 in her bank account. She deposited $190.50. She bought a DVD player for $169.99. How much money did she have left?

$$\begin{array}{lll} \$\,142.12 & \text{starting amount} \\ +190.50 & \text{amount deposited} \\ \hline \$\,332.62 \\ -169.99 & \text{amount deducted} \\ \hline \$\,162.63 & \text{money left} \end{array}$$

First Federal Bank

Deposit $190.50

New
Balance $332.62

Maggie has $162.63 left in her bank account.

Your Turn

f. **FINANCE** Andres had $657.92 in his bank account. He deposited $14.95. Then he paid a bill of $445.29. What is his final balance?

The rules for adding and subtracting positive and negative decimals are the same as for integers.

EXAMPLES Subtract with Negative Decimals

Find each difference.

6 4.63 − 12.99

$$4.63 − 12.99 = 4.63 + (−12.99)$$ Rewrite as a sum using the additive inverse of 12.99.

$$= −8.36$$ Add 4.63 and −12.99.

7 −9.16 − (−1.18)

$$−9.16 − (−1.18) = −9.16 + 1.18$$ Rewrite as a sum using the additive inverse of −1.18.

$$= −7.98$$ Add −9.16 and 1.18.

Your Turn
Find each difference.

g. −8.7 − 34.9

h. 23.45 − (−45.62)

The commutative, associative, and identity properties also apply to decimals. These properties can be used to simplify expressions that include decimals.

EXAMPLES Use Properties with Decimals

Simplify each expression. Give a reason for each step.

8 $1.5x + (2.3 + 4.4x)$

$$1.5x + (2.3 + 4.4x) = 1.5x + (4.4x + 2.3)$$ Commutative Property of Addition

$$= (1.5x + 4.4x) + 2.3$$ Associative Property of Addition

$$= 5.9x + 2.3$$ Add 1.5 and 4.4.

9 $(2.1 + 6.8) − 5.8$

$$(2.1 + 6.8) − 5.8 = 2.1 + (6.8 − 5.8)$$ Associative Property

$$= 2.1 + 1$$ Subtract 5.8 from 6.8.

$$= 3.1$$ Add 2.1 and 1.

Your Turn
Simplify each expression. Give a reason for each step.

i. $1.2z + (3.4 − 4.5z)$

j. $4.6r + (7.9 + 6.6r)$

Examples 8–9
(page 245)

VOCABULARY Name the property used in each equation.

Vocabulary Review
Commutative Property of Addition
Associative Property of Addition
Identity Property of Addition

1. $2.5 + 0 = 2.5$
2. $2.4 + (1.2 - 5.6) = (2.4 + 1.2) - 5.6$
3. $3.8 + 9.1 = 9.1 + 3.8$

Examples 1–2
(page 243)

Add.

4. $7.35 + 19.2$
5. $12.9 + 61 + 0.123$

Examples 3–4
(page 244)

Subtract.

6. $7.40 - 5.23$
7. $21.4 - 18.672$

Example 5
(page 244)

8. **FINANCE** Matt saved $24.76. He earned $25.00 baby-sitting. He spent $21.99 on a new DVD. How much money does he have now?

9. **GEOGRAPHY** From a map, Tavio found two ways to travel between two cities. The first route is on flat land and is 145.8 kilometers long. The second route goes over mountains and is 167.3 kilometers long. How much longer is the mountain route?

Examples 6–7
(page 245)

Find each difference.

10. $6.7 - 9.9$
11. $17.8 - 23.16$

Examples 8–9
(page 245)

Simplify each expression. Give a reason for each step.

12. $4.8f + (7.6 + 5.3f)$
13. $3.67x + (6.78 - 5.87x)$
14. $99x + 21x - 0$

Skills, Concepts, and Problem Solving

HOMEWORK HELP	
For Exercises	**See Example(s)**
15–18	1–2
19–22	3–4
23–26	5
27–30	6–7
31–36	8–9

Add.

15. $56.8 + 97.66$
16. $12.3 + 9.824$
17. $5.4 + 10 + 13.57$
18. $1.23 + 4.5 + 7.983$

Subtract.

19. $23.19 - 7.63$
20. $14.79 - 10.22$
21. $12 - 4.98$
22. $59.4 - 34$

23. **HOBBIES** Sharika makes picture frames. Each side of a frame is 14.5 inches long. She cuts one of these sides from a piece of wood 59.8 inches long. How much wood is left?

24. **COOKING** Dina is using a recipe that calls for 2.5 cups of flour, 1.25 cups of sugar, and 0.375 cup of oil. How many cups of ingredients is she using?

25. **RETAIL** Rohan had $38.29 in his wallet when he got to the shopping mall. When he left, he had $7.41. How much money did Rohan spend at the mall?

26. **CHEMISTRY** Peta is studying an effect of dissolving salt with water. Before adding the salt, the temperature of the water was 36.7 degrees Celsius. After the salt dissolved, the temperature was 33.8 degrees Celsius. How much did the temperature change?

Find each difference.

27. $5.9 - 13.6$

28. $12.9 - 14.7$

29. $-15.2 - 20$

30. $9.1 - (-40.06)$

Simplify each expression. Give a reason for each step.

31. $14.5k + (5.4 + 4.5k)$

32. $9.8b + (17.1 + 6.3b)$

33. $8.0n + (12.3 - 7.8n)$

34. $(6.92 + 0.82w) - 6.03w$

35. $(4.2m + 6.4p) - (4.2p + 6.7m)$

36. $5.61n + (6.82 - 7.34n)$

37. *Writing in Math* Explain the importance of using zeros as placeholders in decimal subtraction problems. Use the problem $42 - 34.8$ as an example.

38. **H.O.T.** Problem Blanca, Dylan, and Jonah came up with three different answers to the following problem.

$$8 + (-8) + a = \,?$$

Blanca said the answer is 0. Dylan gave an answer of 16. Jonah's answer is a. Which student is correct? Explain your answer.

STANDARDS PRACTICE

Use the figure to solve.

39 Ms. Garcia's checkbook shows the entries at right. What is her final balance?

Date	Transaction	Debit (-)	Credit (+)	Balance
				$1425.98
8/12	Morgan Oil	$235.43		
8/14	Deposit		$500.00	
8/14	Farmer Foods	$58.67		
8/16	Hillside Electric	$172.54		
8/18	Jones Plumbing	$234.67		

A $224.67

B $1,224.67

C $1,627.29

D $2,627.29

Spiral Review

Change each fraction to a decimal. (Lesson 4-1)

40. $\dfrac{3}{10}$

41. $\dfrac{11}{50}$

42. $\dfrac{7}{8}$

43. $\dfrac{9}{20}$

Add, subtract, multiply, or divide. Express the answer as a fraction or mixed number in simplest form. (Lessons 3-5, 3-6, 3-7, 3-8, and 3-9)

44. $\dfrac{2}{5} + \dfrac{4}{5}$

45. $5\dfrac{7}{10} - 2\dfrac{3}{4}$

46. $\dfrac{4}{9} \times \dfrac{3}{8}$

47. $5\dfrac{8}{9} \div \dfrac{1}{3}$

48. **BAKING** Doria made $2\dfrac{2}{3}$ dozen granola bars. She gave Alvar half of them. How many bars does Alvar have? (Lesson 3-5)

Math Lab
The Whole Thing

Materials

unifix or centimeter cubes (5 green, 5 red, 5 purple)

5 index cards, labeled 0.1 through 0.5

Standard 7NS1.2 Add, subtract, multiply, and divide rational numbers (integers, fractions, and **terminating decimals**) and take positive rational numbers to whole number powers.

Standard 7MR2.5 Use a variety of methods, such as words, numbers, symbols, charts, graphs, tables, diagrams, **and models, to explain mathematical reasoning.**

 The What: I will learn to add decimals to make a whole.

The Why: Adding decimals helps you know how much change you will receive from a dollar.

Models can help you understand decimals. For example, a row of cubes is a model for a whole made of several parts. If 10 cubes make one whole, each cube is 0.1 of the row. For this activity, you will make whole rows of different colors of cubes.

part, or 0.1

whole, or 1

In this row, 0.5 is green, 0.2 is red, and 0.3 is purple. The equation $0.5 + 0.2 + 0.3 = 1$ represents the model.

ACTIVITY   Part 1: Model a Whole

Make a row of 10 cubes. Make 0.3 of the row green and 0.4 of the row red.

1 **What decimal represents the number of purple cubes needed to complete the row?**

Three purple cubes are needed to make a row of 10. Purple cubes make up 0.3 of the row.

2 **What equation represents the final model?**

$$0.3 + 0.4 + 0.3 = 1$$

Your Turn **Make a row of 10 cubes. Make 0.2 of the row green and 0.4 of the row purple.**

a. What decimal represents the number of red cubes needed to complete the row?

b. What equation represents the final model?

> **Remember!**
> The sum of the decimal parts equals 1 whole, not 10 cubes.

ACTIVITY Part 2: Model a Whole

Make a row of 10 cubes. Take turns drawing index cards to determine what portion of the row will be green and what portion will be red. The first card drawn will represent green, and the second will represent red.

3 What decimal represents the number of purple cubes needed to complete the row?

0.1

0.5

Four purple cubes are needed to make a row of 10.
Purple cubes make up 0.4 of the row.

4 What equation represents the final model?

$$0.1 + 0.5 + 0.4 = 1$$

Your Turn

Make a row of 10 cubes. Draw index cards to determine what portion of the row will be green and what portion will be red. Each partner will draw an index card. The first card drawn will represent green, and the second will represent red.

c. What decimal represents the number of purple cubes needed to complete the row?

d. What equation represents the final model?

Analyze the Results

In a row of 10 cubes, five cubes are purple, two cubes are green.

1. What decimal represents the number of red cubes needed to complete the row?

2. What equation represents this row of cubes?

3. Nickels are worth \$0.05. Dimes are worth \$0.10. Quarters are worth \$0.25. Write three different equations of coin combinations that total \$1.00.

Progress Check 1
(Lessons 4-1 and 4-2)

▷ Vocabulary and Concept Check

Associative Property of Addition (p. 245)	fraction (p. 236)
Commutative Property of Addition (p. 245)	Identity Property of Addition (p. 245)
decimal (p. 236)	repeating decimal (p. 237)
decimal point (p. 236)	terminating decimal (p. 237)

Write the term that *best* completes each sentence. (Lessons 4-1 and 4-2)

1. The equation $64.5 + 0 = 64.5$ is an example of the _____?_____.

2. The number 3.44 . . . is an example of a(n) _____?_____.

3. A(n) _____?_____ is a part of a mixed number.

4. The equation $-0.3 + 5.2 = 5.2 + (-0.3)$ is an example of _____?_____.

5. The name for the period in a decimal is a(n) _____?_____.

▷ Skills Check

Change each terminating decimal to a fraction or mixed number in simplest form. (Lesson 4-1)

6. 0.24   7. 1.37   8. −4.375

Change each fraction or mixed number to a decimal. (Lesson 4-1)

9. $-\dfrac{5}{8}$ 10. $2\dfrac{1}{2}$ 11. $12\dfrac{3}{50}$

Add or subtract. (Lesson 4-2)

12. $23.8 - 12.5$ 13. $4.5 + 3.024$ 14. $12 - 3.56$

Simplify each expression. Give a reason for each step. (Lesson 4-2)

15. $12.9k + 6.7 - 2.3k$ 16. $1.24b + 2.43 + 4.87b$

17. $6.7m + 7.8p - 1.9p + 2.6m$ 18. $4.2n + 7.0w - 7.3w + 6.6n$

▷ Problem-Solving Check

19. **TRAVEL** Rigo drives the same way to work every day. He drives 16.34 miles one way. Next week his usual route will be closed. The detour route is 20.71 miles long. How much farther must he drive using the detour? (Lesson 4-2)

20. **REFLECT** Describe a situation in which it is easier to use a fraction. Describe a situation in which it is easier to use a decimal. (Lesson 4-1)

Multiplying Decimals

Vocabulary

factor (p. 251)

Commutative Property of Multiplication (p. 253)

Associative Property of Multiplication (p. 253)

Identity Property of Multiplication (p. 253)

Multiplicative Property of Zero (p. 253)

Distributive Property (p. 253)

 Standard 7NS1.2 Add, subtract, **multiply,** and divide **rational numbers (integers, fractions, and terminating decimals) and take positive rational numbers to whole-number powers.**

Standard 7NS1.3 Convert fractions to decimals and percents and **use these representations in estimations, computations, and applications.**

Standard 7AF1.3 Simplify numerical expressions by applying the properties of rational numbers (e.g., **identity,** inverse, **distributive, associative, commutative)** and justify the **process used.**

 The What: I will learn to multiply decimals by whole numbers or other decimals.

 The Why: Multiply decimals to calculate weekly earnings.

Jodie volunteers at a food pantry. Last month, four tenths of donations were canned goods. Of donated canned goods, two tenths were canned soup.

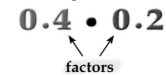

$$0.4 \bullet 0.2$$

factors

$$
\begin{array}{ccccccc}
0.4 & \bullet & 0.2 & = & 0.08 \\
\underbrace{} & & \underbrace{} & & \underbrace{} \\
1 \text{ place} & + & 1 \text{ place} & = & 2 \text{ places}
\end{array}
$$

To multiply numbers in decimal form:

- Count the number of digits to the right of the decimal point in each factor; add these numbers together.

- Place the product's decimal point the same number of places from the right.

To find the amount of all donations that were canned soup, multiply 0.4 by 0.2.

The product $0.4 \bullet 0.2$ can be written as $\frac{4}{10} \bullet \frac{2}{10}$. So, $\frac{8}{100}$ or 0.08 of all donations were canned soup.

MULTIPLYING DECIMALS

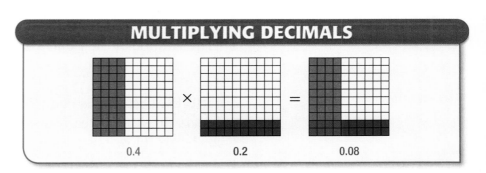

Talk Math

WORK WITH A PARTNER Work together to estimate the product of 2.4 and 3.1. Then multiply the pair of numbers. Compare your exact answer to your estimate.

Study Tip

The signs of the factors tell you the sign of the product. Use these four rules:

$(+) \cdot (+) = (+)$

$(-) \cdot (-) = (+)$

$(+) \cdot (-) = (-)$

$(-) \cdot (+) = (-)$

EXAMPLES Multiply Decimals

Multiply.

① **25 × 12.43**

$$
\begin{array}{r}
12.43 \quad \leftarrow \text{2 places} \\
\times \quad 25 \quad \leftarrow \text{0 places} \\
\hline
6215 \\
24860 \\
\hline
310.75 \quad \leftarrow \text{2 places}
\end{array}
$$

② **−0.23 × 0.044**

$$
\begin{array}{r}
-0.23 \quad \leftarrow \text{2 places} \\
\times \ 0.044 \quad \leftarrow \text{3 places} \\
\hline
92 \\
920 \\
\hline
-0.01012 \quad \leftarrow \text{5 places}
\end{array}
$$

Your Turn

Multiply.

a. 34×6.73 **b.** 1.26×0.321 **c.** $3.2 \times (-5.42)$

Real-World EXAMPLE

❸ **RECREATION** The Wongs have a rectangular pool. Volume is found by multiplying base area and depth. The pool's base area is 28.8 square meters and its depth is 1.8 meters. What is the volume of the pool in cubic meters (m³)?

Area	times	depth	equals	volume
28.8	×	1.8	=	?

Multiply 28.8 and 1.8.

$$
\begin{array}{r}
28.8 \quad \leftarrow \text{1 place} \\
\times \ 1.8 \quad \leftarrow \text{1 place} \\
\hline
2304 \\
2880 \\
\hline
51.84 \quad \leftarrow \text{2 places}
\end{array}
$$

area 28.8 m²

depth 1.8 m

The volume of the pool is 51.84 cubic meters (m³).

Your Turn

d. **TRAVEL** Distance is equal to rate times time traveled. Olivia drove at a steady speed of 54.8 miles per hour for 2.3 hours. How far did she drive?

Remember!

The exponent shows how many times the base is used as a factor.

$3^2 = 3 \times 3$

$4.1^3 = 4.1 \times 4.1 \times 4.1$

EXAMPLE   Use Decimals and Exponents

④ Simplify 1.4^3.

$$1.4^3 = 1.4 \times 1.4 \times 1.4$$
$$= 2.744 \qquad \text{Use the base 1.4 as a factor 3 times.}$$

Your Turn

Simplify.

e. 3.5^2 **f.** 5.61^2 **g.** 2.3^4

The properties that apply to multiplying whole numbers also apply to multiplying decimals.

Commutative Property of Multiplication
$3.1 \times 2.6 = 2.6 \times 3.1$

Associative Property of Multiplication
$(-8.2 \times 3.7) \times 6.5 = -8.2 \times (3.7 \times 6.5)$

Identity Property of Multiplication
$4.3 \times 1 = 4.3$

Multiplicative Property of Zero
$4.3 \times 0 = 0$

Distributive Property
$3.5(4.2 + 5.6) = (3.5 \times 4.2) + (3.5 \times 5.6)$

EXAMPLES   Properties of Multiplication

Simplify each expression. Give a reason for each step.

⑤ $(w \times 1.2) \times 2.4$

$(w \times 1.2) \times 2.4 = w \times (1.2 \times 2.4)$ Associative Property of Multiplication

$= w \times 2.88$ Multiply 1.2 and 2.4.

$= 2.88w$ Commutative Property of Multiplication

⑥ $2.3(t + 5.6)$

$2.3(t + 5.6) = (2.3 \times t) + (2.3 \times 5.6)$ Distributive Property

$= 2.3t + 12.88$ Multiply 2.3 and 5.6.

Your Turn

Simplify each expression. Give a reason for each step.

h. $(r \times 2.6) \times 5.3$ **i.** $(s \times 0) \times 2.1$

j. $3.3(x + 6.1)$ **k.** $8.9(y + 1.3)$

Vocabulary Review
factor
Commutative Property of Multiplication
Associative Property of Multiplication
Identity Property of Multiplication
Multiplicative Property of Zero
Distributive Property

Examples 1–6
(pages 252–253)

VOCABULARY Name the property used in each equation.

1. $4.5(3.6 + 4.8) = (4.5 \times 3.6) + (4.5 \times 4.8)$

2. $5.9 \times 0 = 0$

3. $(4.3 \times 6.5) \times 7.2 = 7.2 \times (4.3 \times 6.5)$

Examples 1–2
(page 252)

Multiply.

4. 56×3.4

5. 1.98×0.017

6. $3.54 \times (-2.3)$

Example 3
(page 252)

7. **RECREATION** Sheryl fills her hot tub at a rate of 200.5 gallons per hour. It takes 2.7 hours to fill the hot tub. How much water does the hot tub hold?

8. **FARMING** What is the total number of kilograms of seeds needed for 68.2 acres of soybeans and 2.3 acres of onions?

Seeds Needed Per Acre		
Seed	Soybeans	Onions
Mass (kg)	27.2	1.4

Example 4
(page 253)

Simplify.

9. 3.6^2

10. 4.52^2

11. 7.1^3

Examples 5–6
(page 253)

Simplify each expression. Give a reason for each step.

12. $(b \times 1) \times 4.5$

13. $(g \times 0) \times 4.4$

14. $8.1(d + 6.2)$

15. $5.2(p + 2.3)$

Skills, Concepts, and Problem Solving

HOMEWORK HELP

For Exercises	See Example(s)
16–21	1–2
22–25	3
26–31	4
32–39	5–6

Multiply.

16. 9×1.89

17. 3.8×90

18. 14×0.046

19. 2.97×1.2

20. -6.2×9.05

21. $14.3 \times (-1.2)$

22. **TRAVEL** Lucy pulled her sister in a wagon along a trail. Lucy pulled at an average rate of 1.8 miles per hour for 2.6 hours. How long was the trail? (distance = rate × time)

23. **FINANCE** In March, Fernando worked 14.5 hours at Gil's Music for $7.85 an hour. In April, Fernando worked 15.6 hours for $8.10 an hour. How much did Fernando earn in March and April?

24. **GARDENS** Sabino planted flowers in a rectangular garden. The area of the garden can be found by multiplying length by width. The garden has a length of 2.3 meters and a width of 1.4 meters. What is the area of the garden in square meters?

25. **COOKING** Holly is baking a ham for a family picnic. The recipe tells her to bake the ham for 16 minutes per pound. The ham weighs 7.5 pounds. How long should she bake the ham?

Simplify.

26. 4.1^2 27. 7.1^2 28. 3.41^2

29. 5.12^2 30. $(-5.6)^3$ 31. 3.3^4

Simplify each expression. Give a reason for each step.

32. $(q \times 1.4) \times 6.7$ 33. $(y \times 1) \times 7.1$

34. $(n \times 0) \times 5.4$ 35. $(m \times 6.7) \times 3.9$

36. $5.8(z + 2.2)$ 37. $4.1(\ell + 4.7)$

38. $6.1(k + 4.2)$ 39. $9.0(f + 7.3)$

40. **H.O.T.** Problem Create two different multiplication problems that have products of 2.86. Use only decimal factors.

41. *Writing in Math* Write a word problem that can be solved by using the expression $(2.3 \times 4.89) + (1.2 \times 2.99)$.

STANDARDS PRACTICE

Refer to the table below for Exercises 42 and 43.

Local Utility Rates	
Utility	**Cost ($)**
Propane	1.69 per gallon
Water	1.52 per 100 gallons
Electricity	0.09 per kilowatt-hour

42 To the nearest penny, how much does Mrs. Bennett pay for 158.9 gallons of propane?

  A $2.42   **C** $268.54

  B $26.85   **D** $524.37

43 Michael's family used 58.7 gallons of propane, 825.3 gallons of water, and 1265 kWH of electricity. How much did they pay for utilities?

  F $118.86   **H** $225.60

  G $182.49   **J** $338.50

Spiral Review

Add or subtract. (Lesson 4-2)

44. $28 + 4.8$ 45. $53.6 + 0.925$ 46. $98.23 - 54.7$ 47. $6.5 - 3.087$

48. **FOOD** Sr. Fernandez bought 2.45 pounds of pears. Mr. Johnson bought $2\frac{2}{5}$ pounds. Who bought more pears? (Lesson 4-1)

Solve each equation. (Lesson 2-7)

49. $5m = 185$ 50. $3x + 6 = 12$ 51. $14p = 182$

Vocabulary

round (p. 258)

 Standard 7NS1.2 Add, subtract, multiply, and **divide rational numbers** (integers, fractions, and **terminating decimals**) and take positive rational numbers to whole-number powers.

Standard 7NS1.3 Convert fractions to decimals and percents and **use these representations in estimations, computations, and applications.**

 The *What*: I will learn how to divide decimals.

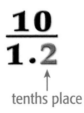 **The** *Why*: A runner's speed or a car's gas mileage can be found by dividing decimals.

Suzie is a sprinter on the track team. The results of a race can be determined by tenths of a second. She ran 10 meters in 1.2 seconds. What was her average speed in meters per second?

To find her speed, Suzie divides the distance by her time.

$$\text{speed} = \frac{\text{distance}}{\text{time}} = \frac{10}{1.2}$$

Suzie renames the fraction so both the numerator and denominator are whole numbers. She identifies the least place value in the fraction. The denominator has a place value of tenths.

$$\frac{\mathbf{10}}{\mathbf{1.2}}$$

↑
tenths place

Remember!

Any number times one equals the original number.

$$\frac{10}{10} = 1$$

Next, Suzie eliminates the decimal place in the fraction. She multiplies the denominator (divisor) and the numerator (dividend) by 10.

$$\frac{10}{1.2} \cdot \frac{10}{10} = \frac{100}{12}$$

Suzie can now divide to find her speed in meters per second.

$$\begin{array}{r} 8.33 \\ 12\overline{)100.00} \\ -96 \\ \hline 40 \\ -36 \\ \hline 40 \\ -36 \\ \hline 4 \end{array}$$

Rewrite the problem and divide.

Place a decimal point in the quotient directly above the decimal point in the dividend.

Divide. Add zeros to the dividend as placeholders.

Divide until there is no remainder or until the quotient can be rounded to the given place value.

The quotient was found to the nearest hundredth.

DIVIDING DECIMALS

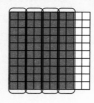

$$0.8 \div 0.2 = 4$$

$$0.8 = \frac{80}{100} \qquad 0.2 = \frac{20}{100}$$

There are 4 groups of 0.2 in 0.8.

EXAMPLES Divide Decimals

Study Tip

In Example 1, the divisor and dividend are multiplied by 10.
In Example 2, each is multiplied by 100.

Divide.

① 24.4 ÷ 6.1

Move each decimal point the same number of places.

$$
\begin{array}{r}
4 \\
6.1\overline{)24.4.} \\
\underline{244} \\
0
\end{array}
$$

The quotient is 4.

② 184.5 ÷ 3.75

Move each decimal point the same number of places. Add zero as a placeholder.

$$
\begin{array}{r}
49.2 \\
3.75\overline{)184.50.0} \\
\underline{1500} \\
3450 \\
\underline{3375} \\
750 \\
\underline{750} \\
0
\end{array}
$$

The quotient is 49.2.

Your Turn

Divide.

a. 50.4 ÷ 7.2

b. 205.6 ÷ 25.7

c. 1.22 ÷ 0.244

Talk Math

WORK WITH A PARTNER Suppose a box holds 10.5 cups of cereal. One cereal serving is 0.75 cup. How do you find the number of cereal servings in the box?

When dividing decimals, you may need to **round** the quotient to a specified decimal place. Suppose you want to find a quotient to the nearest hundredth. Divide to the thousandths place, and then round.

If the thousandths digit is greater than or equal to 5, the hundredths place is rounded up.

$$2.566 \longrightarrow 2.57$$

If the thousandths digit is less than 5, the hundredths place does not round up. It stays the same.

$$2.562 \longrightarrow 2.56$$

Study Tip

Rounded to the nearest tenth:

$6.32 \rightarrow 6.3$ 2 is less than 5.

$6.35 \rightarrow 6.4$ 5 equals 5

$6.38 \rightarrow 6.4$ 8 is greater than 5.

EXAMPLES   Round Quotients

Round each quotient to the nearest tenth.

3 $78.772 \div 9.4 = 8.38$

$$94.\overline{)787.72}$$ 8.38   The digit 8 is greater than 5. Round 3 up to 4.

The quotient to the nearest tenth is 8.4.

4 $16.3 \div 2.5 = 6.52$

$$25.\overline{)163.00}$$ 6.52   The digit 2 is less than 5. Do not round up 5 in the tenths place.

The quotient to the nearest tenth is 6.5.

Your Turn

Round each quotient to the nearest hundredth.

d. $68.912 \div 9.42 = 7.315$ e. $62.94 \div 39.8 = 1.581$

Real-World EXAMPLE

5 SPORTS Suppose that Luke ran 200 meters in 22.9 seconds. What was his average speed in meters per second?

$22.9\overline{)200.}$ Write as a division problem.

$229.\overline{)2000.0}$ Move each decimal point right one place. Add zeros as placeholders.

$\qquad 8.7$
$229.\overline{)2000.0}$ Divide to the tenths place. The number 8.7 rounds to 9.

Luke's average speed was about 9 meters per second.

Your Turn

f. **SWIMMING** It takes 5.3 hours to add 1,200 gallons of water to Chase's swimming pool. What is the rate at which Chase's swimming pool fills to the nearest tenth of a gallon per hour?

Examples 1–5
(pages 257–258)

VOCABULARY

1. You are asked to round a number to the nearest tenth. Explain when you round up the digit in the tenths place and when you do not.

Examples 1–2
(page 257)

Divide.

2. $79.0 \div 15.8$ 3. $352.8 \div 29.4$ 4. $282 \div 23.5$

Examples 3–4
(page 258)

Round each quotient to the nearest tenth.

5. $97.2 \div 21.3 = 4.56$ 6. $21.891 \div 68.4 = 0.32$

Round each quotient to the nearest hundredth.

7. $2.14 \div 5.8 = 0.368$ 8. $450 \div 24.6 = 18.292$

Example 5
(page 258)

9. **NUTRITION** Hallie bought a box of cereal. The box states that each serving is 0.85 ounce and that the box contains 17.0 ounces of cereal. How many servings are in the box?

10. **TRAVEL** Isaac traveled 465.7 miles in 9.25 hours. What was his average speed in miles per hour? Round your answer to the nearest tenth.

Skills, Concepts, and Problem Solving

HOMEWORK HELP	
For Exercises	**See Example(s)**
11–16	1–2
17–24	3–4
25–28	5

Divide.

11. $84.5 \div 6.5$ 12. $277.2 \div 92.4$

13. $422.5 \div 84.5$ 14. $1447.5 \div 96.5$

15. $77.8 \div 3.89$ 16. $215 \div 8.6$

Round each quotient to the nearest tenth.

17. $42.3 \div 32.8 = 1.28$ 18. $56.4 \div 87.9 = 0.64$

19. $68.7 \div 6.79 = 10.11$ 20. $71.2 \div 9.04 = 7.87$

Round each quotient to the nearest hundredth.

21. $807.14 \div 1.3 = 620.876$ 22. $29.6 \div 3.4 = 8.705$

23. $482.5 \div 61.2 = 7.883$ 24. $904 \div 87.2 = 10.366$

TRAVEL Aidan rides the bus 14.3 kilometers to school each day. The trip takes 0.3 hour.

25. In kilometers per hour, what is the average speed of the bus? Round your answer to the nearest tenth.

26. Suppose the bus had an average speed of 58.3 kilometers per hour. How long would the trip take? Round your answer to the nearest hundredth.

27. **INTERIOR DESIGN** Flo covers pillows for sofas. One week she bought a total of 47.75 yards of fabric. Each pillow needs 2.45 yards. How many pillows can she cover?

28. **TRAVEL** Mr. Marc travels each day on business. He made the table to keep track of how far he traveled and how much time it took. On which day was his average speed greater?

Mr. Marc's Business Travel		
Day	Mon	Tues
Miles	434.2	155.3
Hours	8.54	2.90

29. *Writing in Math* Orlando drove at a speed of 75.2 kilometers per hour for 3.4 hours. He says he traveled 255.7 kilometers. His friend Mishka says he traveled 221.2 kilometers. Explain which person was correct, and why.

30. **H.O.T.** Problem Suppose you are asked to solve the following problem.

The Red Cross held a neighborhood blood drive and collected blood from 242 people. The neighborhood has 16 streets. On average, how many people on each street gave blood?

The division does not come out evenly. How do you know the correct place value of the answer?

STANDARDS PRACTICE

Choose the *best* answer.

31 Shelly rolled a toy car down a ramp several times. Each time, she changed the material on the surface of the ramp. She measured and recorded the car's distance and travel time.

Shelly's Data				
Surface	Sandpaper	Waxed paper	Paint	Wood
Distance (cm)	50.6	89.4	80.4	64.2
Time (s)	6.1	7.2	6.7	6.4

On which surface was the car's speed the fastest?

A sandpaper   C paint

B waxed paper   D wood

Spiral Review

Multiply. (Lesson 4-3)

32. 14×3.56 33. $25.3 \times (-15.6)$ 34. -0.94×6.45 35. 0.45×0.03

36. **FINANCE** Sheila starts the month with a balance of $642.40 in her bank account. She deposits $158.90. At the end of the month, her balance is $560.25. How much did she spend that month? (Lesson 4-2)

Find the value of each expression. Show your work. (Lesson 1-3)

37. $4(2 + 3) \times 3(2 - 1)$ 38. $56 \div 8 + 20$ 39. $3^2(5 - 2) + 3(2^3 - 5)$

Progress Check 2
(Lessons 4-3 and 4-4)

▷ Vocabulary and Concept Check

Associative Property of Multiplication (p. 253)	factor (p. 251)
Commutative Property of Multiplication (p. 253)	Identity Property of Multiplication (p. 253)
Distributive Property (p. 253)	Multiplicative Property of Zero (p. 253)
	round (p. 258)

Choose the term that *best* completes each statement. (Lessons 4-3 and 4-4)

1. The equation $-3.21 \times 0 = 0$ is an example of the _____?_____.
2. The equation $29.2 \times 42.3 = 42.3 \times 29.2$ is an example of the _____?_____.
3. The equation $64.5 \times 1 = 64.5$ is an example of the _____?_____.
4. A number multiplied by another number is a(n) _____?_____.

▷ Skills Check

Multiply. (Lesson 4-3)

5. 5.6×7.8 6. $0.423 \times (-56)$ 7. 30.5×6.78

Simplify each expression. Give a reason for each step. (Lesson 4-3)

8. $(p \times 6.2) \times 1.3$ 9. $(u \times 2.3) \times 4.6$
10. $6.6(a + 7.1)$ 11. $15.1(0.7 + x)$

Divide. (Lesson 4-4)

12. $580.5 \div 64.5$ 13. $22.4 \div 3.2$ 14. $5.01 \div 1.67$

Round each quotient to the nearest hundredth. (Lesson 4-4)

15. $24.6 \div 10.7 = 2.299$ 16. $58 \div 12.4 = 4.677$
17. $57.7 \div 0.49 = 117.755$ 18. $0.35 \div 1.69 = 0.207$

▷ Problem-Solving Check

19. **FINANCE** Missy earns $9.50 an hour working in a restaurant. Suppose she works 26.9 hours a week. How much does she earn in a week? (Lesson 4-3)

20. **INDUSTRY** A chemical is made at a rate of 12.34 gallons per hour. To the nearest tenth of an hour, how long will it take to make 607 gallons of the chemical? (Lesson 4-4)

21. **REFLECT** Your friend is absent from class. Explain to your friend how to place the product's decimal point when multiplying two decimals. Use two examples in your explanation. (Lesson 4-3)

Problem-Solving Strategy:
Work Backward

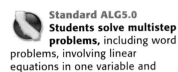

Standard ALG5.0
Students solve multistep problems, including word problems, involving linear equations in one variable and provide justification for each step.

Standard 7NS1.2 Add, subtract, multiply, and divide rational numbers (integers, fractions, and **terminating decimals**) and take positive rational numbers to whole-number powers.

Standard 7MR2.5 Use a variety of methods, such as words, numbers, symbols, charts, graphs, tables, diagrams, and models, **to explain mathematical reasoning.**

Remember!
Use inverse operations. Addition and subtraction undo each other. Multiplication and division undo each other.

Use the Strategy

One month, Courtney made deposits of $26.70, $38.95, and $250.50. She made a withdrawal of $120. At the end of the month, Courtney's account had a balance of $475.93. How much money was in her account at the beginning of the month?

In most problems, you use the information given to find the final answer. In this situation, you will use the final answer and **work backward** to find the original account balance.

Understand **What do you know?**
- She made deposits of $26.70, $38.95, and $250.50.
- She made a withdrawal of $120.
- At the end of the month, her final balance was $475.93.

Plan **What do you want to know? How can you figure it out?**

You want to know how much money Courtney had in her bank account at the beginning of the month. Start with Courtney's final balance. Work backward to find the original number.

Solve **Start with $475.93. Undo each step.**

Courtney has $475.93.

Undo the withdrawal.

$475.93 + $120 = $595.93

Undo the deposits.

$595.93 − $26.70 − $38.95 − $250.50 = $279.78

Courtney had $279.78 at the beginning of the month.

Check **Does your answer make sense?**

Courtney had $279.78 at the beginning of the month. She deposited $26.70, $38.95, and $250.50, giving her $595.93. She withdrew $120, leaving her with $475.93. Your solution makes sense.

Work backward to solve.

1. Angela went shopping for groceries. Her final bill was $29.65. She bought 4.2 pounds of potatoes at $1.95 a pound. She also bought 3.2 pounds of tomatoes at $1.75 a pound and 5.2 pounds of hamburger. What was the cost per pound of the hamburger?

 Understand What do you know?

 Plan What do you want to know? How can you figure it out?

 Solve Work backward. Start with the total cost. Undo each purchase. Calculate the unknown.

 Check Start with your answer, and work forward as a check.

2. *Writing in Math* Explain how the order of operations and working backward are connected.

Solve using the *work backward* strategy.

3. **NUMBER SENSE** Chin picked grapes from his family's vines. He gave half his grapes to Montes. Montes gave a third of the grapes to Myron. Myron gave a fourth of his grapes to David. David has 0.55 pound of grapes. How many pounds of grapes did Chin pick?

4. **FINANCE** After shopping, Julia had the money left shown below. While shopping, she spent $20.45 on a book and $1.69 on graph paper. She ate lunch, which cost $6.99. How much money did she have when she started shopping?

5. **HOBBIES** Willis traded half his baseball cards for two of Gustava's Ryan Howard cards. Then Willis gave Gustava's sister eight of his cards. When Willis left, he had 48 baseball cards. How many cards did Willis have to start with?

Solve using any strategy.

6. **NUMBER SENSE** A certain number is multiplied by 3, and then 5 is added to the result. The final number is 41. Find the number.

7. **FASHION** Antonio has two colors of socks, four colors of shirts, and three colors of pants. How many different outfits are possible?

8. **NUMBER SENSE** The product of two consecutive odd integers is 195. What are the integers?

Decimals in Expressions and Equations

expression (p. 264)

equation (p. 264)

Standard ALG5.0
**Students solve
multistep problems,
including word problems,
involving linear equations in
one variable and provide
justification for each step.**

**Standard 7AF4.1 Solve two-step
linear equations** and inequalities
**in one variable over the rational
numbers,** interpret the solution
or solutions in the context from
which they arose, **and verify the
reasonableness of the results.**

The What: I will learn to simplify expressions and
solve equations that have decimals.

The Why: Simplifying expressions and solving
equations with decimals allow me to
solve problems about rates and money.

Jada bought 3.4 pounds of bananas for $0.59 per pound,
2.5 pounds of pears for $1.29 per pound, and 4.8 pounds of
grapes for $1.69 per pound. How much did Jada spend?

Jada can simplify **expressions** and solve **equations** that
have integers and fractions. She can use the same rules for
expressions and equations that contain decimals.

Jada writes an expression to show the total amount of her
purchase.

cost of bananas	cost of pears	cost of grapes
3.4×0.59 +	2.5×1.29 +	4.8×1.69

To get the total cost of her purchase, Jada simplifies the
expression. She first multiplies all three pairs of factors and
then adds the products.

$3.4 \times 0.59 + 2.5 \times 1.29 + 4.8 \times 1.69$ Multiply the factors.

2.006 + 3.225 + 8.112 Add the products.

13.343

Talk Math

WORK WITH A PARTNER Based on the rounding rules,
what does Jada expect to pay for bananas, pears, and
grapes? Suppose the store does not follow the rounding
rules. What if this store rounds $8.112 to $8.12? Discuss
how this would affect your total bill.

Remember!

Order of operations:
1. grouping symbols and exponents
2. multiply/divide, from left to right
3. add/subtract, from left to right

Estimate the answer to each expression. Then use order of operations to simplify to the nearest hundredth.

❶ $5.8(36.7 - 25.83) \div 14.2$

Estimate.

$5.8(36.7 - 25.83) \div 14.2 \longrightarrow 6(40 - 30) \div 14$

	Round each decimal.
$= 6 \times 10 \div 14$	Subtract 30 from 40.
$= 60 \div 15$	Multiply 6 and 10.
≈ 4	Divide 60 by 15, since 60 and 15 are compatible numbers.

$5.8(36.7 - 25.83) \div 14.2$

$= 5.8(10.87) \div 14.2$	Subtract 25.83 from 36.7.
$= 63.046 \div 14.2$	Multiply 5.8 and 10.87.
≈ 4.44	Divide 63.046 and 14.2.

An estimate of 4 shows that 4.44 is a reasonable answer.

❷ $2.3^2 + 14.1(3.2 - 2.5)$

Estimate.

$2^2 + 14(3 - 2) \longrightarrow 4 + 14 \times 1 \longrightarrow 4 + 14 \longrightarrow 18$

$2.3^2 + 14.1(3.2 - 2.5)$

$= 2.3^2 + 14.1(0.7)$	Subtract 2.5 from 3.2.
$= 5.29 + 14.1(0.7)$	$2.3 \times 2.3 = 5.29$
$= 5.29 + 9.87$	Multiply 14.1 and 0.7.
$= 15.16$	Add 5.29 and 9.87.

An estimate of 18 shows that 15.16 is a reasonable answer.

Your Turn

Estimate the answer to each expression. Then use order of operations to simplify to the nearest hundredth.

a. $2.34(1.4 + 1.5) \div 6.9$ **b.** $1.5^3 - 2.94(0.3 + 0.067)$

3 **RETAIL** Arturo makes jewelry from unpolished gem stones. He bought a 3.4-pound Tiger Eye stone at $6.50 per pound. He bought a 2.1-pound Cat's Eye opal at $7.20 per pound. He also bought a 4.8-pound piece of Tree Agate at $2.70 per pound. How much did Arturo spend? Verify that your answer is reasonable.

Total Cost of Stones

Cost of Tiger Eye	+	Cost of Cat's Eye	+	Cost of Tree Agate	
3.4×6.50	+	2.1×7.20	+	4.8×2.70	Multiply the factors.
22.10	+	15.12	+	12.96	Add the products.
		50.18			

The total cost of the stones was $50.18. This amount is reasonable because the total can be estimated with the expression $(3 \times 7) + (2 \times 7) + (5 \times 3)$. Evaluating this expression gives you 50.

Your Turn

c. **LANDSCAPING** Richard grows vegetables in separate gardens. The tomato garden is 1.4 meters by 2.3 meters. The bean garden is 3.0 meters by 1.9 meters. His pepper garden is 0.5 meter by 7.4 meters. Write and simplify an expression for the total area of Richard's gardens. Verify that your answer is reasonable.

Study Tip

Area of a Rectangle: length × width
Tomato Garden Area: 1.4 × 2.3

EXAMPLE **Simplify Algebraic Expressions**

4 Simplify $3.6b + (9.4c + 4.8b)$. Provide a reason for each step.

$$3.6b + (9.4c + 4.8b) = 3.6b + (4.8b + 9.4c) \quad \text{Commutative Property of Addition}$$

$$= (3.6b + 4.8b) + 9.4c \quad \text{Associative Property of Addition}$$

$$= (3.6 + 4.8)b + 9.4c \quad \text{Distributive Property}$$

$$= 8.4b + 9.4c \quad \text{Order of operations}$$

The expression $8.4b + 9.4c$ consists of unlike terms. It is in simplest form.

Your Turn

Simplify each expression. Provide a reason for each step.

d. $25.4r + 13.8r$ **e.** $6.7g + 2.9 - 4.2g$

Solve each equation. Show your work. Provide a reason for each step.

5 $x - 4.3^2 = 6.23$

$x - 18.49 = 6.23$	Multiply 4.3 by itself.
$x - 18.49 + 18.49 = 6.23 + 18.49$	Add 18.49 to each side.
$x = 24.72$	

6 $2.1z + 54.8 = 59.84$

$2.1z + 54.8 - 54.8 = 59.84 - 54.8$	Subtract 54.8 from each side.
$2.1z = 5.04$	
$2.1z \div 2.1 = 5.04 \div 2.1$	Divide each side by 2.1.
$z = 2.4$	

Your Turn

Solve each equation. Show your work. Provide a reason for each step.

f. $k - 4.54 = 56.7$

g. $6.8c + 45.2 = 86.884$

Real-World EXAMPLE

7 **PACKAGING** Zina needs to send a rectangular piece of art to a friend. The art has an area of 256.28 square centimeters, and its width is 12.6 centimeters. Write and solve an equation to find the length ℓ of the smallest envelope she could use to the nearest tenth of a centimeter.

Area = length × width	
$256.28 = \ell \times 12.6$	
$\dfrac{256.28}{12.6} = \dfrac{\ell \times 12.6}{12.6}$	Divide each side by 12.6.
$20.3 \approx \ell$	

The smallest envelope Zina could use would be about 20.3 centimeters long.

Your Turn

h. **NATURE** Caley is a park ranger. She counts 216 deer living in the state park where she works. In her state, there are about 1.57 deer on every acre a of land. Write and solve an equation to find how many acres there are to the nearest hundredth in the state park.

Examples 1–7
(pages 265–267)

VOCABULARY Match each description with a vocabulary term.

1. contains an equals sign

2. contains terms separated by + and − signs

Examples 1–2
(page 265)

Use order of operations to simplify each expression to the nearest hundredth. Estimate the answer to check.

3. $2.4 - 3.2(6.6 - 4.2) \div 7.1$ 4. $5.1^2 - 3.9 \times 1.2(3.1 - 1.2)$

Example 3
(page 266)

5. **RECREATION** Simon painted squares on his wall. He painted three red squares with sides of 3.6 inches and two yellow squares with sides of 6.8 inches. Write and simplify an expression that shows the total area painted. Verify that your answer is reasonable.

Example 4
(page 266)

Simplify each expression. Provide a reason for each step.

6. $3.8v + 3.4 - 2.98v$ 7. $3.5s + 6.4t - 5.4s$

Examples 5–6
(page 267)

Solve each equation. Show your work. Provide a reason for each step.

8. $2.3u = 12.65$ 9. $1.7j - 4.7 = 2.44$

Example 7
(page 267)

10. **RETAIL** Basillo bought 3.4 pounds of mixed gem stones for $50.15. Write and solve an equation to find the cost per pound for the stones.

Skills, Concepts, and Problem Solving

HOMEWORK HELP

For Exercises	See Example(s)
11–14	1–2
15–16	3
17–20	4
21–24	5–6
25–26	7

Use order of operations to simplify each expression to the nearest hundredth. Estimate the answer to check.

11. $3.4 \times 4.2 + 5.2 \div 2.6$ 12. $3.4 \times 12.9 - 10.4 \times 1.8$

13. $1.2[3.7(2.4 - 1.6) + 5.6]$ 14. $3.9(6.1^2 - 5.4^2 \times 1.1)$

15. **FOOD** Manu bought food for a cookout. He got 4.7 pounds of hamburger at $2.19 per pound and 4.9 pounds of potatoes at $0.79 per pound. Write and simplify an expression that shows the total amount Manu spent. Verify that your answer is reasonable.

16. **SPORTS** Eduardo is a member of a diving team. His score is a product of the sum of the judges' scores and the degree of difficulty. For his first dive, the judges' scores were 6.5, 7.0, 6.5, 5.5, and 7.0 with a 1.5 degree of difficulty. Write and simplify an expression that shows Eduardo's score for his first dive.

Simplify each expression. Provide a reason for each step.

17. $2.3y - 1.5y$ 18. $15.7w + 2w - 8.54w$

19. $4.2q + 3.53q - 7.234$ 20. $4.5p + 3.2t - 3.8p - t$

Solve each equation. Show your work. Provide a reason for each step.

21. $3.4s = 12.58$

22. $7.95m = 11.13$

23. $4.3x - 3.2 = 6.4 - 2.1x$

24. $15.8 - 4.5c = 19.6 - 6.4c$

25. RETAIL Brianna takes inventory at a local hardware store. She measures a stack of table tops. The stack is 45 inches high. Each top is 1.875 inches thick. Write and solve an equation to find how many table tops are in the stack.

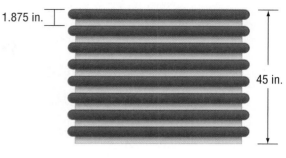

1.875 in.

45 in.

26. COOKING Bruno likes to make his own tomato sauce. He bought 4.5 bushels of tomatoes at a farmer's market. He paid $114.75 for all the tomatoes. Write and solve an equation to find out the cost per bushel for the tomatoes.

27. *Writing in Math* Write a word problem for the following equation. Then solve the equation.

$$\$4.75x + \$13.95 = \$28.20$$

28. H.O.T. Problem Jose and Marianne both solved the following equation.

$$2.5\ell - 1.6 = 1.7\ell + 2.6$$

Jose said the value of ℓ is 5.25. Marianne said that ℓ equals 1.25. Who is correct? Examine the equation and answers. What mistake did the other student make in solving the equation?

STANDARDS PRACTICE

29 The readings on the odometer of Amber's car, at right, show the number of miles when she started her trip and when she came back. How many miles did she drive?

| 52893.6 | 55946.2 |

A 1,057.7

C 3,053.4

B 3,052.6

D 108,839.8

30 Suppose Amber drove for 57.4 hours. What was her average speed?

F 53.18 mi/h

H 55.00 mi/h

G 53.20 mi/h

J 65.00 mi/h

Spiral Review

Solve using the *work backward* strategy. (Lesson 4-5)

31. BIOLOGY In an experiment, bacteria doubles its population every 12.5 hours. After 50.0 hours, there are 1,920 bacteria. How many bacteria were present at the start of the experiment?

Multiply. Write each product in simplest form. (Lesson 3-5)

32. $\frac{4}{5} \times \frac{5}{8}$

33. $3\frac{2}{3} \times \frac{6}{7}$

34. $2\frac{4}{5} \times 1\frac{3}{7}$

Study Guide

Study Tips
Review your notes to:
• note vocabulary words.
• ask questions.
• call out the most important ideas.

Understanding and Using the Vocabulary

After completing the chapter, you should be able to define each term, property, or phrase and give an example of each.

Associative Property of Addition (p. 245)
Associative Property of Multiplication (p. 253)
Commutative Property of Addition (p. 245)
Commutative Property of Multiplication (p. 253)
decimal (p. 236)
decimal point (p. 236)
Distributive Property (p. 253)
equation (p. 264)
expression (p. 264)

factor (p. 251)
fraction (p. 236)
Identity Property of Addition (p. 245)
Identity Property of Multiplication (p. 253)
Multiplicative Property of Zero (p. 253)
order of operations (p. 265)
repeating decimal (p. 237)
round (p. 258)
terminating decimal (p. 237)
work backward (p. 262)

Complete each sentence with the correct mathematical term or phrase.

1. _____?_____ is an example of a problem-solving strategy.

2. A fraction or a _____?_____ is used to represent part of a whole or part of a group.

3. A _____?_____ separates whole units and partial units.

4. A _____?_____ is a decimal with a finite number of digits.

5. A _____?_____ is a decimal that goes on forever in a pattern.

6. In the equation $0.3 \times 0.4 = 0.12$, 0.3 is a _____?_____.

7. The equation $1.5 + (2.5 + 1.6) = (1.5 + 2.5) + 1.6$ is an example of the _____?_____.

8. The equation $3.6 + 0.4 = 0.4 + 3.6$ is an example of the _____?_____.

9. Any terminating or repeating decimal can be written as a _____?_____.

10. The _____?_____ states that multiplications and divisions should be done before additions and subtractions.

Skills and Concepts

Objectives and Examples

LESSON 4-1 pages 236–242

Change fractions to decimals and decimals to fractions.

Change 4.36 to a mixed number in simplest form.

$$4.36 = \left(4\frac{36}{100}\right)$$ Use the place value of the last digit to write 0.36 as a fraction.

$$= 4\frac{36 \div 4}{100 \div 4}$$ Simplify $\frac{36}{100}$ by dividing both numerator and denominator by the GCF.

$$= 4\frac{9}{25}$$

Change $\frac{3}{8}$ to a decimal.

$\frac{3}{8} = 8\overline{)3}$ Rewrite the fraction as a division problem.

$8\overline{)3.000}$ Write the decimal point. Add zeros in the divisor.

$\begin{array}{r} 0.375 \\ 8\overline{)3.000} \end{array}$ Put a decimal point in the quotient above the decimal point in the divisor. Divide until there is no remainder or until you see a pattern.

Review Exercises

Change each terminating decimal to a fraction or a mixed number in simplest form.

11. 0.6   **12.** 0.56

13. 12.625   **14.** 142.006

Change each fraction or mixed number to a decimal.

15. $\frac{3}{4}$ **16.** $5\frac{1}{4}$

17. $\frac{7}{50}$ **18.** $6\frac{15}{16}$

19. **PHOTOGRAPHY** Simone and Lela took pictures at a wedding. Simone took $8\frac{3}{4}$ rolls of film. Lela took 8.375 rolls of film. Both used rolls of film with 36 exposures. Who took more pictures?

LESSON 4-2 pages 243–247

Add and subtract decimals.

$\begin{array}{r} 2.854 \\ + 34.670 \\ \hline 37.524 \end{array}$ Write problem. Align decimal points. Add.

$\begin{array}{r} 20.00 \\ -3.54 \\ \hline 16.46 \end{array}$ Write problem. Add zeros as placeholders. Align decimal points. Subtract.

Add.

20. $2.67 + 6.54$ **21.** $9.6 + 3.7$

22. $0.543 + 1.24$ **23.** $5.432 + 7.1$

Subtract.

24. $4.78 - 3.5$ **25.** $7.423 - 6.5$

26. $4.6 - 2.143$ **27.** $19 - 6.85$

Study Guide

Objectives and Examples

LESSON 4-3   pages 251–255

Multiply decimals.

$$\begin{array}{r} 2.43 \\ \times\ 4.3 \\ \hline 729 \\ 9720 \\ \hline 10.449 \end{array}$$

2.43   2 places right of decimal point
× 4.3   1 place right of decimal point
729   Multiply.
9720
10.449   3 places right of decimal point

LESSON 4-4   pages 256–260

Divide decimals.

$4.5 \div 1.2$

$$\begin{array}{r} 3.75 \\ 1.2.\overline{)4.5.00} \\ 36 \\ \hline 90 \\ 84 \\ \hline 60 \\ 60 \\ \hline 0 \end{array}$$

Rewrite the problem. Move each decimal point the same number of places. Add zeros as placeholders.

Divide 45 by 12.

LESSON 4-6   pages 264–269

Simplify expressions and solve equations that have decimals.

$3.5x + 4.2(2.3x - 1.2)$

$= 3.5x + (4.2 \times 2.3x) - (4.2 \times 1.2)$
                    Distributive Property

$= 3.5x + 9.66x - 5.04$ Order of operations

$= 13.16x - 5.04$ Add like terms.

$5.6y = 25.9$

$5.6y \div 5.6 = 25.9 \div 5.6$ Inverse operations

$y = 4.625$ Divide both sides by 5.6.

Review Exercises

Multiply.

28. 9×6.7 **29.** 8.21×14

30. 4.6×9.03 **31.** 6.42×0.3

32. 14.8×9.32 **33.** 0.042×12.4

Divide. Round each quotient to the nearest hundredth.

34. $9.025 \div 2.5$ **35.** $26.398 \div 6.7$

36. $6.905 \div 4.2$ **37.** $0.469 \div 0.032$

38. $6.8 \div 5.24$ **39.** $4.5 \div 5.78$

40. Kosumi spends $9.43 on grapes. Grapes cost $2.05 per pound. How many pounds of grapes does Kosumi buy?

Simplify each expression.

41. $3.6p + 5.8p$

42. $4.29t - 1.23t$

43. $5.9(5.2r - 3.6) + 6.89r$

44. $2.8s + 1.5(4.6s + 6.81)$

Solve each equation. If necessary, round to the nearest tenth.

45. $7.9z = 2.3$

46. $19.2g = 69.2$

47. $5.4y - 8.7 = 2(y + 4.4)$

48. $5(2.3f - 1.4) = f + 1.5$

Chapter Test

▷ Vocabulary and Concept Check

1. Both fractions and decimals express parts of a whole. Explain how the number of parts in the *whole* is given in a decimal.

▷ Skills Check

Change each fraction or mixed number to a decimal.

2. $\frac{9}{20}$

3. $2\frac{7}{8}$

Change each terminating decimal to a fraction in simplest form.

4. 4.85

5. 34.12

Add or subtract.

6. $14.85 + 6.543$

7. $0.423 + 3.98$

8. $5.98 - 3.8$

Multiply.

9. 4.8×1.23

10. 1.9×2

11. 79.2×0.023

Divide. Round each quotient to the nearest hundredth.

12. $2.3 \div 0.45$

13. $7 \div 2.48$

14. $6.93 \div 2.7$

Simplify each expression.

15. $2.3d + 4.96d$

16. $5.7r - 3.94r$

17. $3.4(5.1q + 1.2r) + 2.34r$

Solve each equation. If necessary, round to the nearest hundredth.

18. $9t(-2.1 \times 6) = -260.82$

19. $0.7p + 6.4 = 6.4p - 0.7$

▷ Problem-Solving Check

20. **BUSINESS** Chelsea makes $6.75 an hour. One week she made $103.95. Write and solve an equation to find the number of hours h she worked.

NUMBER SENSE Damon recycles aluminum cans. He gets cans from his family and friends. His friend Max gave him 45 cans. His friend Marsha gave him three times as many cans as Max. His family gave him the rest of his cans. Now, Damon has 332 cans.

21. How many cans did Damon get from his family?

22. The recycling center paid Damon $28.25 for his cans. At $0.58 per pound, how many pounds of cans did Damon sell? Round your answer to the nearest tenth.

Choose the *best* answer.

1 Marcella has a rectangular swimming pool that is surrounded by a wooden deck. She needs to buy enough paint to cover the area of the deck. (To find the area of a rectangle, multiply length and width.)

Deck

Pool   24.2 m

30.5 m

How many square feet are in the deck if the deck is 3 meters wide?

A 155.1   C 445.9

B 292.2   D 738.1

2 Which property does this equation show?

$5.34x(2.3 + 5.8) = 5.34x \times 2.3 + 5.34x \times 5.8$

F Associative Property

G Commutative Property

H Distributive Property

J Identity Property of Addition

3 What is the difference when you subtract $4.8r$ from $20r$?

A $15.2r$ C $16.8r$

B $16.2r$ D $68r$

4 What is the simplified form of the expression $4.2x + 3.2(1.9x - 4.3)$?

F $-3.48x$ H $10.28x - 4.3$

G $7.4x - 13.76$ J $10.28x - 13.76$

5 Which is the *best* estimate of 24.9×4.12?

A 96   C 120

B 100   D 1000

6 Cari brought all the game tokens she had to Game Gazer. She gave half of them to her friend Heath. Heath gave one-third of these tokens to Neela. Neela has six tokens. How many tokens did Cari bring?

F 1   H 18

G 4   J 36

7 Enrique rode 125.3 kilometers on a bus. The trip took 3 hours 30 minutes. Which of these will give Enrique's average speed in kilometers per hour?

A Divide 125.3 by 3.5.

B Divide 125.3 by 3.3.

C Multiply 125.3 by 3.5.

D Multiply 125.3 by 3.3.

8 Hillary has R roast beef sandwiches in her restaurant. She sells 6 of them and makes $2.5R$ more. Which expression shows how many roast beef sandwiches she now has?

F $1.5R - 6$ H $3.5R - 6$

G $1.5R + 6$ J $3.5R + 6$

PART 2 Short Answer

Record your answers on the answer sheet provided by your teacher or on a separate sheet of paper.

9 Lynn ate $\frac{3}{8}$ of a pizza, and Leigh ate $\frac{1}{2}$. To the nearest hundredth of a pizza, how much more pizza did Leigh eat?

10 The graph below shows the heights in meters of five girls in Mr. Noelle's class.

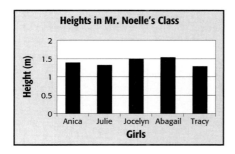

To the nearest tenth of a meter, how much taller is the tallest girl than the shortest one?

11 A number is increased by 5^2 and then divided by 3. The resulting number is 27. What was the original number?

12 Mr. Sanchez's sons are $5\frac{3}{4}$ feet, 5.8 feet, and 60 inches tall. Write their heights in order from shortest to tallest.

13 To the nearest tenth, what is the simplified form of this expression?

$$\frac{6.2x - 2.5x}{2.6 + 6.4 \div 2 \cdot 4.1}$$

PART 3 Extended Response

Record your answers on the answer sheet provided by your teacher or on a separate piece of paper.

14 Mr. Vargas traveled Route A when going to see a friend. Going to his friend's house, he drove 3.4 hours and averaged 52.4 kilometers per hour. Coming home, Mr. Vargas used Route B. He drove 3.8 hours and averaged 49.3 kilometers per hour.

a. Write an equation for the distance he drove on each route. Let D_A be the distance for Route A, and let D_B be the distance for Route B.

b. Which route was longer?

c. How much longer was the longer route?

NEED EXTRA HELP?												
If You Missed Question...	1	2	3	4	5	6	7	8	9, 12	10	11	13	14
Go to Lesson...	4-3	4-3	4-2	4-6	4-3	4-5	4-4	4-6	4-1	4-2	4-5	4-6	4-3
For Help with Algebra Readiness Standard...	7NS1.2	7NS1.2	7NS1.2	7NS1.2	7NS1.2	ALG5.0	ALG5.0	ALG4.0	7NS1.3	7NS1.2	7NS1.2	ALG4.0	7AF4.1

CHAPTER 5

Exponents and Roots

 Here's What I Need to Know

Standard ALG2.0 **Students understand and use such operations as** taking the opposite, finding the reciprocal, **taking a root**, and raising to a fractional power. **They understand and use the rules of exponents.**

Standard 7NS2.1 Understand negative whole-number exponents. Multiply and divide expressions involving exponents with a common base.

Standard 7AF2.1 Interpret positive whole-number powers as repeated multiplication and negative whole-number powers as repeated division or multiplication by the multiplicative inverse. **Simplify and evaluate expressions that include exponents.**

Vocabulary Preview

exponent The number of times a base is used as a factor (pp. 22, 72)

$$2^4 = (2)(2)(2)(2)$$

with "exponent" labeling the superscript 4 and "base" labeling the 2.

square root One of two equal factors of a number (p. 293)

3 is a square root of 9 because $3 \cdot 3 = 9$.

radical sign A symbol that indicates a positive or a nonnegative square root (p. 293)

$$\sqrt{81} = 9$$

with "radical sign" labeling the √ symbol.

 The *What*

I will learn to:

- use exponent rules to simplify expressions.
- use integer exponents and roots to simplify expressions.
- compare and order rational numbers.

The *Why*

Exponents can be used to express large and small numbers, like the population of lions in Serengeti National Park. Roots can be used to calculate the wingspan of a hang glider.

Option 2

Get ready for Chapter 5. Review these skills.

Vocabulary and Skills Review

Match each sentence with the correct vocabulary word.

1. In 2^3, 3 is the ____?____ .
2. In 2^3, 2 is the ____?____ .
3. 3^4 is read 3 raised to the fourth ____?____ .

   A. base
   B. exponent
   C. power
   D. variable

Find the value of each expression.

4. $5 \cdot 7 + 4^2$
5. $13 - 5 + 6 \div 2$
6. $9 \div 3 + 4 \cdot 6$
7. $2(3 + 5) - 2^2$
8. $4[(8 + 10) \div 6]$
9. $\dfrac{27 + 8}{19 - 12}$
10. 10^2
11. 2^5
12. $3^2 + 8 - 9$

Tip
Use an idea web to capture information you read.

Note-taking

Chapter Instruction

Numbers that are expressed using exponents are called **powers.** (p. 278)

To multiply powers that have the same base, add their exponents. (p. 278)

To divide powers that have the same base, subtract their exponents. (p. 280)

To raise a power to a power, multiply the exponents. (p. 281)

My Notes

$4^1 = 4$

$4^5 \cdot 4^6 = 4^{5+6}$

Exponents/Powers

$\dfrac{4^7}{4^2} = 4^{7-2}$

$(4^5)^2 = 4^{5 \cdot 2}$

Vocabulary

powers (p. 278)

Product of Powers (p. 278)

Quotient of Powers (p. 280)

Power of Powers (p. 281)

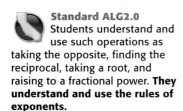 **Standard ALG2.0** Students understand and use such operations as taking the opposite, finding the reciprocal, taking a root, and raising to a fractional power. **They understand and use the rules of exponents.**

Standard 7AF2.1 Interpret positive whole-number powers as repeated multiplication and negative whole-number powers as repeated division or multiplication by the multiplicative inverse. **Simplify and evaluate expressions that include exponents.**

Real-World Challenge·······

A male lion weighs about 450 pounds. A female lion weighs about 300 pounds. Write an algebraic expression to represent the difference in the weights of male and female lions.

The What: I will use the exponent rules to simplify expressions when multiplying powers, dividing powers, and raising powers to powers.

The Why: Powers can be used to calculate large numbers, such as the number of lions in Serengeti National Park.

Numbers expressed using exponents are called **powers.**

$4^1 = 4$ or 4 to the first power is 4.

$4^2 = 16$ or 4 to the second power, or 4 squared, is 16.

$4^3 = 64$ or 4 to the third power, or 4 cubed, is 64.

The expression $4^2 \cdot 4^3$ is an example of multiplying powers with the same base.

$$\underbrace{(4)(4)}_{4^2} \quad \cdot \quad \underbrace{(4)(4)(4)}_{4^3} \quad = 4^5$$

$$4^2 \quad \cdot \quad 4^3 \quad = 4^5$$

·· ·There are about 2^{12} square miles in Serengeti National Park. There are about 2^2 lions per square mile.

To find the number of lions, multiply the number of square miles by the number of lions per square mile.

Number of Lions $= \underbrace{2^{12}}_{\text{square miles}} \quad \cdot \quad \underbrace{2^2}_{\text{lions per square mile}}$

Simplify the expression $2^{12} \cdot 2^2$ by using the **Product of Powers.**

$2^{12} \cdot 2^2 = (2)(2)(2)(2)(2)(2)(2)(2)(2)(2)(2)(2) \cdot (2)(2)$

$= 2^{14}$

PRODUCT OF POWERS		
Numbers	**Words**	**Algebra**
$4^2 \times 4^3 = 4^5$	To multiply powers that have the same base, add their exponents. The base will remain the same.	$x^m \cdot x^n = x^{m+n}$

WORK WITH A PARTNER Name an expression that simplifies to p^7 using the Product of Powers rule. Explain how you created your expression. Listen as your partner names another expression with the same product.

Study Tip
The bases do not change when multiplying exponents. Don't make the mistake of multiplying the bases.

EXAMPLES Multiply Powers

Find each product. Express using exponents.

1 $5^2 \cdot 5^4$

$5^2 \cdot 5^4 = 5^{2+4}$	Use the Product of Powers.
$= 5^6$	Add the exponents 2 and 4.

2 $k^2 \cdot k \cdot k^3$

$k^2 \cdot k \cdot k^3 = k^2 \cdot k^1 \cdot k^3$	Rewrite k as k^1.
$= k^{2+1+3}$	Use the Product of Powers.
$= k^6$	Add the exponents 2, 1, and 3.

Your Turn
Find each product. Express using exponents.

a. $4^2 \times 4^3$ **b.** $p^3 \cdot p^8$ **c.** $h \cdot h^6 \cdot h^5$

Sometimes powers are multiplied by coefficients other than 1.

EXAMPLES Multiply Powers with Coefficients

Find each product. Express using exponents.

3 $(-6y^2)(3y^6)$

$(-6y^2)(3y^6) = (-6 \cdot 3)(y^2 \cdot y^6)$	Use the Commutative and Associative Properties.
$= -18(y^2 \cdot y^6)$	Multiply the coefficients.
$= -18y^{2+6}$	Use the Product of Powers.
$= -18y^8$	Add the exponents 2 and 6.

4 $10x^3y(-2xy^2)$

$10x^3y(-2xy^2) = (10 \cdot -2)(x^3 \cdot x)(y \cdot y^2)$	Use the Commutative and Associative Properties.
$= (-20)(x^3 \cdot x)(y \cdot y^2)$	Multiply 10 and −2.
$= (-20)(x^{3+1})(y^{1+2})$	Use the Product of Powers.
$= -20x^4y^3$	Add the exponents.

Your Turn
Find each product. Express using exponents.

d. $(2b^7)(6b^{20})$ **e.** $(5y^2z^3)(-2y^6z)$

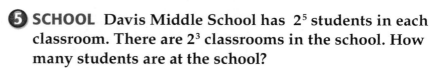

⑤ SCHOOL Davis Middle School has 2^5 students in each classroom. There are 2^3 classrooms in the school. How many students are at the school?

$2^5 \cdot 2^3 = 2^{5+3}$ Use the Product of Powers.

         $= 2^8$ Add the exponents.

Your Turn

f. HOUSES Brady's neighborhood has 3^4 blocks. There are 3^2 houses in each block. How many houses are in Brady's neighborhood?

In 2005, San Francisco had a population of more than 700,000 people.

The expression $8^5 \div 8^2 = 8^3$ is an example of dividing powers with the same base.

$$\frac{8^5}{8^2} = \frac{(\cancel{8} \cdot \cancel{8} \cdot 8 \cdot 8 \cdot 8)}{(\cancel{8} \cdot \cancel{8})} = 8^3$$

QUOTIENT OF POWERS

Numbers	Words	Algebra
$4^9 \div 4^2 = 4^7$	To divide powers that have the same base, subtract their exponents. The base will remain the same.	$x^m \div x^n = x^{m-n}$

EXAMPLES Divide Powers

Find each quotient. Express using exponents.

⑥ $\dfrac{3^5}{3^2}$

$\dfrac{3^5}{3^2} = 3^{5-2}$ Use the Quotient of Powers.

     $= 3^3$ Subtract the exponents.

⑦ $\dfrac{24h^7d^3}{4h^2d}$

$\dfrac{24h^7d^3}{4h^2d} = 6h^{7-2}d^{3-1}$ Divide 24 by 4.

                 Use the Quotient of Powers.

         $= 6h^5d^2$ Subtract the exponents.

Your Turn

Find each quotient. Express using exponents.

g. $\dfrac{10^{13}}{10^8}$ **h.** $\dfrac{a^7}{a^6}$ **i.** $\dfrac{12w^{18}z^{12}}{4w^9z^{12}}$

8 **WILDLIFE** Serengeti National Park has an area of about 2^{12} square miles. There are approximately 2^{17} gazelles in the park. Find the average number of gazelles in each square mile.

To find the average number of gazelles in each square mile, divide the number of gazelles by the park's area. Use the Quotient of Powers.

gazelles ⟶ $\dfrac{2^{17}}{2^{12}} = 2^{17-12}$ ⟵ gazelles per square mile
square miles ⟶

$= 2^5$ Subtract 12 from 17.

There are 2^5 gazelles in each square mile.

Your Turn

j. POPULATION About 5^7 students live in Kenton. It has an area of 5^4 square miles. Find the average number of students in each square mile.

WILDLIFE Grant's gazelles live in Serengeti National Park, Africa.

The expression $(7^2)^4$ is an example of raising a power to a power.

$$(7^2)^4 = 7^2 \cdot 7^2 \cdot 7^2 \cdot 7^2 = 7^{2 \times 4} = 7^8$$

POWER OF POWERS		
Numbers	**Words**	**Algebra**
$(4^2)^3 = 4^6$	To raise a power to a power, multiply the exponents.	$(x^m)^n = x^{m \cdot n}$

EXAMPLES Raise a Power to a Power

Simplify.

9 $(4^2)^3$

$(4^2)^3 = 4^{2 \times 3}$ Use the Power of Powers.

$= 4^6$ Multiply the exponents 2 and 3.

10 $(3fg^5)^4$

$(3fg^5)^4 = 3^4 f^{1 \cdot 4} g^{5 \cdot 4}$ Use the Power of Powers.

$= 3^4 f^4 g^{20}$ Multiply the exponents 1 and 4.
Multiply the exponents 5 and 4.

$= 81 f^4 g^{20}$ Find 3^4. $3 \times 3 \times 3 \times 3 = 81$

Your Turn
Simplify.

k. $(4^9)^{12}$ **l.** $(w^7)^{11}$ **m.** $(6r^5 s^2)^2$

Examples 1–7
(pages 279–280)

VOCABULARY Match each equation to the appropriate vocabulary term.

1. $x^9 \cdot x^4 = x^{13}$

2. $(x^4)^6 = x^{24}$

3. $\dfrac{x^7}{x^4} = x^3$

Find each product. Express using exponents.

Examples 1–2
(page 279)

4. $2^5 \times 2^3$

5. $t^2 \cdot t^9$

6. $r^4 \cdot r \cdot r^7$

Examples 3–4
(page 279)

7. $(10g^4)(-7g^5)$

8. $(3h^7k)(3h^3k^8)$

9. $3t^2 \cdot 4t^9$

Example 5
(page 280)

10. **FITNESS** At a high school, 10^2 students walked to raise money for a local charity. Each student walked 10^4 meters. How many total meters did the students walk?

Examples 6–7
(page 280)

Find each quotient. Express using exponents.

11. $\dfrac{2^{12}}{2^6}$

12. $\dfrac{56d^{14}}{7d^{11}}$

13. $\dfrac{36r^6s^{10}}{4r^4s^3}$

Example 8
(page 281)

14. **POPULATION** The population of the People's Republic of China is about 10^9 people. There are about 10^7 million square kilometers of land. How many people on average are in each square kilometer?

Examples 9–10
(page 281)

Simplify.

15. $(4^2)^6$

16. $(4d^5)^2$

17. $(3f^2g^4)^3$

18. **Talk Math** Explain to a classmate why any number other than zero to the zero power is 1.

Skills, Concepts, and Problem Solving

HOMEWORK HELP

For Exercises	See Example(s)
19–24	1–2
25–30	3–4
31–32	5
33–38	6–7
39–40	8
41–46	9–10

Find each product. Express using exponents.

19. $3^4 \times 3^6$

20. $5^2 \times 5^7$

21. $z^6 \cdot z^8$

22. $n^{14} \cdot n^5$

23. $d^3 \cdot d \cdot d^5$

24. $s^2 \cdot s^4 \cdot s$

25. $(3z^2)(5z^{13})$

26. $(-6p^{16})(4p^3)$

27. $(3d^8f)(5d^9f)$

28. $(-2gh^6)(-7g^4h)$

29. $(x^3y^4)(3xy^5)$

30. $(-5p^2q)(4p^5q^7)$

31. **CONSTRUCTION** A homeowner is replacing the windows in her home. There are 3^2 panes per window. There are 3^3 windows in the home. How many panes are in the home?

32. **FOOD SERVICE** In a serving of chicken fingers, there are 2^3 pieces of chicken. A restaurant sold 2^8 servings of chicken fingers. How many pieces of chicken were sold?

Find each quotient. Express using exponents.

33. $\dfrac{8^4}{8^3}$

34. $\dfrac{12^6}{12^2}$

35. $\dfrac{c^{19}}{c^{11}}$

36. $\dfrac{32w^{24}}{4w^8}$

37. $\dfrac{-6h^{14}}{3h^3}$

38. $\dfrac{18j^6s^9}{2j^3s}$

39. **RECREATION** Pleasant Place has 4^6 campers each summer. There are 4^2 camp sessions. How many campers are in each session?

40. **INDUSTRY** A company makes 12^6 pencils. They put 12^2 pencils in each case. How many cases of pencils do they have to sell?

Simplify.

41. $(5^4)^3$

42. $(2^9)^2$

43. $(-2d^5)^4$

44. $(-3m^6)^3$

45. $(r^6s)^8$

46. $(2t^3u)^6$

47. **H.O.T.** Problem How are these expressions alike and different?

$$\dfrac{x^{15}}{x^3} \qquad x^8 \cdot x^4 \qquad (x^3)^4$$

48. *Writing in Math* Yang said $3^2 \times 3^4$ was equal to 3^6. Trevor said it was equal to 3^8. Who was correct? Explain your thinking.

STANDARDS PRACTICE

49 Choose the *best* answer.

Sound	Intensity
Normal Breathing	10^1
Normal Conversation	10^6
Subway	10^{11}
Jet Engine Taking Off	10^{15}

How many more times intense is the sound of a jet engine than normal conversation?

A 9

C 10^9

B 90

D 10^{21}

50 Which expression is equivalent to $(3x^5)^4(-2x^6)$?

F $79x^{26}$

G $-162x^{26}$

H $-162x^{120}$

J $5182x^{26}$

51 $\dfrac{10x^6}{5x^3} =$

A $2x^2$

C $2x^9$

B $2x^3$

D $2x^{18}$

Spiral Review

Solve. (Lesson 4-6)

52. $3.5x + 1.2 = 18.7$

53. $\dfrac{x}{2.6} = 13.4$

54. **MONEY** Five friends share the cost of lunch equally. Each person pays $6. The total bill is $33.00. How much money is the group short? (Lesson 2-7)

Find the value of each expression. (Lesson 1-3)

55. $7[(12 + 5) - 3(19 - 14)]$

56. $\dfrac{37 + 38}{30 - 5}$

Progress Check 1

(Lesson 5-1)

▷ Vocabulary and Concept Check

Power of Powers (p. 281)
powers (p. 278)
Product of Powers (p. 278)
Quotient of Powers (p. 280)

Give an example of each term. (Lesson 5-1)

1. Product of Powers

2. Quotient of Powers

3. Power of Powers

▷ Skills Check

Find each product. Express using exponents. (Lesson 5-1)

4. $(d^8)(d^{12})$

5. $(3r^3)(4r^5)$

6. $(5w^9x^2)(-6wx^4)$

Find each quotient. Express using exponents. (Lesson 5-1)

7. $\dfrac{e^{13}}{e^8}$

8. $\dfrac{6r^{10}}{2r^2}$

9. $\dfrac{36q^{14}p^{12}}{12q^8p^5}$

Simplify. (Lesson 5-1)

10. $(c^8)^4$

11. $(2r^3)^2$

12. $(2y^4k^9)^5$

▷ Problem-Solving Check

13. **FITNESS** The number of adults who participated in a charity run was 10^3. Each participant ran 10^4 meters. How many meters did they run altogether? (Lesson 5-1)

14. **POPULATION** The population of Los Angeles county is about 10^7 people. The land area of the county is about 10^4 square kilometers. What is the number of people per square kilometer? (Lesson 5-1)

15. **REFLECT** Explain the process you would use to simplify the expression $\dfrac{a^{10} \cdot a^3}{(a^2)^5}$. (Lesson 5-1)

5-2 Integer Exponents

Vocabulary

negative exponent
(p. 285)

Standard 7NS2.1 Understand negative whole-number exponents. Multiply and divide expressions involving exponents with a common base.

Standard 7AF2.1 Interpret positive whole-number powers as repeated multiplication and negative whole-number powers as repeated division or multiplication by the multiplicative inverse. **Simplify and evaluate expressions that include exponents.**

Remember!

Any value not equal to zero raised to the 0 power equals 1.

$3^0 = 1$ and $x^0 = 1$

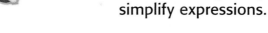

The What: I will use integer exponents to simplify expressions.

The Why: The size of a red blood cell can be expressed using a power with a negative exponent.

A common flea is 2^{-4} inch long. This length is written using a **negative exponent.** Negative exponents are used to express numbers less than one.

Patterns of Powers

$2^4 = 16$ $\quad 2^3 = 8$ $\quad 2^2 = 4$ $\quad 2^1 = 2$ $\quad 2^0 = 1$

$2^{-4} = \frac{1}{16}$ $\quad 2^{-3} = \frac{1}{8}$ $\quad 2^{-2} = \frac{1}{4}$ $\quad 2^{-1} = \frac{1}{2}$

Quotient of Powers

$$\frac{x^2}{x^6} = x^{2-6}$$
$$= x^{-4}$$

Definition of Powers

$$\frac{x^2}{x^6} = \frac{\overset{1}{\cancel{x}} \cdot \overset{1}{\cancel{x}}}{\underset{1}{\cancel{x}} \cdot \underset{1}{\cancel{x}} \cdot x \cdot x \cdot x \cdot x}$$
$$= \frac{1}{x \cdot x \cdot x \cdot x}$$
$$= \frac{1}{x^4}$$

Evaluate the expression for the length of a common flea.

$$\mathbf{2^{-4} = \frac{1}{2^4} = \frac{1}{2 \cdot 2 \cdot 2 \cdot 2} = \frac{1}{16}}$$

The common flea is $\frac{1}{16}$ inch long.

EXAMPLE Negative Exponents

Rewrite using positive exponents, then evaluate.

1 5^{-3}

$5^{-3} = \frac{1}{5^3}$ Rewrite using a positive exponent.

$= \frac{1}{125}$ Find 5^3. $5 \times 5 \times 5 = 125$

Your Turn

Rewrite using positive exponents, then evaluate.

a. 4^{-2}

b. 10^{-4}

Rewrite using positive exponents. Assume that no denominators are equal to 0.

2 b^{-6}

$b^{-6} = \dfrac{1}{b^6}$ Rewrite using a positive exponent.

3 $c^{-9}d^3$

$c^{-9}d^3 = \dfrac{1}{c^9} \cdot d^3$ Rewrite c^{-9} using a positive exponent.

$= \dfrac{d^3}{c^9}$

4 $7x^{-4}y^2$

$7x^{-4}y^2 = 7 \cdot \dfrac{1}{x^4} \cdot y^2$ Rewrite x^{-4} using a positive exponent.

$= \dfrac{7y^2}{x^4}$

Your Turn

Rewrite using positive exponents. Assume that no denominators are equal to 0.

c. t^{-12} **d.** $5g^{-14}$ **e.** $f^{-8}g^2$

Talk Math

WORK WITH A PARTNER Name an expression with a negative exponent. Have a partner write an equivalent expression with only positive exponents.

The exponent rules you studied in Lesson 5-1 also apply to negative exponents.

EXPONENT RULES

Product of Powers		
Numbers	**Words**	**Algebra**
$4^{-2} \times 4^3 = 4^1$ $4^2 \times 4^{-3} = 4^{-1}$	To multiply powers that have the same base, add their exponents. The base will remain the same.	$x^m \cdot x^n = x^{m+n}$
Quotient of Powers		
Numbers	**Words**	**Algebra**
$4^{-9} \div 4^2 = 4^{-11}$ $4^9 \div 4^{-2} = 4^{11}$	To divide powers that have the same base (where the base is not equal to 0), subtract their exponents. The base will remain the same.	$x^m \div x^n = x^{m-n}$
Power of Powers		
Numbers	**Words**	**Algebra**
$(4^{-2})^3 = 4^{-6}$ $(4^2)^{-3} = 4^{-6}$	To raise a power to a power (where the base and the exponents are not equal to 0), multiply the exponents.	$(x^m)^n = x^{m \cdot n}$

EXAMPLES Product of Powers

Find each product. Answers should have only positive exponents. Assume that no denominators are equal to 0.

5 $(a^3)(a^{-7})$

$(a^3)(a^{-7}) = a^{3 + (-7)}$ Use the Product of Powers.

$\qquad = a^{-4}$ Add the exponents 3 and −7.

$\qquad = \dfrac{1}{a^4}$ Rewrite using a positive exponent.

6 $(3x^2y^{-4})(-5x^7y^8)$

$(3x^2y^{-4})(-5x^7y^8) = (3 \cdot -5)(x^2 \cdot x^7)(y^{-4} \cdot y^8)$

$\qquad\qquad\qquad\qquad\qquad\qquad$ Use the Commutative and
$\qquad\qquad\qquad\qquad\qquad\qquad$ Associative Properties.

$\qquad = (-15)(x^2 \cdot x^7)(y^{-4} \cdot y^8)$ Multiply the coefficients 3 and −5.

$\qquad = (-15)x^{2+7}y^{-4+8}$ Use the Product of Powers.

$\qquad = -15x^9y^4$ Add the exponents.

Your Turn

Find each product. Answers should have only positive exponents. Assume that no denominators are equal to 0.

f. $(r^9)(r^{-15})$

g. $(-4x^6y^{-7})(7x^{-2}y^{12})$

EXAMPLES Quotient of Powers

Find each quotient. Answers should have only positive exponents. Assume that no denominators are equal to 0.

7 $\dfrac{h^6}{h^{-10}}$

$\dfrac{h^6}{h^{-10}} = h^{6-(-10)}$ Use the Quotient of Powers.

$\qquad = h^{6+10}$ Rewrite the exponent $6 - (-10)$.

$\qquad = h^{16}$ Add the exponents 6 and 10.

8 $\dfrac{20u^2r^{10}}{4u^5r^9}$

$\dfrac{20u^2r^{10}}{4u^5r^9} = 5u^{2-5}r^{10-9}$ Divide the coefficients 20 and 4.
$\qquad\qquad\qquad\qquad\qquad$ Use the Quotient of Powers.

$\qquad = 5u^{-3}r^1$ Subtract the exponents.

$\qquad = 5u^{-3}r$ Rewrite r^1 as r.

$\qquad = \dfrac{5r}{u^3}$ Rewrite u^{-3} using a positive exponent.

Your Turn

Find each quotient. Answers should have only positive exponents. Assume that no denominators are equal to 0.

h. $\dfrac{m^2}{m^5}$

i. $\dfrac{30c^9d^2}{10c^{-3}d^7}$

Remember!

Subtracting a negative number is the same as adding a positive number.

$6 - (-10) = 6 + 10$

ASTRONOMY The Griffith Observatory in Los Angeles is ranked as one of the top tourist attractions of southern California. It is visited by nearly two million people each year.

Real-World EXAMPLE

9 **ASTRONOMY** The diameter of the sun is approximately 20^9 meters. A model of the sun has a diameter of 20^{-1} meters. How many models would fit across the diameter of the sun?

$20^9 \div 20^{-1} = 20^{9-(-1)}$ Use the Quotient of Powers rule.

$= 20^{10}$ Subtract the exponent −1 from 9.

20^{10} models would fit across the diameter of the sun.

Your Turn

j. BUILDING A framed window is 3^{-1} yard wide. The side of the house is 3^2 yards wide. How many framed windows could fit across the side of the house?

EXAMPLES Raise a Power to a Power

Simplify. Answers should have only positive exponents. Assume that no denominators are equal to 0.

Study Tip

In $(7x)^2$, both 7 and x are raised to the second power. In $7x^2$, only x is squared.

10 $(7x^6y^{-3})^2$

$(7x^6y^{-3})^2 = 7^2x^{6\cdot2}y^{-3\cdot2}$ Use the Power of Powers.

$= 7^2x^{12}y^{-6}$ Multiply the exponents 6 and 2.
                     Multiply the exponents −3 and 2.

$= 49x^{12}y^{-6}$ Find 7^2. $7 \times 7 = 49$

$= \dfrac{49x^{12}}{y^6}$ Rewrite using a positive exponent.

11 $5(c^5y^{-3})^4$

$5(c^5y^{-3})^4 = 5c^{5\cdot4}y^{-3\cdot4}$ Use the Power of Powers.

$= 5c^{20}y^{-12}$ Multiply the exponents.

$= \dfrac{5c^{20}}{y^{12}}$ Rewrite using a positive exponent.

Your Turn
Simplify. Answers should have only positive exponents. Assume that no denominators are equal to 0.

k. $(2p^{-2}m^3)^5$ **l.** $9(w^6z^{-5})^3$

Examples 1–11
(pages 285–288)

VOCABULARY

1. Write an expression with a negative exponent. Then, write the same expression using only positive exponents.

Example 1
(page 285)

Rewrite using positive exponents. Then evaluate.

2. 2^{-5}

3. 13^{-2}

Examples 2–4
(page 286)

Rewrite using positive exponents. Assume that no denominators are equal to 0.

4. c^{-11}

5. $k^9 m^{-5}$

6. $14q^{-6}$

Examples 5–6
(page 287)

Find each product. Answers should have only positive exponents. Assume that no denominators are equal to 0.

7. $(b^6)(b^{-9})$

8. $(2q^5 r^{-3})(-4q^2 r^6)$

Examples 7–8
(page 287)

Find each quotient. Answers should have only positive exponents. Assume that no denominators are equal to 0.

9. $\dfrac{s^3}{s^{-4}}$

10. $\dfrac{80f^2 h^4}{4f^2 h^6}$

Example 9
(page 288)

11. **MEASUREMENT** A millimeter is 10^{-3} meter. How many millimeters are there in 10^0 meters?

12. **SCIENCE** The average platelet is about 3^{-8} centimeter in diameter. How many platelets would fit across a test tube that is 3^3 centimeters long?

Examples 10–11
(page 288)

Simplify. Answers should have only positive exponents. Assume that no denominators are equal to 0.

13. $(4h^8 k^{-2})^2$

14. $7(c^2 g^{-4})^3$

Skills, Concepts, and Problem Solving

HOMEWORK HELP

For Exercises	See Example(s)
15–17	1
18–23	2–4
24–29	5–6
30–35	7–8
36–38	9
39–42	10–11

Rewrite using positive exponents, then evaluate.

15. 8^{-3}

16. 5^{-4}

17. 1^{-100}

Rewrite using positive exponents. Assume that no denominators are equal to 0.

18. d^{-4}

19. y^{-34}

20. $x^{26} w^{-51}$

21. $p^{-50} q^{76}$

22. $14c^{-12}$

23. $32q^{-8}$

Find each product. Answers should have only positive exponents. Assume that no denominators are equal to 0.

24. $(t^5)(t^{-3})$

25. $(r^6)(r^{-1})$

26. $(2x^5 y^{-4})(-6x^2 y^6)$

27. $(3u^3 v^{-3})(-4u^2 v^7)$

28. $(2d^{-6} f^3)(6d^3 f^6)$

29. $(-3p^7 q^2)(-7p^{-8} q^{-4})$

Find each quotient. Answers should have only positive exponents. Assume that no denominators are equal to 0.

30. $\dfrac{d^3}{d^{-9}}$ **31.** $\dfrac{t^4}{t^6}$ **32.** $\dfrac{22r^2t^9}{11r^3t^4}$

33. $\dfrac{5j^6k^{11}}{15j^5k^{13}}$ **34.** $\dfrac{12c^9v^5}{c^7v^6}$ **35.** $\dfrac{36n^8m^3}{24n^4m^6}$

36. **INSECTS** Fire ants are 2^{-3} inch long. The ants are in a line across a porch that is 2^6 inches long. How many fire ants are there?

37. **MEASUREMENT** The diameter of a grain of sand is about 2^{-9} meter. The diameter of a grain of powder is about 2^{-17} meter. How many grains of powder could fit across one grain of sand?

38. **SCIENCE** A virus is about 7^{-9} meter in size. A white blood cell is about 7^{-5} meter in size. How much larger is the white blood cell than the virus?

Simplify. Answers should have only positive exponents. Assume that no denominators are equal to 0.

39. $(2x^4y^{-2})^2$ **40.** $(3r^8s^{-3})^4$ **41.** $2(c^3y^{-5})^3$ **42.** $5(a^2b^{-8})^2$

H.O.T. Problems Study the pattern of exponents in the chart.

43. If you continued the pattern, what is the next exponent for 2?

44. What is the next value in the pattern?

45. *Writing in Math* Explain the difference between x^n and x^{-n}.

$$
\begin{aligned}
2^6 &= 64 \\
2^5 &= 32 \quad \} \div 2 \\
2^4 &= 16 \quad \} \div 2 \\
2^3 &= 8 \quad \} \div 2 \\
2^2 &= 4 \quad \} \div 2 \\
2^1 &= 2 \quad \} \div 2 \\
2^0 &= 1 \quad \} \div 2
\end{aligned}
$$

STANDARDS PRACTICE

Choose the *best* answer.

46 Which is equivalent to $(5f)^{-2}$?

A $-25f^2$ **C** $\dfrac{25}{f^2}$

B $-5f^2$ **D** $\dfrac{1}{25f^2}$

47 The area of the head of a pin is about 2^{-5} square centimeter. It takes about 2^{45} gold atoms to cover the pin head. What is the area of a gold atom?

F 2^{-9} cm^2 **H** 2^{-50} cm^2

G 2^9 cm^2 **J** 2^{50} cm^2

Spiral Review

Simplify. (Lesson 5-1)

48. $(2ab^6)^{10}$ **49.** $7g^3h^5 \cdot 9g^7h$ **50.** $\dfrac{d^8e^9}{d^6e^9}$ **51.** $\dfrac{(3m^5n^2)(6mn^6)^2}{4m^3n^5}$

52. **MONEY** Let x be the total number of dollars raised by a class for charity. Then, $x \div 28$ is the average amount each person donated. If the average donation was \$6.73, what was the total amount donated? (Lesson 4-6)

Problem-Solving Strategy:
Solve a Simpler Problem

Vocabulary

solve a simpler problem
(p. 291)

 Standard 7AF2.1
Interpret positive whole-number powers as repeated multiplication and negative whole-number powers as repeated division or multiplication by the multiplicative inverse. Simplify and evaluate expressions that include exponents.

Standard 7MR2.2 Apply strategies and results from simpler problems to more complex problems.

Use the Strategy

Mrs. Chavez rewards her son Enrico for doing his chores. She gives Enrico a penny on the first day he does his chores. On the second day, Enrico earns 2 pennies. On the third day, he earns 4 pennies.

Each day Enrico does his chores, he earns double the amount of money. Suppose he does his chores for 10 days. How much total money will he have on the 10th day?

Understand ▸ **What do you know?**

- Mrs. Chavez gives 1 penny on Day 1, 2 pennies on Day 2, and 4 pennies on Day 3.
- Each day the amount will double.
- Enrico does his chores for 10 days.

Plan ▸ **What do you want to know?**

You want to know how much total money Enrico will receive. Start by **solving a simpler problem.**

Solve ▸ **How much money will Enrico have in fewer days?**

Day	Pennies	Total
1	1 or 2^0	$2^0 = 1$
2	2 or 2^1	$2^0 + 2^1 = 3$
3	4 or 2^2	$2^0 + 2^1 + 2^2 = 7$
4	8 or 2^3	$2^0 + 2^1 + 2^2 + 2^3 = 15$

Remember!

Any nonzero number raised to the zero power is 1.

$$2^0 = 1.$$

Continue this pattern to find the amount of money Enrico will have on the 10th day.

10-day total: $2^0 + 2^1 + 2^2 + 2^3 + 2^4 + 2^5 + 2^6 + 2^7 + 2^8 + 2^9$
$= 1 + 2 + 4 + 8 + 16 + 32 + 64 + 128 + 256 + 512$
$= 1,023$ pennies or $10.23

Check ▸ **Does your answer make sense?**

Round each day's answer to the nearest dollar (100 pennies) to see if your final answer makes sense.

$$1 + 2 + 4 + 8 + 16 + 32 + 64 + 128 + 256 + 512$$
$$\approx 0 + 0 + 0 + 0 + 0 + 0 + 1 + 1 + 3 + 5$$
$$= 10$$

So, the answer makes sense since $10 is close to $10.23.

MONEY How much total money did Enrico have on the 10th day?

Use a simpler problem to solve the following.

1. A 4-by-4 square is made of 16 smaller squares. What is the total number of squares found on the 4-by-4 square? (**Hint:** There are more than 16 squares.)

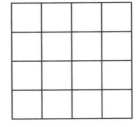

 What do you know?

 How can you figure out how many squares there are?

Solve **Look for a pattern of smaller squares, such as 2-by-2 and 3-by-3 squares, until you can solve for a 4-by-4 square.**

Check **How can you be sure that your answer is correct?**

2. *Writing in Math* There is a soccer tournament with 16 teams. Each team must play every other team. You want to know how many games will be played. Explain how you would create a simpler problem to help solve this problem.

Problem-Solving Practice

Use a simpler problem to solve.

3. **NUMBERS** Find the sum of the whole numbers from 1 to 200.

4. **SHAPES** Fifteen points are marked on a circle. Line segments are drawn connecting every pair of points. How many line segments are drawn?

5. **SHAPES** What is the total number of triangles in the figure?

6. **COMMUNICATION** Juanita heard a funny joke on Sunday. On Sunday, she told the joke to 3 friends. On Monday, each friend told the joke to 3 more friends. On Tuesday, these 3 friends told the joke to 3 more friends. Suppose this pattern continues. How long will it take for 100 people to hear the joke?

Solve using any strategy.

7. **FOOD SERVICE** Chef Martino made a huge pan of lasagna for a banquet. He makes 6 cuts along the length of the rectangular pan and 10 cuts along the width. How many pieces does he have?

8. **RECYCLING** Elroy wants students to recycle containers. After 2 weeks of encouraging recycling, 173 containers are thrown away instead of recycled. After 4 weeks, he counts 168. If this rate continues, what number will he find in 10 weeks?

9. **COMMUNITY** Tamika noticed that her zip code is made of five consecutive digits. If the sum of these digits is 25, what is Tamika's zip code?

10. **MONEY** A penny, a nickel, a dime, and a quarter are in a purse. Elise picks two coins out without looking. How many different amounts of money could she choose?

Roots

Vocabulary

square root (p. 293)

radical sign (p. 293)

radicand (p. 295)

 Standard ALG2.0
Students understand and use such operations as taking the opposite, finding the reciprocal, **taking a root,** and raising to a fractional power. They understand and use the rules of exponents.

Remember!

Inverse operations undo one another.

Addition and subtraction are inverse operations.

Multiplication and division are also inverse operations.

 The *What*: I will use roots to simplify expressions and solve problems.

 The *Why*: Roots can be used to calculate the wingspan of a hang glider.

In the expression s^2, or s squared, the exponent 2 signifies that the variable s will be multiplied by itself. Therefore, $s^2 = s \times s$. The numbers 9, 36, 64, and 81 are square numbers.

$$3 \times 3 = 3^2 = 9 \qquad 6 \times 6 = 6^2 = 36$$
$$8 \times 8 = 8^2 = 64 \qquad 9 \times 9 = 9^2 = 81$$

To solve an equation with a squared number or variable, use the inverse operation. Take the **square root** of both sides.

radical sign
$$\sqrt{81} = \sqrt{9 \cdot 9} = 9$$
radicand $\qquad$ square root

The symbol $\sqrt{}$ is called the **radical sign.** It indicates a positive (nonnegative or 0) square root. Both 9^2 and $(-9)^2$ equal 81.

$$9 \cdot 9 = 9^2 = 81 \qquad\qquad (-9) \cdot (-9) = (-9)^2 = 81$$

The radical sign indicates that the answer is the positive square root. The correct answer to $\sqrt{81}$ is +9, not −9.

SQUARE ROOTS		
Numbers	**Words**	**Algebra**
Since $5 \times 5 = 5^2 = 25$, then $5 = \sqrt{25}$	A square root of a number is one of its two equal factors.	If $x \cdot x = y$, then $x = \sqrt{y}$

Find each value.

① $\sqrt{36}$

$6 \cdot 6 = 36$

So, $\sqrt{36} = 6$.

② $-\sqrt{225}$

$-\sqrt{225} = -1 \cdot \sqrt{225}$

$= -1 \cdot 15$

$= -15$

So, $-\sqrt{225} = -15$.

Your Turn

Find each value.

a. $\sqrt{16}$

b. $\sqrt{49}$

c. $-\sqrt{4}$

d. $-\sqrt{441}$

EXAMPLES Compare Numbers on a Number Line

Order the numbers from least to greatest.

③ $4\frac{3}{10}, \sqrt{25}, 4.6, \sqrt{16}$

Write each number as a decimal. Graph the numbers.

$4\frac{3}{10} = 4.3 \qquad \sqrt{25} = 5.0 \qquad 4.6 = 4.6 \qquad \sqrt{16} = 4.0$

The order from least to greatest is $\sqrt{16}, 4\frac{3}{10}, 4.6, \sqrt{25}$.

④ $\sqrt{64}, 7\frac{5}{8}, 7.37, \sqrt{49}$

Write each number as a decimal. Graph the numbers.

$\sqrt{64} = 8.0 \qquad 7\frac{5}{8} = 7.625 \qquad 7.37 = 7.37 \qquad \sqrt{49} = 7.0$

The order from least to greatest is $\sqrt{49}, 7.37, 7\frac{5}{8}, \sqrt{64}$.

Your Turn

e. Order the numbers from least to greatest.

$\sqrt{9}, 4\frac{1}{4}, \sqrt{16}, 2.3$

> **Remember!**
>
> To change a fraction to a decimal, divide the denominator by the numerator.

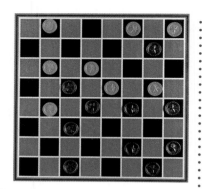

Real-World Challenge

A checkerboard is a large square comprised of 64 smaller squares.

What fraction of smaller squares contain a checker?

EXAMPLE **Use Square Roots**

5 MEASUREMENT The area of a square, A, can be found by using the equation $A = s^2$. A represents the area of the square, and s represents the length of one side. If the area of a square is 25 square inches, what is the length of a side of the square?

$A = s^2$

$25 = s^2$ Substitute 25 for A.

$\sqrt{25} = \sqrt{s^2}$ Take the square root of both sides.

$5 = s$ Simplify $\sqrt{25}$ and $\sqrt{s^2}$.

5 inches   The length of the side is 5 inches.

$A = 25$ in.^{2}

Your Turn Solve using the equation $A = s^2$.

f. MEASUREMENT The area of a square is 16 square feet. What is the length of a side of the square?

Talk Math

WORK WITH A PARTNER Each square on a checkerboard has an area of 9 square inches. What is the area of the checkerboard? Describe how you found the answer.

The radical sign acts as a grouping symbol in the order of operations. Simplify the **radicand** (numbers and/or variables inside the radical sign) before taking the square root.

EXAMPLES **Order of Operations**

Simplify.

6 $\sqrt{3^2 + 4^2}$

$\sqrt{3^2 + 4^2} = \sqrt{9 + 16}$ Find the values of the exponents first.

$= \sqrt{25}$ Add the numbers inside the radical sign.

$= 5$ Find the square root of 25.

7 $3 \cdot 6 + \sqrt{15 - 6}$

$18 + \sqrt{15 - 6} = 18 + \sqrt{9}$ Subtract 6 from 15 in the radical sign.

$= 18 + 3$ Find the square root of 9.

$= 21$ Add 18 and 3.

Your Turn Simplify.

g. $\sqrt{5^2 - 3^2}$

h. $\sqrt{100 - 36} + 18 \div 3$

Lesson 5-4 Roots **295**

Vocabulary Review
square root
radical sign
radicand

Examples 1–7
(pages 294–295)

VOCABULARY

1. Explain why 5 is the square root of 25.

2. Identify the radical sign and the radicand in the expression $\sqrt{49}$.

Examples 1–2
(page 294)

Find each value.

3. $\sqrt{16}$

4. $\sqrt{0}$

5. $-\sqrt{64}$

6. $-\sqrt{144}$

Examples 3–4
(page 294)

Order the numbers from least to greatest.

7. $11\frac{3}{5}, \sqrt{81}, \sqrt{121}, 10.4$

8. $\sqrt{100}, 11.7, 10\frac{1}{8}, \sqrt{144}$

Examples 5
(page 295)

9. **MEASUREMENT** The area of a square is 81 square centimeters. Use the equation $A = s^2$. What is the length of a side of the square?

Examples 6–7
(page 295)

Simplify.

10. $\sqrt{5^2 + 12^2}$

11. $21 \div 7 + \sqrt{21 - 5}$

12. $10 \cdot 3 - \sqrt{10 + 6}$

13. $\sqrt{81 + 19} - 10 \div 2$

Skills, Concepts, and Problem Solving

HOMEWORK HELP

For Exercises	See Example(s)
14–22	1–2
23–30	3–4
31–32	5
33–36	6–7

Find each value.

14. $\sqrt{4}$

15. $\sqrt{1}$

16. $\sqrt{49}$

17. $\sqrt{169}$

18. $-\sqrt{400}$

19. $-\sqrt{25}$

20. $-\sqrt{100}$

21. $-\sqrt{9}$

22. $-\sqrt{121}$

Order the numbers from least to greatest.

23. $\sqrt{64}, \sqrt{49}, 7\frac{3}{8}, \sqrt{100}$

24. $\sqrt{1}, 0.63, \frac{3}{4}, \sqrt{0}$

25. $5\frac{1}{5}, \sqrt{25}, 4.8, 4\frac{5}{8}$

26. $\sqrt{121}, \sqrt{64}, 11\frac{7}{8}, 9.4$

27. $12\frac{1}{8}, \sqrt{169}, \sqrt{144}, 13\frac{3}{4}$

28. $\sqrt{4}, 2\frac{1}{4}, 3.2, \sqrt{9}$

29. $5.25, \sqrt{36}, 5\frac{5}{8}, \sqrt{25}$

30. $\sqrt{1}, 0.8, \sqrt{4}, 1.85$

Solve using the equation $A = s^2$.

31. **MEASUREMENT** The area of a square is 100 square inches. What is the length of a side of the square?

32. **BASEBALL** The area of the square baseball diamond is 8,100 square feet. When it rains, teams cover it with a tarp. How long is the tarp on each side?

BASEBALL Minor league baseball players take the field for one of 10 teams in the California League of Professional Baseball.

Simplify.

33. $\sqrt{2^2 + 5}$

34. $\sqrt{8^2 - 39}$

35. $6 \cdot 8 + \sqrt{55 - 6}$

36. $56 \div 7 + \sqrt{42 + 7}$

37. **H.O.T. Problems** You cannot calculate $\sqrt{-16}$. Why not? Explain your thinking.

STANDARDS PRACTICE

38 The chart on the right shows an expression evaluated for four different values of x.

x	$\sqrt{x}$
4	2
25	5
49	7
121	11

Anna concluded that for all positive values of x, $\sqrt{x}$ produces a prime number. Which value of x serves as a counterexample to prove Anna's conclusion false?

A 1

B 9

C 16

D 169

39 City code requires 4 square feet for each person on a dance floor.

Number of people	9	16	25	36
Number of square feet	36	64	100	144
Side length	6	8	10	12

A square dance floor needs to be large enough for 100 people. How long should it be on each side?

F 5 feet

G 10 feet

H 20 feet

J 40 feet

Spiral Review

Use a simpler problem to solve. (Lesson 5-3)

40. **NUMBERS** Find the sum of the first 100 even whole numbers.

Evaluate each expression if $a = 4$, $b = 2$, and $c = 3$. (Lesson 1-3)

41. $a + b \cdot c$

42. $ab - bc$

43. $7a - (2c + b)$

44. $\dfrac{6(a + b)}{3c}$

45. **GEOMETRY** The surface area of Julio's cubic puzzle is the product of the square of the length of one side and 6. What is an expression that represents the surface area of Julio's cubic puzzle? (Lesson 1-2)

Progress Check 2
(Lessons 5-2 and 5-4)

▷ Vocabulary and Concept Check

> negative exponent (p. 285)   radicand (p. 295)
> radical sign (p. 293)   square root (p. 293)

Choose the term that *best* completes each sentence. (Lessons 5-2 and 5-4)

1. Any number or expression inside a radical sign is called the _____?_____.

2. If $x^2 = y$, then x is the _____?_____ of y.

3. The symbol $\sqrt{}$ is called the _____?_____.

▷ Skills Check

Rewrite using positive exponents, then evaluate. (Lesson 5-2)

4. 9^{-2}

5. 2^{-6}

Rewrite using positive exponents. Assume that no denominators are equal to 0. (Lesson 5-2)

6. $w^{-3}v^{19}$

7. $p^{-3}q^4$

Find the quotient. Answers should have only positive exponents. Assume that no denominators are equal to 0. (Lesson 5-2)

8. $\dfrac{4x^4}{12x^2}$

9. $\dfrac{7g^{14}h^2}{21g^7h^3}$

Simplify. Answers should have only positive exponents. Assume that no denominators are equal to 0. (Lesson 5-2)

10. $(g^8h^{-11})^{-5}$

11. $(x^3y^{-4})^2$

Find each value. (Lesson 5-4)

12. $\sqrt{16}$

13. $-\sqrt{36}$

▷ Problem-Solving Check (Lesson 5-4)

14. **MEASUREMENT** Mrs. Chavez's bedroom is shaped like a square. Use the equation $A = s^2$ to find the length of one side of the room, s, if the area, A, is 64 square feet.

15. **REFLECT** Write a paragraph that explains the difference between squares and square roots.

5-5

Simplifying and Evaluating Expressions

Vocabulary

formula (p. 299)

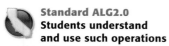

Standard ALG2.0
Students understand
and use such operations
as taking the opposite, finding
the reciprocal, **taking a root,** and
raising to a fractional power. **They**
understand and use the rules of
exponents.

The What: I will evaluate expressions that include powers and roots.

The Why: The velocity of water leaving a firefighter's hose can be found by using equations with powers and roots.

Formulas are equations that describe a relationship among certain quantities. The following formula indicates how fast water leaves the nozzle of a hose.

$$V = 12\sqrt{P}$$

velocity
(feet per second)

pressure
(pounds per square inch—psi)

STEPS FOR USING A FORMULA

❶ Write the formula.

❷ Substitute values for the variables given.

❸ Simplify.

❹ Write the answer with the appropriate label.

EXAMPLE Use Formulas

❶ **FIREFIGHTING** Suppose the fire hose water pressure is 64 psi. What is the water's velocity?

$V = 12\sqrt{P}$	Write the formula.
$V = 12\sqrt{64}$	Substitute 64 for P.
$V = 12(8)$	Simplify $\sqrt{64}$.
$V = 96$	Multiply 12 and 8.
$V = 96$ feet per second	Include the appropriate label.

The water's velocity is 96 feet per second.

Your Turn

a. **FIREFIGHTING** Suppose the fire hose water pressure is 49 psi. What is the water's velocity?

You evaluate an expression by substituting values for variables and simplifying. Remember to use the order of operations.

EXAMPLES Evaluate Expressions Involving Powers

Evaluate each expression for $x = 3$ and for $x = -3$.

2 x^2

$x = 3$

$x^2 = 3^2$ Substitute 3 for x.

$= 3 \cdot 3$ Simplify exponents.

$= 9$ Simplify.

$x = -3$

$x^2 = (-3)^2$ Substitute −3 for x.

$= (-3)(-3)$ Simplify exponents.

$= 9$ Simplify.

3 x^3

$x = 3$

$x^3 = 3^3$ Substitute 3 for x.

$= 3 \cdot 3 \cdot 3$ Simplify exponents.

$= 27$ Simplify.

$x = -3$

$x^3 = (-3)^3$ Substitute −3 for x.

$= (-3)(-3)(-3)$ Simplify exponents.

$= -27$ Simplify.

4 $2x^2$

$x = 3$

$2x^2 = 2 \cdot 3^2$ Substitute 3 for x.

$= 2 \cdot 3 \cdot 3$ Simplify exponents.

$= 18$ Simplify.

$x = -3$

$2x^2 = 2 \cdot (-3)^2$ Substitute −3 for x.

$= 2(-3)(-3)$ Simplify exponents.

$= 18$ Simplify.

5 $3x^3$

$x = 3$

$3x^3 = 3 \cdot 3^3$ Substitute 3 for x.

$= 3 \cdot 3 \cdot 3 \cdot 3$ Simplify exponents.

$= 81$ Simplify.

$x = -3$

$3x^3 = 3 \cdot (-3)^3$ Substitute −3 for x.

$= 3(-3)(-3)(-3)$ Simplify exponents.

$= -81$ Simplify.

Remember!

Simplify exponents before multiplying by any coefficient. This includes expressions with negative signs.

Your Turn

Evaluate each expression for $x = 2$ and for $x = -2$.

b. x^2 c. x^3

d. $3x^4$ e. $-4x^3$

Talk Math

WORK WITH A PARTNER Explain the difference between -2^3 and $(-2)^3$. Describe how simplifying the expressions helps you explain the difference.

EXAMPLE Evaluate Expressions

6 Evaluate $v + n^p$ for $v = 75$, $p = 4$, and $n = 2$.

$v + n^p = 75 + 2^4$ Substitute 75 for v, 4 for p, and 2 for n.

$= 75 + 16$ Simplify exponents.

$= 91$ Add 75 and 16.

Your Turn

f. Evaluate $y^z - x$ for $x = 12$, $y = 4$, and $z = 3$.

Real-World EXAMPLE   Use Formulas

7 **FIREFIGHTING** The formula $Q = 30D^2\sqrt{P}$ indicates the flow rate of water through the hose Q in gallons per minute. The hose nozzle's diameter in inches is represented by D, and the water pressure is represented by P.

Suppose the hose nozzle has a 1-inch diameter and the water pressure is 49 psi. What is the rate of water flow in gallons per minute?

$Q = 30D^2\sqrt{P}$ Write the formula.

$Q = 30(1^2)\sqrt{49}$ Substitute 1 for D and 49 for P.

$Q = 30(1)(7)$ Simplify.

$Q = 210$ gallons per minute   Write the answer and label.

The rate of water flow is 210 gallons per minute.

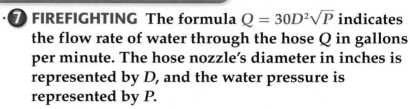

Real-World Challenge· · · · · ·

FIREFIGHTING Fire hoses can have different sizes and types of nozzles.

Your Turn

g. **FIREFIGHTING** Suppose the hose nozzle has a 2-inch diameter and the water pressure is 36 psi. What is the rate of water flow in gallons per minute?

Guided Practice

Vocabulary Review
formula

Examples 1 and 7 (pages 299 and 301)

VOCABULARY

1. Give an example of when you would use a formula.

Example 1 (page 299)

2. **FIREFIGHTING** A pumper can be connected to a fire hydrant so that the water pressure is 144 psi. Use the formula $V = 12\sqrt{P}$, where V represents the velocity in feet per second and P represents the pressure, to find the velocity of the water when a pumper is used.

Examples 2–5 (page 300)

Evaluate each expression for $m = 5$ and for $m = -5$.

3. m^2

4. m^3

5. $2m^2$

6. $-2m^2$

Example 6
(page 301)

Evaluate each expression for $s = 16$, $b = 1$, and $g = 2$.

7. $s^b - 6g$

8. $g^3 + sb$

Example 7
(page 301)

FIREFIGHTING The flow rate of water in gallons per minute is given by the following formula.

$Q = 30D^2\sqrt{P}$ $Q =$ the flow rate

$D =$ the hose nozzle's diameter

$P =$ the water pressure

9. Suppose the hose nozzle has a 1-inch diameter and the water pressure is 64 psi. What is the rate of water flow?

FIREFIGHTING The Santa Ana winds help spread flames from wildfires in southern California.

10. Suppose the hose nozzle has a 2-inch diameter and the water pressure is 25 psi. What is the rate of water flow?

Skills, Concepts, and Problem Solving

HOMEWORK HELP

For Exercises	See Example(s)
11–12	1
13–20	2–5
21–24	6
25–28	7

GEOMETRY The length of the hypotenuse of a right triangle can be found using the Pythagorean Theorem.

$c = \sqrt{a^2 + b^2}$ $c =$ length of hypotenuse

$a =$ length of a leg of the triangle

$b =$ length of the other leg of the triangle

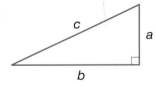

11. What is the hypotenuse of a right triangle with legs measuring 6 inches and 8 inches?

12. What is the hypotenuse of a right triangle with legs measuring 5 inches and 12 inches?

Evaluate each expression for $w = 2$ and for $w = -2$.

13. w^4

14. w^5

15. $4w^4$

16. $-w^4$

Evaluate each expression for $p = 4$ and for $p = -4$.

17. p^2

18. p^3

19. $10p^2$

20. $-p^2$

Evaluate each expression for $x = 3$, $d = 2$, and $u = 9$.

21. $x^2 - u^d$

22. $u^2 + 3dx + 1$

23. $5d^4 + x$

24. $10d^3 + u$

GEOMETRY The approximate radius r of a ball is given by $r = \sqrt{\frac{SA}{12}}$, where SA is its surface area.

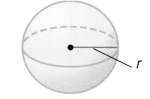

25. A tetherball is made of about 192 square inches of rubber. What is the radius of the ball?

26. A playground ball is made of about 300 square inches of rubber. What is the radius of the ball?

CAREER CONNECTION The flow rate of water in gallons per minute is given by the formula $Q = 30D^2\sqrt{P}$, where Q = the flow rate, D = the hose nozzle's diameter, and P = the water pressure.

27. Suppose the hose nozzle has a 2-inch diameter and the water pressure is 16 psi. What is the rate of water flow?

28. Suppose the hose nozzle has a 1-inch diameter and the water pressure is 81 psi. What is the rate of water flow?

29. **H.O.T.** Problem Use the order of operations to compare $(c - d)^2$ and $c^2 - d^2$. Are they equivalent? Explain. **Hint:** Substitute values for c and d.

30. *Writing in Math* Explain the difference between $2r^3$ and $(2r)^3$. Describe how evaluating the expressions with values may help explain your answer.

STANDARDS PRACTICE

Choose the best answer.

31 Evaluate $\sqrt{\dfrac{3c}{2a + L^2}}$ where $c = 16$, $a = 1$, and $L = 2$.

 A $\sqrt{8}$

 B 16

 C -8

 D $\sqrt{64}$

32 In the formula $R = \dfrac{s^2}{A}$, A is the wing area of a hang glider, s is its wingspan, and R is the ratio that helps determine if the glider will fly. Suppose a glider has a ratio, R, of 1.6 and a wing area, A, of 10 square feet. What is the glider's wingspan?

 F $\sqrt{10}$ feet   **H** 16 feet

 G 4 feet   **J** 160 feet

Spiral Review

Simplify. (Lesson 5-4)

33. **MEASUREMENT** The area of a square tapestry is 121 square inches. What is the length of a side of the tapestry? (Lesson 5-4)

Divide. (Lesson 2-5)

34. $-24 \div 8 \div (-1)$ 35. $18 \div 2 \div (-3)$ 36. $36 \div (-6) \div 3$

Multiply. (Lesson 2-4)

37. $(-3)(-3)(-4)$ 38. $(-1)(5)(-2)$ 39. $(4)(2)(-2)$

Explore 5-6

Math Lab
Locate Rational Numbers on a Number Line

Materials

classroom number line with hash marks from −5 to 5

sticky notes with numbers −5.5, 3², 2⁻¹, $\frac{-9}{2}$, and −√25

 Standard 6NS1.1 Compare and order positive and negative fractions, decimals, and mixed numbers and place them on a number line.

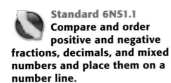 **Study Tip**

Remember that the least numbers on a number line are farthest to the left and the greatest numbers are farthest to the right.

The What: I will compare and order rational numbers.

The Why: Rational numbers are used to represent money and recipe ingredient amounts.

Integers, fractions, decimals, powers, and roots of perfect squares are all rational numbers.

ACTIVITY **Order Rational Numbers**

① **Use the classroom number line to order the numbers on the sticky notes.**

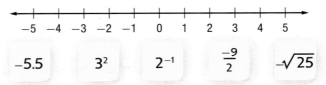

| −5.5 | 3² | 2⁻¹ | $\frac{-9}{2}$ | −√25 |

Step 1: Select a number on a sticky note.
Step 2: Place the number on the classroom number line.
Step 3: Graph all of the numbers on the number line.
Step 4: List all of the numbers from least to greatest.

Your Turn
Graph the numbers on a number line. Then list numbers from *least* to *greatest*.

a. $0.6, -2, \frac{7}{8}, 2^{-3}, -\sqrt{1}$ b. $5^2, 5^{-2}, 0.5, \sqrt{25}, \frac{1}{5}$

Analyze the Results

Graph the numbers on a number line. Then list numbers from *least* to *greatest*.

1. $8^0, \sqrt{16}, \frac{4}{9}, 3^{-2}, -0.66$ 2. $-0.9, -\frac{9}{9}, -3^2, -\sqrt{9}, 9^0$

3. $4.25, 5\frac{58}{100}, 2^4, \sqrt{64}, 12$ 4. $\frac{3}{8}, -\sqrt{4}, -0.5, 2^{-2}, 1.5$

5-6

Comparing and Ordering Rational Numbers

Vocabulary

rational number (p. 305)

 Standard 6NS1.1
Compare and order positive and negative fractions, decimals, and mixed numbers and place them on a number line.

 The What: I will compare and order rational numbers.

The Why: Rational numbers can be used to determine correct distances when traveling to a new place.

Uma was driving to Jessica's house. According to Jessica, Uma should turn right after $3\frac{3}{8}$ miles. The odometer on the car indicated that Uma had traveled 3.75 miles. Did Uma miss the turn?

In this problem, you need to compare **rational numbers.** Using a number line can help when comparing numbers.

$$3\frac{3}{8} < 3.75$$

Remember that on a number line, the greater number is always to the right. So, 3.75 is greater than $3\frac{3}{8}$.

You also can compare by writing both numbers in the same form. It is usually best to write all as decimals.

To determine whether Uma missed the turn, change $3\frac{3}{8}$ to a decimal and then compare decimals. To compare decimals, write numbers so that they have the same number of decimal places.

$$\frac{3}{8} = 3 \div 8 = 0.375$$

Remember!

Divide the numerator by the denominator when changing a fraction to a decimal.
$\frac{1}{4} = 1 \div 4 = 0.25$

Uma was supposed to turn after $3\frac{3}{8}$ or 3.375 miles. Uma had driven 3.75 or 3.750 miles.

$$3.750 \text{ miles} > 3.375 \text{ miles}$$

distance traveled   distance in directions

Since 3.75 is greater than 3.375, Uma had missed the turn.

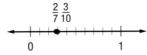

EXAMPLES **Compare Rational Numbers**

Replace each ● with one of the symbols <, >, or = to make a true sentence.

① $\frac{2}{7}$ ● $\frac{3}{10}$

Graph the numbers on a number line to compare.

$\frac{3}{10}$ is to the right of $\frac{2}{7}$.
So, $\frac{2}{7} < \frac{3}{10}$.

Compare by writing in the same form. Change to decimals. Divide the numerator by the denominator.

$\frac{2}{7} = 2 \div 7 = 0.2857\ldots$

$\frac{3}{10} = 3 \div 10 = 0.3$

Write the same number of decimal places.

$0.2857\ldots < 0.3000\ldots$
So, $\frac{2}{7} < \frac{3}{10}$.

② $\frac{3}{4}$ ● $\frac{5}{8}$

Graph the numbers on a number line to compare.

$\frac{3}{4}$ is to the right of $\frac{5}{8}$.
So, $\frac{3}{4} > \frac{5}{8}$.

Compare by writing in the same form. Write both fractions using a common denominator.

$\frac{3}{4} = \frac{6}{8}$ $\frac{6}{8} > \frac{5}{8}$

So, $\frac{3}{4} > \frac{5}{8}$.

③ $\frac{2}{6}$ ● 3^{-1}

Evaluate the power and find a common denominator.
$\frac{2}{6} = \frac{1}{3}$ $3^{-1} = \frac{1}{3}$
So, $\frac{2}{6} = 3^{-1}$.

④ $4\frac{2}{3}$ ● $\sqrt{25}$

Evaluate the root.
$\sqrt{25} = 5$ $4\frac{2}{3} < 5$
So, $4\frac{2}{3} < \sqrt{25}$.

Study Tip
Finding common denominators can make it easier to compare fractions.

Your Turn
Replace each ● with one of the symbols <, >, or = to make a true sentence.

a. $\frac{2}{5}$ ● $\frac{5}{15}$

b. $\frac{4}{9}$ ● $\frac{6}{11}$

c. $\frac{4}{8}$ ● 4^{-2}

d. $5\frac{1}{6}$ ● $\sqrt{25}$

Talk Math

WORK WITH A PARTNER Which is greater: 2.6 or $2\frac{2}{3}$? Explain your thinking to your partner. Then listen as your partner explains whether 3.25 or $3\frac{1}{3}$ is greater.

3 for $4.69

7 for $8.50

5 **RETAIL** Tennis balls sell 3 for $4.69. Another store is selling the same tennis balls 7 for $8.50. Which is the better deal?

Calculate the cost of 1 tennis ball at each store.

$4.69 \div 3 = 1.563 \ldots$ $8.50 \div 7 = 1.214 \ldots$

Compare the costs.

$1.563 \ldots > 1.214 \ldots$ The better deal costs less.

So, 7 for $8.50 is the better deal.

6 **SPORTS** The girls' basketball team has won 16 out of 18 games. The boys' team has a record of 0.875. Which team has the better record?

Calculate the girls' record.

$16 \div 18 = 0.888 \ldots$

Compare the records.

$0.888 \ldots > 0.875$ The better record is more.

So, the girls' team has the better record.

Your Turn

e. **RETAIL** Brandy wants to buy pillows. One store sells 4 for $10.84. Another store sells 7 for $19.65. Which is the better deal?

EXAMPLE   **Order Rational Numbers**

7 Order the numbers from least to greatest. 2^{-2}, 2.3, $\sqrt{4}$

$2^{-2} = \dfrac{1}{2^2} = \dfrac{1}{4} = 0.25$ Express each number as a decimal. Then compare the decimals.

2.3 is already expressed as a decimal.

$\sqrt{4} = 2$

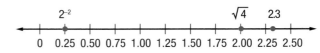

From least to greatest, the order is 2^{-2}, $\sqrt{4}$, 2.3.

Your Turn

Order the numbers from least to greatest.

f. $-5, 3\dfrac{7}{8}, \sqrt{9}$

g. $2^{-2}, \sqrt{1}, 4^{-2}$

Examples 1–7
(pages 306–307)

VOCABULARY

1. Write three examples of rational numbers. Order the numbers from least to greatest.

Examples 1–4
(page 306)

Replace each ● with one of the symbols <, >, or = to make a true sentence.

2. $\frac{8}{10}$ ● $\frac{3}{4}$

3. $-\frac{2}{6}$ ● $-\frac{4}{3}$

4. 6^{-1} ● $\frac{2}{12}$

5. $-4\frac{1}{4}$ ● $-\sqrt{16}$

6. -4.25 ● $-\sqrt{9}$

7. -3^3 ● -26.99

Examples 5–6
(page 307)

8. **SPORTS** One team has a winning record of 0.678. Another has won 15 out of 25 games. Which team has the better record?

9. **CARPENTRY** The directions to build a tree house require 0.375-inch nails. The nail box label reads $\frac{3}{8}$-inch nails. Will the nails in the box work? Explain.

Example 7
(page 307)

Order the numbers from least to greatest.

10. $\sqrt{16}, 9\frac{4}{5}, -4.95$

11. $-5\frac{1}{4}, 8^{-2}, -\sqrt{36}$

Skills, Concepts, and Problem Solving

HOMEWORK HELP	
For Exercises	See Example(s)
12–23	1–4
24–27	5–6
28–29	7

Replace each ● with one of the symbols <, >, or = to make a true sentence.

12. $\frac{1}{4}$ ● $\frac{3}{8}$

13. $\frac{4}{5}$ ● $\frac{7}{9}$

14. $-\frac{4}{9}$ ● $-\frac{5}{12}$

15. $-\frac{1}{3}$ ● $-\frac{2}{9}$

16. -4^{-2} ● $-\frac{4}{2}$

17. 2^5 ● $\frac{100}{4}$

18. $\frac{21}{5}$ ● $\sqrt{16}$

19. $-\sqrt{1}$ ● $-\frac{10}{11}$

20. 3^4 ● 81.01

21. -7.98 ● -2^3

22. $\sqrt{64}$ ● 3.5

23. 3.14 ● $\sqrt{4}$

24. **RETAIL** Oranges are on sale 10 for $1.20. Usually you pay $0.85 for 6 of them. Is this a good sale? Explain.

25. **SPORTS** The high school football team won 8 games out of 11. The middle school team won 7 games out of 9. Which team has the better record?

26. **CARPENTRY** A carpenter has a $\frac{2}{9}$-inch hole in a piece of wood she is using to make a magazine rack. She wants to put a dowel into the wood to hold the magazines. The dowel measures 0.25 inch in diameter. Will the dowel fit into the magazine rack?

27. HEALTH A doctor prescribes $\frac{1}{10}$ gram of medication for a patient. The pills only come in 0.05-gram tablets. The pharmacist calculates that the patient needs two tablets. Is the pharmacist right? Explain.

Order the numbers from least to greatest.

28. $\sqrt{100}$, 10^{-2}, 0.10   **29.** $-\frac{9}{9}$, -3^2, $\sqrt{9}$

30. *Writing in Math*   Explain how you would graph -0.25, $-\frac{1}{2}$, and 2^{-2} on a number line.

31. H.O.T. Problem Sometimes you can compare two powers by rewriting them with the same base. For example, compare 3^8 and 9^3. $9^3 = (3^2)^3 = 3^6$. Then, you can compare their exponents $3^8 > 3^6$. Using this method, which is greater, 2^9 or 4^5?

HEALTH Doctors and pharmacists work together to ensure proper dosage.

STANDARDS PRACTICE

Choose the *best* answer.

32 In what order would these numbers appear on a number line?

$$-1\frac{1}{3}, -3^0, -\sqrt{9}, -0.3$$

A $-3^0, -\sqrt{9}, -1\frac{1}{3}, -0.3$

B $-\sqrt{9}, -1\frac{1}{3}, -3^0, -0.3$

C $-1\frac{1}{3}, -0.3, -3^0, -\sqrt{9}$

D $-0.3, -1\frac{1}{3}, -3^0, -\sqrt{9}$

33 Based on the table of wins and losses so far this season, choose the answer that correctly shows a comparison that would tell you which team has the better record so far.

Team	Wins	Losses
Los Angeles Angels	15	8
Oakland A's	14	7

F $\frac{8}{23} < \frac{7}{21}$ **H** $\frac{15}{23} > \frac{14}{21}$

G $\frac{8}{23} > \frac{7}{21}$ **J** $\frac{15}{23} < \frac{14}{21}$

Spiral Review

34. MEASUREMENT The area of a square is 144 square inches. Use the equation $A = s^2$ where A represents the area and s represents the length of a side of a square. What is the length of a side of the square? (Lesson 5-4)

Divide. (Lesson 4-4)

35. $0.48 \div 0.6$ **36.** $2.4 \div 0.12$

Subtract. (Lesson 3-9)

37. $5 - 2\frac{3}{7}$ **38.** $7\frac{2}{12} - 4\frac{5}{8}$

Study Guide

Study Tips
Use an idea web to capture information you read.
Review all rules and formulas.

Understanding and Using the Vocabulary

After completing the chapter, you should be able to define each term, property, or phrase and give an example of each.

> formula (p. 299)
>
> negative exponent (p. 285)
>
> power (p. 278)
>
> Power of Powers (p. 281)
>
> Product of Powers (p. 278)
>
> Quotient of Powers (p. 280)
>
> radical sign (p. 293)
>
> radicand (p. 295)
>
> rational number (p. 305)
>
> solve a simpler problem (p. 291)
>
> square root (p. 293)

Complete each sentence with the correct mathematical term or phrase.

1. 4^3 is an example of a _____?_____.

2. $\frac{1}{x^{-n}}$ is written using a _____?_____.

3. $\sqrt{}$ is called the _____?_____.

4. According to the _____?_____, $(x^4)^5 = x^{20}$.

5. 9 is a _____?_____ of 81.

6. Any number or expression inside a radical sign is called the _____?_____.

7. According to the _____?_____, $\frac{y^{10}}{y^2} = y^8$.

8. $d = rt$ is an example of a _____?_____.

9. According to the _____?_____, $s^9 \cdot s^5 = s^{14}$.

Skills and Concepts

Objectives and Examples

Review Exercises

LESSON 5-1   pages 278–283

Apply the Product of Powers, the Quotient of Powers, and the Power of Powers to simplify expressions.

$$2r^9 \cdot \frac{6r}{(r^3)^2} = \frac{12r^9r}{(r^3)^2} \quad \text{Multiply 2 and 6.}$$

$$= \frac{12r^{10}}{(r^3)^2} \quad \text{Use the Product of Powers.}$$

$$= \frac{12r^{10}}{r^6} \quad \text{Use the Product of Powers.}$$

$$= 12r^4 \quad \text{Use the Quotient of Powers.}$$

Simplify.

10. $y^6 \cdot y^2 \cdot y$

11. $(-6y^2z^3)(3y^6z)$

12. $\dfrac{g^6h^3}{g^2h}$

13. $\dfrac{12w^{18}z^{12}}{4w^9z^{12}}$

14. $(s^5)^4$

15. $(2h^3m)^5$

LESSON 5-2   pages 285–290

Use integer exponents to simplify expressions.

$$\frac{20x^2y^{10}}{8x^4y^5} = \frac{5x^2y^{10}}{2x^4y^5} \quad \text{Simplify numbers.}$$

$$= \frac{5x^{-2}y^5}{2} \quad \text{Use the Quotient of Powers.}$$

$$= \frac{5y^5}{2x^2} \quad \text{Rewrite using a positive exponent.}$$

Simplify expressions using positive exponents. Assume that no denominators are equal to 0.

16. 5^{-2}

17. 3^{-4}

18. $6d^{-8}$

19. g^5p^{-3}

20. $\dfrac{x^5c^7}{x^{10}c^3}$

21. $\dfrac{3m^8n^6}{21m^3n^{10}}$

LESSON 5-4   pages 293–297

Use roots to simplify expressions.

$$\sqrt{4} = 2 \qquad 2^2 = 2 \times 2 = 4, \text{ so } \sqrt{4} = 2.$$

Find each value.

22. $\sqrt{64}$

23. $-\sqrt{49}$

24. $\sqrt{169}$

25. $-\sqrt{100}$

Objectives and Examples

Review Exercises

LESSON 5-5   pages 299–303

Evaluate expressions that include powers and roots.

Evaluate $p - \sqrt{k + kj^2}$ for $k = 5$, $p = 6$, and $j = 2$.

$p - \sqrt{k + kj^2} = 6 - \sqrt{5 + 5 \cdot 2^2}$

            Substitute 5 for k, 6 for p, and 2 for j.

$= 6 - \sqrt{5 + 5 \cdot 4}$ Simplify exponents.

$= 6 - \sqrt{5 + 20}$ Multiply.

$= 6 - \sqrt{25}$ Add.

$= 6 - 5$ Take the square root of 25.

$= 1$ Subtract.

Evaluate each expression for $z = 2$ and for $m = -2$.

26. z^2

27. z^3

28. $-z^2$

29. $-z^3$

30. m^2

31. m^3

32. $-m^2$

33. $-m^3$

LESSON 5-6   pages 305–309

Compare and order rational numbers.

$2^{-2}, 0.14, \sqrt{4}, \frac{1}{2}$

$2^{-2} = \frac{1}{4} = 0.25$ Change all numbers to decimals.

$0.14 = 0.14$

$\sqrt{4} = 2$

$\frac{1}{2} = 0.5$

$0.14, 0.25, 0.5, 2$ Order the decimals.

$0.14, 2^{-2}, \frac{1}{2}, \sqrt{4}$ Rewrite the numbers in their original form.

Replace each ● with one of the symbols $<$, $>$, or $=$ to make a true sentence.

34. $-\frac{1}{3}$ ● $-\frac{8}{6}$

35. $\frac{12}{16}$ ● $\frac{3}{4}$

36. 4.78 ● $4\frac{7}{10}$

37. 12^{-1} ● $\frac{2}{6}$

Order the numbers from least to greatest.

38. $5.25, -5\frac{1}{5}, -\sqrt{25}$

39. $-6\frac{1}{4}, -7.98, -2^3$

Chapter Test

▷ Vocabulary and Concept Check

1. Rewrite x^{-n} using a positive exponent.

▷ Skills Check

Simplify each expression.

2. $(-7k^3b^7)(2kb^4)$

3. $\dfrac{21u^7z^{13}}{3u^7z^{10}}$

4. $(3pk^3)^4$

5. $\dfrac{(2x^4)^3}{4x^3 \cdot x^2}$

Rewrite using positive exponents. Assume that no denominators are equal to 0.

6. $13m^{-2}$

7. $s^{-9}j^8$

Find the quotient and write with positive exponents. Assume that no denominators are equal to 0.

8. $\dfrac{36d^2e^{10}}{9d^6e^4}$

9. $\dfrac{50r^3s^8}{5r^9s^3}$

Simplify each expression.

10. 10^{-2}

11. $-\sqrt{16}$

Evaluate each expression for $w = 3$ and for $w = -3$.

12. w^2

13. $-w^3$

Evaluate each expression for $q = -3$ and for $w = 3$.

14. $q + \sqrt{w^4 - 17}$

15. $\dfrac{q^3}{w} - 1$

Replace each ● with <, >, or = to make a true sentence.

16. 1^{25} ● $\sqrt{25}$

17. 3^{-2} ● $\dfrac{2}{3}$

18. Order the numbers from least to greatest.

$-\dfrac{1}{3}, \ 3^{-2}, \ -0.3, \ -\sqrt{1}$

▷ Problem-Solving Check

19. **SPORTS** There are 6 basketball teams playing in a tournament. Each team must play every other team. How many total games will be played?

20. **MEASUREMENT** The area of a square is 169 square millimeters. What is the length of a side of the square?

Standards Practice

PART 1 Multiple Choice

Choose the *best* answer.

1 How is the product $7 \times 7 \times 7 \times 7 \times 7 \times 7$ expressed in exponential notation?

  A 6×7

  B 6^7

  C 7^6

  D 7^7

2 Which is equivalent to $n^3 \cdot n^2$?

  F n^1

  G n^5

  H n^6

  J n^9

3 Which is equivalent to $\dfrac{3^{11}}{3^6}$?

  A 3^5

  B 3^6

  C 3^{17}

  D 3^{66}

4 Using the formula $A = \pi r^2$, what is the area (A) of a circle with a radius (r) of 4 cm?

  F 18.85 cm^2

  G 25.13 cm^2

  H 50.27 cm^2

  J 157.91 cm^2

5 Which set of numbers is in order from *least* to *greatest*?

  A $\{-3^{-1}, -0.3, -3\}$

  B $\{-0.3, -3^{-1}, -3\}$

  C $\{-3, -0.3, -3^{-1}\}$

  D $\{-3, -3^{-1}, -0.3\}$

6 Which is equivalent to $(-2d)^3$?

  F $-2d^3$

  G $-6d^3$

  H $-8d^3$

  J d^{-6}

7 What is the value of $3m^3$ if $m = -2$?

  A -216

  B -24

  C 24

  D 216

8 Maria bought 40 shares of stock. She sold her shares at a loss of \$3 per share. Which integer represents the total change in value of the stock?

  F -120

  G -3

  H 3

  J 120

9 Which is equivalent to 4^{-3}?

  A $-4 \times -4 \times -4$

  B $-3 \times -3 \times -3 \times -3$

  C $-\dfrac{1}{4 \times 4 \times 4}$

  D $\dfrac{1}{4 \times 4 \times 4}$

10 Which of the following quotients is simplified completely?

  F $\dfrac{24g^2}{36gh}$

  G $\dfrac{10g^2}{4}$

  H $\dfrac{2kg^2}{3gh^2}$

  J $\dfrac{7g^2}{12kh}$

Record your answers on the answer sheet provided by your teacher or on a separate sheet of paper.

11 A group of 8 students meet each other. They each shake hands with every other student in the group. How many handshakes occurred?

12 The table shows a relationship between the area of a square (A) and its side length (s). Write a formula showing this relationship.

Side Length (s) in inches	3	5	7	11	15
Area (A) in square inches	9	25	49	121	225

13 Simplify the expression.

$$4 \cdot 3a + 9[5 - 3(a - 2)]$$

14 What is the value of the following expression if $a = 4$, $b = 3$, and $c = 2$?

$$\frac{3c + \sqrt{a}}{b^2 - 5}$$

15 Graph the following set of numbers. Order the numbers from least to greatest.

$$\left\{ \frac{3}{8}, 1.3, 3^{-1}, \sqrt{4} \right\}$$

Record your answers on the answer sheet provided by your teacher or on a separate sheet of paper.

16 Jackson heard a song before school, and it stuck in his head. During the first hour of school, he sang it out loud. Samantha and Reese got it stuck in their heads. During the second hour, Samantha and Reese each passed the song on to 2 more of their friends.

Each hour, anyone who heard the song the hour before sang it out loud and made 2 more people sing along.

  a. Make a chart showing the hour and how many people were hearing the song.

  b. How many people heard the song during 6th hour?

  c. How many total people had heard the song by 6th hour?

  d. When would the 150th person hear the song?

NEED EXTRA HELP?												
If You Missed Question...	1–3	4	5	6	7	8–9	10	11	12–13	14	15	16
Go to Lesson...	5-1	5-5	5-6	5-1	5-5	5-2	5-5	5-3	5-5	5-4	5-6	5-3
For Help with Algebra Readiness Standard...	7NS2.1	7AF2.1	ALG2.0	7NS2.1	7AF2.1	7NS2.1	7AF2.1	7MR2.2	7AF2.1	ALG2.0	ALG2.0	7MR2.2

CHAPTER 6

Ratios, Rates, Proportion, and Percent

Here's What I Need to Know

Standard 7NS1.3 Convert fractions to decimals and percents and use these representations in estimations, computations, and applications.

Standard 7AF4.2 Solve multistep problems involving rate, average speed, distance, and time or a direct variation.

Standard ALG5.0 Students solve multistep problems, including word problems, involving linear equations in one variable and provide justification for each step.

Vocabulary Preview

ratio A comparison of two quantities by division (p. 318)

$$3 \text{ girls out of } 8 \rightarrow \frac{3}{8}$$

rate A ratio of two quantities with different units (p. 319)

130 miles in 2 hours

proportion An equation stating that two ratios are equivalent (p. 330)

$$\frac{5}{8} = \frac{15}{24}$$

percent A ratio that compares a number to 100 (p. 323)

76 out of 100 or 76%

 The What

I will learn to:

- find ratios and rates.
- set up and solve proportions, including percent proportions.
- use direct variations.

 The Why

Proportions and percents are used to calculate pay, sale prices, and test scores.

Option 1

Math Online Are you ready for Chapter 6? Take the Online Readiness Quiz at ca.algebrareadiness.com to find out.

Option 2

Complete the Quick Check below.

Vocabulary and Skills Review

Match each sentence with the correct vocabulary word.

1. In $4x = 20$, x is the _____?_____ .   A. equation

2. To _____?_____ $4x = 20$, divide   B. expression
   each side by 4.   C. solve

3. Both sides of a(n) _____?_____   D. variable
   have the same value.

Solve each equation.

4. $5n = 4 \cdot 10$ 5. $20 \cdot 5 = 4p$

Use the relationship between x and y to find the missing number.

6.

x	10	15	20	?
y	2	3	4	5

7.

x	12	15	18	?
y	4	5	6	7

Note-taking

Tips
Use simpler words.
Include examples.

Chapter Instruction	My Notes
A ratio is a comparison of quantities, or numbers, by division. (p. 318)	Ratios compare numbers. Ratios can look like fractions. $\dfrac{18 \text{ boys}}{12 \text{ girls}} = 18 \text{ to } 12 = 18:12$
A ratio of two measurements or quantities with different units is called a rate. (p. 319)	A rate is a ratio with different units. $\dfrac{300 \text{ miles}}{6 \text{ hours}}$

6-1 Ratios and Rates

Vocabulary

ratio (p. 318)

rate (p. 319)

unit rate (p. 319)

**Standard 7NS1.3
Convert fractions to decimals** and percents **and use these representations in estimations, computations, and applications.**

> **The** *What*: I will express relationships as ratios. I will use rates and unit rates to solve real-world problems.
>
> **The** *Why*: Measurements such as miles per hour and price per pound can be represented with rates.

A middle school has 300 students. A high school has 500 students. You can compare the number of middle school students to high school students by using a ratio.

A **ratio** is a comparison of quantities, or numbers, by division. A ratio can be written three ways:

RATIOS

Words	**300** middle school students to **500** high school students
Colon	**300:500**
Fraction	$\dfrac{300}{500}$

Usually a ratio is expressed as a fraction in simplest form or as a decimal.

$$\frac{300}{500} = \frac{300 \div 100}{500 \div 100} = \frac{3}{5} = 0.6$$

There is a 3:5 or 0.6 ratio of middle school to high school students.

Suppose the middle school has 160 boys and 140 girls. The ratio of boys to girls is 160:140, which simplifies to 8:7.

SHAPES Write the answer to Example 4 in colon form.

Write each ratio in the form requested.

1 2 out of 10 students as a fraction in simplest form

$$\frac{2}{10} = \frac{2 \div 2}{10 \div 2} = \frac{1}{5}$$

2 6 out of 12 people as a fraction in simplest form

$$\frac{6}{12} = \frac{6 \div 6}{12 \div 6} = \frac{1}{2}$$

3 10 girls to 4 boys in colon form

10:4

4 5 rectangles to 20 shapes as a decimal

$$\frac{5}{20} = \frac{1}{4} = 0.25$$

Your Turn

Write each ratio in the form requested.

a. 6 out of 18 people as a fraction in simplest form

b. 5 out of 15 people as a fraction in simplest form

c. 8 boys to 11 girls in colon form

d. 9 poodles to 36 dogs as a decimal

A ratio of two measurements or quantities with different units is called a **rate.** For example, the number of miles driven in a certain number of hours is a rate.

$$\text{Rate} \implies \frac{\mathbf{300} \text{ miles}}{\mathbf{6} \text{ hours}}$$

To find the number of miles per hour, you will use a **unit rate.** A unit rate has a denominator of 1. Divide both the numerator and the denominator by the denominator to find the unit rate.

$$\text{Unit Rate} \implies \frac{300 \text{ miles}}{6 \text{ hours}} = \frac{300 \div 6}{6 \div 6} = \frac{50 \text{ miles}}{1 \text{ hour}}$$

The unit rate is 50 miles per hour.

Talk Math

WORK WITH A PARTNER Suppose a school has 700 students and 35 teachers. Find the unit rate of students per teacher.

RATIOS, RATES, AND UNIT RATES

	Ratio	Rate	Unit Rate
Definition	a comparison by division	a comparison involving different units	a rate with a denominator of 1
Examples	Students to Students	Cats to Dogs	Miles per Hour
Words	2 out of 10	10 cats to 4 dogs	50 miles in 1 hour
Symbols	$2:10$ $\frac{2}{10}$ or $\frac{1}{5}$	10 cats $:4$ dogs $\frac{10 \text{ cats}}{4 \text{ dogs}}$ or $\frac{5 \text{ cats}}{2 \text{ dogs}}$	50 miles $:1$ hour $\frac{50 \text{ miles}}{1 \text{ hour}}$ 50 miles per hour

EXAMPLE   Write Unit Rates

5 **Express 360 cookies to 9 boxes as a unit rate.**

$$\frac{360 \text{ cookies}}{9 \text{ boxes}} = \frac{360 \div 9}{9 \div 9} = \frac{40 \text{ cookies}}{1 \text{ box}}$$ Divide each part by 9 to get a denominator of one.

The unit rate is 40 cookies per box.

Your Turn

Express as a unit rate.

e. 185 miles on 5 gallons of gas

> **Remember!**
>
> Unit rates always have a denominator of one. Don't forget to divide both the numerator and denominator by the number in the denominator.

Real-World EXAMPLE

6 **BAKING** Patsy is making cookies. The recipe calls for 8 cups of flour for every 4 teaspoons of baking soda. What is the unit rate of flour to baking soda?

$$\frac{8 \text{ cups flour}}{4 \text{ teaspoons baking soda}} = \frac{8 \div 4}{4 \div 4} = \frac{2 \text{ cups flour}}{1 \text{ teaspoon baking soda}}$$

Divide each part by 4 to get a denominator of one.

The unit rate is 2 cups of flour per teaspoon of baking soda.

Your Turn

f. **FITNESS** Adelina is doing sit-ups in the gym. She does 64 sit-ups in 4 minutes. What is the unit rate at which she does her sit-ups?

Vocabulary Review
ratio
rate
unit rate

Examples 1–6
(pages 319–320)

VOCABULARY

1. A(n) ____?____ describes a comparison by division.

2. The denominator of a(n) ____?____ is always equal to one.

Examples 1–4
(page 319)

Write each ratio in the form requested.

3. 12 out of 20 people as a fraction in simplest form

4. 5 out of 7 people in colon form

5. 20 out of 25 free throws as a decimal

Example 5
(page 320)

Express as a unit rate.

6. 32 ounces for 16 cents

7. 88 feet in 2 seconds

Example 6
(page 320)

8. **COOKING** Alfredo uses nine cups of flour for every three eggs in his dumplings batter. What is the unit rate of flour to eggs?

9. **SURVEY** A group of 450 people was asked their favorite color. Ninety people said blue. Express the ratio of people who prefer blue as a fraction in simplest form.

10. *Talk Math* Have a classmate say a ratio of two different objects in the classroom. Write the information as a unit rate.

Skills, Concepts, and Problem Solving

HOMEWORK HELP

For Exercises	See Example(s)
11–16	1–4
17–20	5
21–24	6

Write each ratio in the form requested.

11. 30 out of 45 as a fraction in simplest form

12. 243 out of 9 as a fraction in simplest form

13. 6 out of 11 in colon form

14. 350 out of 12 in colon form

15. 16 out of 20 as a decimal

16. 84 out of 168 as a decimal

Express as a unit rate.

17. 297 miles on 9 gallons

18. 480 miles in 8 hours

19. $25,000 for 100 employees

20. 96¢ per dozen eggs

21. **MARKET RESEARCH** A group of 459 people was asked which is their favorite amusement park. The results are shown in the table.

Express the ratio of people who prefer River Raft Kingdom to Old West City as a decimal.

Favorite Amusement Parks

Park	Responses
Patrick's Park	9
Old West City	30
Blue Mountain World	75
River Raft Kingdom	210
Adventure Island	135

22. **RETAIL** Grocery store signs display the unit prices of food items. A 14-ounce box of cereal costs $2.66. Find the unit price of the cereal.

23. **FITNESS** Cedro runs 2 miles in 18 minutes. What is the unit rate at which he runs in minutes per mile?

24. **TRANSPORTATION** A commuter train travels a distance of 45 miles in 60 minutes. What is the unit rate in miles per minute?

25. **H.O.T.** Problem The Surfliner can travel from Santa Barbara to Los Angeles in 180 minutes. The unit rate is 30 miles per hour. What is the distance from Santa Barbara to Los Angeles?

26. *Writing in Math* A hybrid automobile can travel 86 miles on 2 gallons of gasoline. Explain how to find the unit rate of miles per gallon.

Spiral Review

Replace each ⬤ with one of the symbols <, >, or = to make a true sentence. (Lesson 5-6)

30. -3 ⬤ -1.5

31. $\frac{7}{9}$ ⬤ 0.6

32. -0.6 ⬤ $-\frac{3}{5}$

33. **INDUSTRY** A local manufacturing company is buying a warehouse that measures 24-by-24 yards. How much larger is this warehouse than their original 15-by-15-yard warehouse? (Lesson 5-6)

6-2 Fractions, Decimals, and Percents

Vocabulary

percent (p. 323)

 Standard 7NS1.3 Convert fractions to decimals and percents and use these representations in estimations, computations, and applications.

Standard 6NS2.0 Students calculate and solve problems involving addition, subtraction, multiplication, and **division**.

> **The What:** I will write percents as fractions and decimals. I will write fractions as decimals and percents.
>
> **The Why:** Test scores are often given as percents.

A **percent** is a ratio that compares a number to 100. Percent means *hundredths* or *out of 100*. The symbol for percent is %.

The same number can be written as a fraction, a decimal, or a percent.

EQUIVALENT REPRESENTATIONS

Fraction	Decimal	Percent	Model
$\frac{20}{100}$ or $\frac{1}{5}$	0.20	20%	
$\frac{75}{100}$ or $\frac{3}{4}$	0.75	75%	

Chapter Test 95%

Suppose you get a score of 95% on a test. This is the same as $\frac{95}{100}$. Write the fraction in simplest form.

$$95\% = \frac{95}{100}$$
$$= \frac{95 \div 5}{100 \div 5}$$
$$= \frac{19}{20}$$

To write the test score as a decimal, remember that 95% is the same as 95 hundredths.

$$95\% = 0.95$$

SCHOOL What percent of students do not ride the bus?

Talk Math

WORK WITH A PARTNER Suppose 80% of the students in your school ride the bus. How can you write this as a fraction in simplest form? How can you write it as a decimal?

A teacher may write a test score as a fraction. Suppose you answer 20 out of 25 questions, or $\frac{20}{25}$, correctly. You can divide to write this fraction as a decimal.

$$\frac{20}{25} = 20 \div 25$$
$$= 0.8$$

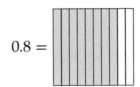

$$\frac{20}{25} =$$

$$0.8 =$$

Remember!

The number 0.8 is the same as 0.80. You can write the decimal 0.80 as $\frac{80}{100}$, or 80%.

FRACTIONS, DECIMALS, AND PERCENTS

Type of Change	What to Do	Example
percent to fraction	• write as number out of 100 • write the fraction in simplest form	$38\% = \frac{38}{100}$ $= \frac{19}{50}$
percent to decimal	• read percent as *per hundred* • divide by 100	$38\% = 38 \div 100$ $= 0.38$
fraction to decimal	• divide the numerator by the denominator	$\frac{19}{50} = 19 \div 50$ $= 0.38$
fraction to percent	• change the fraction to a decimal • read the decimal as hundredths	$\frac{19}{50} = 0.38$ $= 38$ hundredths $= 38\%$

Change a Percent to a Fraction and a Decimal

Write each percent as a fraction in simplest form. Then write it as a decimal.

① 5%

Fraction: $5\% = \dfrac{5}{100} = \dfrac{5 \div 5}{100 \div 5} = \dfrac{1}{20}$

Decimal: $5\% = 5 \div 100 = 0.05$

Write as 5 out of 100. Divide the numerator and denominator by the GCF, 5.

Read percent as *per hundred*. Divide by 100.

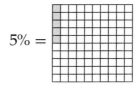

$5\% =$ $\dfrac{1}{20} =$

② 64%

Fraction: $64\% = \dfrac{64}{100} = \dfrac{64 \div 4}{100 \div 4} = \dfrac{16}{25}$

Decimal: $64\% = 64 \div 100 = 0.64$

Write as 64 out of 100. Divide the numerator and denominator by the GCF, 4.

Read percent as *per hundred*. Divide by 100.

Your Turn

Write each percent as a fraction in simplest form. Then write it as a decimal.

a. 7%   **b.** 35%   **c.** 51%

Change a Percent to a Fraction and a Decimal

③ Write 125% as a fraction in simplest form. Then write it as a decimal.

Fraction: $125\% = \dfrac{125}{100}$

$= \dfrac{125 \div 25}{100 \div 25} = \dfrac{5}{4} = 1\dfrac{1}{4}$

Decimal: $125\% = 125 \div 100 = 1.25$

Write as 125 out of 100. Divide the numerator and denominator by the GCF, 25.

Read percent as *per hundred*. Divide by 100.

Your Turn

Write each percent as a fraction in simplest form. Then write it as a decimal.

d. 162%   **e.** 175%   **f.** 114%

Write each fraction as a decimal. Then write it as a percent.

4 $\frac{2}{5}$

Decimal: $\frac{2}{5} = 2 \div 5 = 0.4$ — Divide the numerator by the denominator.

Percent: $0.4 = 0.40 = 40$ hundredths $= 40\%$ — Read the decimal as *hundredths*.

5 $\frac{5}{4}$

Decimal: $\frac{5}{4} = 5 \div 4 = 1.25$ — Divide the numerator by the denominator.

Percent: $1.25 = 125$ hundredths $= 125\%$

6 $\frac{4}{7}$

Decimal: $\frac{4}{7} = 4 \div 7 \approx 0.57$ — Round if there is no exact decimal.

Percent: $0.57 = 57$ hundredths $= 57\%$

Study Tip

Make sure that there are at least 2 places to the right of the decimal point before you change a decimal to a percent. If there are not at least 2, add a 0 to the right end of the deicmal.

Your Turn Write each fraction as a decimal. Round to the nearest hundredth if necessary. Then write it as a percent.

g. $\frac{3}{20}$ **h.** $\frac{21}{20}$ **i.** $\frac{7}{9}$

Real-World EXAMPLE

7 **SCHOOL** Paz has 32 problems to do for math homework. She has completed 24 of the problems. What percent of the problems has she completed?

$\dfrac{\text{number completed}}{\text{total number}} \begin{array}{l} \rightarrow \\ \rightarrow \end{array} \dfrac{24}{32}$ — Write as a fraction.

$24 \div 32 = 0.75$ — Divide the numerator by the denominator.

$0.75 = 75$ hundredths $= 75\%$ — Read the decimal as *hundredths*.

Paz has completed 75% of her math problems.

Your Turn

j. **FITNESS** Wes walked a total of 24 kilometers last week. His friend Dale walked with him for 6 kilometers. What percent of the distance did Dale walk?

Examples 1–7
(page 325–326)

VOCABULARY

1. Explain the meaning of *percent*.

Examples 1–3
(page 325)

Write each percent as a fraction in simplest form. Then write it as a decimal.

2. 4%   3. 9%   4. 56%

5. 85%   6. 145%   7. 265%

Examples 4–6
(page 326)

Write each fraction as a decimal. Then write it as a percent.

8. $\frac{1}{5}$ 9. $\frac{13}{25}$ 10. $\frac{7}{5}$

11. $\frac{13}{10}$ 12. $\frac{9}{11}$ 13. $\frac{14}{15}$

Example 7
(page 326)

14. **TRAVEL** Derek's family lives 216 miles from their vacation site. They have traveled 140 miles. What fraction of the miles have they traveled? About what percent of the miles do they have left to travel?

15. **GARDENING** About 35% of Carlota's flower garden is planted with roses. Write this percent as a decimal.

Skills, Concepts, and Problem Solving

HOMEWORK HELP	
For Exercises	**See Example(s)**
16–27	1–3
28–39	4–6
40–43	7

Write each percent as a fraction in simplest form. Then write it as a decimal.

16. 8%   17. 2%   18. 3%   19. 1%

20. 96%   21. 18%   22. 82%   23. 26%

24. 135%   25. 285%   26. 375%   27. 440%

Write each fraction as a decimal. Then write it as a percent.

28. $\frac{3}{5}$ 29. $\frac{8}{10}$ 30. $\frac{9}{20}$ 31. $\frac{17}{25}$

32. $\frac{3}{2}$ 33. $\frac{6}{5}$ 34. $\frac{8}{5}$ 35. $\frac{7}{4}$

36. $\frac{12}{10}$ 37. $\frac{18}{12}$ 38. $\frac{23}{20}$ 39. $\frac{26}{25}$

Solve.

40. **FARMING** A farmer has 225 acres of farmland. Only 125 of the acres are planted. About what percent of the land is planted?

41. **ARCHEOLOGY** Scientists are studying the site of an ancient city. The site is divided into 25 equal parts. The shaded parts of the figure show areas that have not been studied. About what percent has not been studied?

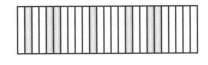

42. **CIVICS** In order to amend, or change, the United States Constitution, at least 75% of the states must agree. Write this percent as a fraction in simplest form.

43. **STATISTICS** The circle graph shows how a sample of people spent their vacations. Write each percent as a fraction in simplest form.

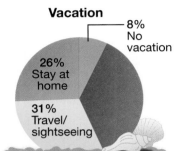

Vacation

- 8% No vacation
- 26% Stay at home
- 31% Travel/sightseeing

44. *Writing in Math* Which is greater: $\frac{3}{5}$ or 65%? Describe the steps you take to decide.

45. **H.O.T.** Problem There are 9 boys and 15 girls in a classroom. Explain how to find the percent of students that are girls.

Choose the *best* answer.

46 Which fraction is the same as 4.15?

  A $\frac{3}{20}$ **C** $\frac{3}{5}$

  B $\frac{83}{20}$ **D** $\frac{77}{5}$

47 Which of the following is equivalent to $\frac{15}{12}$?

  F 0.80   **H** 1.25

  G 4.5   **J** 5.4

48 During football practice, Alvaro made 17 field goal attempts out of 21. What is his field goal kicking percentage?

  A 81%

  B 85%

  C 90%

  D 91%

Spiral Review

Write each ratio as a decimal. (Lesson 6-1)

49. 176 : 220   50. 33 to 15   51. 36 out of 48

52. **FITNESS** Brad lives 0.9 kilometer from a park. He jogs to the park and back 4 days each week. What is the total distance he jogs each week? (Lesson 4-3)

Identify the greatest common factor (GCF) of each pair of numbers. (Lesson 3-3)

53. 42 and 64   54. 8 and 20   55. 24 and 30

Simplify each expression. (Lesson 1-8)

56. $8h + 9j + 7h + 3k$ 57. $3(4x + 9y) + 8(2x + 3y)$

Progress Check 1
(Lessons 6-1 and 6-2)

▷ Vocabulary and Concept Check

> ratio (p. 318)   unit rate (p. 319)
>
> rate (p. 319)   percent (p. 323)

Choose the term that best matches the example. You may use a term more than once. (Lessons 6-1 and 6-2)

1. $18:23$
2. 54%
3. 14.5 km/h
4. 3 gallons in 8 hours
5. 68 to 103
6. $5 per pound

▷ Skills Check

Write each ratio in the form requested. (Lesson 6-1)

7. 12 boys to 42 students as a fraction in simplest form

8. 2 out of 5 doctors in colon form

9. 85 out of 100 points as a decimal

Express as a unit rate. (Lesson 6-1)

10. 360 miles to 9 hours
11. 155 miles on 5 gallons
12. 27 students on 3 teams
13. 16 servings in 4 boxes

Write each percent as a fraction in simplest form. Then write it as a decimal. (Lesson 6-2)

14. 4%
15. 85%
16. 160%

Write each fraction as a decimal. Then write it as a percent. (Lesson 6-2)

17. $\frac{3}{5}$
18. $\frac{9}{20}$
19. $\frac{3}{8}$

▷ Problem-Solving Check

NUTRITION Marcel's granola bar has a mass of 50 grams, 24 of which are carbohydrates. (Lesson 6-2)

20. In simplest form, what fraction of the bar is carbohydrates?

21. What percent of the bar is carbohydrates?

22. **REFLECT** In your own words, explain how a fraction can be changed to a percent. (Lesson 6-2)

Proportions and Proportional Reasoning

Vocabulary

proportion (p. 330)

cross product (p. 330)

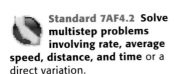

Standard 7AF4.2 Solve multistep problems involving rate, average speed, distance, and time or a direct variation.

Standard 7NS1.2 Add, subtract, **multiply, and divide rational numbers** (integers, fractions, and **terminating decimals**) and take positive rational numbers to whole-number powers.

Standard ALG5.0 Students **solve multistep problems, including word problems,** involving linear equations in one variable and provide justification for each step.

 The What: I will find cross products and solve proportions. I will use proportions to solve problems.

 The Why: Proportions are used when comparing quantities, such as miles per hour.

Average speed is the total distance an object travels divided by the time to move that distance. A car travels 150 miles in 3 hours. Another car travels 2 hours and goes 100 miles. The cars have the same average speed.

A **proportion** is an equation stating that two ratios are equivalent, or equal.

To check whether two ratios are equal, find their cross products. A **cross product** is a product of the numerator of one fraction and the denominator of another fraction.

$$\frac{150 \text{ miles}}{3 \text{ hours}} = \frac{100 \text{ miles}}{2 \text{ hours}}$$

$$150 \cdot 2 \quad = \quad 3 \cdot 100$$

$$300 \quad = \quad 300$$

The cross products of the proportion are equal. This means the ratios of miles to hours for the two cars are equal.

EQUAL RATIOS

Words: The cross products of a proportion are equal.

Numbers: $\frac{3}{4} = \frac{9}{12} \rightarrow 3 \cdot 12 = 4 \cdot 9$

Algebra: If $\frac{a}{b} = \frac{c}{d}$, then $ad = bc$.

Determine whether each pair of ratios is equal. Use cross products. Show your work.

1 $\frac{3}{5}, \frac{9}{15}$

$$\frac{3}{5} \stackrel{?}{=} \frac{9}{15}$$ Set ratios equal to each other.
Place ? over = sign.

$3 \cdot 15 \stackrel{?}{=} 5 \cdot 9$ Find the cross products.

$45 = 45$ ✔   Multiply to simplify each side.

The cross products are equal. The ratios are equal.

2 $\frac{3}{4}, \frac{7}{9}$

$$\frac{3}{4} \stackrel{?}{=} \frac{7}{9}$$ Set ratios equal to each other.
Place ? over = sign.

$3 \cdot 9 \stackrel{?}{=} 4 \cdot 7$ Find the cross products.

$27 \neq 28$ Multiply to simplify each side.

The cross products are not equal, so the ratios do not form a proportion.

Your Turn

Determine whether each pair of ratios is equal. Use cross products. Show your work.

a. $\frac{4}{5}, \frac{16}{20}$ b. $\frac{7}{9}, \frac{5}{7}$ c. $\frac{5}{9}, \frac{7}{9}$

EXAMPLE Solve a Proportion

3 Solve $\frac{m}{24} = \frac{5}{12}$.

$m \cdot 12 = 24 \cdot 5$ Find the cross products.

$12m = 24 \cdot 5$ Use the Commutative Property of Multiplication.

$\dfrac{12m}{12} = \dfrac{24 \cdot 5}{12}$ Divide each side by 12.

$m = \dfrac{^2 24 \cdot 5}{12_1}$ Divide 24 by 12.

$m = 2 \cdot 5$ Multiply 2 and 5.

$m = 10$

Your Turn

Solve each proportion.

d. $\frac{m}{27} = \frac{5}{9}$ e. $\frac{5}{b} = \frac{10}{24}$ f. $\frac{3}{8} = \frac{p}{24}$

Study Tip

Use mental math to solve proportions.

$\frac{2}{3} = \frac{x}{27}$

THINK:

$3 \cdot 9 = 27$
$2 \cdot 9 = 18$

$\overset{\times 9}{\frown}$
$\frac{2}{3} = \frac{x}{27}$
$\underset{\times 9}{\smile}$

So, $x = 18$.

Solve a Proportion

④ **Solve** $\dfrac{1.2}{5} = \dfrac{3.6}{r}$.

$1.2r = 5 \cdot 3.6$ Find the cross products.

$\dfrac{1.2r}{1.2} = \dfrac{5 \cdot 3.6}{1.2}$ Divide each side by 1.2.

$r = 5 \cdot \dfrac{\overset{3}{\cancel{3.6}}}{\underset{1}{\cancel{1.2}}}$ Divide 3.6 by 1.2.

$r = 5 \cdot 3$ Multiply 5 and 3.

$r = 15$

Your Turn

Solve each proportion.

g. $\dfrac{s}{2} = \dfrac{11.1}{3.7}$ h. $\dfrac{4.6}{c} = \dfrac{6.9}{3}$ i. $\dfrac{7}{2.1} = \dfrac{3}{n}$

Real-World EXAMPLE

⑤ **TRANSPORTATION** Kim can drive 150 miles in 3 hours. At the same rate, how far could she drive in 7 hours?

$\begin{array}{l}\text{miles} \rightarrow \\ \text{hours} \rightarrow\end{array} \dfrac{150}{3} = \dfrac{m}{7} \begin{array}{l}\leftarrow \text{miles} \\ \leftarrow \text{hours}\end{array}$ Write a proportion.

$150 \cdot 7 = 3m$ Find the cross products.

$1050 = 3m$ Multiply to simplify.

$\dfrac{1050}{3} = \dfrac{3m}{3}$ Divide each side by 3.

$m = 350$

Kim can drive 350 miles in 7 hours.

TRANSPORTATION The answer to Example 5 can be written two other ways.

$350 : 7$

$\dfrac{350}{7}$

Your Turn

j. **RETAIL** A store sells 3 notebooks for $2.25. At that rate, how many notebooks could you buy for $6.00?

k. **ENGINEERING** Peter draws a diagram of his house. One room 12 feet long is 3 centimeters on his diagram. Each room he draws should have the same ratio. How long should a 16-foot room be on his diagram?

Talk Math

WORK WITH A PARTNER List three ratios equal to $3 : 5$. Listen as your partner lists three ratios equal to $4 : 5$. Discuss how you can tell if two ratios are equal.

Examples 1–5
(pages 331–332)

VOCABULARY

1. A ____?____ is the result of multiplying the numerator of a fraction with the denominator of another fraction.

2. One ratio equal to a second ratio is called a ____?____.

Examples 1–2
(page 331)

Determine whether each pair of ratios is equal. Use cross products. Show your work.

3. $\frac{4}{6}, \frac{30}{45}$

4. $\frac{2}{3}, \frac{9}{11}$

5. $\frac{15}{24}, \frac{10}{16}$

Examples 3–4
(pages 331–332)

Solve each proportion.

6. $\frac{x}{3} = \frac{9}{15}$

7. $\frac{3}{5} = \frac{r}{25}$

8. $\frac{6}{n} = \frac{28}{7}$

9. $\frac{3.8}{2} = \frac{5.7}{c}$

10. $\frac{1.5}{8} = \frac{9}{6.6}$

11. $\frac{2.4}{10} = \frac{h}{1.5}$

Example 5
(page 332)

12. **GARDENING** Mori finds that about six flowers bloom for every 35 seeds she plants. If this ratio continues, about how many flowers will bloom if she plants 175 seeds?

13. **PHYSICAL SCIENCE** A toy car travels 1.3 meters in 2.5 seconds. At this speed, how far will the car travel in 15 seconds?

14. *Talk Math* **WORK WITH A PARTNER** Suppose a recipe uses three cups of flour for every two eggs. Discuss how to set up proportions to double the recipe and to triple the recipe.

Skills, Concepts, and Problem Solving

HOMEWORK HELP

For Exercises	See Example(s)
15–20	1–2
21–26	3
27–32	4
33–36	5

Determine whether each pair of ratios is equal. Use cross products. Show your work.

15. $\frac{11}{13}, \frac{4}{7}$

16. $\frac{4}{12}, \frac{5}{15}$

17. $\frac{6}{7}, \frac{8}{12}$

18. $\frac{5}{8}, \frac{20}{21}$

19. $\frac{12}{18}, \frac{8}{12}$

20. $\frac{6}{8}, \frac{4}{9}$

Solve each proportion.

21. $\frac{n}{5} = \frac{6}{15}$

22. $\frac{7}{2} = \frac{p}{26}$

23. $\frac{14}{21} = \frac{m}{18}$

24. $\frac{36}{24} = \frac{27}{n}$

25. $\frac{x}{25} = \frac{36}{15}$

26. $\frac{28}{t} = \frac{21}{15}$

27. $\frac{5}{s} = \frac{3}{1.8}$

28. $\frac{a}{1.6} = \frac{7}{3.2}$

29. $\frac{1.2}{6} = \frac{c}{3.6}$

30. $\frac{1.8}{4.5} = \frac{4}{r}$

31. $\frac{1.8}{0.05} = \frac{k}{0.5}$

32. $\frac{0.5}{5} = \frac{1.3}{b}$

33. **BUSINESS** The table shows the number of lamps a company can put together by various numbers of employees. If the rate continues, how many lamps will 18 employees assemble?

Number of Employees	Lamps Assembled
8	144
12	216
15	270
18	?

34. **MARKET RESEARCH** A poll shows two out of three people support a candidate. Suppose the city has 45,000 people. Based on this ratio, about how many people support the candidate?

35. **FOOD** If 40 ounces of cranberry juice costs $3.20, how much should 25 ounces cost?

36. **NUTRITION** A poll shows that about two out of five students at a school take a daily vitamin. 850 students attend the school. Based on this ratio, about how many take a daily vitamin?

37. *Writing in Math* Describe an example in your daily life where you could use a proportion.

38. **H.O.T.** Problem If three bakers can prepare 15 cakes in one hour, how long will it take six bakers working at the same rate to prepare 20 cakes?

STANDARDS PRACTICE

Choose the *best* answer.

39 Which pair of ratios would form a proportion?

A $\frac{10}{6}, \frac{6}{10}$

B $\frac{10}{6}, \frac{12}{8}$

C $\frac{10}{6}, \frac{15}{9}$

D $\frac{10}{6}, \frac{3}{5}$

40 Pam designs a rectangular advertisement as shown. She makes a new advertisement that is 6 centimeters wide and has the same height to width ratio. About how high is the new advertisement?

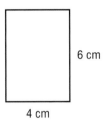

6 cm

4 cm

F 4 cm   H 6 cm

G 9 cm   J 10 cm

Spiral Review

Write each percent as a decimal. (Lesson 6-2)

41. 18%   42. 92%   43. 143%   44. 364%

45. **GARDENING** Elian uses about $\frac{1}{8}$ of his yard for a vegetable garden. He wants to plant squash in $\frac{2}{5}$ of the garden. What fraction of his yard is used to plant squash? (Lesson 3-5)

Divide. (Lesson 2-5)

46. $64 \div (-4)$ 47. $-96 \div (-6)$ 48. $-42 \div 14$

Math Lab
Capture and Recapture

Materials

small bowl

dried lima beans

marker

Standard 7AF4.2 Solve multistep problems involving rate, average speed, distance, and time or a direct variation.

Standard 7MR2.5 Use a variety of methods, such as words, numbers, symbols, charts, graphs, **tables,** diagrams, and **models, to explain mathematical reasoning.**

The What: I will use lima beans to model a study of wild animals.

The Why: Scientists use proportions to monitor animals in the wild.

You can estimate how many salmon are in the Sacramento River using the capture-recapture technique.

Sample	Tagged	Recaptured
A		
B		
C		
D		
E		

Study Tip

Count the number of beans in the bowl. How close was your estimate?

ACTIVITY Capture and Recapture

1. **CAPTURE** Grab a handful of beans. Mark each side of the beans with an X. Record the number *captured.* Return the "tagged" beans to the bowl. Mix well.

2. **RECAPTURE** Grab a handful of beans. Record the number *recaptured.* Record the number of "tagged" beans as sample A. Return the beans to the bowl.

3. Repeat **RECAPTURE** five more times as samples B through E. Find the total tagged and the total recaptured.

4. **ESTIMATE** Use the proportion below to estimate the number of beans in your bowl.

$$\frac{\text{original number captured}}{\text{number in bowl}} = \frac{\text{total tagged in samples}}{\text{total recaptured}}$$

Your Turn

a. Suppose you initially captured 12 beans. After several samples, you count 45 tagged and 150 recaptured beans. About how many total beans did you originally have?

Analyze the Results

1. Suppose some tags wore off. What would happen to your estimate?

2. **REFLECT** Why do estimates and actual totals differ?

The Percent Proportion

Vocabulary

percent proportion
(p. 336)

percentage (p. 336)

base (p. 336)

 Standard 7NS1.3 Convert fractions to decimals and percents and use these representations in estimations, computations, and applications.

Standard 7AF4.2 Solve multistep problems involving rate, average speed, distance, and time or a direct variation.

Standard ALG5.0 Students solve multistep problems, including word problems, involving linear equations in one variable and provide justification for each step.

 The What: I will use percent proportions to solve problems.

The Why: Percent proportions are used to decide how much merchandise to order.

You can use the **percent proportion** to express fractions as percents and to solve percent problems.

$$\underset{\text{base or total}}{\overset{\text{part}}{\frac{P}{B}}} = \overset{\text{percent}}{\frac{r}{100}} \Big\} \text{percentage}$$

The first ratio is the part P divided by the **base** B. The base represents the total amount. The second ratio, $\frac{r}{100}$, is the **percentage.** The value of r is the percent.

Katrina is ordering merchandise for the school store. She decides to order school shirts. A survey finds 50 out of 250 students wear medium size shirts. Katrina uses a percent proportion to find the percent of medium shirts she should order.

$\frac{50}{250} = \frac{r}{100}$	Set up percent proportion.
$250r = 50 \cdot 100$	Find cross products.
$250r = 5000$	Multiply 50 by 100.
$\frac{250r}{250} = \frac{5000}{250}$	Divide each side by 250.
$r = 20$	

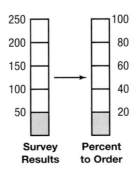

Survey Results   **Percent to Order**

Katrina should order 20% of the school shirts in size medium.

Talk Math

WORK WITH A PARTNER Katrina also wants to order large shirts. Of the 250 students surveyed, 150 students wear large shirts. Work with your partner to find the number of size large shirts she should order.

USING A PERCENT PROPORTION

Direction	Example
Set up the percent proportion.	$\dfrac{P}{B} = \dfrac{r}{100}$
Substitute the two known values.	What number is 12% of 25? $\dfrac{P}{25} = \dfrac{12}{100}$
Find cross products.	$100 \cdot P = 25 \cdot 12$
Solve for the unknown variable.	$P = 3$

EXAMPLES **Express a Fraction as a Percent**

Express each fraction as a percent.

① $\dfrac{2}{5}$

$\dfrac{2}{5} = \dfrac{r}{100}$	Set up the percent proportion.
$2 \cdot 100 = 5r$	Find cross products.
$200 = 5r$	Multiply 2 by 100.
$\dfrac{200}{5} = \dfrac{5r}{5}$	Divide each side by 5.
$r = 40$	Simplify.
$\dfrac{2}{5} = 40\%$	Write the answer as a percent.

② $\dfrac{6}{4}$

$\dfrac{6}{4} = \dfrac{r}{100}$	Set up the percent proportion.
$6 \cdot 100 = 4r$	Find cross products.
$600 = 4r$	Multiply 6 by 100.
$\dfrac{600}{4} = \dfrac{4r}{4}$	Divide each side by 4.
$r = 150$	Simplify.
$\dfrac{6}{4} = 150\%$	Write the answer as a percent.

Your Turn

Express each fraction as a percent.

a. $\dfrac{7}{25}$ b. $\dfrac{9}{24}$ c. $\dfrac{32}{25}$

Remember!

You multiply numerators by denominators to find cross products.

$\dfrac{a}{b} = \dfrac{c}{d}$

$a \cdot d = b \cdot c$

Use a proportion to solve.

3 **What number is 15% of 20?**

$$\frac{P}{20} = \frac{15}{100}$$ Set up the percent proportion.

$$P \cdot 100 = 300$$ Find cross products and multiply.

$$\frac{100P}{100} = \frac{300}{100}$$ Divide each side by 100.

$$P = 3$$ Simplify.

4 **8 is 20% of what number?**

$$\frac{8}{B} = \frac{20}{100}$$ Set up the percent proportion.

$$800 = 20 \cdot B$$ Find cross products and multiply.

$$\frac{800}{20} = \cdot \frac{20B}{20}$$ Divide each side by 20.

$$40 = B$$ Simplify.

5 **30 is what percent of 150?**

$$\frac{30}{150} = \frac{r}{100}$$ Set up the percent proportion.

$$3{,}000 = 150r$$ Find cross products and multiply.

$$\frac{3{,}000}{150} = \frac{150r}{150}$$ Divide each side by 150.

$$20 = B$$ Simplify.

$$20\%$$ Write the answer as a percent.

Your Turn Use a proportion to solve.

d. What number is 70% of 50?

e. 20 is 40% of what number?

f. 7 is what percent of 28?

● **Real-World Challenge**· · · · ·

RETAIL A bike usually costs $240. It is on sale for 30% off. What is the savings?

Real-World EXAMPLE

6 **RETAIL** A camera costs $320. The sales tax is 5%. What is the total cost of the camera?

$$\frac{P}{320} = \frac{5}{100}$$ Set up the percent proportion.

$$P \cdot 100 = 1600$$ Find cross products and multiply.

$$\frac{100P}{100} = \frac{1{,}600}{100}$$ Divide each side by 100.

$$P = 16$$ Simplify.

The tax is $16. The camera costs $320 plus $16, or $336.

Your Turn

g. **RETAIL** A jacket costs $40. The sales tax is 6%. What is the total cost of the jacket?

Examples 1–5
(pages 337–338)

VOCABULARY

Vocabulary Review
percent proportion
percentage
base

1. A number divided by 100 is a ____?____.

2. A ____?____ is an equality between a fraction and a percentage.

Examples 1–2
(page 337)

Express each fraction as a percent.

3. $\frac{3}{5}$ 4. $\frac{7}{20}$ 5. $\frac{15}{4}$ 6. $\frac{8}{5}$

Examples 3–5
(page 338)

Use a proportion to solve.

7. What number is 7% of 400?   8. What number is 8% of 50?

9. 9 is 20% of what number?   10. 14 is 7% of what number?

11. 15 is what percent of 75?   12. 75 is what percent of 250?

Example 6
(page 338)

13. **SALES** A skateboard is on sale for $56. The sale price is 80% of the regular price. What is the regular price?

14. **POPULATION** Today the world population is 6.5 billion people. Experts estimate that by 2050 the population will reach 9 billion. Determine the approximate percent that the number will increase.

15. **Talk Math** Discuss with a classmate the different types of percent proportion problems in this lesson. What are two steps you should always take when solving these problems?

Skills, Concepts, and Problem Solving

HOMEWORK HELP

For Exercises	See Example(s)
16–23	1–2
24–35	3–5
36–39	6

Express each fraction as a percent.

16. $\frac{2}{5}$ 17. $\frac{4}{5}$ 18. $\frac{7}{8}$ 19. $\frac{5}{8}$

20. $\frac{24}{25}$ 21. $\frac{13}{5}$ 22. $\frac{3}{2}$ 23. $\frac{5}{4}$

Use a proportion to solve.

24. What number is 15% of 40?   25. What number is 12% of 25?

26. What number is 60% of 85?   27. What number is 16% of 25?

28. 36 is 45% of what number?   29. 21 is 35% of what number?

30. 48 is 60% of what number?   31. 75 is 12% of what number?

32. 48 is what percent of 60?   33. 56 is what percent of 80?

34. 49 is what percent of 35?   35. 72 is what percent of 25?

36. **FITNESS** Ellen bikes, jogs, and walks a total of 25 kilometers each week. Each week, she rides her bike 8 kilometers and jogs 7 kilometers, she walks the rest of the difference. What percent of the distance does she walk?

37. **SALES** A $240 mountain bike is on sale for 60% of the regular price. What is the sale price?

38. **SPORTS** Forty-eight students try out for the tennis team. Only 12 students are chosen. What percent of students do not make the team?

39. **CAREER CONNECTION** California has more than 7,000 veterinarians. About 75% work in private practice. About how many of the veterinarians work in private practice?

40. *Writing in Math* A skateboard is on sale for $36. This sale price is 80% of the usual price. Write a paragraph describing how a percent proportion can be used to find the usual price.

41. **H.O.T. Problem** The table shows nutrition facts for one serving of macaroni and cheese. The recommended amount of saturated fat is 20 grams. What percent of this recommended amount is the actual amount?

Amount (g)		
Nutrient		**Percent Daily Value**
Total Fat	12	18%
Saturated Fat	3	?

STANDARDS PRACTICE

Use the graph to answer the questions.

U.S. Land Use

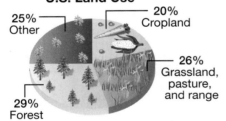

- 25% Other
- 20% Cropland
- 26% Grassland, pasture, and range
- 29% Forest

The total land area of the United States is about 2,263 million acres.

42 Which proportion could you use to determine the amount of forest land?

A $\frac{2,263}{x} = \frac{29}{100}$ C $\frac{29}{2,263} = \frac{x}{100}$

B $\frac{x}{2,263} = \frac{29}{100}$ D $\frac{x}{29} = \frac{2,263}{100}$

43 About how many millions of acres of land is grassland, pasture, and range?

F 453   H 4,530

G 588   J 5,880

Spiral Review

Solve each proportion. (Lesson 6-3)

44. $\frac{y}{5} = \frac{18}{30}$

45. $\frac{n}{12} = \frac{24}{48}$

46. **HOBBIES** If 45 comic books cost $180, how many can be purchased for $20? (Lesson 6-3)

Express each fraction as a decimal. (Lesson 4-1)

47. $\frac{8}{9}$

48. $\frac{12}{15}$

49. $\frac{23}{24}$

50. $6\frac{3}{4}$

Multiply. (Lesson 2-4)

51. $(-x)(-7)$

52. $3(12s)$

53. $6ab(-3)(4)$

54. $7(-m)(n)$

Progress Check 2
(Lessons 6-3 and 6-4)

▶ Vocabulary and Concept Check

> base (p. 336)
> cross products (p. 330)
> percentage (p. 336)
>
> percent proportion (p. 336)
> proportion (p. 330)

Choose the term that *best* matches the example. (Lessons 6-3 and 6-4)

1. A _____?_____ is an equation that shows that two ratios are equal.

2. When you write a _____?_____ , you write an equality between a fraction and a percentage.

3. The _____?_____ of a proportion are equal.

4. In a percent proportion, the _____?_____ is the original amount.

▶ Skills Check

Solve each proportion. (Lesson 6-3)

5. $\dfrac{m}{8} = \dfrac{21}{56}$

6. $\dfrac{12}{39} = \dfrac{n}{13}$

7. $\dfrac{v}{24} = \dfrac{5}{6}$

8. $\dfrac{3.6}{9} = \dfrac{1.2}{n}$

9. $\dfrac{4.9}{28} = \dfrac{0.7}{b}$

10. $\dfrac{0.9}{c} = \dfrac{7.2}{32}$

Express each fraction as a percent. (Lesson 6-4)

11. $\dfrac{6}{8}$

12. $\dfrac{11}{25}$

13. $\dfrac{16}{25}$

14. $\dfrac{17}{50}$

Use a proportion to solve. (Lesson 6-4)

15. What number is 30% of 60?

16. What number is 35% of 80?

17. 24 is what percent of 96?

18. 15 is what percent of 75?

19. 16 is 25% of what number?

20. 32 is 40% of what number?

▶ Problem-Solving Check

21. **SPORTS** Kara Lawson, of the Sacramento Monarchs, made about 92% of the free throws she shot during a recent season. Suppose she attempted 50 free throws. About how many did she make? (Lesson 6-4)

22. **CONSUMER AWARENESS** Suppose 10 cans of cat food cost $4. At this same rate, how much will 25 cans of cat food cost? (Lesson 6-3)

23. **REFLECT** Write a paragraph that explains how to set up a percent proportion. Tell how to decide where to place the variable. Use examples in your explanation. (Lesson 6-4)

Vocabulary

discount (p. 342)

interest (p. 342)

principal (p. 342)

tip (p. 345)

commission (p. 345)

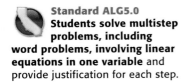

Standard ALG5.0 Students solve multistep problems, including word problems, involving linear equations in one variable and provide justification for each step.

Standard 7AF4.2 Solve multistep problems involving rate, average speed, distance, and time or a direct variation.

Standard 6NS1.4 Calculate given percentages of quantities and solve problems involving discounts at sales, interest earned, and tips.

 The *What:* I will use percents to solve problems.

 The *Why:* Percents are used to find discounted prices, interest on bank accounts, and tips for restaurant servers.

Stores often advertise discounted sales prices. A **discount** is a decrease in the regular price.

$$\frac{\text{discounted amount}}{\text{original amount}} = \frac{r}{100}$$

Interest is money earned on a bank account or money paid toward a bank loan. To find simple interest, you need to know three items.

1. p: the **principal,** or starting amount of money

2. r: the interest rate expressed as a decimal

3. t: the time, in years

You can use the interest equation to solve problems involving interest.

Interest = principal · rate · time

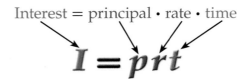

$$I = prt$$

Percents are used to calculate discount prices and interest amounts. Percents are also used to calculate tips and commissions.

30% off   20% tip   6.5% interest

There are two ways to find the discount price of an item.

Method 1—Find Discount and Subtract

Method 2—Find Discounted Price

Both ways use the percent proportion you studied in Lesson 6-4.

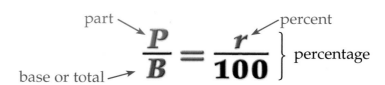

$$\frac{P}{B} = \frac{r}{100} \left.\right\} \text{percentage}$$

part → P (part)
base or total → B
percent → r

Real-World EXAMPLE   Calculate Discount

1 RECREATION The entry fee for a water park is $35. The park gives a 20% discount to groups. What is the discounted group fee?

Method 1—Find Discount and Subtract

$\dfrac{\text{Discounted Amount}}{\text{Original Amount}} = \dfrac{r}{100}$ Use the percent proportion.

$\dfrac{P}{35} = \dfrac{20}{100}$ Substitute for known values.

$100P = 700$ Find cross products and multiply.

$P = \dfrac{700}{100}$ Divide each side by 100.

$P = 7$

The discount is $7. The discounted fee is $35 – $7 or $28.

Method 2—Find Discounted Price

$\dfrac{\text{Discounted Amount}}{\text{Original Amount}} = \dfrac{r}{100}$ Use the percent proportion.

$\dfrac{P}{35} = \dfrac{80}{100}$ Discounted fee will be 80% of original amount.

$100P = 2800$ Find cross products and multiply.

$P = 28$ Divide each side by 100.

The discounted fee is $28.

Study Tip

Use estimation to check your answers.

18% can be rounded to 20%, or $\frac{1}{5}$.

$\frac{1}{5}$ of $50 is $10.

Estimate: $50−$10 or $40

Your Turn

a. **RETAIL** A $50 jacket is on sale for 18% off. What is the sale price of the jacket?

b. **TECHNOLOGY** The cost of a computer is $1,200. It is on sale for 15% off. What is the sale price of the computer?

Use the formula $I = prt$ to find simple interest earned on money in a bank account or money paid toward a bank loan. Remember to change percents to decimals when using the interest formula.

Real-World EXAMPLE Calculate Interest

2 **BANKING** Evan has $660 in a bank account. He earns 5.5% on the money each year. How much will he have in the account after 6 months?

p (principal) = $660
r (interest rate) = 5.5%
t (time) = 6 months $\left(\frac{1}{2} \text{ year}\right)$

$I = prt$	Use the interest equation.
$I = (660)(5.5\%)\left(\frac{6}{12} \text{ year}\right)$	Substitute for p, r, and t.
$I = (660)(0.055)\left(\frac{6}{12} \text{ year}\right)$	Change the percent to a decimal.
$I = (660)(0.055)(0.5)$	Change the fraction to a decimal.
$I = 18.15$	Multiply to find the interest.

Evan will earn $18.15 interest.
The total in his account will be $660 + $18.15 = $678.15.

Remember!

There are 12 months in a year.

3 months = $\frac{3}{12}$ or 0.25

6 months = $\frac{6}{12}$ or 0.5

9 months = $\frac{9}{12}$ or 0.75

Your Turn

c. **TRAVEL** Antonia wants to pay for a school trip. She has $50 in an account that earns 5% each year. How much money will Antonia have in 3 months?

d. **LOANS** Jairo borrowed $1,500 for school tuition at a 3.5% interest rate. How much money will Jairo owe at the end of one year?

Talk Math

WORK WITH A PARTNER Take turns changing percents to decimals. For example, you say $3\frac{1}{2}\%$ and your partner says 0.035. Change a total of six different percents.

A **tip** is money you give someone for doing a job well. A **commission** is money a salesperson earns based on sales.

Real-World EXAMPLE   Calculate a Tip

❸ FOOD SERVICE Tami and her family went to a restaurant. The check was $29.35. Tami's family gave a tip to the server that was 15% of $30. How much money did the server receive?

$\dfrac{\text{Tip Amount}}{\text{Original Amount}} = \dfrac{r}{100}$ Use the percent proportion.

$\dfrac{T}{30} = \dfrac{15}{100}$ Substitute for known values.

$100T = 450$ Find cross products and multiply.

$T = \dfrac{450}{100}$ Divide each side by 100.

$T = 4.5$ Simplify.

The server received a $4.50 tip.

Your Turn

e. **SERVICE** The cost of a haircut is $24.50. You give a tip to the stylist that is 15% of $25. How much money does the stylist receive in all?

Real-World EXAMPLE   Calculate Commission

❹ SALES James is a salesperson at a department store. He earns $5\frac{1}{2}$% commission on all sales. One week, James sold $15,000 worth of clothing. What was his commission?

$\dfrac{\text{Commission}}{\text{Original Amount}} = \dfrac{r}{100}$ Use the percent proportion.

$\dfrac{C}{15{,}000} = \dfrac{5.5}{100}$ Substitute for known values.

$100C = 82{,}500$ Find cross products and multiply.

$C = \dfrac{82{,}500}{100}$ Divide each side by 100.

$C = 825$

James earned a commission of $825.

Your Turn

f. **SALES** A clerk sells $18,000 worth of goods. Suppose the clerk's commission rate is $6\frac{1}{2}$% of all sales. What is the clerk's commission?

Vocabulary Review
interest
principal
discount
tip
commission

Examples 1–4 **VOCABULARY**
(pages 343–345)

1. If an item sells for 80% of the original price, the amount taken off is the ___?___.

2. When you use the formula $I = prt$, to find ___?___, you multiply the ___?___ by the rate and the time.

3. A ___?___ is a percent of sales that a salesperson receives.

Example 1
(page 343)

4. **READING** A bookstore offers a membership that gives a 20% discount on any item purchased to book club members. If a book costs $25, how much will a member pay for the book?

5. **RETAIL** All electronic appliances are sold for 25% off the regular price during a sale at an appliance store. If a 50-inch TV originally sells for $1,800, how much does it cost on the day of the sale?

Example 2
(page 344)

6. **BANKING** Yoko opens a bank account with $900. The money earns 3%. How much money is in the account at the end of 6 months?

Example 3
(page 345)

7. **FOOD SERVICE** Dana and her friends go out for lunch. The bill for the meal is $68.33. If Dana and her friends want to leave a tip that is 15% of $70, how much money should they leave?

Example 4
(page 345)

8. **SALES** A salesperson receives a $4\frac{1}{2}\%$ commission on all sales. If she has sales totaling $7,500 one week, how much commission will she receive?

9. *Talk Math* Discuss with a classmate how finding commission is similar to finding the amount of a tip.

Skills, Concepts, and Problem Solving

HOMEWORK HELP

For Exercises	See Example(s)
10–13	1
14–15	2
16–17	3
18–19	4

Calculate the discount to solve.

10. **RETAIL** A store has a sale on wrapping paper. All paper is on sale for 20% off the regular price. If rolls of wrapping paper regularly sell for $6 each, how much is a roll on sale?

11. **BUSINESS** Freddie's Fruit Stand is selling all its fruit for 20% off. Tansy purchases fruit that regularly sells for $8. How much did Tansy pay for the fruit?

12. **FASHION** All dresses in the juniors department are on sale for 25% off on Saturday. If Zoe wants a dress that originally sells for $88, how much will she pay for the dress on sale?

13. **SPORTS** The Sports Equipment Store is selling all sports equipment for 25% off the regular price. Prem buys equipment that originally sold for $96. How much does he pay for the equipment?

Calculate the interest to solve.

14. **TECHNOLOGY** Emil is saving to buy a laptop computer. He opens a savings account which offers 5% interest. If he puts $1,200 in the account, how much interest does the money earn after 6 months?

15. **BANKING** Ofelia has an account at the bank. If the bank gives 4.5% interest and Ofelia has $1,400 in the bank, how much will she receive in interest at the end of 6 months?

Calculate the tip to solve.

16. **CARS** Gilberto has his car detailed at the local car wash. The total bill comes to $157.31. If he plans to give a tip that is 20% of $160, how much should he give?

17. **FOOD SERVICE** A waitress served dinner to a group of 25 people. The bill came to $259.98. If she receives a tip that is 15% of $260, how much money will she receive?

Calculate the comission to solve.

Real-World Challenge· · · · · · · ·
In 1998, 60% of salespeople in the car industry worked on commission only. By 2000, 49.5% of the salespeople worked on commission only.

18. **SALES** Many car salespeople receive only commission as their pay. Mr. Evans is a salesperson at a car dealership. He receives a $6\frac{1}{2}$% commission on each car he sells. In one month, he sells $75,000 worth of cars. How much money does Mr. Evans earn that month?

19. **REAL ESTATE** Ms. Feurtado is a real estate agent. She receives a $5\frac{1}{2}$% commission on each property she sells. Suppose she sells a plot of land for $90,000. How much money does she earn?

20. **SPORTS** Last year, the Cougar Soccer Team scored 30 goals. Their coach wants to increase the number of goals scored by 20%. How many goals does the coach want the Cougars to score this year?

21. **COMPUTERS** A personal computer is currently running at a speed of 8 MHz. A new processor increases the speed by 9%. How fast will the computer operate with the new processor?

22. *Writing in Math* Explain how you can use estimation to check your answer to Exercise 19.

23. **H.O.T. Problem** Cheep's Discount House sells items for 10% off the original price the first week of a sale. That price is discounted 10% the second week. Each week the item does not sell, it is discounted 10% off the previous week's price.

The original price of a chair is $80. Jan figures the price the third week will be $58.32. Patrick figures the price will be $56. Who is correct? Explain your answer.

Jan

$(80)(1 - 0.10) = 72$

$(72)(1 - 0.10) = 64.80$

$(64.80)(1 - 0.10) = 58.32$

Patrick

$10\% + 10\% + 10\% = 30\%$

$80(1 - 0.30) = 56$

STANDARDS PRACTICE

Choose the *best* answer.

24 Justine took a test in Spanish. She got 10% of the questions wrong. If there were 40 questions on the test, how many did Justine get right?

A 4

B 30

C 10

D 36

25 The Levenson family went to dinner. The bill came to $96.21. If they leave a 15% tip, how much money will they leave? Round your answer to the nearest dollar.

F $14

G $16

H $15

J $17

Spiral Review

Use a proportion to solve. (Lesson 6-4)

26. What number is 40% of 90?   27. 36 is 30% of what number?

28. 12 is what percent of 48?   29. What number is 10% of 220?

30. **RECREATION** Students at Central High School were asked the number of hours they watch TV each day. What percent of the students surveyed reported that they watched 2 or more hours of TV?

(Lesson 6-4)

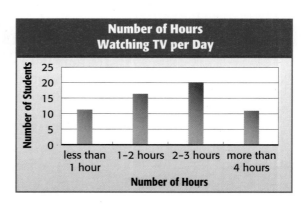

Number of Hours Watching TV per Day

Direct Variation

Vocabulary

direct variation (p. 349)

constant of variation (p. 349)

 Standard ALG5.0
Students solve multistep problems, including word problems, involving linear equations in one variable, and provide justification for each step.

Standard 7AF4.2 Solve multistep problems involving rate, average speed, distance, and time **or a direct variation.**

 The What: I will find constants of variation and use direct variations.

The Why: Direct variations are used to calculate earnings for jobs paying hourly rates.

Galeno has a part-time job at a restaurant. One week, he earned $70 for 10 hours work. The next week, he earned $140 for 20 hours work.

The amount Galeno earns depends directly on the hours he works.

$$\frac{\text{Earnings}}{\text{Hours Worked}} = \frac{70}{10} = \frac{140}{20}$$

Galeno makes $7 per hour. The equation $y = 7x$ can be used to find his earnings y when he works x hours. This equation is an example of a direct variation. Galeno's hourly rate of pay is the constant of variation.

DIRECT VARIATION

A **direct variation** is an equation of the form

$$y = kx \quad \text{or} \quad \frac{y}{x} = k,$$

where $k \neq 0$. The variable k is a ratio called the **constant of variation.**

variable quantities   variable quantities

$$y = k \cdot x \qquad\qquad \frac{y}{x} = k$$

constant of variation   constant of variation

To determine if a table of values represents a direct variation, find the ratio of the dependent and independent values. When each ratio is the same, or constant, the table shows a direct variation. Use a direct variation's input-output values to determine the constant of variation.

EXAMPLES Identify Direct Variations

Determine whether each table shows a direct variation. If a table shows a direct variation, find the constant of variation.

 1

x	1	2	3	4
y	5	10	15	20

$$\frac{y}{x} = \frac{5}{1} = \frac{10}{2} = \frac{15}{3} = \frac{20}{4}$$

The ratios are equal. The table shows a direct variation. The constant of variation is 5.

 2

hours worked	1	2	3
amount paid	6	10	18

$$\frac{y}{x} = \frac{6}{1} = \frac{18}{3} \neq \frac{10}{2}$$

The ratios are not equal. The table does not show a direct variation.

Your Turn

Determine whether each table shows a direct variation. If a table shows a direct variation, find the constant of variation.

a.

x	−1	−2	−3
y	−5	−10	−15

b.

time in weeks	1	2	3
height in centimeters	12	15	18

EXAMPLE Use Direct Variations

3 **Assume y varies directly as x. If $y = 32$ when $x = 8$, find x when $y = 48$.**

$\dfrac{y_1}{x_1} = \dfrac{y_2}{x_2}$ Set up a proportion.

Use subscripts to indicate the two sets of values.

$\dfrac{32}{8} = \dfrac{48}{x}$ Substitute known values.

$32x = 8 \cdot 48$ Find cross products.

$32x = 384$ Multiply 8 by 48.

$\dfrac{32x}{32} = \dfrac{384}{32}$ Divide each side by 32.

$x = 12$ Simplify.

Your Turn

c. Assume h varies directly as g. If $h = 24$ when $g = 6$, find h when $g = 44$.

d. The weight of a rope varies directly with the length. A six-foot length of rope weighs 2 pounds. How much will a 15-foot length weigh?

Study Tip

You can simplify ratios before finding cross products.

$\dfrac{32}{8} = \dfrac{48}{x}$

$\dfrac{4}{1} = \dfrac{48}{x}$

$4x = 48$

$x = 12$

Talk Math

WORK WITH A PARTNER Name a set of three ordered pairs with a constant of variation equal to 6. Listen as your partner names a set of three ordered pairs with a constant of variation equal to 7.

Real-World EXAMPLES   Apply Direct Variations

Use direct variation to solve. Give a reason for each step.

④ ELECTRICITY In a transformer, voltage varies with the number of turns of wire on the coil. Suppose 100 volts come from 55 turns. How many volts will come from 66 turns?

$$\frac{y_1}{x_1} = \frac{y_2}{x_2}$$ Set up a proportion.
Use subscripts to indicate the two sets of values.

$$\frac{100}{55} = \frac{y}{66}$$ Substitute known values.

$100 \cdot 66 = 55y$ Find cross products.

$6{,}600 = 55y$ Multiply 100 by 66.

$$\frac{6{,}600}{55} = \frac{55y}{55}$$ Divide each side by 55.

$y = 120$ Simplify.

66 turns produce 120 volts of electricity.

⑤ SPACE The weight of an object on Earth varies directly as the object's weight on the moon. A 120-pound person would weigh 20 pounds on the moon. How much would a 180-pound person weigh on the moon?

$$\frac{y_1}{x_1} = \frac{y_2}{x_2}$$ Set up a proportion.
Use subscripts to indicate the two sets of values.

$$\frac{120}{20} = \frac{180}{x}$$ Substitute known values.

$120x = 20 \cdot 180$ Find cross products.

$120x = 3{,}600$ Multiply 20 by 180.

$$\frac{120x}{120} = \frac{3{,}600}{120}$$ Divide each side by 120.

$x = 30$ Simplify.

A 180-pound person would weigh 30 pounds on the moon.

SPACE An astronaut weighs $\frac{1}{6}$ of his Earth weight on the moon.

Your Turn

Use direct variation to solve. Give a reason for each step.

e. **TRAVEL** Serefina drives 550 miles in 10 hours. How far will she travel if she drives for 12 hours?

Examples 1–5
(pages 350–351)

VOCABULARY

1. Write three equations that represent direct variations.

2. Explain the difference between a constant and a variable.

Examples 1–2
(page 350)

Determine whether each table shows a direct variation. If a table shows a direct variation, find the constant of variation.

3.

hours	1	3	4
distance	43	129	172

4.

x	−3	−5	−6
y	−90	−120	−180

Example 3
(page 350)

5. Assume d varies directly as t and $d = 180$ when $t = 3$. Find t when $d = 390$.

Examples 4–5
(page 351)

6. **FINANCE** The number of servings of popcorn varies directly with the number of bags of popcorn. There are 2.5 servings in 1 bag. How many servings are there in 6 bags?

7. **COMMUNITY SERVICE** Lacie recycles newspapers. The amount she recycles varies directly with the number of weeks during which she recycles. In four weeks, she recycles 20 pounds of paper. How many pounds of paper will she recycle in 10 weeks?

Skills, Concepts, and Problem Solving

HOMEWORK HELP	
For Exercises	**See Example(s)**
8–10	1–2
11–12	3
13–15	4–5

Determine whether each table shows a direct variation. If a table shows a direct variation, find the constant of variation.

8.

x	12	16	20
y	2	4	5

9.

x	9	7	3
y	−1350	−1050	−450

10.

minutes on parking meter	15	30	60
money put in meter	30	50	75

11. Suppose y varies directly as x. Let $y = -20$ when $x = -10$. Find x when $y = -60$.

12. Suppose d varies directly as t. Let $d = 400$ when $t = 8$. Find d when $t = 10$.

Use direct variation to solve. Give a reason for each step.

13. **FINANCE** Nodin earns $150 for 20 hours work. How much will he earn for 24 hours work?

14. **ASTRONOMY** The weight of an object on the moon varies directly as its weight on Earth. With all his gear, Neil Armstrong weighed 360 pounds on Earth and 60 pounds on the moon. Marisa weighs 108 pounds on Earth. How much would she weigh on the moon?

15. **GEOMETRY** The height of an object varies directly as its shadow's length. Suppose a 7-foot high sign casts a shadow 3 feet long. How tall is a tree that casts a shadow 12 feet long?

16. *Writing in Math* Explain why the equation $y = 2x + 5$ does not represent a direct variation.

17. **H.O.T.** Problem Complete the table. Write an equation that represents this direct variation relationship.

x	2	3	4	?	6
y	25	37.5	50	62.5	?

Use the table to answer the questions.

Buying Cherries

Number of Pounds	Cost ($)
3.4	6.77
5.7	11.34
1.2	2.39

18 How much does a pound of cherries cost?

A $0.50

B $2.30

C $1.99

D $2.87

19 In Exercise 18, what term describes the cost of a pound of cherries?

F constant of variation

G total amount

H direct variation

J variable

Spiral Review

20. **SPORTS** Santos Sporting Goods offered a 25% discount on all baseball equipment. How much did Ines pay for baseball equipment originally priced at $42.56? (Lesson 6-5)

21. **BANKING** Bruce earns 3% interest on the money in his savings account. Suppose he has $245.90 in his account. About how much interest will his account earn in one year? (Lesson 6-5)

Change each fraction to a decimal. (Lesson 4-1)

22. $\frac{7}{8}$

23. $\frac{11}{4}$

24. $\frac{9}{2}$

Progress Check 3

(Lessons 6-5 and 6-6)

▷ Vocabulary and Concept Check

> commission (p. 345)   interest (p. 342)
> constant of variation (p. 349)   principal (p. 342)
> direct variation (p. 349)   tip (p. 345)
> discount (p. 342)

Choose the term that *best* matches the example. You may use a term more than once. (Lessons 6-5 and 6-6)

1. A decrease in the regular price is called a(n) _____?_____.

2. Someone who earns money based on percent of sales receives a(n) _____?_____.

3. In the equation $y = kx$, k is called the _____?_____.

▷ Skills Check

4. Suppose y varies directly as x. Let $y = 45$ when $x = -9$. Find y when $x = 4$. (Lesson 6-6)

5. Assume h varies directly as g. If $h = 77.4$ when $g = 4.3$, find h when $g = 12$. (Lesson 6-6)

▷ Problem-Solving Check

6. **FASHION** Dionne bought jeans for 75% of the original price. The original price was $38. How much did Dionne pay for the jeans? (Lesson 6-5)

7. **FOOD SERVICE** Eloy and his friends go out to lunch. The bill totals $38.34. The group wants to leave 15% of the bill for a tip. How much money should they leave? (Lesson 6-5)

8. **SALES** Mrs. Jacobs sells cars. She earns 5% of the total sale price of every car. Mrs. Jacobs sold a car for $34,900. How much money did she earn from the sale of that car? (Lesson 6-5)

Use direct variation to solve. (Lesson 6-6)

9. **TRAVEL** The Montoya family travels from Sacramento to Los Angeles, a distance of 351 miles. They travel at an average speed of 54 miles per hour. How long will the trip take?

10. **REFLECT** Explain two ways you can solve a problem involving discount. Use an example to show the two ways. (Lesson 6-5)

Problem-Solving Strategy: Make a Table

Vocabulary

make a table (p. 355)

 Standard 7AF4.2 Solve multistep problems involving rate, average speed, distance, and time or a direct variation.

Standard 7MR2.5 Use a variety of methods, such as words, numbers, symbols, charts, graphs, **tables,** diagrams, and models, **to explain mathematical reasoning.**

Standard 6NS1.2 Interpret and use ratio in different contexts to show the relative sizes of 2 quantities using appropriate notations.

Beng made a salad. He used a 4:6 ratio of cucumber slices to tomato slices. Suppose he used a total of 30 cucumber and tomato slices. How many of each did he use?

Use the Strategy

 What do you know?

- Beng used cucumber and tomato slices.
- Beng used a total of 30 slices.
- The ratio of cucumbers to tomatoes is 4:6.
- The ratio simplifies to 2:3.

Plan **What do you want to know?**

You want to know the number of both cucumber and tomato slices. You can **make a table** to find out. Make a table of ratios and extend the table until you have 30 total slices.

Solve **Make a table.**

List 2 cucumber and 3 tomato slices, for a total of 5 slices. Multiply both 2 and 3 by 2 to find an equivalent ratio. Use multiples to list equivalent ratios for cucumber and tomato slices. List total slices until your table shows 30 total slices.

COOKING You can use ratios to increase and decrease recipes.

Cucumber Slices	Tomato Slices	Total Slices
2	3	5
4	6	10
6	9	15
8	12	20
10	15	25
12	18	30	✔

Check **Check that your answers meet both conditions.**

You check that 12:18 is equal to 4:6.

$$\frac{12}{18} \div \frac{3}{3} = \frac{4}{6}$$

You check that 12 and 18 add to 30.

Guided Practice

Use the *make-a-table* strategy to solve.

1. Eric wants to drive from San Diego to Sacramento. He travels a total distance of almost 480 miles. Eric travels at a rate of 60 miles per hour. How long will it take him to get to Sacramento?

   Understand **What do you know?**

   Plan **What do you want to know?**

   Solve **Make a table of the distance he traveled after each hour.**

   Check **Check that your answer is correct.**

2. *Writing in Math* Two numbers are in a ratio of 6 : 5. If their sum is 77, explain how making a table can help you find the two numbers.

Problem-Solving Practice

Use the *make-a-table* strategy to solve.

3. **AGES** The ratio of the ages of two women is 4:3. The total of their ages is 84 years. What is the age difference of the women?

4. **COOKING** Talia is preparing for a barbecue. She plans to cook hamburgers and hot dogs in a ratio of 3:2. The total number of items she will barbecue is 70. How many hamburgers and hot dogs will she cook?

5. **NUMBER SENSE** Two numbers have a ratio of 7:5. Suppose the sum of the numbers is 72. What is the difference between the two numbers?

6. **BAKING** A pastry recipe calls for 3 cups of flour for every 1.5 cups of sugar. Suppose you use a total of 27 cups of flour and sugar when making pastries. How many of each do you use?

7. **COOKING** A pizza recipe calls for 9 cups of cheese to make 4 pizzas. How many cups of cheese are needed for 14 pizzas?

Use any strategy to solve.

8. **GEOMETRY** Look at the following diagram. How many dots will it take to make the 7th triangle?

9. **FINANCE** Elvio went to the mall. He spent half of his money on a pair of shoes. Then, he spent $15 on a shirt. After shopping, he spent half of the money he had left on dinner. After dinner, he had $12 left. How much money did Elvio have when he went to the mall?

10. **RECREATION** In Rebecca's backyard, there are bicycles and tricycles. Rebecca sees 17 seats and 43 wheels. How many tricycles are in the backyard?

Study Guide

Study Tips
• Use simpler words.
• Include examples.

Understanding and Using the Vocabulary

After completing the chapter, you should be able to define each term or phrase and give an example of each.

base (p. 336)	percent proportion (p. 336)
commission (p. 345)	percentage (p. 336)
constant of variation (p. 349)	principal (p. 342)
cross product (p. 330)	proportion (p. 330)
direct variation (p. 349)	rate (p. 319)
discount (p. 342)	ratio (p. 318)
interest (p. 342)	tip (p. 345)
percent (p. 323)	unit rate (p. 319)

Complete each sentence with the correct mathematical term or phrase.

1. A(n) _____?_____ is an equation of two equal fractions.

2. The amount or percent off the original price is the _____?_____.

3. A(n) _____?_____ has a denominator of one.

4. The amount of money you put into an account is called the _____?_____.

5. A(n) _____?_____ is a comparison of two quantities.

6. In $\frac{a}{b} = \frac{c}{d}$, $a \cdot d$ is called a(n) _____?_____.

7. Salespeople may work on _____?_____. This means that they receive a percent of the amount of the sale.

8. It is common to leave at least a 15% _____?_____ for a server at a restaurant.

9. In a percent proportion, the _____?_____ is the total amount or the original price.

10. A(n) _____?_____ is a ratio that compares a number to 100.

Study Guide

LESSON 6-1 — pages 318–322

Express ratios, rates, and unit rates as fractions and decimals.

25 to 40

$\dfrac{25}{40} = \dfrac{25 \div 5}{40 \div 5} = \dfrac{5}{8}$ Write as a fraction. Divide by the GCF to simplify.

$\dfrac{5}{8} = 5 \div 8 = 0.625$ Write as a decimal.

$\dfrac{54 \text{ miles}}{6 \text{ hours}} = \dfrac{54 \div 6}{6 \div 6}$ Write as a unit rate. Divide to get a denominator of one.

$= 9$ miles per hour

Write each ratio in the form requested.

11. 12 out of 144 as a fraction in simplest form
12. 13 out of 25 as a decimal
13. 36 boys to 40 girls as a fraction in simplest form
14. 27 robins out of 30 birds in colon form

Express as a unit rate.

15. 250 miles for 10 gallons
16. 24 feet in 6 seconds
17. 50 boxes for 10 cards
18. 60 books for 4 shelves

LESSON 6-2 — pages 323–328

Write percents as fractions and decimals.

$45\% = \dfrac{45}{100}$ Write as a fraction.

$= \dfrac{45 \div 5}{100 \div 5} = \dfrac{9}{20}$ Write in simplest form.

$35\% = 35 \div 100 = 0.35$ Divide by 100.

Write fractions as decimals and percents.

$\dfrac{3}{4} = 3 \div 4 = 0.75$ Divide the numerator by the denominator.

$\dfrac{3}{4} = 3 \div 4$ Divide the numerator by the denominator.

$= 0.75 = 75\%$ Read the decimal as *per hundred*.

Write each percent as a fraction in simplest form. Then write it as a decimal.

19. 30%   20. 48%

21. 55%   22. 76%

Write each fraction as a decimal. Then write it as a percent.

23. $\dfrac{3}{5}$ 24. $\dfrac{9}{12}$

25. $\dfrac{3}{20}$ 26. $\dfrac{17}{25}$

Skills and Concepts

Objectives and Examples

Review Exercises

LESSON 6-3   pages 330–334

Find cross products.

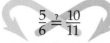

$$\frac{5}{6} \overset{?}{=} \frac{10}{11}$$ Set up the proportion.

$5 \cdot 11 \overset{?}{=} 6 \cdot 10$ Find the cross products.

$55 \neq 60$ The two ratios do not form a proportion.

Solve proportions.

$$\frac{5}{6} = \frac{m}{20}$$

$6m = 5 \cdot 20$ Find the cross products.

$$\frac{6m}{6} = \frac{100}{6}$$ Divide each side by 6.

$m = 16.\overline{6}$ Simplify.

Determine whether each pair of ratios is equal. Use cross products. Show your work.

27. $\frac{4}{5}, \frac{9}{10}$ **28.** $\frac{3}{8}, \frac{12}{32}$

29. $\frac{5}{7}, \frac{25}{35}$ **30.** $\frac{7}{9}, \frac{27}{21}$

Solve each proportion.

31. $\frac{3}{5} = \frac{n}{15}$ **32.** $\frac{9}{10} = \frac{b}{40}$

33. $\frac{30}{45} = \frac{2}{h}$ **34.** $\frac{72}{81} = \frac{8}{z}$

LESSON 6-4   pages 336–340

Solve percent proportions.

What number is 30% of 20?

$$\frac{P}{20} = \frac{30}{100}$$ Set up the percent proportion.

$100 \cdot P = 20 \cdot 30$ Find the cross products.

$100 \cdot P = 600$ Multiply 20 by 30.

$$\frac{100 \cdot P}{100} = \frac{600}{100}$$ Divide each side by 100.

$P = 6$ Simplify.

5 is what percent of 80?

$$\frac{5}{80} = \frac{r}{100}$$ Set up the percent proportion.

$5 \cdot 100 = 80 \cdot r$ Find the cross products.

$500 = 80 \cdot r$ Multiply 5 by 100.

$$\frac{500}{80} = \frac{80 \cdot r}{80}$$ Divide each side by 80.

$r = 6.25$ Simplify.

Use a proportion to solve.

35. 9 is what percent of 45?

36. 12 is 40% of what number?

37. 16 is what percent of 64?

38. What number is 12.5% of 88?

39. 6 is what percent of 96?

40. 18 is what percent of 60?

Skills and Concepts

Objectives and Examples

LESSON 6-5 pages 342–348

Use percents to solve problems.

A sweater costs $120. It is on sale for 20% off the original price. What is the cost of the sweater on sale?

$\dfrac{\text{Discounted Amount}}{\text{Original Amount}} = \dfrac{r}{100}$ Use the percent proportion.

$\dfrac{P}{120} = \dfrac{20}{100}$ Substitute for known values.

$100P = 2400$ Find cross products. Multiply. Divide each side by 100.

$P = 24$

The discount is $24. The discounted fee is $120 − $24 or $96.

Review Exercises

41. **RETAIL** Camden bought a television set on sale for 20% off. Suppose the original price was $540. How much did Camden pay for the television set?

42. **FOOD SERVICE** A local restaurant automatically adds a 15% tip to bills for eight people or more. If a bill for 10 people came to $220, how much of a tip would be added to the bill?

43. **BANKING** Leo puts $1,300 in the bank at a rate of 6% interest. How much money will he have in the bank at the end of 6 months?

44. **SALES** A salesperson sells $27,000 worth of merchandise. Their commission rate on all sales is $5\frac{1}{2}\%$. What is the salesperson's commission?

LESSON 6-6 pages 349–353

Find constants of variation and use direct variations.

x	5	6	7
y	15	18	21

$\dfrac{15}{5} = \dfrac{18}{6} = \dfrac{21}{7}$

The ratios are equal. The table shows a direct variation. The constant of variation is 3.

Use direct variation to solve. Assume y varies directly as x.

45. If $y = 10.5$ when $x = 3$, find y when $x = 7$.

46. If $y = 50$ when $x = 2$, find x when $y = 75$.

Chapter Test

▷ Vocabulary and Concept Check

1. Write 40 people to 5 tables as a unit rate.

2. Write a fraction. Then, convert the fraction to a decimal and a percent.

▷ Skills Check

Write each ratio in the form requested.

3. 8 out of 12 in colon form

4. 12 out of 15 as a fraction in simplest form

5. 7 out of 14 as a decimal

Write each percent as a fraction in simplest form. Then write it as a decimal.

6. 12%   7. 95%   8. 165%   9. 150%

Write each fraction as a decimal. Then write it as a percent.

10. $\frac{4}{5}$ 11. $\frac{5}{8}$ 12. $\frac{8}{5}$ 13. $\frac{7}{4}$

Solve each proportion.

14. $\frac{4}{5} = \frac{20}{n}$ 15. $\frac{n}{7} = \frac{15}{35}$ 16. $\frac{5}{6} = \frac{m}{48}$ 17. $\frac{3}{m} = \frac{2}{12}$

Use a proportion to solve.

18. 13 is what percent of 65?   19. 21 is what percent of 28?

20. What number is 25% of 72?   21. 18 is 20% of what number?

Determine whether each table shows a direct variation. If a table shows a direct variation, find the constant of variation.

22.

x	3	4	5
y	30	48	50

23.

x	2	3	4
y	6	9	12

▷ Problem-Solving Check

24. **SPORTS** Lisa Leslie, of the Los Angeles Sparks, made 40% of her 3-point baskets during a recent season. Suppose she attempted 60 3-point shots. About how many did she make?

25. **MUSIC** Lucas bought a CD that has 15 songs. Suppose 20% of the songs are instrumentals. How many songs are instrumentals?

Standards Practice

PART 1 Multiple Choice

Choose the *best* answer.

1 Which shows 80% written as a fraction in simplest form?

A $\frac{2}{25}$

B $\frac{8}{100}$

C $\frac{4}{5}$

D $\frac{5}{4}$

2 Find the interest earned in one year on a deposit of $7,500 if the rate is 6%.

F $420

G $450

H $4,500

J $7,950

3 Doneeka works as a salesperson at a shoe store. She earns 8% commission on her sales. In one month, Doneeka's sales were $24,000. Which shows the amount earned on commission?

A $1,920

B $4,800

C $19,200

D $22,080

4 A radio originally cost $54. Ian bought the radio for 20% off. Which shows the amount deducted from the original price?

F $5.40

G $10.80

H $43.20

J $52.92

5 What fraction is the same as 0.04?

A $\frac{1}{50}$

B $\frac{1}{25}$

C $\frac{1}{5}$

D $\frac{2}{5}$

6 Natasha had lunch in a restaurant. The bill totaled $16.67. Natasha plans to leave a 15% tip. Which amount shows the tip left by Natasha?

F $2.40

G $2.50

H $2.55

J $19.17

7 The table shows the temperature drop each hour for a 2 hour time period.

Time	Temperature
8 P.M.	40°
8:30 P.M.	38°
9 P.M.	36°
9:30 P.M.	33°
10 P.M.	30°

Which shows the percent the temperature dropped from 8 P.M. to 10 P.M.?

A 4%

B 10%

C 20%

D 25%

8 What is the fraction $\frac{7}{8}$ written as a percent?

F 1.14%

G 11.5%

H 87.5%

J 875%

Record your answers on the answer sheet provided by your teacher or on a separate sheet of paper.

9 The McCormick family went out to dinner. The bill totaled $152.93. They want to leave a 10% tip. To the nearest dollar, how much tip should the McCormicks leave?

10 The table shows Tiesha's sales at the department store. Tiesha earns 7% commission for her sales. What was the total Tiesha earned for the months January through June?

Month	Total Sales
JANUARY	$14,000
FEBRUARY	$19,000
MARCH	$16,000
APRIL	$17,000
MAY	$13,000
JUNE	$16,000

11 Ilya bought a jacket on sale for $54. Suppose the jacket originally cost $60. What was the percent of discount on the jacket?

12 There are 15 boys to every 12 girls at a local high school.

  a. Write this ratio in simplest form.

  b. Suppose there are a total of 360 boys in the high school. Find the number of girls in the high school.

13 To bake 3 dozen cookies, Darien needs 4 cups of flour. He has 7 cups of flour. How many cookies can he bake?

  a. Write a proportion to find the number of cookies.

  b. Solve the proportion.

Record your answers on the answer sheet provided by your teacher or on a separate sheet of paper.

14 U-SAVE sells all items at a 20% discount. The salespeople earn a 6% commission on all sales.

  a. Sonya buys a refrigerator that sells for $850. What is the U-SAVE discounted price?

  b. How much commission will the salesperson earn for selling the refrigerator?

  c. How much more money would the salesperson have earned if the refrigerator was sold at the undiscounted price?

NEED EXTRA HELP?														
If You Missed Question...	1	2	3	4	5	6	7	8	9	10	11	12	13	14
Go to Lesson...	6-1 6-2	6-5	6-5	6-5	6-2	6-5	6-5	6-2	6-5	6-5	6-4 6-5	6-1	6-1	6-5
For Help with Algebra Readiness Standard...	7NS1.3	7AF4.2	6NS1.4	6NS1.4	7NS1.3	6NS1.4 ALG5.0	6NS1.4 ALG5.0	7NS1.3	6NS1.4	7NS1.3 ALG5.0	7NS1.3 ALG5.0	7NS1.3	7NS1.3	6NS1.4 ALG5.0

CHAPTER 7

Algebra on the Coordinate Plane

Doug Menuez/Getty Images

Here's What I Need to Know

Standard 7AF3.3 Graph linear functions, noting that the vertical change (change in *y*-value) per unit of horizontal change (change in *x*-value) is always the same and know that the ratio ("rise over run") is called the slope of a graph.

Standard 7AF3.4 Plot the values of quantities whose ratios are always the same. Fit a line to the plot and understand that the slope of the line equals the ratio of the quantities.

Standard 7MG1.3 Use measures expressed as rates and measures expressed as products to solve problems; check the units of the solutions; and use dimensional analysis to check the reasonableness of the answer.

Vocabulary Preview

coordinate plane   A plane in which a horizontal number line and a vertical number line intersect at the point where each line is zero (p. 366)

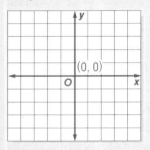

slope   The ratio of the change in *y* to the change in *x*. (p. 388)

The What

I will learn to:

- graph linear functions.
- solve problems using measures expressed as rates or products.
- use the Pythagorean Theorem to solve problems.

The Why

Linear functions can be used to show the relationship between two sets of data in many areas, including sports like football and car racing.

Option 1

Math Online Are you ready for Chapter 7? Take the Online Readiness Quiz at ca.algebrareadiness.com to find out.

Option 2

Complete the Quick Check below.

Vocabulary and Skills Review

Match each description with the correct vocabulary word.

1. a three-sided polygon   A. coordinate plane

2. a statement in which two   B. equation
   expressions are equal
        C. ordered pair

3. (x, y) D. triangle

Identify the coordinates of the given points.

4. A

5. B

6. C

7. D

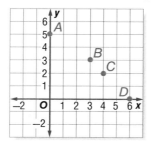

Solve each equation.

8. $x^2 = 36$ 9. $x^2 = 100$

Tips

- Give graphs a title and label each axis as needed.
- Copy example graphs carefully.
- Include any hints about how to make the graph.
- Use your notes to study for your tests.

Note-taking

Chapter Instruction	My Notes
The numbers in an ordered pair are called **coordinates.** (p. 367)	Coordinates name an ordered pair. The x-coordinate comes first. The y-coordinate comes second. (x, y)

The Coordinate Plane

Vocabulary

ordered pair (p. 366)

coordinate plane (p. 366)

x-axis (p. 366)

y-axis (p. 366)

origin (p. 366)

quadrant (p. 366)

coordinate (p. 367)

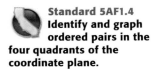 **Standard 5AF1.4 Identify and graph ordered pairs in the four quadrants of the coordinate plane.**

SPORTS In organized football games, a touchdown is worth 6 points.

Study Tip

The origin and the two axes do not lie in any quadrant.

"Axes" is the plural for axis.

 The What: I will plot points on a coordinate plane.

The Why: Plotting points shows the relationship between the number of touchdowns and the number of points in a football game.

The neighborhood football field does not have goalposts. So, each team earns 7 points every time they make a touchdown.

In math, graphs can provide a picture of this relationship.

$$x = \text{touchdowns} \qquad y = \text{points}$$

The relationship between the number of touchdowns and the number of points can be expressed by **ordered pairs** (x, y).

1 touchdown = 7 points   (1, 7)

2 touchdowns = 14 points   (2, 14)

3 touchdowns = 21 points   (3, 21)

Ordered pairs are graphed on a **coordinate plane.** In a coordinate plane, a horizontal number line, called the **x-axis,** intersects a vertical number line, called the **y-axis.** The two number lines intersect at the point where each line is zero. This point is the **origin.**

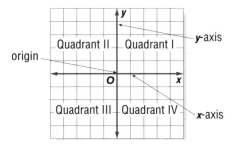

The number lines divide the coordinate plane into four **quadrants.** Each quadrant is numbered with a Roman numeral.

The numbers in an ordered pair are called **coordinates.**

x-coordinate   y-coordinate

(2, 14)

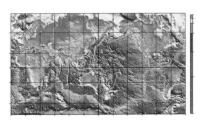

MAPS World seismicity maps are coordinate planes that show the magnitude of earthquakes.

When you graph a point, start by putting your pencil at the origin, (0, 0). If the x-coordinate is positive, move that many units to the right of the origin. If the y-coordinate is positive, move that many units up the grid. Draw a dot to graph the coordinate.

If the value for x is a negative number, move to the left of the origin. If the value for y is a negative number, move below the origin.

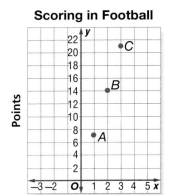

Scoring in Football

Touchdowns

GRAPHING POINTS

x-coordinate	y-coordinate	Quadrant
+	+	I
−	+	II
−	−	III
+	−	IV

EXAMPLE   **Find the Ordered Pair**

Find the ordered pair for the given point.

1 *A*

Think of a vertical line and a horizontal line passing through point *A*.

Since the vertical line intersects the x-axis at −2, the x-coordinate is −2. Since the horizontal line intersects the y-axis at 3, the y-coordinate is 3.

The ordered pair for point *A* is (−2, 3).

Your Turn
Find the ordered pair for each point.

a. *D*   **b.** *E*

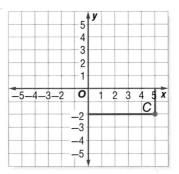

EXAMPLE Graph Points

Graph the point on a coordinate plane.

2 *C* (5, −2)

Start at (0, 0). Move 5 units to the right along the *x*-axis. Move 2 units down along the *y*-axis to the point. Label *C*.

Your Turn

Graph each point on a coordinate plane.

c. *F* (0, 3)

d. *G* (−6, −1)

Real-World EXAMPLE

Baskets	Points	(x, y)
1	12	(1, 12)
2	14	(2, 14)
3	16	(3, 16)
4	18	(4, 18)

3 **SPORTS** In a neighborhood game of basketball, each team scores 2 points for every basket. One team starts with a 10-point lead because its players are younger. Graph the younger team's points on a coordinate plane.

Graph the point for each ordered pair.

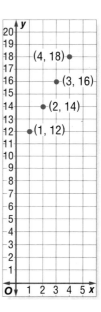

Your Turn

e. **SPORTS** A basketball team with older players takes only 3-point shots. Graph the team's points on a coordinate plane.

Baskets	Points	(x, y)
1	3	(1, 3)
2	6	(2, 6)
3	9	(3, 9)
4	12	(4, 12)

Talk Math

WORK WITH A PARTNER Discuss how you would graph the points *R* (1, 8), *S* (8, 1), *T* (−1, 8), and *U* (1, −8). Identify the quadrant for each point.

Guided Practice

Examples 1–3
(pages 367–368)

VOCABULARY Draw a coordinate plane and label each item.

1. *x*-axis

2. *y*-axis

3. Quadrants I, II, III, and IV

4. origin

Example 1
(page 367)

Find the ordered pair for each point.

5. *A*

6. *B*

7. *C*

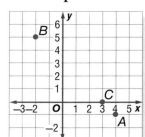

Example 2
(page 368)

Graph each point on a coordinate plane.

8. *E* (0, 6)　　　　9. *F* (9, 1)　　　　10. *G* (−6, −8)

Example 3
(page 368)

11. **BUSINESS** The cost of setting up a lemonade stand is $20. Each cup of lemonade sells for $2. Graph the points on a coordinate plane.

Cups	Profit	(x, y)
0	−20	(0, −20)
5	−10	(5, −10)
10	0	(10, 0)
15	10	(15, 10)

Skills, Concepts, and Problem Solving

HOMEWORK HELP

For Exercises	See Example(s)
12–17	1
18–23	2
24–25	3

Find the ordered pair for each point.

12. *I*

13. *J*

14. *K*

15. *L*

16. *M*

17. *N*

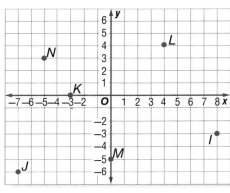

Graph each point on a coordinate plane.

18. *Q* (8, 3)　　　　19. *R* (−4, 1)　　　　20. *S* (−12, −2)

21. *T* (2, −9)　　　　22. *U* (0, 1)　　　　23. *V* (5, 5)

24. **RETAIL** The math club decides to sell T-shirts as a fundraiser. They buy 100 T-shirts for $200. They sell the T-shirts for $10 each. Graph the points on a coordinate plane.

T-shirts	Profit	(x, y)
0	−200	(0, −200)
10	−100	(10, −100)
20	0	(20, 0)
30	100	(30, 100)

25. **TRANSPORTATION** A taxi charges a $5 pickup fee and then $1 per mile. Graph the points on a coordinate plane.

Miles	Dollars	(x, y)
1	6	(1, 6)
2	7	(2, 7)
3	8	(3, 8)
4	9	(4, 9)

26. **H.O.T. Problem** Refer to Exercise 24. The break-even point is when the club makes enough money to cover their expenses without making a profit. Find the break-even point. How many T-shirts must they sell to make a profit?

27. *Writing in Math* Explain why (−2, 3) could never be on the graph in Exercise 25.

28 Dariq needs $40 for a new video game. He has $15. He can earn $5 an hour raking leaves. The equation $y = 5x + 15$ shows the relationship between how many hours he must work and how much money he has. How many hours must he work to buy the game?

A 3   C 8

B 5   D 11

29 Francisco wants to buy 4 shirts. The cost of each shirt ranges from $19.99 to $35.99. What is a reasonable total cost for the shirts?

F $60   H $120

G $70   J $160

Spiral Review

Tell whether each is an example of direct variation. (Lesson 6-6)

30. **BUSINESS** The number of T-shirts sold decreases with an increase in price.

31. **FINANCE** The monthly cost of a cell phone service plan increases with the amount of time talked.

Simplify. (Lesson 5-2)

32. $x^5 \cdot x^3$

33. $\dfrac{e^{13}}{e^{15}}$

34. $(m^4)^9$

Progress Check 1
(Lesson 7-1)

▷ **Vocabulary and Concept Check**

> coordinate (p. 367)   quadrant (p. 366)
> coordinate plane (p. 366)   x-axis (p. 366)
> ordered pair (p. 366)   y-axis (p. 366)
> origin (p. 366)

Choose the term that *best* completes each sentence. (Lesson 7-1)

1. The vertical number line on the coordinate plane is the _____?_____.

2. The horizontal number line on the coordinate plane is the _____?_____.

3. (x, y) is a(n) _____?_____.

4. $(0, 0)$ is called the _____?_____.

5. The coordinate plane is divided into four _____?_____.

▷ **Skills Check**

Find the ordered pair for each given point. (Lesson 7-1)

6. A

7. B

8. C

9. D

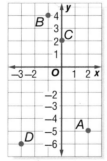

Graph each point on a coordinate plane. (Lesson 7-1)

10. F (5, 3)   11. G (−6, −1)   12. H (−2, 4)   13. I (1, −7)

▷ **Problem-Solving Check**

14. **FINANCE** Abby buys $10 worth of fruit. Each piece of fruit sells for $1. Write this data as ordered pairs. Graph the points on a coordinate plane. (Lesson 7-1)

# Fruit Sold	Profit
0	−10
5	−5
10	0
15	5

15. **REFLECT** Describe the steps you take to graph a point. (Lesson 7-1)

7-2 Problem-Solving Strategy: Draw a Graph

Vocabulary

draw a graph (p. 372)

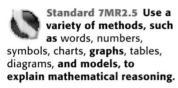 **Standard 7MR2.5 Use a variety of methods, such as** words, numbers, symbols, charts, **graphs,** tables, diagrams, **and models, to explain mathematical reasoning.**

Standard 7AF3.4 Plot the values of quantities whose ratios are always the same. Fit a line to the plot and understand that the slope of the line equals the ratio of the quantities.

BOTANY Redwood trees are some of the largest trees in the world. They can live to be over 2,000 years old.

Use the Strategy

Some trees are sampled to measure growth. After five years, a sample tree is 2 meters tall. Six years later, the same tree is 3.2 meters tall. The tree is harvested when it is 7 meters tall.

If the tree grew at the same rate, how many years after planting was the tree harvested?

Understand **What do you know?**

• At 5 years, the tree is 2 meters tall.
• At 11 years, the tree is 3.2 meters tall.
• When the tree is harvested, it is 7 meters tall.
• The tree grew at the same rate.

Plan **What do you want to know?**

You want to know how many years after planting the tree was harvested.

Solve **Draw a graph of the given information.**

Use your graph. Find the number of years that corresponds to a height of 7 meters.

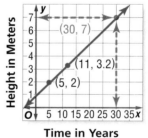

Tree Growth

• Let the horizontal axis represent time in years.
• Let the vertical axis represent the height of the tree in meters.
• Plot ordered pairs to represent the data.
   (5, 2)   (5 years, 2 meters)
   (11, 3.2)   (11 years, 3.2 meters)
• Draw a line through these points.

The graph shows that 7 meters corresponds to 30 years.

The tree was harvested 30 years after planting.

Check **Check to see if your answer makes sense.**

At five years, the sample tree was 2 meters tall. During the next 6 years, the tree grew about 1 meter. At this rate, the tree needed 24 years to grow 4 meters.

Since $11 + 24 = 35$, a solution of 30 years makes sense.

Draw a graph to solve the problem.

1. Ivan deposited $80 in his bank account. He earned $4 interest after a year. If he had deposited $100, he would have earned $5 interest. How much interest would he earn for $120?

> **Understand** **What do you know?**

> **Plan** **What are you trying to find? How can you figure it out?**

> **Solve** **Draw a graph. See what point corresponds to the answer you are seeking.**

> **Check** **How can you be sure your answer makes sense?**

2. *Writing in Math* The math club started with 10 students. The next year, there were 15 students. After five years, there were 35 students. Explain how you would set up a graph of the data.

Problem-Solving Practice

Solve by drawing a graph.

3. **TRAVEL** Mr. McCarthy drives 100 miles in 2 hours. He drives 250 miles in the next 5 hours. How many miles does Mr. McCarthy drive after 8 hours?

4. **RETAIL** All items in a store are reduced in price by the same percent. What is the sale price for a DVR that usually costs $500?

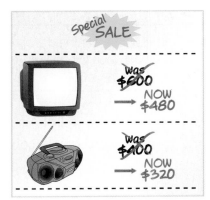

Special SALE

was $600
NOW → $480

was $400
NOW → $320

5. **MEASUREMENT** David measures the height of the steps going into his house. The 2nd step is 1 foot above ground. The 5th step is $2\frac{1}{2}$ feet above ground. What is the height of the 11th step?

Solve using any strategy.

6. **SCHOOL** The lockers in one hall are numbered 256 to 375. The hall is 180 feet long. Another hall is 120 feet long and has lockers numbered 376 to 455. How wide is each locker?

7. **MEASUREMENT** What is the missing measurement in the pattern?

$$\ldots, \underline{\quad ? \quad}, \frac{1}{4} \text{ in.}, \frac{1}{8} \text{ in.}, \frac{1}{16} \text{ in.}, \ldots$$

8. **RECREATION** SouWei is making lemonade. She needs 1 package to make 2 gallons and 4 packages to make 8 gallons. How many packages does she need to make 6 gallons?

9. **NUMBER SENSE** One number is twice another. Their product is 32. Find the numbers.

10. **NUMBER SENSE** Two consecutive even integers add to −38. Their product is 360. Find the numbers.

7-3 Relationships Involving Equal Ratios

Vocabulary

scatter plot (p. 374)

trend (p. 374)

linear relationship (p. 376)

> The **What:** I will use scatter plots to show the relationship between two quantities.
>
> The **Why:** A graph can show a relationship between study time and test scores.

 Standard 7AF3.4 Plot the values of quantities whose ratios are always the same. Fit a line to the plot and understand that the slope of the line equals the ratio of the quantities.

A **scatter plot** is a graph that shows the relationship between two sets of data. Scatter plots also show consistent changes over time, known as **trends.**

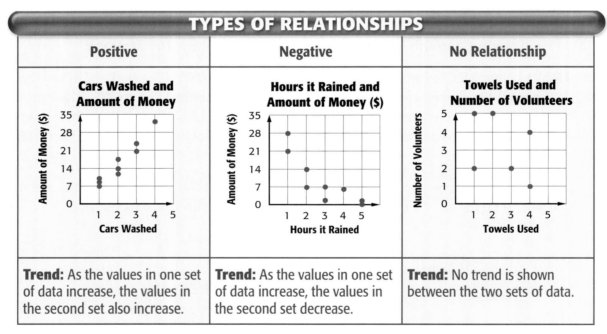

Louise wants to know if there is a relationship between test scores and the amount of time spent studying. She collected the following data.

Student	Doug	Rebecca	Bradley	Justine	Allison	Tami	Mick	Montega	Christy
Study Time (minutes)	10	15	70	60	45	90	60	30	120
Test Score	65	68	87	92	73	95	84	78	98

Analyze Data

1 **Describe the relationship shown in the scatter plot.**

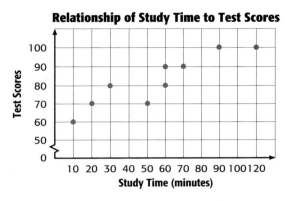

Relationship of Study Time to Test Scores

In general, students who spent more time studying earned higher test scores.

The scatter plot shows a positive relationship between the time spent studying and the test score.

Your Turn

Describe the relationship shown in the scatter plot.

a.

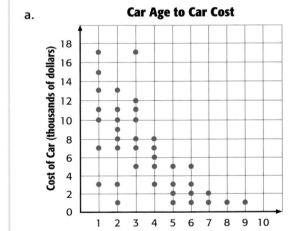

Car Age to Car Cost

> **Remember!**
> The distance between each number along an axis should be the same.

Talk Math

WORK WITH A PARTNER Describe data that has a positive relationship. Listen as your partner describes data with a negative relationship. Work together to describe a situation with no relationship.

Some data have **linear relationships.** This means that the ratio between the two quantities is always the same, or constant. A linear relationship has a straight-line graph.

 EXAMPLE **Linear Relationships**

Make a graph to find the answer.

② **FOOD** Jessie is making snack bars using cereal and marshmallows. How many ounces of marshmallows does she need for 10 cups of cereal?

Cups of Cereal	3	6	12	14
Ounces of Marshmallows	6	12	24	28

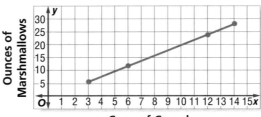

Make a scatter plot of the data. Title the graph.

Label the *x*-axis and number intervals 1 to 15.

Label the *y*-axis and number intervals 0 to 30.

Graph the ordered pairs (3, 6), (6, 12), (12, 24), and (14, 28).

Draw a line connecting the points. Determine the relationship.

She needs 20 ounces of marshmallows for 10 cups of cereal.

Your Turn
Make a graph to find the answer.

b. FOOD Kenya needs 100 limes to make limeade. She knows what some limes cost, but she does not know the cost of 100 limes. How much will she spend?

Number of Limes	20	50	70	40
Cost in Dollars	2	5	7	4

Guided Practice

Examples 1–2
(pages 375–376)

VOCABULARY Write the term that *best* completes each sentence.

1. A _____?_____ shows a relationship between two sets of data.

2. As temperature increases, humidity increases. This shows a _____?_____.

Analyze data.

Example 1
(page 375)

3. **FARMING** Farmer Garcia makes a scatter plot to show the relationship between bushels of apples harvested and amount of fertilizer used. Describe the relationship shown in his scatter plot.

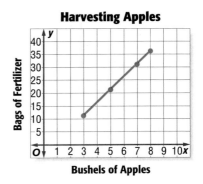

Harvesting Apples

4. **MARKET RESEARCH** A survey about a new cereal produced the following data. Is there a positive, negative, or no relationship? Explain.

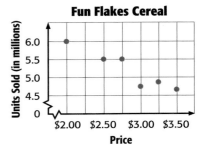

Fun Flakes Cereal

Make a graph to find the answer.

Example 2
(page 376)

5. **SHOPPING** How much does one box of cereal cost?

Cereal Boxes	4	7	9	2
Cost in Dollars	10	17.50	22.50	5

6. **MEASUREMENT** Sabrina needs to find out if a frame with a perimeter of 30 inches is long enough for her 11-inch picture. She knows the measurements of some similar frames. The information is in the chart below. What is the perimeter of a frame with a length of 11 inches?

Length	3	4	6	5
Perimeter	7.6	10.1	15.2	12.7

Skills, Concepts, and Problem Solving

HOMEWORK HELP

For Exercises	See Example(s)
7–10	1
11–13	2

Describe the relationship shown in each scatter plot.

7. **STATISTICS** Mila takes a poll of the girls in her class. She records each girl's height and shoe size. What is the relationship between a girl's height and her shoe size?

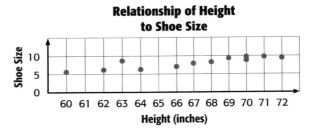

Relationship of Height to Shoe Size

Describe the relationship shown in each scatter plot.

8.

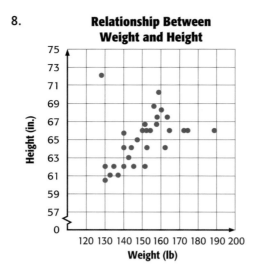

Relationship Between
Weight and Height

Height (in.) vs. Weight (lb)

9.

Relationship Between Profit
and DVD Release Date

Profits (thousands of dollars) vs. Month, Year

10.

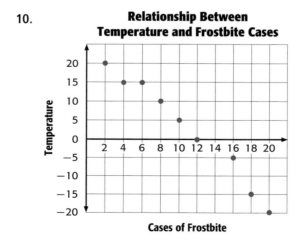

Relationship Between
Temperature and Frostbite Cases

Temperature vs. Cases of Frostbite

Make a graph to find the answer.

11. MEASUREMENT How many pints
are in a gallon?

Gallons	7	2	4	9
Pints	56	16	32	72

12. MEASUREMENT How many farthings are in a penny?

Farthings	24	44	100	68
Pennies	6	11	25	17

13. How much does one pound of grapes cost?

Pounds of Grapes	4.4	3.3	8.8	1.1
Cost in Dollars	2	1.50	4	0.50

14. H.O.T. Problem The table shows the diameters and circumferences of several round items. Suppose the data have a linear relationship. One item has a diameter of 140 centimeters. What is its circumference? An item's diameter was d centimeters. What is its circumference?

Diameter in cm	10	100	50	25
Circumference in cm	31.4	314	157	78.5

15. *Writing in Math* How can you tell if two sets of data have a positive, negative, or no relationship?

16 TRANSPORTATION The scatter plot below shows the average vehicle speed and average gas mileage of a certain car for 50 days.

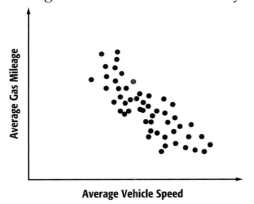

Average Vehicle Speed

Which statement *best* describes the relationship between average vehicle speed and average gas mileage shown on the scatter plot?

A As vehicle speed increases, gas mileage increases.

B As vehicle speed increases, gas mileage decreases.

C As vehicle speed increases, gas mileage increases at first, then decreases.

D As vehicle speed increases, gas mileage decreases at first, then increases.

Spiral Review

Solve by drawing a graph. (Lesson 7-2)

17. TRANSPORTATION Mrs. Madison drives at a constant rate for 6 hours. After $\frac{1}{2}$ hour, she has driven 25 miles. After 2 hours, she has driven 100 miles. How many miles does she drive in 6 hours?

Solve each equation. (Lesson 2-7)

18. $3x + 5 = 23$ **19.** $5 - x = 9$ **20.** $4x - 12 = 20$ **21.** $8 + 10x = 48$

Extend 7-3

Math Lab
Model Linear Relationships

Materials

clock/watch with a second hand

 Standard 7AF3.4 Plot the values of quantities whose ratios are always the same. Fit a line to the plot and understand that the slope of the line equals the ratio of the quantities.

 The What: I will plot data on a coordinate plane.

The Why: Graphs can be useful tools for predicting outcomes.

You can model linear relationships by graphing data on a coordinate plane.

ACTIVITY **Part 1: Collect Data**

Step 1 Begin marking "X's" on your paper when your teacher says "start." Continue until your teacher says "stop."

Step 2 Count the "X's" you made.

Step 3 Record your results.

Your Turn

a. Collect data for the time periods below.
   5, 10 15, 20, 25, and 30 seconds

b. Record your results in a table like the one shown. Label the left column as "Time in Seconds" and the right column as "Number of X's".

MEASUREMENT You can use a stopwatch to measure each time interval.

Time in Seconds	Number of X's
5	?
10	?
15	?
20	?
25	?
30	?

Part 2: Graph Data and Make Predictions

Step 1 Graph your results.

Step 2 Look for patterns. Predict the number of "X's" you would make in 18, 27, and 60 seconds.

Step 3 Make "X's" for 18 seconds. Count your "X's". Record your number.

Step 4 Make "X's" for 60 seconds. Count your "X's". Record your number.

Step 5 Compare your actual numbers with your predictions. How close were your predictions?

Your Turn

c. Predict the number of "X's" you would make in 45, 75, and 90 seconds. Collect data for these time periods. Graph your data on a coordinate plane. Check your predictions.

Analyze the Results

Graph the data on a coordinate plane. Make the prediction.

1.

Time in Seconds	Number of Bounces
5	4
10	8
15	11
20	15
25	18
60	?

2.

Time in Seconds	Number of Feet
10	20
20	40
30	60
35	?
40	78
50	95

3. **TALK ABOUT IT** What factors could contribute to the accuracy of predictions?

Measures as Rates and Products

Vocabulary

dimensional analysis
(p. 383)

density (p. 383)

 Standards 7MG1.3 Use measures expressed as rates and measures expressed as products to solve problems; check the units of the solutions; and use dimensional analysis to check the reasonableness of the answer.

 The What: I will use measures expressed as rates and products to solve problems.

The Why: Calculating someone's speed requires measures expressed as rates.

Speed measures how fast someone or something is going. Speed is equal to distance divided by time. It is usually expressed as a unit rate.

$$\frac{\text{distance}}{1 \text{ unit of time}} = \frac{50 \text{ miles}}{1 \text{ hour}} = 50 \text{ miles per hour}$$

Remember!

A ratio is a relationship between two quantities.

300 to 6 or 300:6

or $\frac{300}{6}$

A rate is a ratio of two measurements or quantities with different units.

$\frac{300 \text{ miles}}{6 \text{ hours}}$

A unit rate has a denominator of 1.

$\frac{50 \text{ miles}}{1 \text{ hour}}$

or 50 miles per hour

EXAMPLE Find Measures Expressed as Rates

Solve using rates.

① **SPORTS** Karen and three friends are running a relay. Each person runs one lap. Each lap is 400 meters. The team finished in 10 minutes. How fast was Karen's team in meters per minute?

Find the total distance.
They ran 4 laps, each 400 meters.

$$4 \text{ laps} \cdot \frac{400 \text{ meters}}{1 \text{ lap}} = 1{,}600 \text{ meters}$$

The total distance is 1,600 meters.

Divide the distance in meters by the time in minutes.

$$\frac{1{,}600 \text{ meters}}{10 \text{ minutes}} = \frac{160 \text{ meters}}{1 \text{ minute}} \text{ or } 160 \text{ m/min}$$

The team's speed is 160 meters per minute. The second amount, 1 minute, is 1. This means it is a unit rate.

Your Turn
Solve using rates.

a. **FITNESS** Andre jogged 50 yards in 16 seconds. How many yards did he jog per second?

Dimensional analysis is a process that includes units of measurement when you compute. You can use dimensional analysis to change from one set of units to another, or to check your answer.

EXAMPLE Find Measures Expressed as Rates

2 **SPORTS** Karen's team had a speed of 160 meters per minute. What was their speed in meters per hour?

$$\frac{160 \text{ m}}{1 \text{ min}}$$ Write the units you have as a fraction.

$$\frac{60 \text{ min}}{1 \text{ h}}$$ Write a new fraction for the unit you have and the unit you want.

$$\frac{160 \text{ m}}{1 \text{ min}} \cdot \frac{60 \text{ min}}{1 \text{ h}}$$ Write the two fractions as a multiplication problem to change units.

$$\frac{160 \text{ m}}{1 \text{ min}} \cdot \frac{60 \text{ min}}{1 \text{ h}} = 9{,}600 \text{ m/h}$$ Divide by units. Multiply.

Your Turn

b. TRANSPORTATION Manuel is driving his car at 30 miles per hour. What is his speed in miles per minute?

Study Tip

Use dimensional analysis to check your answer. You should be able to cancel all units except the ones you want in your final answer. In Example 2, divide by "minutes" in both the numerator and denominator.

Density is a rate used in population studies. For example, the rate 150 people per square mile compares the number of people to the amount of land (density = people/square mile). In science, density can tell you how much mass is in a given space (density = mass/volume).

EXAMPLE Use Density as a Rate

Solve using density as a rate.

3 **POPULATION** In the year 2000, California had about 34,000,000 people in about 156,000 square miles. What is the population density in people per square mile (to the nearest whole number)?

$$\text{density} = \frac{\text{people}}{\text{square mile}}$$ Use a formula for density.

$$= \frac{34{,}000{,}000 \text{ people}}{156{,}000 \text{ square miles}}$$ Substitute values for people and square miles.

$$\approx \frac{218 \text{ people}}{\text{square mile}}$$ Divide 34,000,000 by 156,000. Label the answer.

Your Turn

c. SCIENCE A piece of wood has a volume of 100 cubic centimeters and a mass of 70 grams. What is the density of the piece of wood?

Rates can be found by dividing. Some rates are found by multiplying. Passenger-miles is a rate that tells you the total number of miles traveled by all the passengers as if only one person traveled those miles.

One person traveling 6 miles is 1 · 6, or 6 passenger-miles. Two people traveling 3 miles is 2 · 3, or 6 passenger-miles.

RATES AND PRODUCTS		
	Rates	**Products**
Operation	Division	Multiplication
Examples	meters/minute grams/milliliter	passenger-miles person-days

EXAMPLES Use Measures Expressed as Products

Solve using measures expressed as products.

4 TRAVEL A group of 6 friends travel 300 miles together. How many passenger-miles is that?

Multiply to find the product.

6 passengers · 300 miles = 1,800 passenger-miles

5 TRAVEL The Namkenas go on a family vacation. There are 8 people in their family. The vacation is 10 days long. How many person-days is that?

Multiply to find the product.

8 people · 10 days = 80 person-days

Your Turn
Solve using measures expressed as products.

d. TRAVEL A family of 4 travels to the zoo that is 20 miles away. How many passenger-miles do they travel?

e. SCHOOL There are 20 students. They each attend school 15 days. How many student-days do they attend?

Study Tip

You can treat units like numbers. You can divide them just as you divide numbers.

$$\frac{1 \text{ mi}}{1 \text{ hr}} \cdot \frac{1 \text{ hr}}{60 \text{ min}} \cdot \frac{1 \text{ min}}{60 \text{ s}}$$

$$= \frac{1 \text{ mi}}{3,600 \text{ s}}$$

Talk Math

WORK WITH A PARTNER How can looking at units help you decide if your answer makes sense?

Examples 1–5
(pages 382–384)

VOCABULARY Complete the following sentence.

1. You can use dimensional analysis to _____?_____.

Examples 1–2
(pages 382–383)

Solve using rates.

2. **HOBBY** A toy truck is rolled across the floor. It goes 108 centimeters in 6 seconds. What is its speed in centimeters per second?

3. **FITNESS** Paquito's running stride is about 1 yard long. How many strides does he take during a 2-mile run? (There are 1,760 yards per mile.)

Example 3
(page 383)

4. **POPULATION** In the year 2000, about 980,000 people under the age of 18 lived in Los Angeles, which has an area of about 470 miles. What is the population density of young people, to the nearest whole number?

Examples 4–5
(page 384)

Solve using measures expressed as products.

5. **TRANSPORTATION** A class of 22 students travels 10 miles to a museum. How many passenger-miles are traveled?

6. **COMMUNITY SERVICE** Twenty volunteers clean the beach. They each worked two days. How many person-days did they work?

Skills, Concepts, and Problem Solving

HOMEWORK HELP

For Exercises	See Example(s)
7–11	1–2
12–13	3
14–19	4–5

Solve using rates.

HISTORY An early airplane flight lasted for 12 seconds. The plane flew 144 feet.

7. What was the plane's speed in ft/s?

8. What is the airplane's speed in ft/min?

SPORTS Arnold can swim 400 meters in 5 minutes.

9. What is his speed in m/min?

10. What is Arnold's speed in m/h?

11. **WATER USE** Low-flow showerheads use $2\frac{1}{2}$ gallons of water per minute. If a family showers for a total of $2\frac{1}{3}$ hours per week, about how many gallons of water does the family use each week?

12. **MEASUREMENT** A rock has a mass of 100 grams and a volume of 50 cubic centimeters. What is its density?

13. **GEOGRAPHY** California's Inyo County had about 19,000 people in 10,200 square miles in the year 2005. What was the population density, to the nearest whole number?

Solve using measures expressed as products.

14. **TRAVEL** An airplane carries 46 passengers on a 400-mile trip. How many total passenger-miles are on the trip?

15. **TRANSPORTATION** There are 30 people on a train. The train travels 33 miles. What is the total number of passenger-miles?

16. **SCHOOL** There are 500 students and teachers at a school. There are 180 school days in a year. How many person-days are in a school year?

17. **HEALTH** There are 130 people who break for lunch. Lunch is two hours. How many person-hours are spent at lunch?

18. *Writing in Math* Explain the meaning of "person-days" and how you calculate them.

19. **H.O.T.** Problem Pedro and Ruben drive 100 miles to pick up Nathan. Then, they all travel 50 miles together to the farm. Ruben says the total number of passenger-miles is 150. Pedro says it is 350. Who is right? Explain your answer.

STANDARDS PRACTICE

Choose the *best* answer.

20 The chart below describes the speed of toy race cars.

Car	Speed
Dragster	2 laps per minute
Radical Racer	1 lap every 2 minutes
Speedy	160 laps in 2 hours
Track Titan	100 laps per hour

Which car is the fastest?

A Dragster

B Radical Racer

C Speedy

D Track Titan

21 Beverly ran 6 miles at the speed of 4 miles per hour. How long did it take her to run that distance?

F $\frac{2}{3}$ h

G $1\frac{1}{2}$ h

H 4 h

J 6 h

Spiral Review

Describe the relationship shown in each scatter plot. (Lesson 7-3)

22.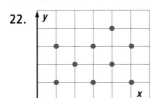

23.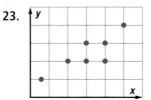

24.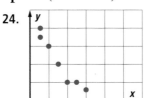

25. **SPORTS** A football team loses 10 yards. Then it gains 4 yards on the next play. What integer represents the final result from the two plays? (Lesson 2-3)

Progress Check 2
(Lessons 7-3 and 7-4)

▷ Vocabulary and Concept Check

density (p. 383)
dimensional analysis (p. 383)
linear relationship (p. 376)
scatter plot (p. 374)
trend (p. 374)

Choose the term that *best* completes each statement. (Lessons 7-3 and 7-4)

1. A _____?_____ is a graph that shows the relationship between two sets of data.

2. _____?_____ is a consistent change over time.

▷ Skills Check

3. Describe the relationship shown in the scatter plot. (Lesson 7-3)

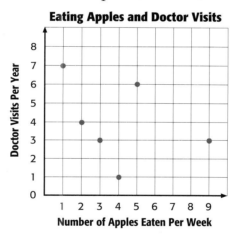

Eating Apples and Doctor Visits

4. Use the table. How are yards and feet related? (Lesson 7-3)

Yards	3	5	7	10
Feet	9	15	21	30

▷ Problem-Solving Check

Solve using rates. (Lesson 7-4)

TRANSPORTATION Lucia drives 120 miles in 2 hours.

5. What is Lucia's speed in miles per hour?

6. What is Lucia's speed in miles per minute?

Solve using measures expressed as products. (Lesson 7-4)

7. **SCHOOL** A class has 20 students and 1 teacher. They all work on science-fair projects for 5 days. How many person-days did they work?

8. **REFLECT** How can you tell if two sets of data have a positive, negative, or no relationship? (Lesson 7-3)

7-5 Slope

Vocabulary

slope (p. 388)

rise (p. 388)

run (p. 388)

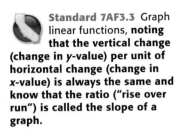

Standard 7AF3.3 Graph linear functions, **noting that the vertical change** (change in y-value) per unit of horizontal change (change in x-value) is always the same and know that the ratio ("rise over run") is called the slope of a graph.

> **The** What**:** I will understand slope as a constant rate of change. I will find slope using graphs and tables.
>
> **The** Why**:** Slope is used in many everyday situations, such as determining the steepness of access ramps and ski runs.

An access ramp from the sidewalk to a school's entrance rises 2 inches for every 24 inches in length. The ramp's steepness is called the **slope.**

Slope is the ratio of the vertical change, called the **rise,** to the horizontal change, called the **run.**

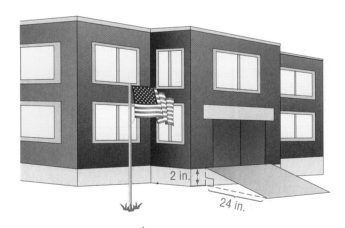

2 in.

24 in.

$$\text{Slope of ramp} = \frac{\text{vertical change}}{\text{horizontal change}} = \frac{\text{rise}}{\text{run}} = \frac{2 \text{ in.}}{24 \text{ in.}} = \frac{1}{12}$$

Study Tip

y_2 is read "y sub 2." The 2 is called a subscript.

SLOPE

$$\text{Slope} = m = \frac{\Delta y}{\Delta x} = \frac{\text{change in } y}{\text{change in } x} = \frac{\text{rise}}{\text{run}} = \frac{y_2 - y_1}{x_2 - x_1}$$

$$m = \frac{y_2 - y_1}{x_2 - x_1}, \text{ where } x_2 \neq x_1$$

1 SPORTS Suppose a ski run drops 18 feet for every 24 feet of horizontal change. Find the slope.

$$\text{slope} = \frac{\text{vertical change}}{\text{horizontal change}}$$ Definition of slope
vertical change = −18 ft
horizontal change = 24 ft

$$= \frac{-18 \text{ ft}}{24 \text{ ft}}$$

$$= -\frac{3}{4}$$ Simplify.

The slope of the ski ramp is $-\frac{3}{4}$.

Your Turn

a. **EXERCISE** A treadmill is 48 inches long. Suppose it is raised 10 inches. Find the slope.

You can find slope from a graph. Choose two points on the line. Find the rise and the run. Then find the $\frac{\text{rise}}{\text{run}}$ ratio.

EXAMPLE   **Slope Using a Graph**

2 Use the graph to find the slope of the line.

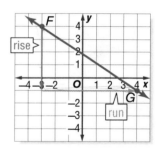

$$\text{slope} = \frac{\text{rise}}{\text{run}}$$ Definition of slope

$$= -\frac{5}{7}$$ rise = −5, run = 7

The slope of the line is $-\frac{5}{7}$.

Your Turn
Find the slope of each line.

b.

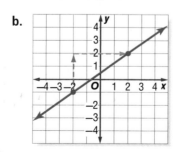

c.

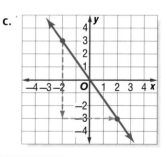

Study Tip

Recall the directions you used to plot points.
Use these to find rise and run values.

Positive changes are up and right.
Negative changes are down and left.

positive   negative
↑   ↓

→   ←

For Example 2, you could use the coordinates of points *F* and *G* to find the slope. Let (−3, 4) be (x_1, y_1) and (4, −1) be (x_2, y_2).

$$\text{Slope} = \frac{y_2 - y_1}{x_2 - x_1} = \frac{-1 - 4}{4 - (-3)} = \frac{-5}{7}$$

You can also find slope from a table. Choose ordered pairs representing two points on a line. Find the changes in the *x*- and *y*-values to set up the slope ratio.

EXAMPLE Slope Using a Table

The points given in the table lie on a line. Find the slope of the line.

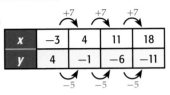

$$\text{slope} = \frac{\text{change in } y}{\text{change in } x}$$

$$= \frac{-1 - 4}{4 - (-3)}$$

$$= -\frac{5}{7}$$

The slope is $-\frac{5}{7}$.

> **Study Tip**
>
> Use two different ordered pairs to find the slope. You should get the same slope.
>
> You can use any two ordered pairs or points when finding slope.

Your Turn

The points given in each table lie on lines. Find the slope of each line.

d.

x	2	6	10
y	0	1	2

e.

x	−4	0	4	8
y	−2	−5	−8	−11

The graph in Example 2 and the table in Example 3 represent the same line.

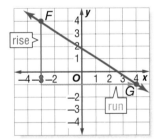

Whether you are given information in a graph or in a table, the line's slope will be the same.

Notice the slope is the same going from point *F* to point *G* or from point *G* to point *F*.

A rise of −5 and a run of 7 is the same as a rise of 5 and a run of −7.

Talk Math

WORK WITH A PARTNER Tell your partner how to find slope using a graph. Listen as your partner tells you how to find slope using a table.

Vocabulary Review
slope
rise
run

Examples 1–3
(pages 389–390)

VOCABULARY

1. What term describes the constant rate of change?

2. The vertical change is called the _____?_____.

3. The _____?_____ describes the horizontal change.

Example 1
(page 389)

4. **SAFETY** A fire truck uses a moveable ladder to reach high locations. Suppose the ladder's base is 50 feet from a building and is raised 100 feet. Find the ladder's slope.

5. **RECREATION** A biking path rises 5 feet for every horizontal change of 100 feet. Find the path's slope.

Example 2
(page 389)

Find the slope of each line.

6.

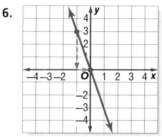

7.

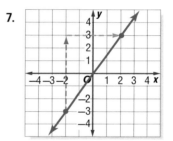

Example 3
(page 390)

The points given in each table lie on a line. Find the slope of each line.

8.

x	8	6	4	2
y	0	5	10	15

9.

x	4	8	12	16
y	−12	0	12	24

10.

x	5	6	7	8
y	15	12	9	6

11.

x	2	4	6	8
y	3	0	−3	−6

Skills, Concepts, and Problem Solving

HOMEWORK HELP

For Exercises	See Example(s)
12–14	1
15–18	2
19–22	3

12. **SAFETY** A ladder is placed 3 feet from a house. The ladder is extended to a vertical height of 16 feet. Find the slope of the ladder.

13. **RECREATION** A snowboarding hill decreases 20 feet vertically for every 25 feet of horizontal distance. Find the slope of the hill.

14. An access ramp increases 1 inch vertically for every foot of horizontal distance. Find the slope of the access ramp.

Use the graphs to find the slope of each line.

15.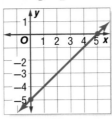

16.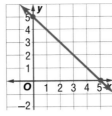

17.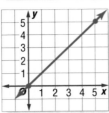

18.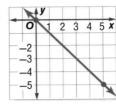

The points given in each table lie on a line. Find the slope of each line.

19.

x	2	3	4	5
y	5	10	15	20

20.

x	2	4	6	8
y	7	14	21	28

21.

x	10	8	6	4
y	4	8	12	16

22.

x	4	7	10	13
y	20	18	16	14

23. *Writing in Math* Sashi says that a slope of $-\frac{2}{3}$ indicates a vertical change of 2 down and a horizontal change of 3 left. Do you agree? Why or why not?

24. **H.O.T.** Problem Compare and contrast finding slopes using graphs and tables. How are the procedures alike? How are they different?

STANDARDS PRACTICE

25 The price of pizzas with different toppings is shown in the graph. What ordered pair shows the price in dollars of a pizza with one topping?

  A $(0, 9)$ **C** $(1, 10.25)$

  B $(9, 0)$ **D** $(10.25, 1)$

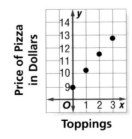

26 What is the slope of the line that passes through the points shown on the graph?

  F $\frac{1}{2}$ **H** 1

  G $\frac{4}{5}$ **J** $\frac{5}{4}$

Spiral Review

27. **RECREATION** A sailboat carries 6 people. They sail for 8 hours. What is the total number of passenger-hours? (Lesson 7-4)

Rewrite using only positive exponents. (Lesson 5-2)

28. $a^3 g^{-5}$ 29. $b^{-1} d^{-3}$ 30. $\dfrac{r^5}{t^{-2}}$

Linear Functions

Vocabulary

function (p. 393)

nonlinear (p. 396)

 Standard 7AF3.3 Graph linear functions, noting that the vertical change (change in *y*-value) per unit of horizontal change (change in *x*-value) is always the same and know that the ratio ("rise over run") is called the slope of a graph.

 The What: I will graph linear functions.

 The Why: Graphing a linear function can show how far a car has traveled at the California Speedway in Fontana.

The California Speedway track is 2 miles long.

If a car travels 1 lap, it goes 2 • 1 or 2 miles.
If a car travels 2 laps, it goes 2 • 2 or 4 miles.
If a car travels 5 laps, it goes 2 • 5 or 10 miles.
If a car travels x laps, it goes 2 • x or $2x$ miles.

To make an input-output table, substitute values of x to find values of y. Write at least three ordered pairs in your table.

An Indy race car can reach speeds of more than 230 miles per hour.

Laps, x (input)	$2x$	Distance, y (output)	(x, y)
1	2 • 1	2	(1, 2)
2	2 • 2	4	(2, 4)
5	2 • 5	10	(5, 10)

Notice how each value of x is related to a unique value of y. There is exactly one distance traveled for each number of laps. If a driver drives 2 laps, the only distance related to that is 4 miles. A driver can not travel both 4 miles and 10 miles in 2 laps.

This is an example of a function. A **function** is a relation that assigns exactly one output for each input. The equation $y = 2x$ represents the California Speedway track input-output table.

Linear functions can be represented by graphs, tables, and equations.

The equation $y = x + 2$ is another example of a linear function. It has exactly one output, or y-value, for every input, or x-value.

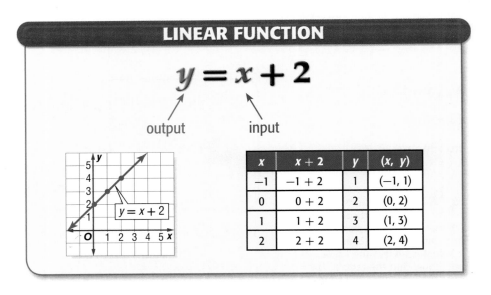

LINEAR FUNCTION

$$y = x + 2$$

output   input

x	x + 2	y	(x, y)
−1	−1 + 2	1	(−1, 1)
0	0 + 2	2	(0, 2)
1	1 + 2	3	(1, 3)
2	2 + 2	4	(2, 4)

Remember $y = x + 2$ can also be written as $x + 2 = y$. In $x + 2 = y$, x is still the input and y is still the output.

EXAMPLE Graph Linear Functions

1 Use an input-output table to graph the linear function $y = 2x$.

Choose any 3 values for the inputs, or x-values.

Substitute each value to find the outputs, or y-values.

Graph each ordered pair.

x	0	1	2
2x	0	2	4
y	0	2	4
(x, y)	(0, 0)	(1, 2)	(2, 4)

Study Tip

It takes only two points to graph a line. Use a third point to check your graph. If your three points do not make one line, check your calculations.

Draw a line that passes through all 3 points.

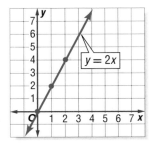

Your Turn

Use an input-output table to graph each linear function.

a. $y = x + 4$

b. $y = 3x$

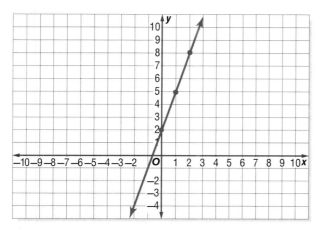

WORK WITH A PARTNER Which graph better shows $y = -3x + 2$? Discuss your choice with your partner.

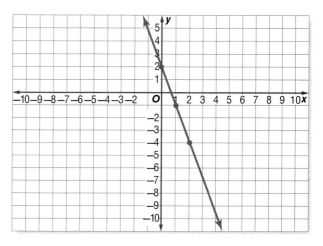

EXAMPLE Graph Linear Functions

2 Use an input-output table to graph the linear function $y = 3x + 1$.

Choose any three values for the inputs, or *x*-values.

Substitute each value to find the outputs, or *y*-values.

Graph each ordered pair.

x	0	1	2
3x + 1	1	4	7
y	1	4	7
(x, y)	(0, 1)	(1, 4)	(2, 7)

Draw a line that passes through all three points.

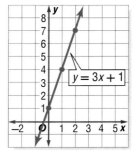

Your Turn

Use an input-output table to graph each linear function.

c. $y = 4x + 2$ **d.** $y = 2x - 1$

When a function is linear, the x and y values each change by a constant amount. This is called slope. It is a constant rate of change. If the rate of change is not constant, the function is **nonlinear.**

EXAMPLES **Identify Linear Functions**

Determine whether each table shows a linear or nonlinear function.

3

	+1	+1	
x	1	2	3
y	5	9	13
	+4	+4	

The rate of change is constant.
The table shows a linear function.

4

x	y	
1	5	+4
2	9	+5
3	14	+6
4	20	

(+1 on each x step)

The rate of change is not constant.
The table shows a nonlinear function.

Your Turn

Determine whether each table shows a linear or nonlinear function.

e.

x	0	3	7
y	10	15	20

f.

x	y
0	0
5	−4
10	−8
15	−12

Real-World Challenge·····
Determine linear costs for 1, 2, and 3 sandwiches.

Real-World EXAMPLE

5 FOOD At a restaurant, one sandwich costs $2.00, two sandwiches cost $3.50, and three sandwiches cost $4.75. Determine if these prices represent a linear or nonlinear function. Explain.

x	y	
1	2.00	+1.50
2	3.50	+1.25
3	4.75	

(+1 on each x step)

The rate of change is not constant.
The sandwich pricing is nonlinear.

Your Turn

g. **SPORTS** An athlete runs 1 mile in 6 minutes, 2 miles in 12 minutes, and 3 miles in 18 minutes. Does the athlete run at a linear rate? Explain.

Examples 1–4
(pages 394–396)

VOCABULARY

1. Define function.

Examples 1–2
(pages 394–395)

Use an input-output table to graph each linear function.

2. $y = x + 1$

3. $y = 3x - 2$

Examples 3–4
(page 396)

Determine whether each table shows a linear or nonlinear function.

4.

x	3	5	7	9
y	20	17	13	8

5.

x	12	10	8	6
y	4	8	12	16

Example 5
(page 396)

6. **FINANCE** Jasmine earns $50 when she works 10 hours. She earns $80 when she works 15 hours. She earns $140 when she works 20 hours. Is Jasmine's pay scale linear? Explain.

Skills, Concepts, and Problem Solving

HOMEWORK HELP

For Exercises	See Example(s)
7–10	1–2
11–14	3–4
15–16	5

Use an input-output table to graph each linear function.

7. $y = x + 5$

8. $y = 4x$

9. $y = 2x + 4$

10. $y = 3x - 1$

Determine whether each table shows a linear or nonlinear function.

11.

x	2	6	10	14
y	20	16	12	8

12.

x	12	10	8	6
y	4	8	12	16

13.

x	y
3	2
6	5
9	9
12	14

14.

x	y
−5	0
0	−3
5	−6
10	−9

15. **TRANSPORTATION** A car travels 50 miles in 1 hour, 100 miles in 2 hours, 150 miles in 3 hours, and 200 miles in 4 hours. Does the car travel at a linear rate?

16. **FOOD** At a restaurant, six bagels cost $6.00. Twelve bagels cost $9.00. Eighteen bagels cost $12.60. Determine if these prices represent a linear or nonlinear function. Explain.

17. *Writing in Math* Explain the difference between linear and nonlinear functions.

18. **H.O.T. Problem** Determine the missing values in the table below. Write a linear equation that would produce the table's input and output values.

x	2	5	8	11		17
y	5	11	17	23	29	

STANDARDS PRACTICE

19 A line is graphed on a coordinate plane. The line passes through the points (0, 0), (5, −6), and (−5, 6). What is the equation of the line?

A $y = -\frac{5}{6}x$ C $y = \frac{6}{5}x$

B $y = \frac{5}{6}x$ D $y = -\frac{6}{5}x$

20 Which equation represents the linear function shown in the table below?

x	−5	0	5	10
y	0	10	20	30

F $y = x + 5$ H $y = \frac{1}{2}x - 5$

G $y = x - 5$ J $y = 2x + 10$

Spiral Review

The points given in the table lie on a line. Find the slope of the line. (Lesson 7-5)

21.

x	0	−5	−10	−15
y	2	4	6	8

Use the graph to find the slope of the line. (Lesson 7-5)

22.

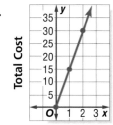

Number of Days

23. **FOOD** Stephen orders box lunches for a picnic. One lunch costs $10. Eleven lunches cost $110. Twenty-one lunches cost $210. Find the slope of the line that represents the box lunch prices. (Lesson 7-5)

Simplify. (Lesson 4-2)

24. $1.35 + 21.9$ 25. $3.1 - 2.78$ 26. $4.02 - 3.19$

27. **WEATHER** One winter evening, the temperature was 2°C at 8:00 P.M. By 11:00 P.M., the temperature dropped to −7°C. By how much did the temperature drop from 8:00 to 11:00? (Lesson 2-3)

28. Solve $4 + 5w = 14$. (Lesson 2-7)

Progress Check 3

(Lessons 7-5 and 7-6)

▷ Vocabulary and Concept Check

> function (p. 393)   nonlinear (p. 396)
> linear function (p. 376)   slope (p. 388)

Complete each sentence with the correct mathematical term or phrase.
(Lessons 7-5 and 7-6)

1. The rise over run is called the _____?_____ .

2. A relationship that assigns exactly one output for each input is called a _____?_____ .

▷ Skills Check

Use the graphs to find the slope of each line. (Lesson 7-5)

3.   4.

Use an input-output table to graph each linear function. (Lesson 7-6)

5. $y = -5x + 5$ 6. $y = \frac{1}{2}x + 1$

Determine whether each table shows a linear or nonlinear function. If linear, give the constant rate of change. (Lesson 7-6)

7.

x	y
11	4
13	13
15	4
26	56
12	32

8.

# days	$
1	2
2	4
3	6
4	8
5	10

▷ Problem-Solving Check

9. A store charges $2.00 for twelve muffins, $3.50 for eighteen muffins, and $4.75 for twenty-four muffins. Determine if the prices represent a linear function. Explain. (Lesson 7-6)

10. **REFLECT** How can you find the slope of a line from its graph?
(Lesson 7-5)

The Pythagorean Theorem

Vocabulary

Pythagorean Theorem
(p. 400)

converse (p. 403)

 Standard 7MG3.3 Know and understand the Pythagorean Theorem and its converse and use it to find the length of the missing side of a right triangle and the lengths of other line segments and, in some situations, **empirically verify the Pythagorean Theorem by direct measurement.**

 The What: I will solve problems using the Pythagorean Theorem.

The Why: The Pythagorean Theorem can be used to determine if a softball diamond is designed correctly.

When groundskeepers lay out a softball diamond, they measure the distances between the bases and the diagonals to be sure the diamond is a square.

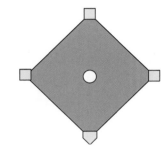

A right triangle is a triangle with one right angle. A Greek mathematician named Pythagoras lived about 2,500 years ago. The Pythagorean Theorem is named after him. The **Pythagorean Theorem** states that, for any right triangle, the square of the length of the hypotenuse is equal to the sum of the squares of the lengths of the legs.

THE PYTHAGOREAN THEOREM

$$a^2 + b^2 = c^2$$

legs: a, b

hypotenuse: c

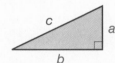 **Talk Math**

WORK WITH A PARTNER A triangle has side measures of 6 feet, 10 feet, and 8 feet. Identify the measures of the legs and the hypotenuse.

EXAMPLE Verify the Pythagorean Theorem

Measure the right triangle. Use the measures to verify the Pythagorean Theorem.

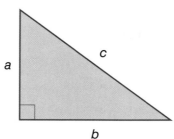

$a = 3$ cm
$b = 4$ cm
$c = 5$ cm

$a^2 + b^2 = c^2$	Use the Pythagorean Theorem.
$3^2 + 4^2 = 5^2$	Substitute lengths a, b, and c.
$9 + 16 = 25$	Simplify.
$25 = 25$	

Your Turn

Measure the right triangle. Use the measures to verify the Pythagorean Theorem.

a.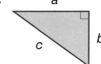

You can use the Pythagorean Theorem to find the length of any side of a right triangle.

EXAMPLE Find the Length of the Hypotenuse

Use the Pythagorean Theorem to find the length of the hypotenuse.

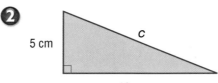

Remember!

A number has both a positive and a negative square root.
The square roots of 9 are 3 and –3.

$3 \cdot 3 = 9$
$-3 \cdot (-3) = 9$

Use only positive square roots when finding lengths.

$a^2 + b^2 = c^2$	Use the Pythagorean Theorem.
$5^2 + 12^2 = c^2$	Substitute lengths a and b.
$25 + 144 = c^2$	Simplify 5^2 and 12^2.
$c^2 = 169$	Add 25 to 144.
$c = 13$ or $c = -13$	Find the square root of each side.

The hypotenuse is 13 centimeters long.

Your Turn

Use the Pythagorean Theorem to find the length of the hypotenuse.

b.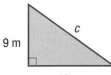

EXAMPLES **Find the Missing Length**

Use the Pythagorean Theorem to find each missing length.

3

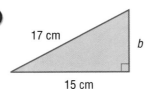

$a^2 + b^2 = c^2$	Use the Pythagorean Theorem.
$15^2 + b^2 = 17^2$	Substitute lengths a and c.
$225 + b^2 = 289$	Simplify.
$b^2 = 289 - 225$	Subtract 225 from each side.
$b^2 = 64$	Subtract 225 from 289.
$|b| = \sqrt{64}$	Find the square root of each side.
$b = 8$ or $b = -8$	Use the positive root for length.

The missing length is 8 centimeters.

Remember!

The hypotenuse is the side of a right triangle that is opposite the right angle.

4 What is the distance between *L* and *S*?

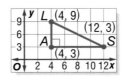

$a^2 + b^2 = c^2$	Use the Pythagorean Theorem.
$6^2 + 8^2 = c^2$	Substitute lengths a and b.
$36 + 64 = c^2$	Simplify 6^2 and 8^2.
$c^2 = 100$	Add 36 to 64.
$c = 10$ or $c = -10$	Find the square root of each side.

The hypotenuse is 10 centimeters long.

Use the positive root for length.

Your Turn

Use the Pythagorean Theorem to find each missing length.

c.

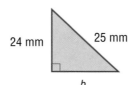

d.

Converse means an if/then statement in which terms are expressed in reverse order. The converse of the Pythagorean Theorem can be used to identify right triangles.

CONVERSE OF THE PYTHAGOREAN THEOREM

If a, b, and c are the lengths of the sides of a triangle and $a^2 + b^2 = c^2$, then the triangle is a right triangle.

EXAMPLE Identify a Right Triangle

5 Determine if a triangle with sides of 7 inches, 9 inches, and 6 inches is a right triangle. Use the converse of the Pythagorean Theorem.

$a^2 + b^2 = c^2$ Use the Pythagorean Theorem.

$6^2 + 7^2 \overset{?}{=} 9^2$ Substitute the greatest value for c and the other values for a and b.

$36 + 49 \overset{?}{=} 81$ Simplify.

$85 \neq 81$ Add 36 to 49.

The triangle is *not* a right triangle.

Your Turn

e. Determine if a triangle with sides of 12 meters, 20 meters, and 16 meters is a right triangle. Use the converse of the Pythagorean Theorem.

Real-World EXAMPLE

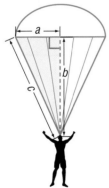

6 SKY DIVING The radius of a parachute canopy is 12 feet. The suspension lines are 20 feet long. How far below the canopy does the person hang?

$a^2 + b^2 = c^2$ Use the Pythagorean Theorem.

$12^2 + b^2 = 20^2$ Substitute lengths a and c.

$144 + b^2 = 400$ Simplify.

$b^2 = 400 - 144$ Subtract 144 from each side.

$b^2 = 256$ Subtract 144 from 400.

$|b| = \sqrt{256}$ Find the square root of each side.

$b = 16$ or $b = -16$ Use the positive root for length.

The person hangs 16 feet below the canopy.

Your Turn

Use the Pythagorean Theorem to find the missing length.

f. A television screen is 20 inches long. It has a diagonal measurement of 25 inches. How high is the screen?

Examples 1–6
(pages 401–403)

VOCABULARY Complete each sentence.

Vocabulary Review
Pythagorean Theorem
converse

1. The Pythagorean Theorem says that _____?_____.

2. The word _____?_____ means opposite.

Example 1
(page 401)

3. Measure the right triangle. Use the measures to verify the Pythagorean Theorem.

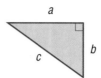

Examples 2–4
(pages 401–402)

Use the Pythagorean Theorem to find each missing length.

4.

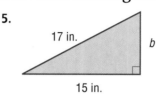

5.

Example 5
(page 403)

6. Determine if a triangle with sides of 4 meters, 3 meters, and 5 meters is a right triangle. Use the converse of the Pythagorean Theorem.

Skills, Concepts, and Problem Solving

HOMEWORK HELP	
For Exercises	**See Example(s)**
7–8	1
9–14	2–4
15–16	5
17	6

Measure each right triangle. Use the measures to verify the Pythagorean Theorem.

7.

8.

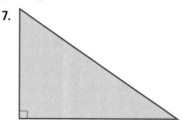

Use the Pythagorean Theorem to find each missing length.

9.

10.

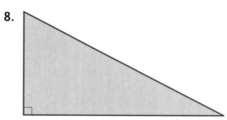

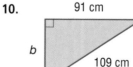

11. How long is the ramp?

12. How high is the kite?

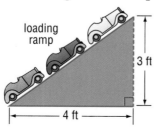

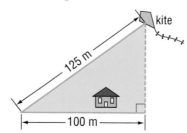

13. What is the distance between A and C?

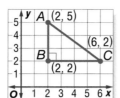

14. What is the distance between X and Z?

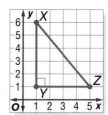

Determine if each triangle is a right triangle. Use the converse of the Pythagorean Theorem.

15. The sides measure 8 km, 15 km, and 17 km.

16. The sides measure 5 inches, 8 inches, and 13 inches.

17. CONSTRUCTION The diagonal brace on a gate is 5 feet long. The height of the gate is 4 feet. How wide is the gate?

18. H.O.T. Problem Anessa says a triangle with sides of 3 centimeters, 4 centimeters, and 7 centimeters is a right triangle. Do you agree? Why or why not?

19. *Writing in Math* For safety reasons, the base of a 24-foot ladder should be at least 8 feet from the wall. Can a 24-foot ladder be used to reach a window 22 feet above the ground? Explain.

STANDARDS PRACTICE

20

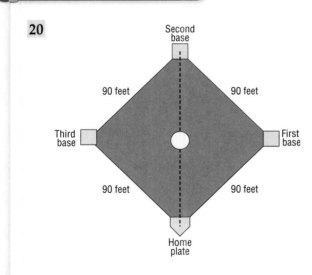

A catcher throws a baseball from home plate to second base. Approximately, how many feet does the ball travel?

A 90 feet

B 127 feet

C 135 feet

D 180 feet

Spiral Review

Use an input-output table to graph each line. (Lesson 7-6)

21. $y = 2x + 4$

22. $y = -3x + 2$

23. CONSTRUCTION Mrs. Blair wants to make shelves that are each $\frac{3}{4}$-yard long. She has 3 yards of lumber. How many shelves can Mrs. Blair make? (Lesson 3-6)

Study Guide

Understanding and Using the Vocabulary

After completing the chapter, you should be able to define each term, property, or phrase, and give an example of each.

converse (p. 403)	linear relationship (p. 376)	rise (p. 388)
coordinate (p. 367)	nonlinear (p. 396)	run (p. 388)
coordinate plane (p. 366)	ordered pair (p. 366)	scatter plot (p. 374)
density (p. 383)	origin (p. 366)	slope (p. 388)
dimensional analysis (p. 383)	Pythagorean Theorem	trend (p. 374)
draw a graph (p. 372)	(p. 400)	x-axis (p. 366)
function (p. 393)	quadrant (p. 366)	y-axis (p. 366)

Complete each sentence with the correct mathematical term or phrase.

1. The _____?_____ is the rise over the run.

2. The _____?_____ states that if a triangle is a right triangle, then $a^2 + b^2 = c^2$.

3. A coordinate plane is divided into four sections called _____?_____.

4. A(n) _____?_____ is a relation that assigns exactly one output for each input.

5. The horizontal axis is called the _____?_____.

6. A graph that shows the relationship between two sets of data is a(n) _____?_____.

7. The _____?_____ is located at (0, 0).

Objectives and Examples

LESSON 7-1 pages 366–370

Plot A (4, −3) on a coordinate plane.

Start at the origin, (0, 0). Move 4 units right. Then move 3 units down. Mark and label the point.

Review Exercises

Graph each point on a coordinate plane.

8. B (1, 4)

9. C (−2, 3)

10. D (−3, −4)

11. E (4, −3)

Objectives and Examples

LESSON 7-3 pages 374–379

Make a graph to find the answer.

Minutes read	20	6	16	12	4
Number of pages read	10	3	8	6	2

How many minutes does it take to read one page?

Make a scatter plot of the data. Title the graph. Label the *x*-axis and *y*-axis. Graph the points. Draw a line connecting them.

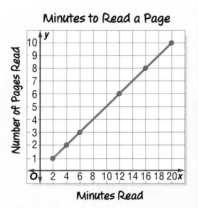

Minutes to Read a Page

It takes two minutes to read one page.

LESSON 7-4 pages 382–386

Use measures expressed as rates and products to solve problems.

Lance and his friend drove 20 miles in 16 minutes. How fast did they travel in miles/minute? How many passenger-miles did they travel?

Divide distance by time to calculate their speed.
20 miles/16 minutes = 1.25 mi/min

Multiply the number of people by the number of miles to find the passenger-miles.
2 people × 20 miles = 40 passenger-miles

Review Exercises

Make a graph to find each answer.

12.

Gallons	5	3	6	4	2
Miles driven	115	69	138	92	46

How many miles can Terri drive on one gallon of gas?

13.

Number of science kits	10	15	12	9	8
Number of students	30	45	36	27	24

How many students use one science kit?

14.

Number of coins	10	2	12	4	6
Value in dollars	35	7	42	14	21

How much is each coin worth?

Solve using rates.

15. **TOYS** A toy car went 400 yards in 5 minutes. What is its speed in yards/minute?

16. **RETAIL** Apples cost $3.20 for 8 apples. What is the cost per apple?

Solve using measures expressed as products.

17. **SCHOOL** A holiday is 12 days long. There are 450 people at the school. How many person-days is that?

18. **TRAVEL** A family of 9 travels 2,100 miles together. How many passenger-miles is that?

Objectives and Examples

Review Exercises

LESSON 7-5   pages 388–392

Use tables and graphs to find the slopes of lines.

Days worked (x)	2	4	6	8	10
Money earned (y)	200	400	600	800	1000

Every time *x* increases by 2, *y* increases by $200.

The relation is linear. The constant rate of change is (the slope) $\frac{\$200}{2}$ or $100.

The points given in the table lie on a line. Find the slope of the line.

19.

Number of pizzas	1	3	5	7	9
Cost	8	24	40	56	72

LESSON 7-6   pages 393–398

Graph linear functions.

$y = 2x + 1$

Make an input-output table. Plot the points.

Draw a line through the points.

x	0	1	2
y	1	3	5

Use an input-output table to graph each line.

20. $y = 3x + 2$ **21.** $y = 4x + 6$

22. $y = -2x + 1$ **23.** $y = -7x + 3$

LESSON 7-7   pages 400–405

Use the Pythagorean Theorem to find the missing length.

$$a^2 + b^2 = c^2$$
$$3^2 + 4^2 = c^2$$

3 cm

4 cm

c

$25 = c^2$ Simplify.

$|c| = \sqrt{25}$ Find the square root of each side.

$c = 5$ or $c = -5$ Use the positive root for length.

The hypotenuse is 5 centimeters long.

Use the Pythagorean Theorem to find each missing length.

24.

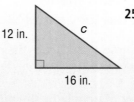

12 in.

16 in.

c

25.

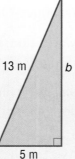

13 m

5 m

b

26. A gate is 6 feet high and 8 feet long. What is the measure of its diagonal brace?

Chapter Test

▷ Vocabulary and Concept Check

Identify the point that lies on, in, or at the given part of the coordinate system.

1. Quadrant I

2. y-axis

3. origin

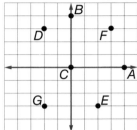

▷ Skills Check

4. Find the ordered pairs for the points shown in the graph above.

5. Graph $y = 2x - 5$ on a coordinate plane.

6. Determine whether the function is linear or nonlinear. If linear, give the constant rate of change.

x	y
2	16
4	12
6	8
8	4
10	0

7. Use the Pythagorean Theorem to find the missing length.

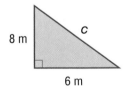

▷ Problem-Solving Check

8. **TRAVEL** Mrs. Elton drove at a constant rate. She drove 120 miles in 3 hours. How many total miles would Mrs. Elton have driven after 8 hours?

Solve using rates.

9. **FINANCE** Chen has $300 in a savings account. At the end of the year his balance has grown to $312. What interest rate does the bank pay?

10. **SPORTS** Faith ran 400 meters in 4 minutes. Find her speed in kilometers per minute.

PART 1 Multiple Choice

Choose the *best* answer.

1 Which point lies on the line $y = x + 2$?

  A $(-2, 0)$ **C** $(0, -2)$

  B $(2, 0)$ **D** $(0, 2)$

2 The gas tank in Alejandra's car holds 15 gallons of fuel. She can drive 300 miles on one tank of gas. How many miles per gallon does her car get?

  F 20   **H** 315

  G 285   **J** 4,500

3 The mileage reading on the Flores' car was 256 before they left on vacation. When they returned home, the reading was 740. There are five members in the Flores family. How many passenger-miles was the trip?

  A 484   **C** 3,700

  B 2,420   **D** 4,980

4

Wins	4	10	12	16
Losses	2	5	6	8

Suppose this pattern continues. If a team wins 30 games, how many losses would you expect?

  F 10   **H** 30

  G 15   **J** 60

5 Triangle ABC is a right triangle. What is the length of the hypotenuse?

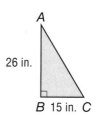

  A 30 in.   **C** 130 in.

  B 41 in.   **D** 139 in.

6 Which ordered pair is represented by point S?

  F $(-3, -2)$ **H** $(-2, -3)$

  G $(-3, 2)$ **J** $(3, -2)$

7 Which *best* represents $y = 3x - 6$?

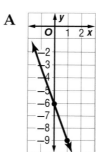

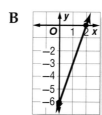

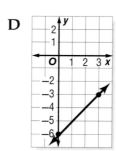

Record your answers on the answer sheet provided by your teacher or on a separate sheet of paper.

8 Graph $y = -2x + 4$ on a coordinate plane.

9 Make a scatter plot for the following data.

Age	8	20	12	24	32
Length of hair (cm)	32	4	8	16	4

What is the relationship between age and length of hair?

10 Find the slope of the line.

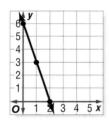

11 A triangle has side lengths 4 meters, 7 meters, and 12 meters. Use the Pythagorean Theorem to determine whether or not this is a right triangle.

12 A right triangle has legs that measure 8 meters and 6 meters. What is the length of the hypotenuse?

Record your answers on the answer sheet provided by your teacher or on a separate sheet of paper.

13 Booker decides to sell bottled water at his brother's soccer games. He spends $20 to buy bottles of water. He sells them for $2 each.

a. Write an equation to find Booker's profit y based on the number of bottles sold x.

b. Use an input-output table to graph your equation.

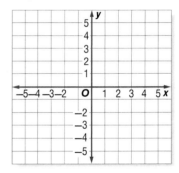

c. Identify the slope of the line. What does it represent?

d. How much money does he have after he sells 15 bottles of water?

e. The break-even point is when he has sold enough water to cover his expenses. At that point, his profit is $0. How many bottles of water must he sell to reach his break-even point?

NEED EXTRA HELP?												
If You Missed Question...	1	2	3	4	5	6	7	8	9	10	11, 12	13
Go to Lesson...	7-5	7-4	7-4	7-3	7-7	7-1	7-6	7-6	7-3	7-7	7-6	7-5, 7-6
For Help with Algebra Readiness Standard...	7AF3.3	7MG1.3	7MG1.3	7AF3.4	7MG3.3	7AF3.3	7AF3.3	7AF3.3	7AF3.4	7MG3.3	7AF3.3	7AF3.3

CHAPTER 8

Geometry Basics

Here's What I Need to Know

Standard 7MG2.1 Use formulas for finding the perimeter and area of basic two-dimensional figures and the surface area and volume of basic three-dimensional figures, including rectangles, parallelograms, trapezoids, squares, triangles, circles, prisms, and cylinders.

Standard 7MR1.1 Analyze problems by **identifying relationships**, distinguishing relevant from irrelevant information, identifying missing information, sequencing and prioritizing information, **and observing patterns.**

Standard 7MR2.0 Students use strategies, skills, and concepts in finding solutions.

Vocabulary Preview

perimeter the distance around a plane figure (p. 434)

area the number of square units needed to cover the inside of a region or plane figure (p. 440)

volume the number of cubic units needed to fill a three-dimensional figure (p. 452)

 The What

I will learn to:

- recognize and identify plane and solid figures.
- use and understand congruent and similar relationships.
- use formulas to find perimeter, area, volume, and surface area.

 **The** Why

Everyday tasks such as wrapping presents, determining the distance around the basketball court, and painting a mural on a wall use geometric concepts.

Option 2

Complete the Quick Check below.

Vocabulary and Skills Review

Match each description with the correct figure.

1. figure with four congruent sides

   A.

2. two rays that create an acute angle

   B.

3. polygon with three sides and a right angle

   C.

Find the perimeter *P* or the area *A*.

4. $P = 4s$
   $s = 2$ inches

    s

5. $A = \ell w$
   $\ell = 8$ cm, $w = 3$ cm

   ℓ w

Solve.

6. $\dfrac{3}{7} = \dfrac{12}{x}$

7. $\dfrac{4}{5} = \dfrac{x}{35}$

Note-taking

Tips
Use page clues to find important information.
- bold words
- special colors

Chapter Instruction	My Notes
The What: I will recognize and name **plane figures** and calculate the measure of their angles.	Plane figures are flat and have two dimensions. Examples: triangle, quadrilateral
The sum of the three interior angles of a triangle is 180°.	Angles of a triangle add up to 180°. $m\angle A + m\angle B + m\angle C = 180°$

8-1

Triangles and Quadrilaterals

Vocabulary

triangle (p. 414)

plane figure (p. 414)

polygon (p. 414)

vertex (p. 414)

quadrilateral (p. 416)

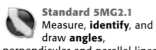 **Standard 5MG2.1** Measure, **identify**, and draw **angles**, perpendicular and parallel lines, **rectangles, and triangles** by using appropriate tools.

Standard 5MG2.2 Know that the sum of the angles of any triangle is 180° and the sum of the angles of any quadrilateral is 360° and use this information to solve problems.

Standard 7MR1.1 Analyze problems by identifying relationships, distinguishing relevant from irrelevant information, identifying missing information, sequencing and prioritizing information, **and observing patterns**.

The What: I will recognize and name plane figures and calculate angles in them.

The Why: Plane figures are used in many designs, including traffic signs.

Triangles and quadrilaterals are two-dimensional, or flat, **plane figures.** **Polygons** are closed figures formed by three or more line segments called sides. The sides of a polygon intersect at a point called a **vertex.** **Triangles** are polygons with three sides and three angles.

TRIANGLES			
Type	**Angles**	**Sides**	**Sum of the Interior Angles**
right △ABC	one right angle	two sides of equal length or no sides of equal length	m∠A + m∠B + m∠C = 180°
isosceles △DEF	two angles of equal measure m∠D = m∠F	two sides of equal length $\overline{DE} = \overline{EF}$	m∠D + m∠E + m∠F = 180°
equilateral △GHI	three angles of equal measure m∠G = m∠H = m∠I	three sides of equal length $\overline{GH} = \overline{HI} = \overline{IG}$	m∠G + m∠H + m∠I = 180°
scalene △JKL	no angles of equal measure m∠J ≠ m∠K ≠ m∠L	no sides of equal length $\overline{JK} \neq \overline{KL} \neq \overline{LJ}$	m∠J + m∠K + m∠L = 180°

The symbol for an angle is ∠. The letter "m" indicates the measure of the angle. For example, ∠*F* is the symbol for angle *F*, and m∠*F* is the measure of angle *F*.

Find the value of x.

① In △*RST*, m∠*R* = *x*, m∠*S* = 57°, and m∠*T* = 48°.

m∠*R* + m∠*S* + m∠*T* = 180°	The sum of the angles of a triangle is 180°.
x + 57 + 48 = 180	Substitute with angle values.
x + 105 = 180	Add 57 and 48.
x + 105 − 105 = 180 − 105	Subtract 105 from each side.
x = 75°	Simplify.

②

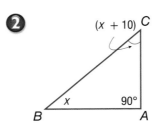

m∠*A* + m∠*B* + m∠*C* = 180°	The sum of the angles of a triangle is 180°.
90 + *x* + (*x* + 10) = 180	Substitute with angle values.
100 + 2*x* = 180	Combine like terms, and add 90 and 10.
100 + 2*x* − 100 = 180 − 100	Subtract 100 from both sides.
$\frac{2x}{2} = \frac{80}{2}$	Divide both sides by 2.
x = 40°	Simplify.

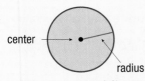

Study Tip

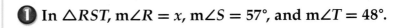

center → radius

A *circle* is a closed figure in a plane, so it is a plane figure. A circle does not have line segments, so it is not a polygon.

Study Tip

Use the corner of an index card to quickly check whether an angle is right, acute, or obtuse. Place the card's corner at the vertex and line up its edge with one of the angle's rays.

❸

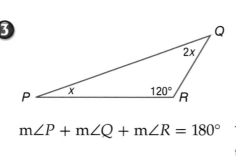

$$m\angle P + m\angle Q + m\angle R = 180°$$ The sum of the angles of a triangle is 180°.

$$x + 2x + 120 = 180$$ Substitute with angle values.

$$3x + 120 = 180$$ Combine like terms. Add x and $2x$.

$$3x + 120 - 120 = 180 - 120$$ Subtract 120 from each side.

$$\frac{3x}{3} = \frac{60}{3}$$ Divide both sides by 3.

$$x = 20$$ Simplify.

$$m\angle P = x = 20°$$ Use the value of x to find each missing angle.

$$m\angle Q = 2x = 20 \times 2 = 40°$$

Your Turn Find the value of x.

a.

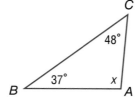

b.

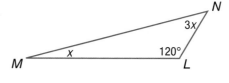

A **quadrilateral** is a four-sided polygon. Quadrilaterals, like triangles, are classified by their sides and angles. The sum of the interior angles of a quadrilateral is always 360°.

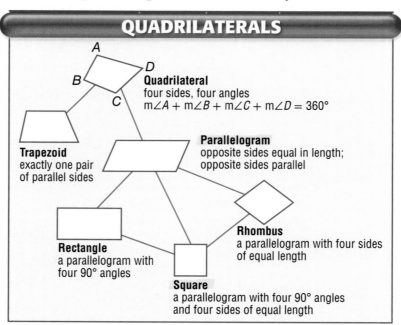

QUADRILATERALS

Quadrilateral
four sides, four angles
$m\angle A + m\angle B + m\angle C + m\angle D = 360°$

Trapezoid
exactly one pair of parallel sides

Parallelogram
opposite sides equal in length; opposite sides parallel

Rectangle
a parallelogram with four 90° angles

Rhombus
a parallelogram with four sides of equal length

Square
a parallelogram with four 90° angles and four sides of equal length

Find the value of x.

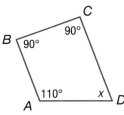

4

$m\angle A + m\angle B + m\angle C + m\angle D = 360°$

The sum of the angles of a quadrilateral is 360°.

$110° + 90° + 90° + x = 360°$

Substitute with angle values.

$x + 290 = 360°$

Combine like terms. Add 110 + 90 + 90.

$x + 290 - 290 = 360 - 290$

Subtract 290 from each side.

$x = 70°$

Simplify.

$m\angle D = 70°$

Use the value of x to find the missing angle.

Your Turn Find the value of x.

c.

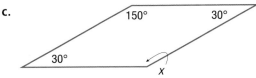

Talk Math

WORK WITH A PARTNER Take turns naming, describing, and drawing different quadrilaterals.

Real-World EXAMPLE

5 LANDSCAPING Gilbert clears an area of the yard that has one pair of parallel sides. What shape did he clear?

He cleared an area in the shape of a trapezoid.

Your Turn

d. **REALTY** A horse corral shows it has four sides of equal length and four right angles. What is the shape of the corral?

Examples 1–5
(pages 415–417)

VOCABULARY Choose the term that *best* completes each sentence.

1. A four-sided polygon is a ___?___.

2. Two sides of a ___?___ intersect at a point called a ___?___.

3. A plane figure with three sides is a ___?___.

Examples 1–4
(pages 415–417)

Find the value of x.

4. In $\triangle GHP$, $m\angle G = x$, $m\angle H = 79°$, and $m\angle P = 92°$.

5.

6.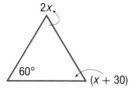

7.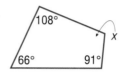

Example 5
(page 417)

8. **SPORTS** A practice field is a parallelogram with four sides. Both pairs of sides are parallel, but all of the sides are not the same length. The angle at each corner is 90 degrees. What shape is the field?

9. **EARTH SCIENCE** To find the location of an earthquake, scientists use data from three seismic stations. Each station is a vertex of a large plane figure. What is the figure?

Skills, Concepts, and Problem Solving

HOMEWORK HELP	
For Exercises	**See Example(s)**
10–16	1–3
17–20	5

Find the value of x.

10. In $\triangle TUV$, $m\angle T = x$, $m\angle U = 108°$, and $\angle V = 67°$.

11. In $\triangle ABC$, $m\angle A = x$, $m\angle B = 62°$, and $\angle C = 39°$.

12.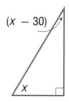

13.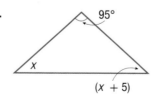

14.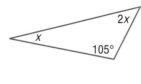

15.

16.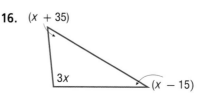

17. **TRAFFIC SIGNS** A speed limit sign is a parallelogram with four 90° angles, but the sides are not all the same length. What shape is a speed limit sign?

18. **FOOD SERVICE** A pizzeria cuts their square pizzas into eight pieces. Each piece is a polygon with three sides. What shape is each pizza slice?

19. **RECREATION** A swimming pool is a polygon with four sides. Only one pair of opposite sides is parallel. What shape is the pool?

20. **CARPENTRY** A carpenter made a table top with four sides. All the sides are the same length, but the angles are not 90 degrees. What shape is the tabletop?

21. **H.O.T.** Problem Look at the figures to the right. How are they the same and how are they different?

22. *Writing in Math* Amado says that a square is always a rectangle, but a rectangle is not always a square. Do you agree? Why or why not?

STANDARDS PRACTICE

Choose the *best* answer.

23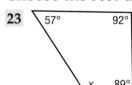

57°   92°

x   89°

What is the value of *x* in the figure above?

A   102°   C   238°

B   122°   D   360°

24 What is the sum of the measures of the angles of a triangle?

F   180°

G   360°

H   540°

J   720°

Spiral Review

25. **LANDSCAPING** A garden is a right triangle. The legs of the triangle are 24 meters and 7 meters long. How much edging is needed to line the other side? (Lesson 7-7)

Find each root. (Lesson 5-4)

26. $-\sqrt{64}$

27. $\sqrt{121}$

Add or subtract. (Lesson 4-2)

28. $90.47 + 14.3$

29. $53.72 - 7.093$

Congruency and Similarity

Vocabulary

congruent (p. 420)

corresponding parts (p. 420)

similar (p. 422)

Standard 7MG3.4 Demonstrate an understanding of conditions that indicate two geometrical figures are congruent and what congruence means about the relationships between the sides and angles of the two figures.

Standard 7MR1.0 Students make decisions about how to approach problems.

Standard 7MR1.1 Analyze problems by identifying relationships, distinguishing relevant from irrelevant information, identifying missing information, sequencing and prioritizing information, **and observing patterns.**

 The What: I will identify corresponding parts of polygons to find missing measures of congruent and similar figures.

The Why: Photographs that are reduced or enlarged are similar to the originals.

Alex ordered reprints of two photographs. The reprints were the same size and shape as the original photographs. Each reprint was **congruent** to its original.

When shapes are congruent, their **corresponding parts** are congruent. Corresponding angles have the same measure. Corresponding sides have the same length. The symbol $\cong$ indicates congruent figures.

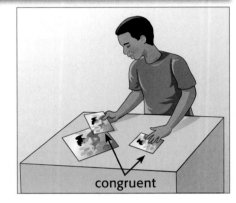

congruent

Study Tip

You can rotate triangles to find corresponding parts.

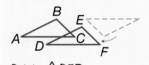

Rotate △DEF.

IDENTIFYING CORRESPONDING PARTS

Match corresponding angles by comparing the number of markings at each angle and the order each vertex is given in naming the triangle.

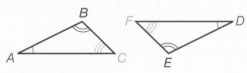

△ABC ≅ △DEF

∠A and ∠D have one marking. ∠A corresponds to ∠D.

∠B and ∠E have two markings. ∠B corresponds to ∠E.

∠C and ∠F have three markings. ∠C corresponds to ∠F.

Match corresponding sides by comparing the number of markings on each side.

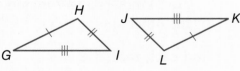

△GHI ≅ △KLJ

$\overline{GH}$ corresponds to $\overline{KL}$.

$\overline{HI}$ corresponds to $\overline{LJ}$.

$\overline{IG}$ corresponds to $\overline{JK}$.

Name Congruent Angles and Sides

Name the congruent angles and sides.

① △LMN ≅ △RST

First, name the three pairs of congruent angles by looking at the order of the vertices.

∠L ≅ ∠R because these angles are named first,
∠M ≅ ∠S because these angles are named second,
∠N ≅ ∠T because these angles are named last.

Now, name the three pairs of corresponding sides by comparing the order each vertex is given in naming the triangle.

$\overline{LM} ≅ \overline{RS}$, $\overline{MN} ≅ \overline{ST}$, $\overline{NL} ≅ \overline{TR}$

Your Turn **Name the congruent angles and sides.**

a. △ABC ≅ △DEF

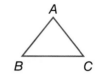

EXAMPLE **Name Corresponding Parts**

Name the corresponding parts. Then complete the congruence statement.

②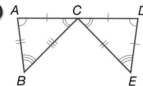

△ABC ≅ △____?____

Name the congruent angles by using the number of markings at each angle.

∠A ≅ ∠D, ∠ACB ≅ ∠DCE, ∠E ≅ ∠B

Name the congruent sides by using the number of markings on each side.

$\overline{AC} ≅ \overline{CD}$, $\overline{AB} ≅ \overline{DE}$, $\overline{CB} ≅ \overline{CE}$

Match the vertices of congruent angles to complete the congruence statement.

△ABC ≅ △DEC

Your Turn **Name the corresponding parts. Then complete the congruence statement.**

b.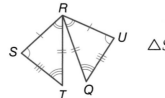

△SRT ≅ △____?____

Remember!

A vertex is the point where two sides of a polygon intersect.

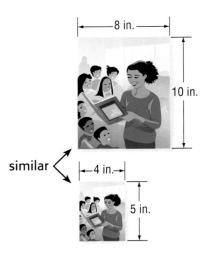

8 in.

10 in.

similar ⟨

4 in.

5 in.

Alex realized that one photograph needed to be reduced. The reduced photograph will have the same shape as the original photograph. The reduced photograph is **similar** to the original.

CONGRUENT AND SIMILAR POLYGONS

Relationship	Meaning	Angles and Sides	Symbol
congruent	same shape, same size	congruent angles, congruent sides	≅
similar	same shape, different size	congruent angles, proportional sides	~

Real-World **EXAMPLE**

❸ **CONSTRUCTION** A pinwheel has two similar triangles connected at a vertex. What is the missing length?

$\triangle ABC \sim \triangle DEC$

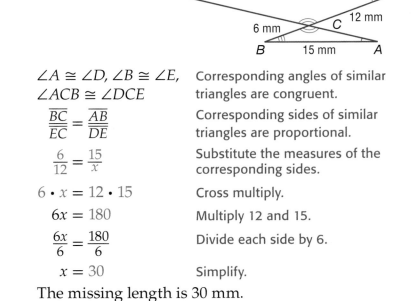

$\angle A \cong \angle D, \angle B \cong \angle E,$ Corresponding angles of similar
$\angle ACB \cong \angle DCE$ triangles are congruent.

$\dfrac{\overline{BC}}{\overline{EC}} = \dfrac{\overline{AB}}{\overline{DE}}$ Corresponding sides of similar triangles are proportional.

$\dfrac{6}{12} = \dfrac{15}{x}$ Substitute the measures of the corresponding sides.

$6 \cdot x = 12 \cdot 15$ Cross multiply.

$6x = 180$ Multiply 12 and 15.

$\dfrac{6x}{6} = \dfrac{180}{6}$ Divide each side by 6.

$x = 30$ Simplify.

The missing length is 30 mm.

Your Turn

c. **TOYS** The roof for a house on a felt board is a triangle. Someone cuts off the top of the roof. What was the side length of the original roof?

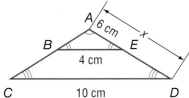

Talk Math

WORK WITH A PARTNER Explain to your partner what congruent shapes are. Have your partner tell you what similar shapes are.

Guided Practice

Examples 1–3
(pages 421–422)

VOCABULARY Write the term that *best* completes each sentence.

1. Two polygons with the same shape and with corresponding sides that are the same length are said to be ___?___.

2. Polygons with the same shape, equal angle measures, and proportional side lengths are said to be ___?___.

Example 1
(page 421)

Name the congruent angles and sides.

3. $\triangle ABC \cong \triangle LMN$

Example 2
(page 421)

Name the corresponding parts. Then complete the congruence statement.

4.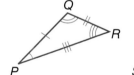

$\triangle PQR \cong \triangle$___?___

Example 3
(page 422)

5. **MEASUREMENT** Sekithia made a triangle smaller to fit on a poster. It is similar to the original one. What is the missing measurement?

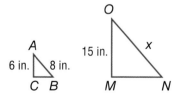

6. **MEASUREMENT** Mario needs to reduce a picture in the shape of a right triangle. The original picture has sides of 10 inches, 15 inches, and 18 inches. What is the length of the hypotenuse of the reduced picture?

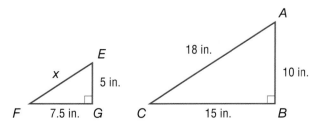

HOMEWORK HELP

For Exercises	See Example(s)
7–10	1
11–12	2
13–14	3

Name the congruent angles and sides.

7. $\triangle CDE \cong \triangle FGH$

8. $\triangle QRS \cong \triangle MNO$

9. $\triangle WXY \cong \triangle TUV$

10. $\triangle LMN \cong \triangle HIJ$

Name the corresponding parts. Then complete the congruence statement.

11.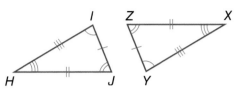

$\triangle MLK \cong \triangle$ ___?___

12.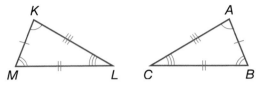

$\triangle HIJ \cong \triangle$ ___?___

13. **ART** Tobias drew pictures of similar houses. Each one had a triangular roof. What is the missing measurement on the smaller triangle?

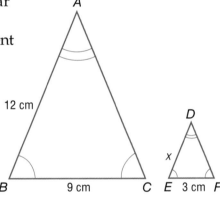

14. **MEASUREMENT** Luna and her doll have matching triangular pillows. What is the missing measurement on Luna's pillow?

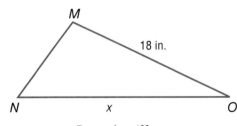

The doll's pillow   Luna's pillow

15. **H.O.T.** Problem A photo negative is 1.5 centimeters wide by 2.2 centimeters long. You want to make three enlargements that are similar to the original photo. Give the measures for the length and width of these three enlargements.

16. *Writing in Math* Draw two congruent triangles. Label the triangles and mark corresponding parts. Use your drawing to write a congruence statement.

Choose the *best* answer.

17 $\triangle ABC \sim \triangle DEF$
Which of the following statements must be true?

A $\overline{AB} = \overline{DE}$ **C** $m\angle A = m\angle D$

B $m\angle C = m\angle B$ **D** $\overline{CA} = \overline{FD}$

18 $\triangle PQR \cong \triangle TVU$
Which of the following statements must be true?

F $m\angle P = m\angle U$ **H** $\overline{RP} = \overline{UT}$

G $m\angle R = m\angle T$ **J** $\overline{VU} = \overline{TU}$

19

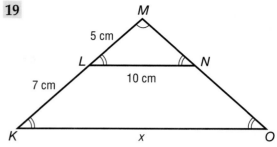

Austin has two similar triangles. He places the smaller triangle on top of the larger one. What is the length of the missing base?

A 8 cm   **C** 14 cm

B 12 cm   **D** 24 cm

Spiral Review

Find the value of *x*. (Lesson 8-1)

20.

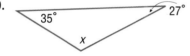

21.

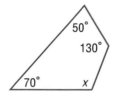

22. Use the Pythagorean Theorem to find the hypotenuse. (Lesson 7-7)

6 cm
8 cm

23. Find the slope of the line segment. (Lesson 7-5)

24. **GRAPHING** Flor has a piece of graph paper that is 5 inches long. It is divided into 0.25-inch squares. How many squares long is the paper?
(Lesson 4-4)

Math Lab
Make Congruent and Similar Figures

Materials

rectangular grid geoboards and rubber bands

overhead geoboard

geoboard blackline master

colored markers

 Standard 7MG1.2
Construct and read drawings and models made to scale.

Standard 7MR1.0 Students make decisions about how to approach problems:

Standard 7MR1.1 Analyze problems by identifying relationships, distinguishing relevant from irrelevant information, identifying missing information, sequencing and prioritizing information, **and observing patterns.**

> **The What:** I will use geoboards to create congruent and similar figures.
>
> **The Why:** Enlargements and original photographs are similar polygons. The dimensions of each will be proportional.

EXAMPLE **Make Congruent and Similar Rectangles**

① Use a rubber band to make each rectangle.

- Make two congruent rectangles on the geoboard using rubber bands.

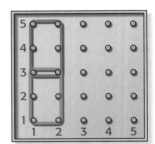

- Make a larger third rectangle that is similar to the first two.

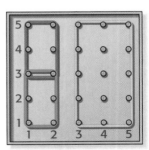

- Draw your rectangles on dot paper.

2 **Use a rubber band to make each triangle.**

- Make two congruent triangles on the geoboard using rubber bands. Make a larger third triangle that is similar to the first two.

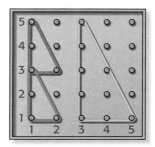

- Draw your triangles on dot paper.

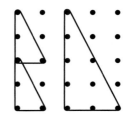

Your Turn

a. Use a rubber band to make the shape of your choice.
   - Make a shape that is congruent to the one you chose.
   - Make a shape that is similar to the first two.
   - Draw your shapes on dot paper.

Analyze the Results

1. **REFLECT** Explain why your first two shapes are congruent. Explain why your third shape is similar to the first two shapes.

2. Suppose you draw a right triangle with legs measuring 3 and 4 units. Find the leg measurements of the two different similar right triangles.

Progress Check 1

(Lessons 8-1 and 8-2)

▷ Vocabulary and Concept Check

> congruent (p. 420)   quadrilateral (p. 416)
> corresponding parts (p. 420)   similar (p. 422)
> plane figure (p. 414)   triangle (p. 414)
> polygon (p. 414)   vertex (p. 414)

Write the term that *best* completes each sentence. (Lessons 8-1 and 8-2)

1. If two shapes are _____?_____ , then their corresponding angles and sides have the same measure.

2. If two shapes are _____?_____ , then their corresponding angles are congruent and their sides are proportional.

3. A(n) _____?_____ is a four-sided polygon.

▷ Skills Check

Find the value of *x*. (Lesson 8-1)

4.

5.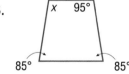

Name the congruent angles and sides. (Lesson 8-2)

6. △BCD ≅ △MNO

7. △ABC ≅ △HIJ

▷ Problem-Solving Check

8. **INTERIOR DESIGN** A trianglar table has three angles of equal measure. What is the measure of each angle of the table? (Lesson 8-1)

9. **MEASUREMENT** Julius made two similar triangles for an art project. What is the missing measurement on △ABC? (Lesson 8-2)

10. **REFLECT** How are congruent and similar triangles alike? How are they different?

    (Lessons 8-1 and 8-2)

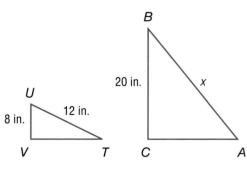

Coordinate Geometry

Vocabulary

coordinate geometry
(p. 429)

Distance Formula (p. 429)

Standard 7MG3.2
Understand and use coordinate graphs to plot simple figures, **determine lengths** and areas related to them, and determine their image under translations and reflections.

Standard 7MR2.3 Estimate unknown quantities graphically and **solve** for them **by using** logical reasoning and **arithmetic and algebraic techniques.**

Study Tip

Ordered pairs, like (x, y), represent points in the coordinate plane. To tell points apart, we can use subscripts, like (x_1, y_1) and (x_2, y_2).

> **The What:** I will use the Distance Formula to solve measurement problems.
>
> **The Why:** Coordinate geometry techniques can be used to find distances.

Alita and Hao live close to school. Alita lives 3 blocks due east of school. Hao lives 4 blocks due south of school. They want to find out how far apart their houses are when they walk the shortest distance between them.

They decide to use **coordinate geometry** to represent their neighborhood. They put the school at the origin. They mark Alita's house at coordinates (3, 0). They mark Hao's house at coordinates (0, −4).

They will use the **Distance Formula** to find out how far apart their houses are.

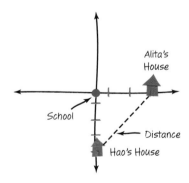

You can use the Distance Formula to find the distance between any two points in the coordinate plane.

DISTANCE FORMULA
$$\text{distance} = \sqrt{(x_2 - x_1)^2 + (y_2 - y_1)^2}$$

Talk Math

WORK WITH A PARTNER Use a coordinate plane to find the distance between the points at (6, 4) and (1, 4). Explain how you can use the coordinates to find the length of a line segment when a coordinate plane is not provided.

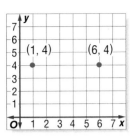

EXAMPLE Use the Distance Formula

1 What is the distance between points at (3, 0) and (0, −4)?

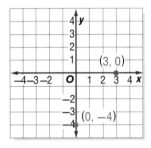

$x_1 = 3, x_2 = 0, y_1 = 0,$ and $y_2 = -4$ Identify the variables.

$\sqrt{(x_2 - x_1)^2 + (y_2 - y_1)^2}$ Use the Distance Formula.

$= \sqrt{(0 - 3)^2 + (-4 - 0)^2}$ Substitute values.

$= \sqrt{(-3)^2 + (-4)^2}$ Simplify inside the parentheses.

$= \sqrt{9 + 16}$ Simplify the exponents.

$= \sqrt{25}$ Simplify inside the radical.

$= 5$ Find the square root.

The shortest distance between the two points is 5 units.

Your Turn

a. What is the distance between points at (5, 5) and (1, −2)?

2 DISTANCE Two houses are on opposite sides of the lake. What is the distance between the two houses in miles if each unit equals one mile?

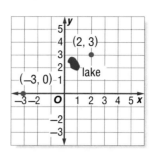

$x_1 = -3$, $x_2 = 2$, $y_1 = 0$, and $y_2 = 3$ Identify the variables.

$$\sqrt{(x_2 - x_1)^2 + (y_2 - y_1)^2}$$ Use the Distance Formula.

$$= \sqrt{[2 - (-3)]^2 + (3 - 0)^2}$$ Substitute values.

$$= \sqrt{5^2 + 3^2}$$ Simplify inside the parentheses.

$$= \sqrt{25 + 9}$$ Simplify the exponents.

$$= \sqrt{34}$$ Simplify inside the radical.

$$\approx 5.8$$ Find the square root and label the answer.

There is approximately 5.8 miles between the houses.

Study Tip

The number 34 is not a perfect square. The number $\sqrt{34}$ lies between $\sqrt{25}$ and $\sqrt{36}$. So, $\sqrt{34}$ is about 5.8.

Your Turn

b. **DISTANCE** Coty walks home from school each day. About how many blocks is the school from his home if each unit equals one block?

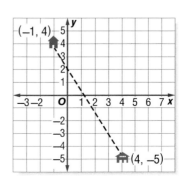

Vocabulary Review
coordinate geometry
Distance Formula

Examples 1–2
(pages 430–431)

VOCABULARY **Write the term that completes each sentence.**

1. $\sqrt{(x_2 - x_1)^2 + (y_2 - y_1)^2}$ is the _____?_____.

2. The study of geometry using the principles of algebra is known as _____?_____.

Example 1
(page 430)

Use the Distance Formula.

3. What is the distance between the points $(-5, -8)$ and $(3, 7)$?

Example 2
(page 431)

4. **DISTANCE** Sarita wants to visit her grandfather. What is the distance between their houses in miles if each unit equals one mile?

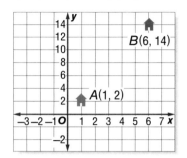

Skills, Concepts, and Problem Solving

HOMEWORK HELP

For Exercises	See Example(s)
5–6	1
7–8	2

5. What is the distance between points at $(4, 2)$ and $(1, 6)$?

6. What is the distance between points at $(-2, 0)$ and $(0, 1)$?

7. **DISTANCE** Mr. Guerra is traveling to a store at the opposite end of Cedar Lane. What is the length of Cedar Lane in miles if each unit equals one mile?

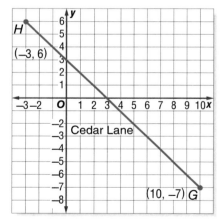

8. **DISTANCE** A farmer plants corn in a field. What is the distance across the field in yards if each unit equals 2 yards?

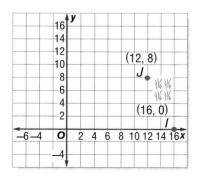

9. *Writing in Math* How could you use the Distance Formula to determine if *LMNO* is a parallelogram?

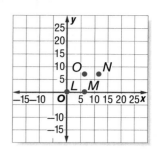

10. **H.O.T.** Problem Find the perimeter of △*FGH* with vertices *F*(3, 4), *G*(25, 0), and *H*(9, 12). Round to the nearest whole number.

STANDARDS PRACTICE

Choose the *best* answer.

11 What is the distance between the two points?

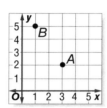

A 3 units   **C** 13 units

B 5 units   **D** $\sqrt{13}$ units

12 Which is the approximate length of line segment *CD*?

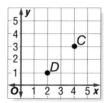

F 16   **H** 2.8

G 4.3   **J** 1.5

Spiral Review

13. **ART** Fina has a picture in the shape of a right triangle. It has sides of 8 inches, 6 inches, and 10 inches. She wants to reduce the picture to a similar triangle. What is the length of the reduced picture's hypotenuse? (Lesson 8-2)

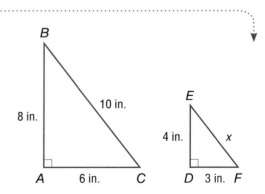

Find the value of *x*. (Lesson 8-1)

14.

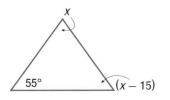

15.

Vocabulary

perimeter (p. 434)

circumference (p. 436)

circle (p. 436)

center (p. 436)

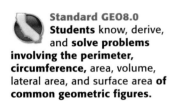

Standard GEO8.0 Students know, derive, and **solve problems involving the perimeter, circumference,** area, volume, lateral area, and surface area **of common geometric figures.**

Standard 7MR1.1 Analyze problems by identifying relationships, distinguishing relevant from irrelevant information, **identifying missing information,** sequencing and prioritizing information, and observing patterns.

Standard 7MR2.6 Express the solution clearly and logically by using the appropriate mathematical notation and terms and clear language; support solutions with evidence in both verbal and symbolic work.

 The What: I will find the perimeter of plane figures.

 The Why: The amount of fencing needed to enclose a space can be calculated using perimeter.

Candice's family decides to fence 12 square yards of space for a new puppy. They want to create a rectangular space no wider than 3 feet. Candice plans to use the least amount of fencing possible.

The **perimeter** can be used to determine how much fencing is needed. Perimeter is the distance around the outside of a closed figure.

$$P = \ell + w + \ell + w = (\ell + w) + (\ell + w) = 2(\ell + w)$$

Perimeter   length   width

Since we are measuring a distance, the units are linear. Two possible dimensions are 2 by 6 and 3 by 4.

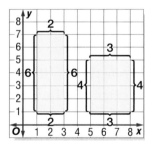

Candice calculates the perimeter of each space.

2 by 6	3 by 4	
$P = 2(\ell + w)$	$P = 2(\ell + w)$	Use the formula.
$P = 2(2 + 6)$	$P = 2(3 + 4)$	Substitute length and width values.
$P = 2(8)$	$P = 2(7)$	Simplify inside the parentheses.
$P = 16$ yards	$P = 14$ yards	Write and label the answer.

The family decides to make a 3-yard by 4-yard space. It has the smallest perimeter and requires the least fencing.

Study Tip

To find the perimeter of any polygon, add the lengths of all its sides.

EXAMPLE   Perimeter of Polygons

1 **Find the perimeter of the polygon.**

5 cm

3 cm

7 cm

4 cm

9 cm

$$P = 3 + 5 + 7 + 9 + 4 \qquad \text{Add the lengths of the sides.}$$
$$P = 28 \qquad \text{Add. Write and label the answer.}$$

The perimeter is 28 centimeters.

Your Turn

a. Find the perimeter of the polygon.

10 cm

4 cm

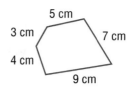

Real-World EXAMPLE

2 **PAINTING** Susan is painting her room. She wants to tape around the rectangular window. The window has a length of 5 feet and she uses 16 feet of tape. What is the width of the window?

5 ft

w

You may want to sketch the shape first. Label the width as *w*.

$$P = 2(\ell + w) \qquad \text{Use the perimeter formula.}$$

$$16 = 2(5 + w) \qquad \text{Substitute the length of the tape used for the perimeter and 5 for } \ell.$$

$$16 = (2 \cdot 5) + 2w \qquad \text{Distributive Property}$$
$$16 = 10 + 2w \qquad \text{Multiply 2 and 5.}$$
$$16 - 10 = 10 - 10 + 2w \qquad \text{Subtract 10 from each side.}$$
$$6 = 2w$$
$$6 \div 2 = 2w \div 2 \qquad \text{Divide each side by 2.}$$
$$w = 3 \qquad \text{Solve for } w. \text{ Label the answer.}$$

The width of the window is 3 feet.

Your Turn

b. **INTERIOR DESIGN** A rectangular mirror is 8 inches wide. Takara uses 36 inches of ribbon around the border. How long is the mirror?

A **circle** is a set of points that are equal distance from a point called the **center.**

Remember!

The diameter is the distance across the circle, passing through its center.

diameter = 2 × radius

The distance around a circle is called the **circumference.** The formula for circumference is $C = \pi d$, where C is the circumference and d is the diameter.

EXAMPLE **Find Circumferences**

❸ **Find the circumference of the circle. Use π = 3.14.**

6 yd

Since the radius is 6 yards, the diameter is 12 yards.

$C = \pi d$	Use the formula.
$C = 3.14(12)$	Substitute 3.14 for π and 12 for d.
$C = 37.68$	Multiply.

The circumference is 37.68 yards.

Your Turn

Find the circumference of the circle. Use π = 3.14.

c.

10 mi

Real-World EXAMPLE

❹ **LIFE SCIENCE** A tree trunk has an 8-inch radius. What is the circumference of the trunk?

Since the radius is 8 inches, the diameter is 16 inches.

$C = \pi d$	Use the formula.
$C = 3.14(16)$	Substitute 3.14 for π and 16 for d.
$C = 50.24$	Multiply. Write and label the answer.

The circumference is 50.24 inches.

Your Turn

d. **MEASUREMENT** A flag pole has a radius of one foot. What is the circumference of the flag pole?

FORMULAS		
perimeter	all polygons	$P =$ sum of side lengths
perimeter	rectangles	$P = 2(\ell + w)$
perimeter	squares	$P = 4s$
circumference	circles	$C = \pi d$

Talk Math

WORK WITH A PARTNER Tell your partner how perimeter and circumference are alike and different.

Examples 1–4
(pages 435–436)

VOCABULARY Write the term that completes each sentence.

1. The distance around a closed figure is called its _____?_____.

2. The distance around a circle is called the _____?_____.

Example 1
(page 435)

3. Find the perimeter of the polygon shown at the right.

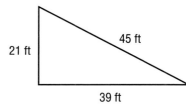

Example 2
(page 435)

Find the missing side.

4. **ART** A teacher has a rectangular bulletin board in his room. It is 4 feet high. He uses 48 feet of border to go around the bulletin board. How long is the bulletin board?

Example 3
(page 436)

Find the circumference of the circle. Use π = 3.14.

5.

Example 4
(page 436)

6. **INTERIOR DESIGN** A circular table has a diameter of 15 inches. What is the circumference of the table?

For Exercises	See Example(s)
7–8	1
9–10	2
11–12	3
13–14	4

HOMEWORK HELP

Find the perimeter of each polygon.

7.

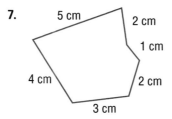

8.

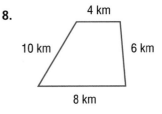

Find the missing side.

9. **GARDENING** Marta used 39 feet of railroad ties to enclose her garden. Her garden is a quadrilateral. Three of the sides have measures of 6 feet, 10 feet, and 12 feet. How long is the fourth side?

10. **CONSTRUCTION** A house has a window that is in the shape of a pentagon. Three sides measure 1 yard. The other two side lengths are equal. A builder puts 6.5 yards of molding around the window. How long are the other two sides?

Find the circumference of each circle. Use π = 3.14.

11.
5 in.

12.
9 cm

13. **SPORTS** A bike tire has a radius of 13 inches. What is the circumference of the tire?

14. **FOOD** A pizza has a diameter of 12 inches. What is the circumference of the pizza?

15. **H.O.T.** Problem Which figure has a greater distance around— a triangle with sides of 12 centimeters each or a circle with a diameter of 12 centimeters? Explain.

16. *Writing in Math* Compare the formulas for the perimeter of rectangle and the perimeter of a square. How are they different? How are they the same.

STANDARDS PRACTICE

Choose the *best* answer.

17 A semi-circular window needs to be sealed around its edges. How many inches of sealant are needed? Use π = 3.14.

14 in.

A 21.98 in.   **C** 43.96 in.

B 35.98 in.   **D** 57.96 in.

18 Find the perimeter of the polygon.

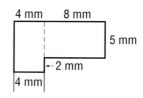
4 mm   8 mm
5 mm
←2 mm
4 mm

F 19 mm   **H** 38 mm

G 36 mm   **J** 44 mm

Spiral Review

19. What is the distance between points at (16, 14) and (7, 2) on a coordinate plane? (Lesson 8-3)

20. **MAPS** On a map, Elmwood is at (22, 50) and Oak City is at (10, 47). What is the distance between these two points? (Lesson 8-3)

Complete each equation using the Distributive Property. (Lesson 1-5)

21. $3(x + 6) = ($ _?_ $\cdot$ _?_ $) + ($ _?_ $\cdot$ _?_ $)$

22. $-4(p - 5) = ($ _?_ $\cdot$ _?_ $) + ($ _?_ $\cdot$ _?_ $)$

Progress Check 2
(Lessons 8-3 and 8-4)

▷ Vocabulary and Concept Check

> center (p. 436) coordinate geometry (p. 429)
> circle (p. 436) Distance Formula (p. 429)
> circumference (p. 436) perimeter (p. 434)

Choose the term that *best* completes each statement. (Lessons 8-3 and 8-4)

1. ____?____ uses the principles of algebra to study geometry.

2. The formula for calculating the ____?____ of a circle is $C = \pi d$.

3. The ____?____ can be used to find the length of a line segment.

4. ____?____ is the distance around the outside of a closed figure.

▷ Skills Check

5. What is the distance between points at (5, 5) and (1, 2)? (Lesson 8-3)

6. What is the distance between points at (0, 6) and (0, 4)? (Lesson 8-3)

Find the perimeter of each polygon. (Lesson 8-4)

7.

6 in.

12 in.

8.

16 ft   12 ft

20 ft

Find the circumference of each circle. Use $\pi = 3.14$. (Lesson 8-4)

9.

4 cm

10.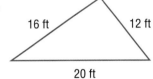

← 10 cm →

▷ Problem-Solving Check

11. **FITNESS** Paul runs on a circular jogging track. The track has a diameter of 170 feet. Paul runs once around the track. How many feet did he run? Use $\pi = 3.14$. (Lesson 8-4)

12. **REFLECT** Marcos wants to line the bottom of his bedroom wall with a wallpaper border. He will not paper under the two doors. How can he figure out how much wallpaper border to buy?

8-5 Area

Vocabulary

area (p. 440)

altitude (p. 442)

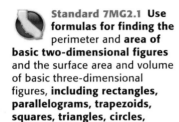 **Standard 7MG2.1 Use formulas for finding the** perimeter and **area of basic two-dimensional figures** and the surface area and volume of basic three-dimensional figures, **including rectangles, parallelograms, trapezoids, squares, triangles, circles,** prisms, and cylinders.

Standard 7MR2.0 Students use strategies, skills, and concepts in finding solutions.

Standard 7MR2.3 Estimate unknown quantities graphically and solve for them by using logical reasoning and arithmetic and algebraic techniques.

 **The What:** I will learn to find the area of some quadrilaterals, triangles, and circles.

The Why: Finding the area of a wall determines the amount of paint to buy when painting a room.

Andrew is making some changes to his room. He needs to find the area of his closet in order to add new carpet. **Area** is the measure of the surface enclosed by a geometric figure. It is measured in square units.

You can use a coordinate graph in order to visualize the area of Andrew's closet. If the closet has a length of 5 feet and a width of 3 feet, what is the area of his closet? The coordinate graph below shows a sketch of Andrew's closet. Each square unit on the graph represents one square foot.

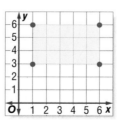

One way to find the area of Andrew's closet is simply to count the number of square units that are shaded. The yellow rectangle indicates the size of his closet. By counting the square units, you can see that the area of his closet is 15 units, or 15 square feet.

Instead of counting squares, you can find the area by multiplying the measures of the length and width.

$$\text{Area} = \text{length} \cdot \text{width}$$

$$A = \ell \cdot w$$

$A = \ell \cdot w$	Write the formula.
$A = 5 \cdot 3$	Substitute 5 for ℓ and 3 for w.
$A = 15$	Multiply 5 and 3.

The area of Andrew's closet is 15 square feet.

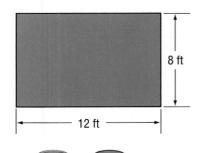

8 ft

12 ft

Coverage:
175 sq ft
per can

Andrew also decides to paint part of his bedroom a bright blue. He needs to find the area of his walls. Two of the walls in Andrew's room are rectangles that measure 12 feet long by 8 feet wide. To find each wall's square footage, Andrew uses the formula for the area of a rectangle.

$A = \ell \cdot w$	Write the formula.
$A = 12 \cdot 8$	Substitute 12 for ℓ and 8 for w.
$A = 96$	Multiply 12 and 8.

The area of one of the walls is 96 square feet.

Since Andrew is painting two of the walls in his bedroom, the total area he needs to paint is 2 · 96 or 192 square feet.

At the store, Andrew learns that a can of paint covers 175 square feet. Since his wall area is greater than 175 square feet, he buys two cans of paint.

Talk Math

WORK WITH A PARTNER Kathy's walls measure 8 feet long and 9 feet wide. If she paints three walls, how many cans of paint does she need?

There are several formulas used to find area. The shape of the figure determines which formula to use.

Notice the similarities in the area formulas for a square, a rectangle, and a parallelogram.

AREA FORMULAS

rectangle	$A = \ell \cdot w$	
square	$A = s \cdot s = s^2$	
parallelogram	$A = b \cdot h$	
trapezoid	$A = \frac{1}{2}h(b_1 + b_2)$	
triangle	$A = \frac{1}{2}b \cdot h$	
circle	$A = \pi r^2$	

Study Tip

The height, or altitude, can be shown inside or outside the figure. Either way, the area is the same.

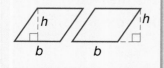

The formulas for finding the areas of triangles, parallelograms, and trapezoids include the height, or **altitude,** of each figure. The altitude is the line segment from a vertex that is perpendicular to the opposite side or opposite extended side.

Notice that you need to use π to find the area of a circle. Recall that $\pi \approx 3.14$.

Find the area of each figure.

1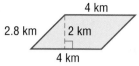

parallelogram

$A = bh$ Use the formula for the area of a parallelogram.

$A = 4(2)$ Substitute the base and height.

$A = 8$ Multiply.

The area is 8 square kilometers.

2

trapezoid

$A = \frac{1}{2}h(b_1 + b_2)$ Use the formula for the area of a trapezoid.

$A = \frac{1}{2}(6)(15 + 5)$ Substitute the bases and height.

$A = \frac{1}{2}(6)(20)$ Add 15 and 5.

$A = 60$ Multiply.

The area is 60 square meters.

3

triangle

$A = \frac{1}{2}bh$ Use the formula for the area of a triangle.

$A = \frac{1}{2}(7)(6)$ Substitute the base and height.

$A = 21$ Multiply $\frac{1}{2}$, 7, and 6.

The area is 21 square centimeters.

Your Turn Find the area of each figure.

a.

parallelogram

b.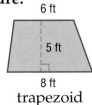

trapezoid

c.

triangle

④ Find the area of the circle. Use π = 3.14.

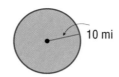

10 mi

$A = \pi r^2$	Use the formula for the area of a circle.
$A = 3.14(10^2)$	Substitute 3.14 for π and 10 for r.
$A = 3.14(100)$	Find 10^2.
$A = 314$	Multiply 3.14 and 100.

The area is 314 square miles.

Your Turn

Find the area of the circle. Use π = 3.14.

d.

6 ft

e.

11 cm

Real-World EXAMPLE

⑤ CONSTRUCTION   A square deck is added to the back of a house. Each side of the deck is 21 feet. What is the area of the deck?

$A = s^2$	Use the formula for the area of a square.
$A = 21^2$	Substitute the side length.
$A = 441$	Find 21^2.

The area is 441 square feet.

Your Turn

f. RECREATION   A square game board is 15 inches on each side. What is the area of the game board?

Examples 1–5
(pages 440–444)

VOCABULARY Write a definition of each term.

1. area

2. altitude

Examples 1–4
(pages 443–444)

Find the area of each figure.

3.
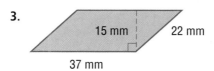
15 mm | 22 mm
37 mm

parallelogram

4.
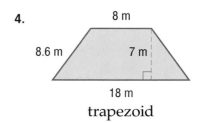
8 m
8.6 m | 7 m
18 m

trapezoid

5.
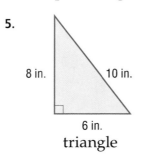
8 in. | 10 in.
6 in.

triangle

6.

4 cm

circle   Use π = 3.14.

Example 5
(page 444)

7. LANDSCAPING A garden has a circular fountain. The radius of the fountain is 2 meters. How much area in the garden does the fountain take? Use π = 3.14.

8. PHOTOGRAPHY A photograph is 5 inches by 7 inches. What is the area of the photograph?

For Exercises	See Example(s)
9–14	1–3
15–16	4
17–20	5

HOMEWORK HELP

Find the area of each figure.

9.
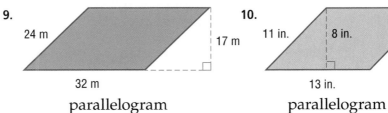
24 m
17 m
32 m

parallelogram

10.
11 in. | 8 in.
13 in.

parallelogram

11.
5 mm
5.1 mm | 4 mm | 5.1
3.5 mm

trapezoid

12.

8 km
7 km
11 km

trapezoid

Find the area of each figure.

13.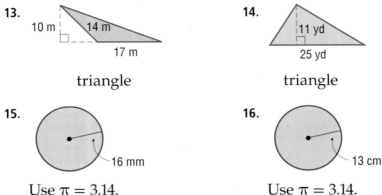

10 m | 14 m
17 m

triangle

14.

11 yd
25 yd

triangle

15.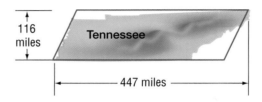

16 mm

Use π = 3.14.

16.

13 cm

Use π = 3.14.

17. **GEOGRAPHY** The state of Tennessee is shaped almost like a parallelogram. Its base is 447 miles and its height is 116 miles. What is the approximate area of the state?

116 miles

Tennessee

447 miles

18. **HOME ECONOMICS** Elena wants to make a pillow. She starts with a piece of fabric that is 28 inches wide and 36 inches long. How many square inches of fabric does she have?

19. **RECREATION** The bottom of a pool has the shape of a trapezoid. One base is 40 feet. The other base is 18 feet. The altitude is 50 feet. What is the area of the bottom of the pool?

20. **LANDSCAPING** A sprinkler sprays water in a circle with a radius of 5 meters. What is the area of the yard that is watered? Use π = 3.14.

21. **H.O.T.** Problem Toba, in Indonesia, is a large volcanic crater. It covers 1,080 square miles. Suppose the crater is circular. About how many miles across the center would it be?

22. *Writing in Math* Explain how you would find the area of the shaded region.

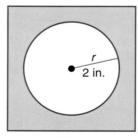

r
2 in.

5 in.

Choose the *best* answer.

23 What is the area of trapezoid *PQRS* in square units?

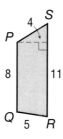

A 22

B 27

C 38

D 48

24 Cherie is making placemats. She cut four triangles off the corners of a rectangle to make an octagon as shown below.

What is the area of the shaded octagon?

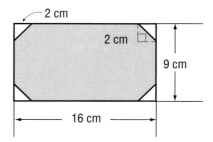

F 128 cm²

G 136 cm²

H 140 cm²

J 152 cm²

Spiral Review

Find the perimeter of each polygon. (Lesson 8-4)

25.

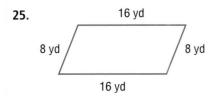

26. a trapezoid with sides 2 cm, 2.5 cm, 3 cm, and 4.5 cm

Find the circumference of the circle. Use π = 3.14. (Lesson 8-4)

27.

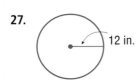

28. SHOPPING A sweater costs $26. It goes on sale for 20% off the regular price. What is the sale price of the sweater? (Lesson 6-5)

29. MUSIC Last month, a CD cost $12. It is on sale for 25% off. What is the discounted price? (Lesson 6-5)

Problem-Solving Strategy: Make a Model

Vocabulary

make a model (p. 448)

Standard GEO8.0 Students know, derive, and **solve problems involving the** perimeter, circumference, **area,** volume, lateral area, and surface area **of common geometric figures.**

Standard 7MR2.0 Students use strategies, skills, and concepts in finding solutions.

Standard 7MR2.5 Use a variety of methods, such as words, numbers, symbols, charts, graphs, tables, diagrams, and **models, to explain mathematical reasoning.**

Remember!

Area of rectangle
$$A = \ell \cdot w$$
A is area.
ℓ is length.
w is width.

Use the Strategy

Mikal has a poster that is 15 centimeters wide and 25 centimeters long. The mat around it is 3 centimeters wide. The frame around the mat is 2 centimeters wide. He wants to know how much wall space he needs to display his poster.

Understand **What does Mikal know?**

- The poster is 15 centimeters by 25 centimeters.
- The mat is 3 centimeters wide around the poster.
- The frame around the mat is 2 centimeters wide.

Plan **What does Mikal want to know?**

Mikal wants to know the total area of the framed poster. He can **make a model** and use it to calculate the area.

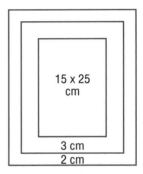

Solve **Mikal draws a model.**

Mikal notes that the entire framed poster is still a rectangle. He uses the formula $A = \ell \cdot w$ to find the area.

First, Mikal finds the total length.

$$25 + 3 + 3 + 2 + 2 = 35$$ The total length is 35 centimeters.

poster   mat   frame

Next, he looks at the width.

$$15 + 3 + 3 + 2 + 2 = 25$$ The total width is 25 centimeters.

poster   mat   frame

Last, Mikal calculates the area.

$A = \ell \cdot w$ Use the formula for area of a rectangle.

$A = 35 \cdot 25$ Substitute the length and the width values.

$A = 875$ Multiply.

Mikal needs 875 square centimeters of wall space to display his poster.

Check **Mikal finds the area another way.**

Mikal drew a proportional model. He then measured his poster to check his calculations. He finds that his answer is correct.

Make a model to solve the problem.

1. There are two tiles. One is a square, and the other is a triangle with three congruent sides. The sides of the square are congruent to the sides of the triangle. The shapes are put together to create a pentagon. If one side of the triangle is 20 centimeters, what is the perimeter of the pentagon?

Understand What do you know?

Plan What are you trying to find? How can you figure it out?

Solve Make a model. Calculate your answer based on the model.

Check How do you know your answer is reasonable?

2. *Writing in Math* You are trying to figure out how many fence posts you will need to surround your yard. How can a model help you decide how many posts you need?

Problem-Solving Practice

Solve using the *make a model* strategy.

3. **GAMES** Bernard has identical circular pieces in his board game (diameter = 1 inch). He wants to lay the pieces flat in a single layer. Which surface will hold more pieces—one 15 inches by 15 inches or one 12 inches by 18 inches?

4. **LANDSCAPING** A playground's perimeter is in the shape of a pentagon. Two sides measure 40 feet, two sides measure 60 feet, and one side measures 70 feet. Bushes will be planted around the playground every 5 feet. How many bushes will be planted?

5. **GARDEN** Sweet corn is planted in one-half of a garden. Strawberries are planted in half of the remaining garden. A third of the remaining space is planted in tomatoes. The tomatoes take up 6 square meters. What is the corn's area?

Solve using any strategy.

6. **SPORTS** Five teams are playing in a tournament. Each team plays every other team. How many total games are played?

7. **INTERIOR DESIGN** A living room, hallway, and bedroom are to be carpeted. Carpet is $12.89 per square yard. How much will it cost?

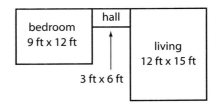

8. **TRANSPORTATION** Vito bought his car 3 years ago. It has lost $\frac{1}{3}$ of its original value. In 2 years it will have lost another $\frac{1}{4}$ of its original value. If he paid $6,000 for the car, what will it be worth after 5 years?

Progress Check 3

(Lessons 8-5 and 8-6)

▷ Vocabulary and Concept Check

> altitude (p. 442)   area (p. 440)

Choose the term that *best* completes each sentence. (Lesson 8-5)

1. The space inside a plane figure is the _____?_____.

2. The height of a triangle is the length of its _____?_____.

▷ Skills Check

Find the area of each figure. (Lesson 8-5)

3.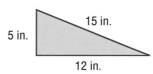

    parallelogram

4.
    15 in.

    5 in.

    12 in.

    triangle

5.

    circle   Use π = 3.14.

6.
    11 cm

    square

▷ Problem-Solving Check (Lessons 8-5 and 8-6)

7. **RECREATION** A pool has the shape of a trapezoid. One base is 40 feet. The other base is 28 feet. The length of the pool is 50 feet. What is the area of the pool?

8. **LANDSCAPING** A circular piece of lawn has an area of 30 square meters. What is the radius? Use π = 3.14.

9. **REFLECT** What are the dimensions of a square for which the measure of

Math Lab
Find Volume and Surface Area

Materials

centimeter cubes

centimeter grid blackline master

scissors

tape

Standard 7MG2.0 Students compute the perimeter, area, and **volume of common objects** and use the results to find measures of less common objects. They know how perimeter, area, and volume are affected by changes of scale.

Standard 7MR1.0 Students make decisions about how to approach problems.

Standard 7MR1.1 Analyze problems by identifying relationships, distinguishing relevant from irrelevant information, identifying missing information, sequencing and prioritizing information, **and observing patterns**.

 The *What*: I will use centimeter cubes and grid paper to find volume and surface area.

 The *Why*: Surface area measures the exterior of a box. Volume measures the amount that a box can contain.

EXAMPLE **Find Volume and Surface Area**

1 **Make a box without a lid using centimeter grid paper. Use centimeter cubes to find the volume and surface area of the box.**

Step 1 Cut a 10-cm by 10-cm square from grid paper.

Step 2 Cut a 3-cm by 3-cm square from each corner. Fold up and tape the sides to make a box.

Step 3 Write the length, width, height on the box.

Step 4 Fill the box with centimeter cubes. Count the cubes needed to fill the box. This is the box's volume in cubic centimeters, cm³.

Step 5 Unfold the box. Count the total number of grid squares. This is the box's surface area in square centimeters, cm².

Step 6 Make a table to record your measures.

Your Turn **Make boxes without lids using centimeter grid paper. Use centimeter cubes to find the volume and surface area of each box.**

a. Cut a 10-cm by 10-cm rectangle from the grid paper. Cut a 4-cm by 4-cm square from each corner. Repeat steps 3–6.

b. Create a box that is 4-cm by 4-cm by 4-cm. Repeat steps 3–6.

c. What is volume and how do you find it?

d. What is surface area and how do you find it?

Vocabulary

solid figure (p. 452)

prism (p. 452)

pyramid (p. 452)

cube (p. 452)

volume (p. 452)

cylinder (p. 453)

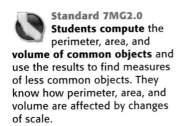

Standard 7MG2.0
Students compute the perimeter, area, and **volume of common objects** and use the results to find measures of less common objects. They know how perimeter, area, and volume are affected by changes of scale.

 The What: I will find the volume of solid figures.

 The Why: Volume can be used to determine which food item is the better buy.

Two kinds of cakes are being sold at a festival. Both cakes cost the same amount. Donna wants to buy the bigger cake.

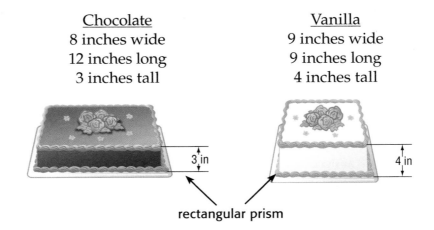

Chocolate
8 inches wide
12 inches long
3 inches tall

Vanilla
9 inches wide
9 inches long
4 inches tall

3 in

4 in

rectangular prism

The cakes are **solid figures** or solids. Solids are three-dimensional shapes. **Prisms** and **pyramids** are examples of solids. A prism has two parallel, congruent faces called bases. When all six faces of a prism are congruent squares, the figure is a **cube.** A pyramid has one base that is a polygon and faces that are triangles.

Volume can be measured by the number of cubic units needed to fill a space.

VOLUME FORMULAS		
rectangular prism		Volume = area of the base • height $= B \cdot h = \ell \cdot w \cdot h$
cube		Volume = area of the base • height $= B \cdot h = s \cdot s \cdot s$
square pyramid		Volume $= \frac{1}{3}$(area of the base • height) $= \frac{1}{3}(B \cdot h) = \frac{1}{3}(\ell \cdot w \cdot h)$

Donna uses the formula for the volume of a rectangular prism.

Chocolate	Vanilla	
$V = \ell \cdot w \cdot h$	$V = \ell \cdot w \cdot h$	Use the formula.
$V = 8 \cdot 12 \cdot 3$	$V = 9 \cdot 9 \cdot 4$	Substitute.
$V = 288$	$V = 324$	Multiply.

The chocolate cake is 288 cubic inches. The vanilla cake is 324 cubic inches. Donna decides to buy the vanilla cake.

EXAMPLES Find the Volume of Solids

Find the volume of each solid.

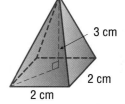

① $V = \frac{1}{3}B \cdot h$ Use the formula for volume of a pyramid.

$V = \frac{1}{3}(\ell \cdot w) \cdot h$ Use the area of a square for B.

$V = \frac{1}{3}(2 \cdot 2) \cdot 3$ Substitute the values.

$V = \frac{1}{3} \cdot 12$ Multiply.

$V = 4$ Simplify. Write and label the answer.

The volume is 4 cubic centimeters.

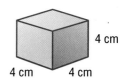

② $V = s \cdot s \cdot s$ Use the formula for volume of a cube.

$V = 4 \cdot 4 \cdot 4$ Substitute.

$V = 64$ Multiply.

The volume is 64 cubic centimeters.

Your Turn Find the volume of each solid.

a.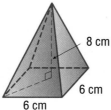

b.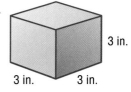

Study Tip

The formula for the volume of a cylinder is the same formula you used for rectangular prisms.

You can also use the formula to find the volume of a triangular prism.

The key is finding the area of the base.

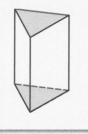

A **cylinder** is a solid with curved sides. To find the volume of a cylinder, you can use the formula $V = B \cdot h$.

VOLUME FORMULA

cylinder		Volume = area of the base • height $= B \cdot h = \pi r^2 \cdot h$

EXAMPLE Find the Volume of Cylinders

③ Find the volume. Use π = 3.14.

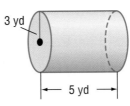

$V = B \cdot h$ Use the formula for volume of a cylinder.

$V = (\pi r^2) \cdot h$ Use the area of a circle for B.

$V = (3.14 \cdot 3^2) \cdot 5$ Substitute.

$V = 141.3$ Multiply.

The volume is 141.3 cubic yards.

Your Turn

c. Find the volume. Use π = 3.14.

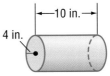

Real-World EXAMPLE

④ PETS A rectangular fish tank is 30 inches long, 16 inches wide, and 12 inches high. How much water can the tank hold?

$V = B \cdot h$ Use the formula for the volume of a rectangular prism.

$V = \ell \cdot w \cdot h$ Use the area of a rectangle for B.

$V = (30 \cdot 16) \cdot 12$ Substitute.

$V = 5,760$ Multiply.

The tank can hold 5,760 cubic inches of water.

Your Turn

d. AGRICULTURE A farmer stores grain in a cylinder. The cylinder has a radius of 6 feet and a height of 3 feet. How much grain does the cylinder hold? Use π = 3.14.

Talk Math

WORK WITH A PARTNER Talk with your partner about the formula to find the area of a triangle. If the formula for the volume of a pyramid is $V = \frac{1}{3}(B \cdot h)$, how could you find the volume of this triangular pyramid?

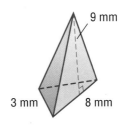

Vocabulary Review
solid figure
prism
pyramid
cube
volume
cylinder

Examples 1–4
(pages 452–454)

VOCABULARY Match the figure with its name.

___?___ **1.** prism

a.

___?___ **2.** cylinder

b.

___?___ **3.** cube

c.

Examples 1–2
(page 453)

Find the volume of each solid figure.

4.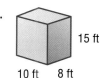
15 ft
10 ft   8 ft

5.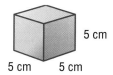
5 cm
5 cm   5 cm

Example 3
(page 454)

6. Find the volume of the cylinder. Use π = 3.14.
18 in.
6 in.

Example 4
(page 454)

7. FOOD SERVICE A soup can has a diameter of 6 centimeters and a height of 8 centimeters. How much soup will the can hold? Use π = 3.14.

8. CONSTRUCTION An animal crate in the shape of a prism has a square base with sides 24 inches long. The crate is 10 inches tall. What is the volume of the crate?

Skills, Concepts, and Problem Solving

HOMEWORK HELP

For Exercises	See Example(s)
9–12	1–2
13–14	3
15–18	4

Find the volume of each solid figure.

9.
9 mm
8 mm
3 mm

10.
5 cm
4 cm   3.5 cm

11.
4 ft
4 ft   4 ft

12.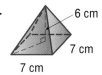
6 cm
7 cm
7 cm

Find the volume of each cylinder. Use π = 3.14.

13.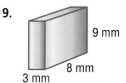
10 yd
6 yd

14.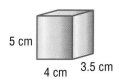
4 ft
6 ft

15. **ENTERTAINMENT** A clown's hat is in the shape of a cylinder. The hat has a radius of 4 inches and is 20 inches tall. How much confetti can he hide in his hat? Use $\pi = 3.14$.

16. **TRANSPORTATION** A truck carries water in a cylindrical tank. The tank is 20 feet long and 6 feet in diameter. How much water does the tank hold? Use $\pi = 3.14$.

17. **CHEMISTRY** A chemistry kit comes in a box that measures 10 inches by 8 inches by 2 inches. How much can the box hold?

18. **HOBBIES** A model of the Great Pyramid in Egypt comes in a box that is a cube. The side of the base measures 7 inches, and the height of the box is 7 inches. How much does the box hold?

19. **H.O.T.** Problem Draw three rectangular prisms. Find the volume of each. Then double the length, width, and height of each prism. Find the volumes of the new prisms. What pattern do you see in the volumes?

20. *Writing in Math* Write a paragraph that explains how to find the volume of a cylinder.

Choose the *best* answer.

21 The formula for the volume of a cube is $V = s^3$. What is the volume of a cube with side length 6 centimeters?

  A 18 cm^3 **C** 42 cm^3

  B 36 cm^3 **D** 216 cm^3

22 Jacinto poured sand into the cylinder. How many cubic inches of sand will the cylinder hold? Use $\pi = 3.14$.

  F 8 in^3 **H** 50.2 in^3

  G 25.1 in^3 **J** 201 in^3

Spiral Review

Find the area of the circle. Use $\pi = 3.14$. (Lesson 8-5)

23.
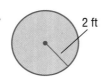
2 ft

24. Find the circumference of the circle in Exercise 23. (Lesson 8-4)

25. **GARDENING** A garden has the shape of a right triangle. One leg measures 8 meters and is against the wall. The hypotenuse is 10 meters. How much fence is needed for the other leg? (Lesson 7-7)

Surface Area

Vocabulary

surface area (p. 457)

Standard 7MG2.1 Use formulas routinely for finding the perimeter and area of basic two-dimensional figures and **the surface area** and volume **of basic three-dimensional figures, including** rectangles, parallelograms, trapezoids, squares, triangles, circles, **prisms, and cylinders.**

Standard 7MR2.0 Students use strategies, skills, and concepts in finding solutions.

Standard 7MR2.3 Estimate unknown quantities graphically and solve for them by using logical reasoning and arithmetic and algebraic techniques.

 The What: I will find the surface area of solid figures.

The Why: Surface area can be used to determine how much wrapping paper covers a package.

Sarai is wrapping a box 9 inches wide, 12 inches long, and 3 inches high. She can find how much wrapping paper it takes to cover the box by calculating the **surface area** of the box. The surface area is the sum of the areas of its faces.

The chart below summarizes what Sarai did. Since each face is a rectangle, she uses the formula $A = \ell w$.

congruent pair: top (and bottom)	area of one face: $A = 9 \cdot 12 = 108$ in²	
congruent pair: right (and left)	area of one face: $A = 9 \cdot 3 = 27$ in²	
congruent pair: front (and back)	area of one face: $A = 12 \cdot 3 = 36$ in²	
surface area = $(2 \cdot 108) + (2 \cdot 27) + (2 \cdot 36) = 342$ in²		

Sarai calculated the area of one of each pair of congruent faces. To find the total surface area, she multiplied the area of each face by 2 and added.

Sarai will need at least 342 square inches of wrapping paper.

Talk Math

WORK WITH A PARTNER How is finding a solid's surface area different than finding its volume?

SURFACE AREA FORMULAS

cubes	$SA = 6s^2$	
rectangular prisms	$SA = 2\ell w + 2\ell h + 2wh$	
cylinders	$SA = 2\pi r^2 + 2\pi rh$	

EXAMPLE Find the Surface Area of Cubes

1 Find the surface area of the cube.

$SA = 6s^2$ Use the formula for surface area.
$SA = 6 \cdot 4^2$ Substitute.
$SA = 6 \cdot 16$ Simplify.
$SA = 96 \text{ in}^2$ Multiply.

4 in.
4 in.   4 in.

The surface area is 96 square inches.

Your Turn Find the surface area of the cube.

a.

5 m

EXAMPLE Find the Surface Area of Rectangular Prisms

2 Find the surface area of the rectangular prism.

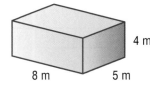

4 m
8 m   5 m

$SA = 2\ell w + 2\ell h + 2wh$
　　Use the formula for surface area.
$SA = 2(8 \cdot 5) + 2(8 \cdot 4) + 2(5 \cdot 4)$
　　Substitute.
$SA = 2(40) + 2(32) + 2(20)$ Simplify.
$SA = 80 + 64 + 40$ Simplify.
$SA = 184$ Add.

The surface area is 184 square meters.

Your Turn

b. Find the surface area of the rectangular prism.

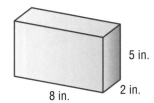

5 in.

8 in.   2 in.

3 Find the surface area of the cylinder. Round to the nearest whole number.

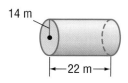

14 m

22 m

$$SA = 2\pi r^2 + 2\pi rh$$ Use the formula for surface area.

$$SA = 2(3.14 \cdot 14^2) + 2(3.14 \cdot 14 \cdot 22)$$ Substitute.

$$SA = 1934.24 + 1230.88$$ Simplify.

$$SA = 3165$$ Add and round.

The surface area is 3,165 square meters.

Remember!

$\pi = $ pi ≈ 3.14

Your Turn

c. Find the surface area of the cylinder. Round to the nearest whole number.

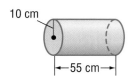

10 cm

55 cm

Real-World EXAMPLE

4 **ART** A red number cube has a side length of 2 centimeters. Paula wants to paint the cube green. How many square centimeters will she paint?

$$SA = 6s^2$$ Use the formula for surface area.

$$SA = 6 \cdot 2^2$$ Substitute the length of the side.

$$SA = 6 \cdot 4$$ Simplify using order of operations.

$$SA = 24$$ Multiply.

Paula will paint 24 square centimeters.

Your Turn

d. **PACKAGING** A box is shaped like a rectangular prism. It is 13 inches long, 6 inches high, and 3 inches wide. How much cardboard was used to make the box?

Examples 1–4
(pages 458–459)

VOCABULARY

1. Write the definition of surface area.

Find the surface area of each solid figure. Round to the nearest whole number if necessary.

Examples 1–2
(page 458)

2.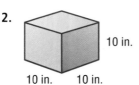

10 in.

10 in.   10 in.

3.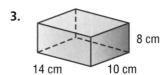

8 cm

14 cm   10 cm

Example 3
(page 459)

4.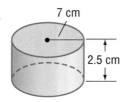

7 cm

2.5 cm

Example 4
(page 459)

5. **ENGINEERING** A water storage tank has a cylindrical shape. The radius of the base is 20 feet, and the tank is 60 feet high. What is the area of the surface to the nearest whole number?

6. **CONSTRUCTION** Paula is building a rectangular chest that is 3 feet long, 2 feet wide, and 1 foot high. She wants to stain all the surfaces. How many square feet will she need to stain?

Skills, Concepts, and Problem Solving

Find the surface area of each solid figure. Round to the nearest whole number if necessary.

HOMEWORK HELP	
For Exercises	**See Example(s)**
7–8	1
9–10	2
11–12	3
13–16	4

7.

15 m

15 m   15 m

8.

3 km

3 km   3 km

9.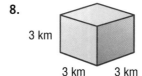

1 mm

4 mm   2 mm

10.

4 cm

6 cm   5 cm

11. 4 ft

2 ft

12.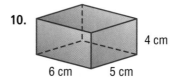

3 yd

1.5 yd

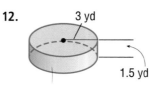

13. **SPORTS** A soccer ball is packaged in a box that is 8 inches on each side. How much plastic will it take to wrap the box?

14. **SPORTS** A tennis ball has a diameter of 2.5 inches. What is the surface area of a can that holds four tennis balls? Round to the nearest whole number.

15. **INTERIOR DESIGN** A cube has sides of 2 feet. How much cloth is needed to cover it?

16. **SPORTS** A water bottle is a cylinder with a 5-centimeter radius and a 12-centimeter height. What is the surface area?

17. *Writing in Math* Write a problem in which you need to know the surface area of a figure.

18. **H.O.T.** Problem A sheet of paper is rolled into a cylinder so that the circumference is 11 inches and the height is 8.5 inches. Another sheet of paper is rolled the other direction to create a cylinder with a circumference of 8.5 inches and a height of 11 inches. Suppose both cylinders have closed ends. How do their surface areas compare?

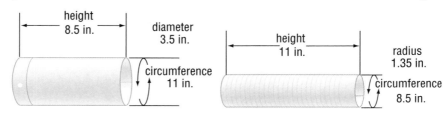

Choose the *best* answer.

19 What is the surface area of the rectangular prism?

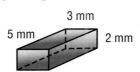

3 mm
5 mm
2 mm

A 30 mm²   C 46 mm²

B 36 mm²   D 62 mm²

20 Shardei peels off the label on a soup can and paints all of the surfaces of the can. How much area does she paint? Use π = 3.14.

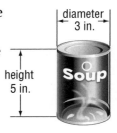

diameter 3 in.
height 5 in.

F 47.1 in²   H 75.36 in²

G 61.23 in²   J 94.2 in²

Spiral Review

21. Find the volume of the soup can in Exercise 20. Use π = 3.14. (Lesson 8-7)

22. **FARMING** A farmer stores water for irrigation in cylinders in different fields. How much water can this cylinder hold? (Lesson 8-7)

3 ft
7 ft

Add or subtract. (Lesson 3-9)

23. $1\frac{6}{7} + 2\frac{1}{2}$ 24. $2\frac{2}{5} + 3\frac{3}{4}$ 25. $6\frac{1}{3} - 2\frac{3}{4}$ 26. $9\frac{3}{8} - 4\frac{3}{5}$

Study Guide

Understanding and Using the Vocabulary

After completing the chapter, you should be able to define each term, property, or phrase and give an example of each.

altitude (p. 442)
area (p. 440)
center (p. 436)
circle (p. 436)
circumference (p. 436)
congruent (p. 420)
coordinate geometry
  (p. 429)

corresponding parts
  (p. 420)
cube (p. 452)
cylinder (p. 452)
Distance Formula (p. 429)
make a model (p. 448)
perimeter (p. 434)
plane figure (p. 414)
polygon (p. 414)

prism (p. 452)
pyramid (p. 452)
quadrilateral (p. 416)
similar (p. 422)
solid figure (p. 452)
surface area (p. 457)
triangle (p. 414)
vertex (p. 414)
volume (p. 452)

Complete each sentence with the correct mathematical term or phrase.

1. A(n) _____?_____ is an enclosed plane figure made of line segments.

2. To find the _____?_____ of a prism, multiply the area of the base times the height of the prism.

3. Two figures that have the same shape but are different sizes are said to be _____?_____.

4. The distance around a circle is the _____?_____

5. A three-sided polygon is called a(n) _____?_____.

6. Add the measures of all sides of a polygon to find its _____?_____.

7. A four-sided polygon is called a(n) _____?_____.

8. Two figures that have the same size and shape are called _____?_____.

9. _____?_____ and _____?_____ are measured in square units.

10. Another word for height is _____?_____.

Skills and Concepts

Objectives and Examples

Review Exercises

LESSON 8-1   pages 414–419

Calculate measures of angles.

Find the value of x.

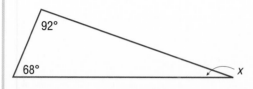

$$68 + 92 + x = 180$$

$$160 + x = 180$$

$$160 + x - 160 = 180 - 160$$

$$x = 20°$$

The sum of the angles of a triangle is 180°.

Add 68 and 92.

Subtract 160 from each side.

Simplify.

Find the value of x.

11.

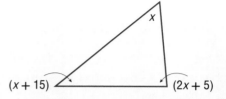

12.

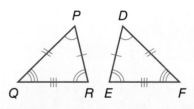

LESSON 8-2   pages 420–425

Identify and use congruent and similar figures.

Name the corresponding parts. Then, complete the congruence statement.

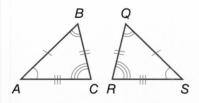

$$\triangle ABC \cong \triangle \underline{\ ?\ }$$

$$\angle A \cong \angle S, \angle B \cong \angle Q, \angle C \cong \angle R;$$

$$\overline{AB} \cong \overline{QS}, \overline{BC} \cong \overline{QR}, \text{ and } \overline{CA} \cong \overline{RS}$$

$$\triangle ABC \cong \triangle QSR$$

Name the corresponding parts. Then, complete the congruence statement.

13.

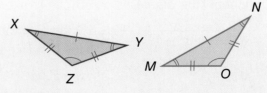

$$\triangle PQR \cong \triangle \underline{\ ?\ }$$

14.

$$\triangle XYZ \cong \triangle \underline{\ ?\ }$$

Skills and Concepts

Objectives and Examples

Find the missing measurement.

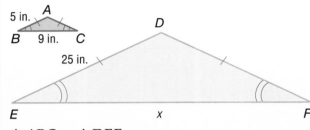

$\triangle ABC \sim \triangle DEF$

$$\frac{AB}{DE} = \frac{BC}{EF}$$
Since the triangles are similar, their sides are proportional.

$$\frac{5}{25} = \frac{9}{x}$$
Substitute the values of the measures.

$5x = 25 \cdot 9$ Cross multiply.

$5x = 225$ Multiply 25 and 9.

$5x \div 5 = 225 \div 5$ Divide each side by 5.

$x = 45$ in.   Simplify.

Review Exercises

Find the missing measurement.

15.

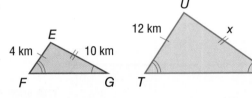

$\triangle EGF \sim \triangle UVT$

16.

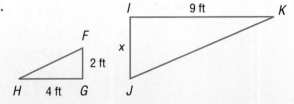

$\triangle FHG \sim \triangle JIK$

LESSON 8-3   pages 429–433

Use the Distance Formula.

What is the distance between points at (4, 3) and (10, 11)?

$x_1 = 4$, $x_2 = 10$, $y_1 = 3$, and $y_2 = 11$
Identify the variables.

$\sqrt{(10 - 4)^2 + (11 - 3)^2}$ Substitute values into the Distance Formula.

$\sqrt{6^2 + 8^2}$ Simplify.

$\sqrt{36 + 64}$

$\sqrt{100} = 10$ Find the square root.

The distance is 10 units.

Use the Distance Formula.

17. What is the distance between points at $(-1, 7)$ and $(5, -14)$?

18. What is the distance between P and Q?

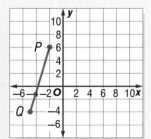

Skills and Concepts

Objectives and Examples

Review Exercises

LESSON 8-4   pages 434–438

Find the perimeter of plane figures.

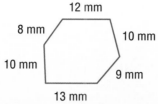

$P = $ sum of side lengths

Use the perimeter formula.

$P = 13 + 10 + 8 + 12 + 10 + 9$

Substitute the side values.

$P = 62$ mm

Add and write the answer with a label.

The perimeter is 62 millimeters.

Find the perimeter of each polygon.

19.

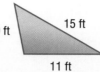

20. a square with each side length equal to 4 miles

21. a rectangle with length 12 inches and width 9 inches

Find the circumference of the circle. Use π = 3.14.

22.

LESSON 8-5   pages 440–447

Find the area of plane figures.

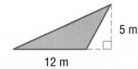

$A = \frac{1}{2}bh$ Use the formula for the area of a triangle.

$A = \frac{1}{2}(12)(5)$ Substitute the base and height.

$A = 30$ Multiply $\frac{1}{2}$, 12, and 5.

The area is 30 square meters.

Find the area of each figure.

23.

24.

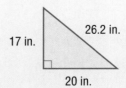

25. Use π = 3.14.

Study Guide

Objectives and Examples

Find the volume of solid figures.

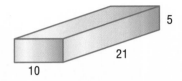

$V = B \cdot h$ Volume of a prism.

$V = (\ell \cdot w)h$ Use the area of a rectangle for *B*.

$V = (10 \cdot 21) \cdot 5$ Substitute the values.

$V = 1{,}050$ Multiply.

The volume is 1,050 cubic inches.

Find the surface area. Round your answer to the nearest whole number. Use π = 3.14.

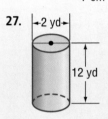

$SA = 2\pi r^2 + 2\pi rh$

Use the formula for surface area.

$SA = 2(3.14 \cdot 3^2) + 2(3.14 \cdot 3 \cdot 20)$

Substitute the values of the dimensions and π.

$SA = 56.52 + 376.8$ Simplify.

$SA = 433$ Add and round.

The surface area is 433 square meters.

Review Exercises

Find the volume of each solid figure.

26.

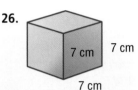

27.

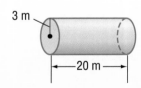

Find the surface area of each solid figure. Round to the nearest whole number. Use π = 3.14.

28.

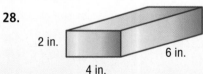

29.

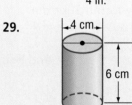

30.

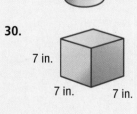

Chapter Test

▷ Vocabulary and Concept Check

1. Explain how a triangle is both a polygon and a plane figure.
2. Explain the difference between similar and congruent.

▷ Skills Check

Find the value of x.

3.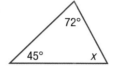

4.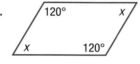

5.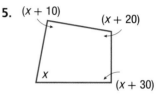

Name the corresponding parts. Then, complete the congruence statement.

6.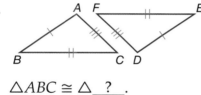

$\triangle ABC \cong \triangle\underline{\ ?\ }$.

7.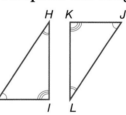

$\triangle GHI \cong \triangle\underline{\ ?\ }$.

Find the perimeter or circumference of each figure, then find the area.

8.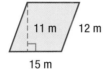

9.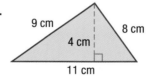

10. Use $\pi = 3.14$.

Find the volume of each solid figure. Use $\pi = 3.14$. Then find the surface area.

11.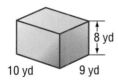

12.

▷ Problem-Solving Check

13. **MEASUREMENT** To show movies, the room must be very dark. Kiva decides to cover her window with black paper. The window is 5 feet long and 3 feet wide. How much paper does it take to cover the window?

PART 1 Multiple Choice

Choose the *best* answer.

1 What is the perimeter of a rectangle with length 9 meters and width 12 meters?

   A 21 meters   **C** 46 meters

   B 42 meters   **D** 108 meters

2 What is the approximate length of the line segment with endpoints at (4, 6) and (−2, −4)?

   F 4   **H** 16

   G 12   **J** 136

3 Find x.

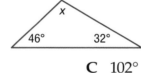

   A 2°   **C** 102°

   B 78°   **D** 282°

4 Find the area of the figure.

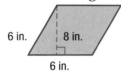

   F 14 in²   **H** 28 in²

   G 24 in²   **J** 48 in²

5 Find the surface area of a cube with side length 10 millimeter.

   A 60 mm²   **C** 600 mm²

   B 400 mm²   **D** 1,000 mm²

6 $\triangle ABC \cong \triangle DEF$. Which side is congruent to $\overline{CA}$?

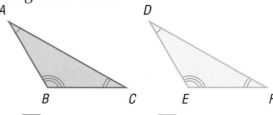

   F $\overline{FE}$ **H** $\overline{EF}$

   G $\overline{DE}$ **J** $\overline{FD}$

7 What is the perimeter of a polygon with six sides all measuring 10 cm?

   A 50 cm   **C** 70 cm

   B 60 cm   **D** 100 cm

8 Find the volume.

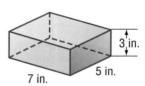

   F 15 in³   **H** 38 in³

   G 30 in³   **J** 105 in³

9 What is the sum of the angles in a trapezoid?

   A 180°   **C** 540°

   B 360°   **D** 720°

10 Find the area.

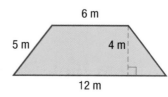

   F 36 m²   **H** 72 m²

   G 45 m²   **J** 90 m²

Record your answers on the answer sheet provided by your teacher or on a separate sheet of paper.

11 A cake is 2 inches tall. A piece of cake measures 1 inch by 3 inches.

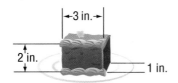

a. What is the volume of the cake?

b. The corner piece of cake has icing on two sides and the top. What area is covered by icing?

12 For a circle with a radius of 2 inches, the following relationships are true.

Area = Circumference

x in^2 = x in.

Explain how this is possible. Find the measure for the area and circumference.

13 Find the surface area.

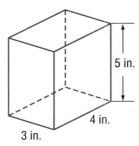

14 What is the distance between W and E?

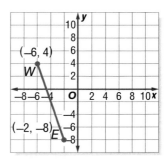

Record your answers on the answer sheet provided by your teacher or on a separate sheet of paper.

15 A puppy lives in a triangular yard. The yard has a base of 50 feet and a height of 30 feet. A storage shed in the yard has a rectangular base 10 feet by 8 feet.

a. What is the area of the yard?

b. What is the area of the yard on which the shed sits?

c. What is the area of the yard where the dog can play? (He has the whole yard except where the storage shed is.)

NEED EXTRA HELP?														
If You Missed Question...	1	2	3	4	5	6	7	8	9	10	11	12	13	14
Go to Lesson...	8-6	8-3	8-1	8-4	8-8	8-2	8-6	8-7	8-1	8-4	8-7	8-6	8-2	8-8
For Help with Algebra Readiness Standard...	GEO8.0	7MG3.2	7MR1.1	7MG2.1	7MG2.1	7MG3.4	GEO8.0	7MG2.0	7MR1.1	7MG2.1	7MG2.0	7MG2.1	7MG3.4	7MG2.1

What's in the Student Handbook?

Extra Practice

The Extra Practice section contains problems similar to those found at the end of each lesson. These problems provide extra practice on the skills you learn in each chapter.

Mixed Problem Solving

The Mixed Problem-Solving section contains a collection of word problems. Although organized by chapter, they are all mixed up—so you will need to figure out which skills to apply in order to solve them.

California Standards Practice

The California Standards Practice section contains problems like those you would find on most California state standardized tests. There is a mix of multiple choice, short answer, and extended response questions. A chart at the end of each practice test will let you know which lesson to review for additional help.

Glossary **528**

The glossary is a list of all the words from every lesson in this book, along with their definitions.

Selected Answers **543**

The Selected Answers contain answers to some, but not all, of the problems from the lessons.

Photo Credits **552**

Index **553**

The index indicates the page(s) on which the major topics appear.

TEST YOURSELF

Find the answers to these questions by using the Student Handbook on the following pages.

1. What is the variable used in Exercise 1 in Extra Practice 1-2?
2. What is the definition of *equation?*
3. Where can I find more word problems that involve fractions?
4. How many multiple-choice problems appear in the California Standards Practice section for Chapter 8?
5. What is the first page on which you'll find information about *variables?*

Test Yourself Answers
1. *n*
2. An *equation* is a mathematical sentence stating that two expressions are equal.
3. p. 502
4. 16
5. p. 58

Extra Practice

Prerequisite Skills Extra Practice

P-1 Whole Numbers

Identify the whole number in each set.

1. $12, 3.4, \frac{5}{6}$
2. $\frac{1}{4}, 1\frac{2}{3}, 11$
3. $7, 15.1, \frac{2}{5}$
4. $23\frac{2}{3}, 23.23, 23$
5. $2.3, 5, \frac{7}{9}$
6. $6\frac{7}{8}, 0.2, 13$
7. $36, 1.245, \frac{7}{8}$
8. $\frac{2}{7}, 3, 9\frac{1}{2}$
9. $0.8, 0, \frac{3}{10}$

Compare the numbers, using the symbols < and >.

10. 1 and 0
11. 6 and 9
12. 17 and 12
13. 18 and 15
14. 21 and 20
15. 3 and 7
16. 19 and 26
17. 14 and 8
18. 29 and 27

Order the numbers from least to greatest.

19. 5, 8, 9
20. 1, 9, 3
21. 23, 19, 32
22. 12, 2, 37
23. 3, 0, 11
24. 41, 37, 34
25. 14, 96, 85
26. 28, 32, 15
27. 53, 35, 47

P-2 Place Value

Write the value of the underlined digit.

1. <u>4</u>
2. <u>1</u>2
3. <u>2</u>3
4. 7<u>0</u>
5. 1<u>0</u>1
6. 76<u>9</u>
7. 2<u>3</u>4
8. 10<u>4</u>
9. <u>5</u>37
10. 8<u>1</u>2
11. <u>3</u>00
12. 1,<u>8</u>76

Write each number in expanded form.

13. 50
14. 89
15. 101
16. 123
17. 253
18. 427
19. 1,987
20. 1,234
21. 2,357
22. 7,849
23. 10,001
24. 5,280

P-3 Addition and Subtraction

Add.

1. $71 + 27$
2. $14 + 94$
3. $46 + 207$
4. $28 + 496$
5. $123 + 234$
6. $456 + 321$
7. $387 + 610$
8. $946 + 101$
9. $654 + 146$
10. $3,333 + 1,111$
11. $2,856 + 386$
12. $5,386 + 57$

Subtract.

13. $432 - 131$
14. $544 - 332$
15. $867 - 358$
16. $4,296 - 77$
17. $3,335 - 900$
18. $1,453 - 145$
19. $7,893 - 946$
20. $1,192 - 1,002$
21. $3,476 - 1,876$
22. $8,410 - 6,936$
23. $2,276 - 2,087$
24. $5,461 - 2,949$

P-4 Multiplication and Division

Multiply.

1. $22 \cdot 34$
2. $10 \cdot 38$
3. $76 \cdot 16$
4. $400 \cdot 75$
5. $232 \cdot 18$
6. $958 \cdot 34$
7. $1,010 \cdot 21$
8. $4,183 \cdot 12$
9. $2,320 \cdot 87$
10. $3,206 \cdot 42$
11. $5,162 \cdot 35$
12. $6,485 \cdot 24$

Divide.

13. $123 \div 3$
14. $459 \div 3$
15. $672 \div 4$
16. $996 \div 4$
17. $3,970 \div 5$
18. $4,564 \div 7$
19. $4,096 \div 2$
20. $5,280 \div 10$
21. $2,772 \div 9$
22. $1,988 \div 4$
23. $5,635 \div 7$
24. $5,913 \div 9$

P-5 Exponents

Simplify.

1. 5^3
2. 2^3
3. 7^2
4. 2^2
5. 4^2
6. 4^3
7. 2^7
8. 3^3

Write the product using exponents.

9. $3 \cdot 3 \cdot 3 \cdot 3$
10. $7 \cdot 7 \cdot 7 \cdot 7$
11. $13 \cdot 13 \cdot 13 \cdot 13 \cdot 13 \cdot 13$
12. $n \cdot n$
13. $p \cdot p \cdot p \cdot p$
14. b

P-6 Prime Factorization

Write the prime factorization of each number.

1. 6   2. 14

3. 8   4. 18

5. 32   6. 63

7. 81   8. 15

9. 20   10. 42

11. 52   12. 12

13. 28   14. 22

15. 50   16. 24

17. 36   18. 35

19. 54   20. 44

P-7 Fractions

Write the fraction.

1. What fraction of the pie has been eaten?

2. What fraction of the mosaic is red?

3. What fraction of the trees is taller than 5 feet?

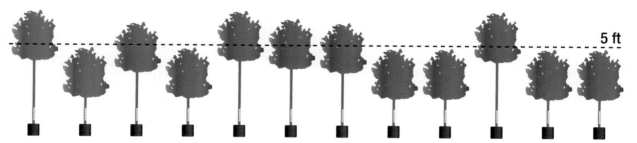

5 ft

4. What fraction of the goal has been reached?

Fundraising Goal

12,000

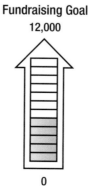

0

P-8 Decimals

Identify the decimal as terminating or repeating.

1. 0.12
2. 0.1333 . . .
3. 0.57
4. 0.111 . . .
5. 0.13
6. 0.3535 . . .
7. 0.040404 . . .
8. 0.69 . . .
9. 0.69

Rewrite each terminating decimal as a fraction.

10. 1.33
11. 0.007
12. 0.57
13. 0.13
14. 5.25
15. 1.143
16. 7.25
17. 2.16
18. 1.125

P-9 Least Common Multiple

Find the LCM of the numbers.

1. 4 and 11
2. 2 and 5
3. 4 and 7
4. 6 and 7
5. 3 and 5
6. 2 and 4
7. 11 and 13
8. 2 and 7
9. 5 and 8
10. 2, 3, and 7
11. 2, 5, and 9
12. 3, 5, and 6

P-10 Estimation

Round to the nearest ten to estimate.

1. $123 + 746$
2. $34 \cdot 12$
3. $198 + 453$
4. $344 + 127$
5. $236 - 103$
6. $28 \cdot 41$
7. $349 \div 7$
8. $692 - 388$
9. $237 \div 4$
10. $398 + 857$
11. $987 - 611$
12. $3,593 + 2,846$
13. $5,959 - 2,022$
14. $3,927 - 2,154$
15. $2,109 + 9,309$
16. $19 \cdot 98$

Use compatible numbers to estimate.

17. $456 \div 3$
18. $234 \div 5$
19. $930 \div 5$
20. $464 \div 9$
21. $109 \div 4$
22. $928 \div 6$
23. $1,029 \div 5$
24. $3,497 \div 7$
25. $7,176 \div 8$
26. $1,189 \div 6$

P-11 Angles

Classify each indicated angle as acute, obtuse, or right.

1.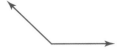

2.

3.

4.

5.

6.

7. angles in a speed-limit sign

8.

Angle A: _____ Angle B: _____ Angle C: _____

P-12 Probability

Find the probability of each event.

1. a coin landing with "heads" up   2. your new friend was born on a Monday

3. the spinner landing on green   4. the spinner landing on blue

5. the spinner landing on red   6. the spinner landing on yellow or green

Describe the event as likely, unlikely, certain, or impossible.

7. drawing a blue marble from a bag of red and green marbles

HOBBIES Amelia folded 11 paper airplanes, all with the same design. She went into the hallway and tossed 10 of them. The distance of each flight is recorded in this chart. What is the likelihood of each?

Distance	Number of Airplanes
< 15 feet	3
15–20 feet	6
> 20 feet	1

8. the remaining plane will fly less than 15 feet

9. the remaining plane will fly more than 20 feet

10. the remaining plane will fly 15 feet to 20 feet

Chapter 1 Extra Practice

1-1 A Plan for Problem Solving

Use the four-step plan to solve the following problems.

1. **PETS** Laramie feeds his dog 5 ounces of dog food every morning and 7 ounces of dog food every evening. He buys dog food in 30-pound bags. How many days will one bag of dog food last?

2. **SPORTS** The Lawrence Tigers scored 26 points in a football game. They scored 2 touchdowns, worth 6 points each, and 2 PAT kicks, worth 1 point each. They also scored some field goals, which are worth 3 points each. How many field goals did they score?

3. **SCHOOL** At Jacobsen Middle School 70 students study only algebra and 45 students study only French. One hundred-thirty students study at least one of these courses. How many students study both?

4. **FINANCE** Damian works 40 hours a week. He earns $12 per hour. He saves $\frac{1}{4}$ of his weekly earnings to buy a scooter worth $960. How many weeks will Damian have to work to buy the scooter?

1-2 Expressions and Equations

Write an algebraic expression for each verbal expression.

1. the sum of 14 and n

2. the product of 10 and x

3. the quotient of 7 and p

4. the difference between h and 3 divided by g

Write an equation for each sentence.

5. Seven divided by x is 12.

6. x plus 7 times y is equal to 42.

7. Eleven more than y is 20.

8. Five multiplied by n equals 30.

1-3 Order of Operations

Find the value of each expression. Show your work.

1. $3 \cdot 5 + 8 \div 2$

2. $32 \div 4 \cdot 3 + 15$

3. $12 + 10 \div 2$

4. $3 \cdot 7 + 10 \cdot 2$

5. $6 \cdot 3 \div 9 - 1$

6. $14 + 50 \div 5 - 3 \cdot 2$

7. $3(12 - 2) \cdot 4$

8. $5^2 + 30 \div (5 - 2)$

9. $20 - 4(3 + 1)$

10. $6^2 \div 2(8 + 1)$

11. $7 + (9 - 1)^2$

12. $5[(10 \div 5)^3 + (3 - 1)^4]$

1-4 Commutative and Associative Properties

Complete the equation using the Commutative Property.

1. $9 \cdot 3 = 3 \cdot \underline{}$

2. $11 + 12 = 12 + \underline{}$

3. $4 \cdot 6 = \underline{} \cdot \underline{}$

4. $3 + 5 = \underline{} + \underline{}$

Complete the equation using the Associative Property.

5. $3 \cdot (9 \cdot 2) = (3 \cdot \underline{}) \cdot 2$

6. $(2 + 3) + 4 = \underline{} + (3 + \underline{})$

7. $8 \cdot (1 \cdot 2) = (\underline{} \cdot \underline{}) \cdot \underline{}$

Simplify each expression. Name the property used in each step.

8. $23 + (7 + 8)$

9. $2 \cdot (5 \cdot 4)$

10. $(9 + 23) + 41$

11. $(8 \cdot 4) \cdot 25$

1-5 Distributive Property

Complete the equations using the Distributive Property.

1. $12(8 + 3) = (12 \cdot \underline{}) + (12 \cdot \underline{})$

2. $7(10 + 5) = (\underline{} \cdot \underline{}) + (\underline{} \cdot \underline{})$

3. $(4 \cdot 2) + (4 \cdot 8) = 4(\underline{} + \underline{})$

4. $(3 \cdot 8) + (3 \cdot 7) = \underline{} (\underline{} + \underline{})$

1-7 Other Properties

Complete each equation. Name the property used in each equation.

1. $98 + \underline{} = 98$

2. $33 \cdot \underline{} = 0$

3. $3y \cdot 1 = \underline{}$

4. $2x + \underline{} = 2x$

1-8 Simplifying Expressions

Determine if each expression has like terms. If so, state them.

1. $7y - 2y$

2. $3x + 2x^2$

3. $9xy + 2x + 3y$

4. $2b - 8b$

Simplify the expression by combining like terms. Provide a reason for each step.

5. $2s + 7s$

6. $9p - 4p$

7. $3x + (4 + x)$

8. $3 + 2(d + 4)$

Simplify the expression.

9. $3f + 2g - f + 4g$

10. $2(h + 4) + 2h + 4$

11. $2(a + b + 3) + 6 + 2a$

12. $3(2 + x) + 5(4 + 3y)$

Chapter 2 Extra Practice

2-1 Equations

Which of the numbers 0, 1, 2, or 3 is the solution for each equation?

1. $3x + 3 = 9$
2. $5n - 3 = 2$
3. $2r - 6 = 0$
4. $4y + 4 = 4$

Solve each equation using mental math.

5. $p + 11 = 17$
6. $3q = 15$
7. $17 - g = 8$
8. $d \div 4 = 4$

Write an equation. Solve your equation.

9. **SPORTS** Adelfo won two games more than Don in a set of tennis. If Adelfo won seven games, how many games did Don win?

10. **SHOES** Lydia has twice as many pairs of shoes as Delia. If Lydia has 24 pairs of shoes, how many pairs does Delia have?

2-2 Integers

Answer each question.

1. What is the opposite of -6?
2. What is the opposite of 5?
3. What is the absolute value of -3?
4. What is the absolute value of 3?

5. Name the integers represented by *D, O,* and *G.*

6. Name the integers represented by *T, M,* and *R.*

7. Graph the integers $-4, -2, 2,$ and 4 on a number line.
8. Graph the integers $-5, -3, 1,$ and 6 on a number line.

Use the number line to complete each statement with < or >.

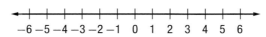

9. $-2 \bigcirc 2$
10. $0 \bigcirc -3$
11. $-4 \bigcirc -2$
12. $-5 \bigcirc 1$

2-3 Adding and Subtracting Integers

Draw a number line. Use the number line to find each sum.

1. $-6 + 1$
2. $8 + (-7)$
3. $-9 + 3$
4. $-2 + 2$

Find each sum.

5. $4 + (-7)$
6. $-11 + (-6)$
7. $-6 + 3$
8. $6 + (-6)$
9. $-7 + 7$
10. $-3 + (-9)$
11. $-8 + 12$
12. $-9 + 9$

Find each difference by using additive inverses.

13. $-6 - (-2)$
14. $-9 - 10$
15. $3 - 5$
16. $-2 - (-2)$

Simplify each expression.

17. $3r + (-4r)$
18. $3n - 5n$
19. $2x + (-4x)$
20. $p - 6p$

2-4 Multiplying Integers

Find each product.

1. $-1 \cdot 8$
2. $4 \cdot 2$
3. $2 \cdot (-6)$
4. $6 \cdot (-5)$
5. $(-1)(3)(-5)$
6. $(-2)(1)(-3)(-1)$
7. $(-2)(3)(-5)(7)$
8. $(-4)^2(3)(-1)(-2)$

Find the value of each expression.

9. $2x$, if $x = -8$
10. $-5a$, if $a = 3$
11. $-7rs$, if $r = 4$ and $s = -3$
12. $3mn$, if $m = -2$ and $n = -5$

2-5 Dividing Integers

Find each quotient.

1. $-25 \div 5$
2. $-32 \div 4$
3. $32 \div (-8)$
4. $36 \div (-4)$

Find the value of each expression.

5. $\left(\frac{-6}{-2}\right)^3$
6. $\left(\frac{-12}{4}\right)^4$
7. $\left(\frac{4}{-2}\right)^5$
8. $\left(\frac{-8}{2}\right)^2$
9. $\frac{k}{-6}$, if $k = 42$
10. $\frac{a}{b}$, if $a = -32$ and $b = -4$
11. $x \div 2$, if $x = 20$
12. $\frac{f}{g}$, if $f = 39$ and $g = -13$

2-7 Solving Equations

Solve each equation.

1. $r + 3 = 9$
2. $a + 4 = 32$
3. $f - 3 = 12$
4. $n - 7 = 14$
5. $2m = 24$
6. $4t = -24$
7. $\frac{y}{5} = 6$
8. $n \div 3 = 6$
9. $2n + 1 = 11$
10. $7y + 5 = 19$
11. $3y - 3 = 21$
12. $5n + 7 = -23$

Chapter 3 Extra Practice

3-1 Fractions

Graph each fraction on a number line.

1. $\frac{1}{4}$

2. $\frac{2}{5}$

3. $-\frac{1}{6}$

4. $-\frac{3}{10}$

Write the additive inverse of each number.

5. $\frac{2}{7}$

6. $\frac{1}{3}$

7. $-\frac{3}{5}$

8. $-\frac{7}{8}$

Write the additive inverse of each number. Then, graph the number and its additive inverse.

9. $-\frac{1}{2}$

10. $\frac{3}{5}$

11. $-\frac{3}{7}$

12. $\frac{3}{4}$

Use a number line to determine which number is greater.

13. $\frac{3}{4}$ or $\frac{1}{2}$

14. $\frac{1}{5}$ or $\frac{3}{8}$

15. $-\frac{1}{5}$ or $-\frac{1}{3}$

16. $-\frac{2}{3}$ or $-\frac{5}{8}$

Order each set of fractions from least to greatest.

17. $\frac{2}{3}, \frac{1}{7}, \frac{5}{9}$

18. $\frac{3}{4}, -\frac{4}{7}, \frac{2}{7}$

19. $-\frac{3}{5}, \frac{5}{6}, -\frac{2}{3}$

20. $-\frac{7}{10}, -\frac{1}{4}, -\frac{5}{8}$

3-2 Fractions and Mixed Numbers

Identify each fraction as proper or improper.

1. $\frac{9}{10}$

2. $-\frac{3}{5}$

3. $\frac{7}{3}$

4. $-\frac{25}{18}$

Write each improper fraction as a mixed number.

5. $\frac{7}{6}$

6. $\frac{9}{2}$

7. $-\frac{4}{3}$

8. $-\frac{22}{7}$

Write each improper fraction as a mixed number.

9. $-\frac{39}{8}$

10. $-\frac{5}{3}$

11. $\frac{23}{6}$

12. $\frac{11}{4}$

Graph each improper fraction and mixed number on a number line.

13. $1\frac{1}{4}$

14. $2\frac{3}{5}$

15. $-4\frac{3}{7}$

16. $-5\frac{2}{3}$

Determine which number is greater.

17. $-3\frac{7}{8}$ or $-2\frac{7}{8}$

18. $-1\frac{9}{13}$ or $-5\frac{6}{11}$

19. $2\frac{1}{6}$ or $1\frac{5}{6}$

20. $5\frac{3}{4}$ or $5\frac{3}{8}$

3-3 Factors and Simplifying Fractions

Identify each number as prime or composite.

1. 3

2. 12

3. 17

4. 21

List the factors for each number.

5. 10

6. 16

7. 7

8. 19

Find the common factors of each set of numbers.

9. 5 and 25

10. 9 and 27

11. 24 and 32

12. 14 and 28

Identify the greatest common factor (GCF) of the numbers.

13. 8 and 12

14. 10 and 15

15. 21 and 28

16. 6 and 16

17. 14 and 18

18. 7 and 20

Write each fraction in simplest form.

19. $\frac{3}{9}$

20. $\frac{6}{18}$

21. $\frac{5}{25}$

22. $\frac{8}{12}$

23. $\frac{7}{27}$

24. $\frac{21}{28}$

3-5 Multiplying Fractions

Multiply. Write the product in simplest form.

1. $\frac{1}{2} \cdot \frac{3}{4}$

2. $\frac{4}{5} \cdot \frac{2}{3}$

3. $\frac{7}{8} \cdot \frac{3}{8}$

4. $\frac{1}{3} \cdot \frac{5}{7} \cdot \frac{1}{4}$

5. $\frac{1}{2} \cdot \frac{1}{4} \cdot \frac{3}{4}$

6. $-\frac{3}{5} \cdot \frac{7}{8}$

7. $\frac{9}{10} \cdot -\frac{5}{6}$

8. $-\frac{1}{6} \cdot -\frac{3}{4}$

9. $\frac{1}{2} \cdot \frac{4}{5}$

10. $\frac{3}{4} \cdot \frac{3}{4}$

11. $\frac{1}{2} \cdot \frac{2}{3}$

12. $\frac{3}{4} \cdot \frac{3}{8}$

13. $\frac{3}{5} \cdot \frac{4}{3}$

14. $\frac{5}{2} \cdot \frac{3}{4}$

15. $2 \cdot -\frac{7}{3}$

16. $-\frac{7}{16} \cdot -\frac{5}{2}$

Evaluate.

17. $\left(\frac{1}{5}\right)^2$

18. $\left(\frac{3}{4}\right)^3$

19. $\left(\frac{2}{3}\right)^4$

20. $\left(\frac{1}{2}\right)^3$

3-6 Dividing Fractions

Write the reciprocal of each fraction.

1. $\frac{2}{3}$

2. $\frac{7}{6}$

3. 5

4. $\frac{5}{8}$

Divide. Write each quotient in simplest form.

5. $\frac{3}{5} \div \frac{2}{5}$

6. $\frac{3}{8} \div \frac{7}{12}$

7. $\frac{2}{3} \div \frac{8}{9}$

8. $\frac{2}{7} \div \frac{18}{35}$

9. $\frac{3}{4} \div \left(-\frac{2}{3}\right)$

10. $-\frac{2}{3} \div \frac{5}{8}$

11. $\frac{1}{4} \div \left(-\frac{7}{10}\right)$

12. $\frac{1}{6} \div \left(-\frac{3}{6}\right)$

13. $\frac{5}{4} \div \frac{3}{4}$

14. $\frac{17}{6} \div \frac{5}{3}$

15. $-\frac{7}{4} \div \frac{3}{2}$

16. $-\frac{7}{2} \div -\frac{11}{4}$

3-7 Adding and Subtracting Fractions with Like Denominators

Add. Write the sum in simplest form.

1. $\frac{1}{3} + \frac{1}{3}$
2. $\frac{3}{14} + \frac{9}{14}$
3. $\frac{2}{7} + \frac{5}{7}$
4. $\frac{3}{4} + \frac{2}{4}$
5. $\frac{67}{16} + \frac{37}{16}$
6. $\frac{9}{4} + \frac{3}{4}$
7. $\frac{3}{2} + \frac{1}{2}$
8. $\frac{17}{12} + \frac{31}{12}$

Subtract. Write each difference in simplest form.

9. $\frac{3}{4} - \frac{1}{4}$
10. $\frac{7}{8} - \frac{5}{8}$
11. $\frac{9}{10} - \frac{1}{10}$
12. $\frac{7}{18} - \frac{5}{18}$
13. $4\frac{1}{6} - 2\frac{1}{6}$
14. $5\frac{3}{4} - 3\frac{1}{4}$
15. $3 - 2\frac{1}{5}$
16. $2 - \frac{3}{7}$

3-8 Adding Fractions with Unlike Denominators

Find the LCM of each pair of numbers.

1. 4 and 10
2. 14 and 21
3. 3 and 12
4. 16 and 24

Rename the fraction with the given denominator.

5. $\frac{1}{3} = \frac{?}{6}$
6. $\frac{3}{5} = \frac{?}{15}$
7. $\frac{2}{7} = \frac{?}{21}$
8. $\frac{1}{3} = \frac{?}{18}$

Find the LCD of each pair of fractions.

9. $\frac{2}{3}$ and $\frac{5}{6}$
10. $\frac{7}{8}$ and $\frac{2}{3}$
11. $\frac{3}{5}$ and $\frac{1}{10}$
12. $\frac{3}{8}$ and $\frac{1}{6}$

Add. Write each sum in simplest form.

13. $\frac{1}{2} + \frac{1}{4}$
14. $\frac{2}{3} + \frac{5}{6}$
15. $\frac{4}{7} + \frac{5}{14}$
16. $\frac{8}{15} + \frac{3}{10}$
17. $\frac{19}{8} + \frac{19}{4}$
18. $\frac{11}{3} + \frac{28}{9}$
19. $\frac{11}{2} + \frac{19}{8}$
20. $\frac{38}{5} + \frac{12}{7}$

3-9 Subtracting Fractions with Unlike Denominators

Subtract. Write each difference in simplest form.

1. $\frac{7}{16} - \frac{3}{8}$
2. $\frac{5}{6} - \frac{1}{3}$
3. $\frac{5}{8} - \frac{1}{4}$
4. $\frac{3}{4} - \frac{2}{3}$
5. $3\frac{5}{6} - 1\frac{2}{3}$
6. $2\frac{2}{3} - 1\frac{2}{3}$
7. $8 - 3\frac{7}{12}$
8. $7 - 1\frac{3}{8}$

3-10 Fractions in Expressions and Equations

Simplify each expression.

1. $\frac{5}{8}g + \frac{2}{3} - \frac{3}{8}g$
2. $\frac{7}{12}p - \frac{3}{8}p - \frac{3}{7}$
3. $\frac{1}{5} + \frac{2}{3} - \frac{2}{3}k$
4. $-\frac{3}{10}s + \frac{7}{12} + \frac{5}{6}s + \frac{1}{4}$

Solve each equation.

5. $y + \frac{1}{2} = \frac{3}{4}$
6. $2r = \frac{2}{3}$
7. $s - \frac{5}{6} = 1$
8. $h \div \frac{2}{3} = \frac{6}{7}$
9. $3k + \frac{3}{4} = 2$
10. $\left(a \div \frac{3}{2}\right) - \frac{3}{4} = 3$
11. $\frac{2}{3}g + \frac{5}{6} = \frac{5}{6}$
12. $\frac{1}{3}x + \frac{2}{5} = 1\frac{1}{2}$

Chapter 4 Extra Practice

4-1 Fractions and Decimals

Change each terminating decimal to a fraction in simplest form.

1. 0.9

2. 0.43

3. 0.1

4. 0.21

5. 0.11

6. 0.23

Change each terminating decimal to a mixed number in simplest form.

7. 3.6

8. 100.85

9. 28.06

10. 2.2

11. 17.25

12. 1.125

Change each fraction or mixed number to a decimal.

13. $\frac{5}{8}$

14. $5\frac{21}{40}$

15. $1\frac{1}{4}$

16. $66\frac{11}{20}$

17. $\frac{3}{4}$

18. $\frac{1}{1,000}$

4-2 Adding and Subtracting Decimals

Add.

1. $5.3 + 7.8$

2. $23 + 2.3$

3. $84.8 + 45.6$

4. $31.301 + 9.378$

5. $43.1 + 22$

6. $1.2 + 3.45 + 6.789$

Subtract.

7. $5.4 - 1.2$

8. $4.86 - 2.42$

9. $17.9 - 6.7$

10. $31.4 - 11.54$

11. $32 - 15.99$

12. $34.1 - 33$

Find the difference.

13. $51.24 - 75.56$

14. $-6.48 - 13.48$

15. $10.9 - 23$

16. $43 - 56.8$

17. $28.3 - 72.14$

18. $12.06 - (-19.5)$

Simplify each expression. Provide a reason for each step.

19. $5.7r + 9.2r + 8.3$

20. $18.8s + (3.9 - 15.1s)$

21. $(0.72t - 14.6) + 10.67t$

22. $(2.7v + 11.4w) - (6.1w - 4.5v)$

4-3 Multiplying Decimals

Multiply.

1. $28 \cdot 1.7$
2. $2.25 \cdot 20$
3. $2.87 \cdot (-1.1)$
4. $43.1 \cdot 2.4$
5. $0.79 \cdot (-2.23)$
6. $2.5 \cdot 10$
7. $4.5 \cdot 4.5$
8. $-24 \cdot (-1.5)$
9. $22.66 \cdot 1.5$

Simplify.

10. 6.2^2
11. 1.9^4
12. 8.14^3
13. 7.32^2
14. $(-12.1)^2$
15. $(-4.32)^3$

Simplify each expression. Provide a reason for each step.

16. $(a \cdot 3.8) \cdot 2.7$
17. $(b \cdot 9.2) \cdot 4.3$
18. $6.1(c + 7.4)$
19. $8.3(f + 1.9)$

4-4 Dividing Decimals

Divide.

1. $75.0 \div 12.5$
2. $96.75 \div 22.5$
3. $0.05 \div 2.5$
4. $517.92 \div 8.3$
5. $84.78 \div 3.14$
6. $36.11 \div 11.5$

Round the quotient to the nearest hundredth.

7. $87.24 \div 9.95 = 8.767$
8. $24 \div 11.6 = 2.068$
9. $237.72 \div 2.2 = 108.055$
10. $100 \div 26.8 = 3.731$
11. $3.14 \div 4 = 0.785$
12. $0.55 \div 0.76 = 0.723$

4-6 Decimals in Expressions and Equations

Use order of operations to simplify the expression to the nearest hundredth. Estimate the answer to check.

1. $8.36 - 3.2(5.5 - 5.2)$
2. $10.1^2 - 3.2 \cdot 1.5(3.34 + 4.66)$

Simplify each expression. Provide a reason for each step.

3. $2.2s - 2.1s$
4. $3.4g - 1.9g$
5. $2.2d + 2.1d$
6. $3.5n + 4.8n$
7. $1.23h - 3.12h$
8. $1.46k + 14.6k$
9. $t + (3.22 + 5.12t)$
10. $(w + 3.5w) + 0.5$

Solve each equation. Show your work. Provide a reason for each step.

11. $3.2n = 14.08$
12. $2.25x + 2.25 = 13.5$
13. $5.7y = 17.67$
14. $1.7p - 51.4 = 86.64$

Chapter 5 Extra Practice

5-1 Exponents

Find each product. Express using exponents.

1. $3^2 \cdot 3^2$

2. $5^4 \cdot 5$

3. $r^2 \cdot r^3$

4. $s^4 \cdot s^3$

5. $n \cdot n \cdot n$

6. $w^2 \cdot w^3 \cdot w^7$

7. $h \cdot h^2 \cdot h^3$

8. $k^2 \cdot k \cdot k^7$

9. $(p^2)^2$

10. $(g^5)^3$

11. $(m^2n)(m^2n)$

12. $(ab)(a^2b^2)$

13. $(5s^2r)(4s^3r^2)$

14. $(4a^3)(3a^2)$

15. $(2k^2)(-3k^2)$

16. $(-2d^2)(-2d^2)$

Find each quotient. Express using exponents.

17. $\dfrac{3^{10}}{3^5}$

18. $\dfrac{7^8}{7^7}$

19. $\dfrac{b^{14}}{b^2}$

20. $\dfrac{h^5}{h^3}$

21. $\dfrac{4a^5}{2a^3}$

22. $\dfrac{16d^{16}}{8d^8}$

23. $\dfrac{2g^6h^5}{3g^4h^4}$

24. $\dfrac{24x^7y^2}{6x^3y}$

Simplify.

25. $(3^2)^3$

26. $(4^3)^2$

27. $(h^4)^5$

28. $(3s^3)^3$

29. $(5r^{11})^7$

30. $(-3h^3g^3)^2$

31. $(10c^6d^4)^3$

32. $(-2r^2s^3)^2$

33. $(-2m^5n^3)^3$

34. $(4a^6b^2)^3$

35. $(5c^2d^6)^4$

36. $(-9r^5p^4)^2$

37. $(7g^3h)^2$

38. $(-2x^4y^6)^5$

39. $(-2k^6m^5)^4$

40. $(5c^3d^2)^2$

5-2 Integer Exponents

Rewrite using positive exponents. Then evaluate.

1. 3^{-3}

2. 2^{-3}

3. 7^{-2}

4. 5^{-1}

Rewrite using positive exponents. Assume that no denominators are equal to 0.

5. r^{-2}

6. d^{-12}

7. $a^{-3}b^4$

8. h^3n^{-3}

9. $9s^{-1}$

10. $7g^{-4}$

Find each product. Answers should have only positive exponents. Assume that no denominators are equal to 0.

11. $(a^3)(a^{-4})$

12. $(c^3)(c^{-3})$

13. $(-3x^{-2}y^{-2})(5x^3y^6)$

14. $(-2m^{-2}n^{-2})(3m^2n^{-2})$

Find each quotient. Answers should have only positive exponents. Assume that no denominators are equal to 0.

15. $\dfrac{s^3}{s^{-4}}$

16. $\dfrac{b^5}{b^9}$

17. $\dfrac{5p^4d^2}{15p^3d^6}$

18. $\dfrac{80f^2h^4}{4f^2h^6}$

Simplify. Answers should have only positive exponents. Assume that no denominators are equal to 0.

19. $(2x^8y^{-2})^2$

20. $3(m^3n^{-2})^3$

5-4 Roots

Find each value.

1. $\sqrt{25}$

2. $\sqrt{0}$

3. $\sqrt{1}$

4. $\sqrt{100}$

5. $-\sqrt{49}$

6. $-\sqrt{121}$

7. $-\sqrt{36}$

8. $\sqrt{529}$

9. $\sqrt{900}$

10. $\sqrt{10,000}$

11. $-\sqrt{729}$

12. $-\sqrt{2,025}$

Simplify.

13. $\sqrt{2^2 + 21}$

14. $\sqrt{4^2 + 8^2 + 1}$

15. $21 + 9 - \sqrt{17 - 8}$

16. $13 - 4 + \sqrt{16 + 20}$

5-5 Simplifying and Evaluating Expressions

GEOMETRY The formula for the area of an equilateral triangle is given by the formula $A = \frac{\sqrt{3}}{4} s^2$, where A = area and s = side length of the triangle.

1. What is the area of an equilateral triangle with side lengths of 3 centimeters?
2. What is the area of an equilateral triangle with side lengths of 4 inches?
3. What is the area of an equilateral triangle with side lengths of 15 meters?

Evaluate each expression for $n = 2$ and for $n = -2$.

4. $3n^3$
5. n^2
6. $4(n-1)^2$
7. $-2n^2$

Evaluate each expression for $p = 3$ and for $p = -3$.

8. $-2p^2$
9. $3p$
10. $3(2-p)^{-1}$
11. $4p^2$

Evaluate each expression for $a = 1$, $b = 4$, and $c = -3$.

12. $2a^2 - c$
13. $-bc - 3a^3$
14. $ab - bc$
15. $a^b - b^c$

5-6 Comparing and Ordering Rational Numbers

Replace each ⬤ with <, >, or = to make a true sentence.

1. $\frac{2}{3}$ ⬤ $\frac{3}{4}$
2. $-\frac{1}{5}$ ⬤ $-\frac{2}{9}$
3. 5^{-1} ⬤ $\frac{1}{5}$
4. $3\frac{1}{2}$ ⬤ $\sqrt{16}$
5. $-3\frac{1}{2}$ ⬤ $-\sqrt{16}$
6. 6^{-2} ⬤ $\frac{3}{18}$
7. -4 ⬤ $-\sqrt{9}$
8. -2^2 ⬤ -4.01
9. $\sqrt{361}$ ⬤ -7
10. 1^{25} ⬤ $\sqrt{25}$
11. 0^{16} ⬤ $-\sqrt{16}$
12. -4^3 ⬤ -63

Chapter 6 Extra Practice

6-1 Ratios and Rates

Write the ratio in the form requested.

1. 3 out of 8 as a fraction in simplest form

2. 10 hits in 25 at-bats as a decimal

3. 28 out of 49 in colon form

4. 9 out of 12 as a fraction in simplest form

5. 8 out of 16 in colon form

6. 132 out of 150 free throws as a decimal

Express as a unit rate.

7. 2 kilometers in 50 hours

8. 6 gallons drained in 36 minutes

9. 22 yards in 48 minutes

10. 6 miles ran in 46 minutes

11. 50 kilometers in 2 hours

12. 888 holes drilled in 4 minutes

13. 30 cookies baked in 15 minutes

14. 92 miles driven on 4 gallons of gas

6-2 Fractions, Decimals, and Percents

Write each percent as a fraction in simplest form. Then write it as a decimal.

1. 12%

2. 50%

3. 28%

4. 75%

5. 13%

6. 10%

7. 32%

8. 1%

9. 333%

10. 450%

11. 222%

12. 105%

Write each fraction as a decimal. Then write it as a percent.

13. $\frac{2}{5}$

14. $\frac{11}{20}$

15. $\frac{7}{10}$

16. $\frac{16}{25}$

17. $\frac{17}{10}$

18. $\frac{26}{25}$

19. $\frac{2}{3}$

20. $\frac{7}{9}$

6-3 Proportions and Proportional Reasoning

Determine whether each pair of ratios is equal. Use cross products. Show your work.

1. $\dfrac{2}{3}, \dfrac{30}{45}$

2. $\dfrac{3}{4}, \dfrac{17}{22}$

3. $\dfrac{5}{32}, \dfrac{20}{96}$

4. $\dfrac{3}{11}, \dfrac{12}{44}$

Solve each proportion. Round to the nearest tenth, if necessary.

5. $\dfrac{d}{2} = \dfrac{7}{14}$

6. $\dfrac{4}{7} = \dfrac{p}{28}$

7. $\dfrac{12}{26} = \dfrac{8}{13}$

8. $\dfrac{5}{17} = \dfrac{k}{34}$

9. $\dfrac{7}{4} = \dfrac{n}{24}$

10. $\dfrac{11}{22} = \dfrac{m}{6}$

11. $\dfrac{7}{35} = \dfrac{x}{5}$

12. $\dfrac{15}{75} = \dfrac{y}{25}$

13. $\dfrac{1.2}{8} = \dfrac{7}{4.4}$

14. $\dfrac{1.1}{h} = \dfrac{4}{2.3}$

15. $\dfrac{2.4}{5} = \dfrac{h}{2.5}$

16. $\dfrac{5}{6} = \dfrac{m}{1.2}$

17. $\dfrac{7.2}{5} = \dfrac{4.8}{x}$

18. $\dfrac{1.5}{8} = \dfrac{10}{5.8}$

19. $\dfrac{3.4}{4.6} = \dfrac{1.7}{n}$

20. $\dfrac{2.3}{5.2} = \dfrac{y}{6.4}$

6-4 The Percent Proportion

Express each fraction as a percent.

1. $\dfrac{3}{10}$

2. $\dfrac{3}{5}$

3. $\dfrac{12}{3}$

4. $\dfrac{13}{10}$

Use a proportion to solve.

5. What number is 13% of 200?

6. What number is 12% of 150?

7. 32 is 25% of what number?

8. 21 is 10% of what number?

9. 78 is 13% of what number?

10. 40 is what percent of 50?

11. 35 is what percent of 700?

12. 120 is what percent of 375?

6-5 Problems Involving Percents

1. **BANKING** Yolanda has $1,500 in her savings account. Yolanda's account earns 5% interest annually. Suppose she does not make any additional deposits. How much money will be in the account after 18 months?

2. **ART** Membership to a local art museum costs $75 per year. All members receive a 25% discount on any prints they purchase. Suppose prints cost $50 each. How many prints would a member have to buy for the total discount to equal the cost of membership?

3. **FOOD SERVICE** At a business lunch, Jakob spent $23.98. His company policy allows him to leave a 15% tip. How much should he leave?

4. **SALES** A car salesperson receives a 3% commission on all sales. Suppose she sells a car for $22,400. How much commission will she receive?

6-6 Direct Variation

Determine if the numbers in the table show direct variation. If they do, find the constant of variation.

1.

cost of meal	14	19	22
tips earned	2	5	4

2.

hours	1	3	10
minutes	60	180	600

3.

acres surveyed	2	6	13
number of trees counted	5	220	75

4.

items received	20	35	50
total cost	200	500	100

5.

hours traveled	4	6	9
number of miles traveled	200	300	450

6.

month of the year	2	5	7
savings account balance	240	600	840

Suppose y varies directly as x.

7. Let $y = 22$ when $x = 1$. Find x when $y = 242$.

8. Let $y = -30$ when $x = -1$. Find y when $x = -10$.

9. Let $y = 175$ when $x = 5$. Find x when $y = -140$.

10. Let $y = 65$ when $x = 13$. Find y when $x = 272$.

Chapter 7 Extra Practice

7-1 The Coordinate Plane

Find the ordered pair for the given point.

1. A
2. B
3. C
4. D

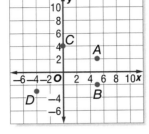

Graph each point on a coordinate plane.

5. $E\ (2, 2)$
6. $F\ (0, 6)$
7. $G\ (9, -1)$
8. $H\ (-6, -8)$
9. $J\ (0, -7)$
10. $K\ (-2, -3)$
11. $L\ (-5, 15)$
12. $M\ (-1, 5)$

7-3 Relationships Involving Equal Ratios

Describe the relationship shown in the scatter plot.

1.

2. **SPORTS** The scatter plot at the right shows the number of minutes played and the number of points scored by each member of the Los Angeles Lakers during the 2004–05 season. Describe the relationship.

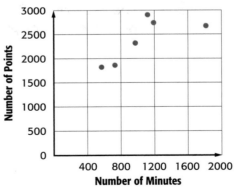

Relationship beween Points and Number of Minutes Played

Make a graph to find the answer.

3. **FITNESS** The table below compares the maximum heart rate of an athlete to his or her age. Write an equation for the maximum heart rate based on age.

Age	20	28	33	45	50
Maximum Heart Rate	200	192	187	175	170

4. **JEWELRY** Gem stones are weighed in units called *carats*. Precious metals are weighed using *troy weight*. The relationship between troy pounds and carats is shown in the table. How many carats are in one troy pound?

Troy Pounds	$\frac{3}{4}$	$\frac{1}{2}$	1.5	2	3	5
Carats	1,080	720	2,160	2,880	4,320	7,200

7-4 Measures as Rates and Products

Solve using rates.

1. **TIME** There are 7 days in 1 week. How many days are in 40 weeks?

2. **TRAVEL** Alexia's car gets 24 miles per gallon of gas. Suppose her gas tank holds 15 gallons. How far can she travel on a full tank of gas?

3. **MEASUREMENT** A gallon contains 4 quarts. A quart contains 32 ounces. How many ounces are in 3 gallons?

Solve using density as a rate.

4. **MEASUREMENT** A subdivision has an area of 12 square mile and a population of 7,500 people. What is its population density?

5. **ECOLOGY** A cave scientist estimated that a cave houses a population of about 120,000 bats. The scientist found the cave to have one room in which bats roost on the ceiling. Suppose the area of the ceiling is 400 square meters. What is the population density of the bats?

Solve using measures expressed as products.

6. **TRAVEL** A plane carrying 50 passengers flew 240 miles between airports. How many passenger-miles were traveled?

7. **MECHANICS** Work is calculated by multiplying force and distance. A force of 16 pounds was applied to move an object a distance of 5 feet. How many foot-pounds of work were done on the object?

8. **CONSTRUCTION** A crew of 8 people worked 7.5 hours a day for 4.25 days to complete a construction project. How many person-hours were worked?

7-5 Slope

Find the slope of each line.

1.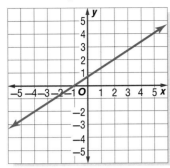

2.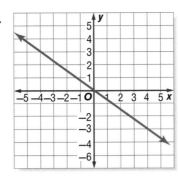

The points given in each table lie on a line. Find the slope of each line.

3.
x	−8	−4	0	4
y	−4	−2	0	2

4.
x	1	6	11	16
y	−4	−3	−2	−1

5.
x	0	2	4	8
y	4	0	−4	−16

6.
x	18	16	10	−2
y	−10	−12	−18	−30

7-6 Linear Functions

Use an input-output table to graph each linear function.

1. $y = -2x + 4$ 2. $y = 3x - 1$

3. $y = 2x - 3$ 4. $y = -3x + 1$

Determine whether each table shows a linear or nonlinear relationship.

5.
x	y
20	16
36	27
52	38
68	49
84	60

6.
x	y
2	4
3	2
3	4
5	8
7	6

7.
x	y
1	2
3	5
4	6
7	2
8	10

7-7 The Pythagorean Theorem

Use the given measures to verify the Pythagorean Theorem.

1.
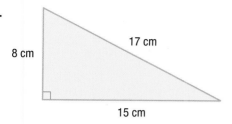
8 cm   17 cm   15 cm

2.
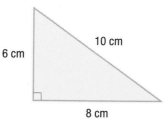
6 cm   10 cm   8 cm

Use the Pythagorean Theorem to find the length of each hypotenuse.

3.
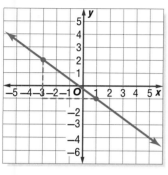

4.
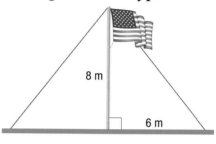
8 m   6 m

Use the Pythagorean Theorem to find each missing length.

5.
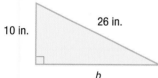
10 in.   26 in.   *b*

6.
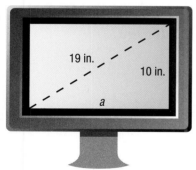
19 in.   10 in.   *a*

Use the converse of the Pythagorean Theorem to determine if each triangle is a right triangle.

7. The sides measure 7 inches, 24 inches, 25 inches.

8. The sides measure 9 cm, 14 cm, 15 cm.

Chapter 8 Extra Practice

8-1 Triangles and Quadrilaterals

Find the value of *x*.

1. In △ABC, m∠A = x°, m∠B = 30°, and m∠C = 60°.

2. In △EFG, m∠E = 29°, m∠F = 66°, and m∠G = x°.

3.

4.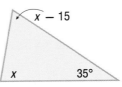

5.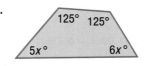

6.

7.

8.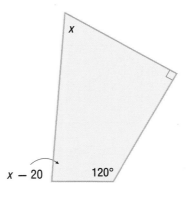

8-2 Congruency and Similarity

Name the congruent angles and sides.

1. △KLM ≅ △CDE

2. △ABC ≅ △IJK

Name the corresponding parts. Then complete the congruence statement.

3. 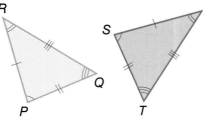   △TUV ≅ △ __?__

4. 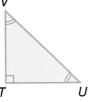   △PQR ≅ △ __?__

5. 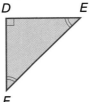   △STU ≅ △ __?__

8-3 Coordinate Geometry

Use the Distance Formula to find the distance between the given points.

1.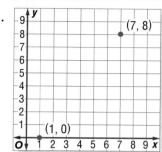

2.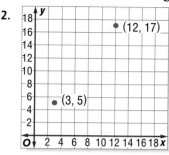

Use the Distance Formula.

3. What is the distance between the points (−5, 15) and (−13, 9)?

4. What is the distance between the points (−3, −4) and (5, 6)?

8-4 Perimeter

Find the perimeter of each polygon.

1.

2.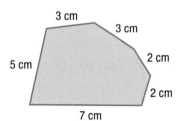

3.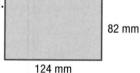

4.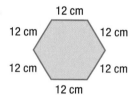

5. a triangle with sides 4 inches long, 5 inches long, and 6 inches long

6. a rhombus with each side 10 inches long

7. a square with each side 1.5 meters long

8. a parallelogram in which two sides are 13 meters long and the other two sides are 10 meters long

Find the circumference of each circle. Use π = 3.14.

9.

10. a circle with a diameter of 11 kilometers

8-5 Area

Find the area of each figure.

1.
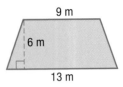
2 cm
3 cm
4 cm

2.
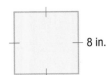
8 in.

3.

80 mm
20 mm

4.
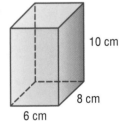
9 m
6 m
13 m

5.

4 cm

Use π = 3.14.

6.

7 cm

Use π = 3.14.

8-7 Solid Figures and Volume

Find the volume of each solid.

1.
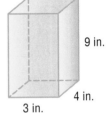
9 in.
4 in.
3 in.

2.
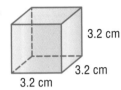
3.2 cm
3.2 cm
3.2 cm

3.

12 mm
10 mm
10 mm

4.
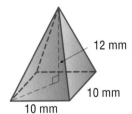
3 ft
5 ft

Use π = 3.14.

8-8 Surface Area

Find the surface area of each solid. Round your answer to the nearest whole number if necessary.

1.
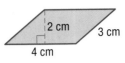
10 cm
8 cm
6 cm

2.
7 in.
7 in.
7 in.

3.
$3\frac{1}{4}$ in.
$1\frac{1}{2}$ in.
$2\frac{1}{2}$ in.

4.
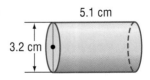
5.1 cm
3.2 cm

Use π = 3.14.

5.
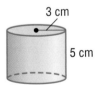
3 cm
5 cm

Use π = 3.14.

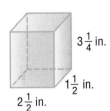
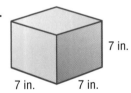

Mixed Problem Solving

Prerequisite Skills

ART A tile pattern on the wall at Hotel Atlantico has 3 green squares for every 2 blue squares.

1. What fraction of the squares are green? (Lesson P-7)

2. What fraction of the squares are blue? (Lesson P-7)

3. **SCHOOL** The lock on Rene's locker has a combination of 27-7-17. Put these numbers in order from least to greatest. (Lesson P-1)

4. **FASHION** Patel has 5 pairs of socks that are a solid color. He also has 6 pairs of socks with a pattern. He pulls out a pair of socks without looking. What is the probability that the socks will have a pattern? (Lesson P-12)

5. **GEOMETRY** Classify the angles in this figure as acute, obtuse, or right. (Lesson P-11)

6. **PROBABILITY** The chance of rain next weekend is 0.40. Is this a terminating or repeating decimal? Explain. (Lesson P-8)

7. **GARDENING** Mrs. Jenkins waters one plant every 6 days. She waters another plant every 10 days. Suppose she waters both plants today. How long will it be before she again waters both plants on the same day? (Lesson P-9)

BAKING Lara makes 264 cookies a day. She puts 12 cookies in each box.

8. How many boxes of cookies will she fill in one day? (Lesson P-4)

9. Suppose Lara bakes every day this week. How many cookies will she make? (Lesson P-4)

10. **FITNESS** Daria rides the stationary bike at her gym for 45 minutes. The bike shows that she burns about 12 calories per minute. Estimate the number of calories she will burn during one workout. (Lesson P-10)

BUSINESS An adventure company is planning a weekend trip. To make a profit, at least 55 people must sign up.

11. Currently, 29 people are signed up for the trip. How many more people are needed to make a profit? (Lesson P-3)

12. Suppose 38 more people sign up for the trip. How many will attend? (Lesson P-3)

HOBBIES Alvin is building a fort out of wooden cubes. All four walls contain the same number of cubes.

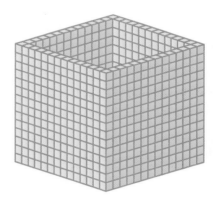

13. Using an exponent, how many cubes will Alvin need for one wall? (Lesson P-5)

14. Suppose the volume of a cube-shaped box is 8 ft³. Find its length, width, and height. (Lesson P-6)

15. **RETAIL** Paul buys marbles in bags of 1, 10, or 100. He just got 2 bags with 1 marble each, 8 bags with 10 marbles each, and 4 bags with 100 marbles each. In total, how many marbles did Paul buy? (Lesson P-2)

Mixed Problem Solving

Chapter 1

BUSINESS Gustaro has two employees who work for his construction company. Both are paid $14 per hour. Last week, Maurice worked 42 hours, and Amanda worked 48 hours.

1. Write two different expressions that represent the total wages earned by Maurice and Amanda last week. (Lesson 1-5)

2. What is the total amount of wages earned by Maurice and Amanda last week? (Lesson 1-5)

SPORTS In a basketball game, Elisha made five shots worth 3 points each and four shots worth 2 points each.

3. Write an expression that shows the total number of points Elisha scored. (Lesson 1-2)

4. Find the total number of points Elisha scored. (Lesson 1-3)

NATURE A forester counted the trees in each of the three equal-sized areas of a forest, as shown in the map.

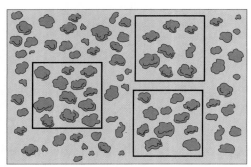

5. Write an expression that shows the total number of trees counted by the forester. (Lesson 1-2)

6. Use the Commutative Property of Addition to write a different expression. (Lesson 1-4)

7. **COOKING** The weight of a turkey determines its cooking time. The table below shows the recommended times.

If the weight of the turkey (w) is . . .	Cook it . . .
Less than or equal to 13 pounds	$12w + 22$ minutes
More than 13 pounds	$10w + 48$ minutes

Use the four-step plan to find how long you should cook an 18-pound turkey. Then find how long you should cook a 12-pound turkey. (Lesson 1-1)

TRAVEL Brenda's charge to rent a car was $0.20 per mile. Let m represent the number of miles that Brenda drove the rental car on Saturday. On Sunday, she drove the car three times as far.

8. Write an expression in simplest form to represent the total cost of the rental car for Saturday and Sunday. (Lesson 1-8)

9. Suppose Brenda drove 45 miles the first day. What was the total cost of the rental car for Saturday and Sunday? (Lesson 1-8)

10. **MANUFACTURING** Sit On It makes stools by adding 3 legs to a seat. Workers make chairs by adding 4 legs to a seat. They have 8 seats and 27 legs. How many stools and chairs should they make to use all of their supplies? (Lesson 1-6)

NUTRITION Morey bought a bag of frozen vegetables. The nutrition label says that the bag contains 16 servings and that each serving has 0 grams of trans fat.

11. Write an equation that represents this information. (Lesson 1-2)

12. Name the property that your equation represents. (Lesson 1-7)

Mixed Problem Solving

Chapter 2

1. **INDUSTRY** A booklet is made by stapling sheets of paper together and folding along the line of staples. Each sheet of paper becomes four different pages in the booklet.

   How many sheets are needed to make a 28-page booklet? (Lesson 2-5)

   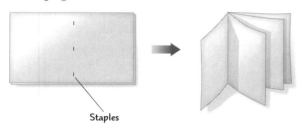

   Staples

2. **TEMPERATURE** At the South Pole, the temperature was 2°F at 8:00 A.M. By 9:00 A.M., the temperature had fallen to −4°F. By 10:00 A.M., the temperature had risen 11°F.

   On a number line, indicate the temperatures at 8:00 A.M., 9:00 A.M., and 10:00 A.M. (Lesson 2-2)

3. **AGES** Jana is twice as tall as her sister. Suppose Jana is 64 inches tall. How tall is her sister? (Lesson 2-1)

4. **SPORTS** The sum of the measures of the five angles in a softball home plate is 540°. What is the measure of each angle indicated in the figure below? (Lesson 2-7)

   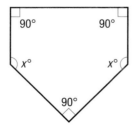

5. **GAMES** When playing a game called "Positively Negative," Phu had consecutive scores of −6, 10, and −15. What was his total score for these three turns? (Lesson 2-3)

6. **RETAIL** Alison and Jemi sold 6 shirts at a profit of $7 each. They also sold 3 pairs of shorts on sale at a loss of $2 each. They split the profits and losses.

   Write an expression that could be used to show how much money Alison should receive. Then simplify the expression to find the amount. (Lesson 2-4)

7. **ARCHITECTURE** An elevator goes down to the first floor. It took 8 seconds to go from floor to floor. It stopped once for 15 seconds to pick up passengers. The entire trip took 79 seconds. On what floor did the elevator start? (Lesson 2-7)

8. **SPORTS** In golf, 0 represents par. Leonardo has a miniature-golf score of 2 over par. Tina has a miniature-golf score of 8 under par. Write their scores as integers. Then, use < or > to compare the scores. (Lesson 2-2)

9. **NUMBER SENSE** A sequence begins 1, 4, 2, . . . The rule for the sequence is:

   • Multiply an odd number by 3 and add 1 to find the next number.

   • Divide an even number by 2 to find the next number.

   What are the first 15 numbers in this sequence? (Lesson 2-6)

Mixed Problem Solving

Chapter 3

1. **ARCHITECTURE** A construction worker places a piece of drywall on either side of a 2-inch × 4-inch stud. The diagram shows the width of these materials. What is the total width of the wall? (Lesson 3-4)

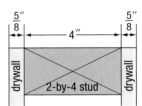

2. **COOKING** A recipe for waffles calls for $2\frac{3}{4}$ cups of flour. The recipe will serve four people. Nadine only wants to make waffles for 3 people. How much flour should she use? (Lesson 3-6)

3. **SCIENCE** The equation $\frac{9}{5}C + 32 = F$ shows how to change temperature in degrees Celsius (C) to degrees Fahrenheit (F). Suppose that the temperature outside is 50°F. What is the temperature in degrees Celsius? (Lesson 3-10)

4. **FITNESS** Coach Fisher measured the time it took two players to do 100 push-ups. The results are shown below. How much less time did it take Abraham to do 100 push-ups than it took Armando? (Lesson 3-9)

Time Taken to Do 100 Push-ups	
Player Name	**Time (minutes)**
Abraham	$6\frac{1}{2}$
Armando	$7\frac{1}{4}$

5. **SCIENCE** During the first hour of an experiment, the temperature of a liquid changes by $-\frac{3}{4}$°C. During the second hour, the temperature of a liquid changes by $-\frac{5}{8}$°C. Graph these fractions on a number line, and then compare them. (Lesson 3-1)

6. **TRACK AND FIELD** Martha and Wilma begin running at the starting line of a circular track. Martha runs one lap in the same time that it takes Wilma to run $\frac{6}{7}$ of a lap. How many laps will Wilma run before she and Martha meet again at the starting line? (Lesson 3-3)

7. **RECREATION** Billy caught a fish in Lake Hooladonga. The picture compares his fish with the longest fish caught in the lake.

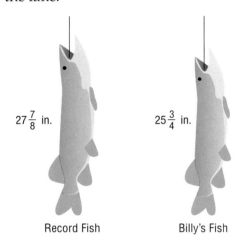

Record Fish   Billy's Fish

How many inches short of the record was Billy's fish? (Lesson 3-2)

8. **HORTICULTURE** Denzel inspected half of a citrus orchard. Patrick worked in the other half of the orchard, where he inspected $\frac{3}{4}$ acre. Combined, they inspected $1\frac{2}{3}$ acres. How many acres are in the orchard? (Lesson 3-8)

9. **BUSINESS** Meg wanted to sell $\frac{1}{3}$ of a company's stock. However, she was only able to sell $\frac{1}{2}$ of the amount she wanted to sell. How much of the company's stock did she sell? (Lesson 3-5)

10. **FOOD** When Dad baked a pie, he cut it into 9 pieces. Rudy ate 2 of them, and Misha ate 5 of them. Combined, how much of the pie did they eat? (Lesson 3-7)

Mixed Problem Solving

Chapter 4

1. **RETAIL** At the grocery store, Nara filled a bag with fresh string beans. The scale in the produce department listed the weight as $1\frac{3}{8}$ pounds.

   At the check-out counter, the weight was shown as a decimal. What decimal value did the display show? (Lesson 4-1)

2. **BANKING** The balance in Marcia's bank account was $1,447.88. She deposited her paycheck for $187.98. Then she withdrew $423.55 for her car payment.

   What was the balance of Marcia's account? (Lesson 4-2)

3. **RECREATION** The Longwells just built a sandbox in their backyard. The area of the sandbox can be found by multiplying length by width. The picture shows the dimensions of the sandbox.

   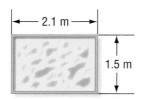

   What is the area of the sandbox in square meters? (Lesson 4-3)

4. **FITNESS** Hernando runs in his races of 6.2 miles. His fastest time was 41 minutes. At the same rate, how long will it take him to run a mile? Round your answer to the nearest tenth. (Lesson 4-4)

5. **TRANSPORTATION** A new hybrid sedan is able to travel 306.72 miles on 7.2 gallons of gas. In miles per gallon, what is the fuel efficiency of this car? (Lesson 4-4)

6. **NUMBER SENSE** A number is divided by 1.4. Then 7.1 is subtracted from this quotient. The result is 3.9. Find the original number. (Lesson 4-5)

RETAIL At the farmer's market, Marita bought two pounds of green beans and some tomatoes. The total cost of her purchases was $4.88.

Cost of Market's Vegetables	
Type of Vegetables	**Cost per Pound**
Peas	$0.99
Green Beans	$1.09
Tomatoes	$1.50
Onions	$1.19
Squash	$1.75

7. Write an equation that could be used to determine the number of pounds of tomatoes Marita bought. (Lesson 4-6)

8. How many pounds of tomatoes did she buy? (Lesson 4-6)

9. How much would it cost to buy 1.5 pounds of onions and 2.5 pounds of squash? (Lesson 4-3)

MAIL DELIVERY City law requires that there is one mailbox for every 2.4 blocks. In the neighborhood of Federal Hill, there are 216 mailboxes.

10. Write an equation to find the number of blocks b in which 216 mailboxes would be found. (Lesson 4-6)

11. Suppose the Federal Hill neighborhood consists of 86 blocks. Are there enough mailboxes to meet the requirement? Explain. (Lesson 4-6)

12. Suppose a neighborhood has 96 blocks. What is the least number of mailboxes that the neighborhood can have? (Lesson 4-6)

Mixed Problem Solving

Chapter 5

DESIGN An interior designer created a square mosaic with five squares on a side. Each square was further divided into a 5×5 arrangement of squares, as shown below.

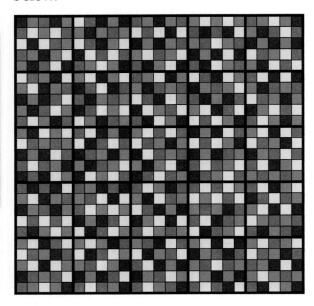

1. How many small squares are in this arrangement? (Lesson 5-1)

2. If the total number of small squares in this arrangement is expressed as 5^n, what is the value of n? (Lesson 5-1)

MEASUREMENT A millimeter is 10^{-3} meters. A kilometer is 10^3 meters.

3. How many millimeters are in 10 meters? (Lesson 5-2)

4. How many millimeters are in a kilometer? (Lesson 5-2)

5. **GEOMETRY** Line segments were used to divide a square into 64 smaller squares. How many line segments were used? (Lesson 5-3)

6. **DISTANCE** An estimate of the distance in miles that a person near the coast can see to the horizon is $\sqrt{1.5h}$. In this expression, h is the height in feet that the person is above sea level. Suppose a person is 54 feet above sea level. How far can this person see to the horizon? (Lesson 5-4)

BIOLOGY The number of bacteria in a petri dish doubles every hour.

7. Write an expression that includes exponents to find the number of bacteria in the dish after 8 hours if there is one bacterium in the dish to start. (Lesson 5-5)

8. Simplify the expression to find the number of bacteria after 8 hours. (Lesson 5-5)

URBAN PLANNING The diagrams show the dimensions of two circular fountains.

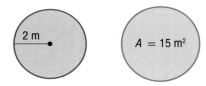

The formula for the area of a circle is $A = \pi r^2$, where A is the area and r is the radius. Let $\pi = 3.14$.

9. What is the area of the fountain that has a radius of 2 meters? (Lesson 5-5)

10. Which of the fountains is larger? (Lesson 5-6)

11. **FOOD SERVICE** In a serving of pasta, there are 2^2 meatballs. A restaurant sold 2^5 servings of pasta. How many meatballs did the restaurant serve? (Lesson 5-1)

Mixed Problem Solving

Chapter 6

1. **PACKAGING** Each box of CDs shipped from a warehouse contains 8 rock CDs and 7 country CDs. A local music store places an order for 300 CDs. How many will be rock CDs? (Lesson 6-7)

2. **FOOD SERVICE** After eating at a restaurant, Alex tells his father that the tip should be $6.63. Suppose the dinner bill is $45.98. Is Alex's amount more than, less than, or exactly equal to a 15% tip? Explain. (Lesson 6-5)

COMMUNITY Six of the 80 apartments in a new complex are being reserved for teachers in local schools.

3. Write a proportion that could be used to determine the percent of apartments that will be reserved for teachers. (Lesson 6-4)

4. Use your proportion to determine the percent. (Lesson 6-4)

INTERIOR DECORATING The table below shows the number of gallons of paint needed to cover surfaces of different sizes.

Number of Gallons	Area To Be Painted (square feet)
1	350
5	1750
10	3500

5. Explain why this situation represents a direct variation. (Lesson 6-6)

6. Determine how many square feet can be painted with 3 gallons. (Lesson 6-6)

FITNESS Mekala ran 4 miles in 32 minutes.

7. What is her unit rate, in minutes per mile? (Lesson 6-1)

8. Suppose Mekala runs at the same speed. How far could Marissa run in an hour? (Lesson 6-3)

FORESTRY Marvin is conducting a study in a forest. She has divided the forest into 20 sections. The map below shows the sections in which trees are dying from disease.

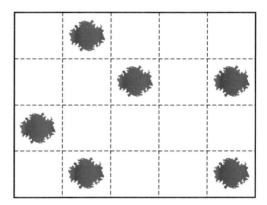

9. What fraction of the sections contains a tree dying from disease? Express your answer as a fraction in simplest form. (Lesson 6-2)

10. Write the number of sections with trees dying from disease as a percent and as a decimal. (Lesson 6-2)

11. **RETAIL** Both Hurrah's Closet and Mary's Boutique sell the same jeans for $50. At Hurrah's, there is a 20% discount on the jeans. Mary's had been giving a 10% discount on the jeans. Then, Mary's offered an additional 10% discount on all jeans. After the discounts are applied, which store has the better price on these jeans? Explain. (Lesson 6-5)

Mixed Problem Solving

Chapter 7

MEASUREMENT Originally, a *cubit* was the distance from the thumb to the elbow. Because arms of different people vary in length, cubits are not standard units. Today, a *natural cubit* has a standard length.

The chart below shows the length in inches for various lengths in natural cubits.

Natural Cubits	Inches	(x, y)
2	36	(2, 36)
3.5	63	(3.5, 63)
$\frac{5}{6}$	15	$\left(\frac{5}{6}, 15\right)$
4	72	(4, 72)

1. Graph the points on a coordinate plane. (Lesson 7-1)

2. Describe the relationship shown on your graph. (Lesson 7-3)

3. How many inches are in a natural cubit? (Lesson 7-3)

4. **SPORTS** The infield for softball, shown below, is a square.

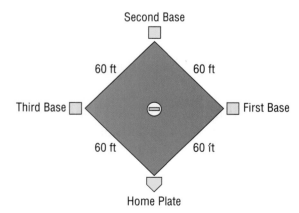

A player throws a ball from third base to first base. How far does the ball travel? (Lesson 7-7)

GRAPHING A line passes through the points (0, −7), (4, 5), and (8, 17).

5. Graph the line. (Lesson 7-6)

6. Find the slope of the line. (Lesson 7-5)

BUSINESS Rolando is getting shirts printed for his employees. The cost is $55 for set-up fees plus $8 for each shirt. He uses this graph to find the cost of different numbers of shirts.

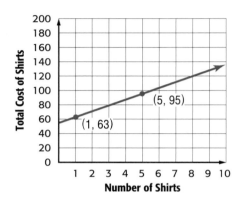

7. Find the slope of this line. (Lesson 7-5)

8. How much will 35 shirts cost? (Lesson 7-3)

MARKETING The sales team at XYZ Jackets conducted a survey. They wanted to know how many jackets they would sell at various prices. The table shows the results of the survey.

Results of Jacket Pricing Survey	
Price per Jacket	Number of Jackets Sold
$20	4,800
$30	4,200
$50	3,000

9. Graph the data in the table above. (Lesson 7-1)

10. Use your graph to predict the number of jackets they will sell at $40. (Lesson 7-2)

11. Use your graph to predict the price at which they sell no jackets. (Lesson 7-2)

12. **CURRENCY** In Rome, a currency exchange used to give you 1,500 Italian lira for every $1. Suppose Bobbie received 31,500 lira. How many dollars did Bobbie exchange? (Lesson 7-4)

Mixed Problem Solving

Chapter 8

1. **CONSTRUCTION** Ramiro and Salvador build houses. Both of them use a "speed square." A speed square is a metal triangle that can be used to check for right angles. The two speed squares are similar triangles. What is the length of the missing side? (Lesson 8-2)

Ramiro's Speed Square   Salvador's Speed Square

TRAFFIC Stop signs, construction signs, and yield signs are common geometric shapes.

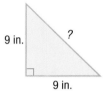

2. By how many degrees is the sum of all angle measures of a dead end sign greater than the sum of all angle measures of a yield sign? (Lesson 8-1)

SCHOOL Ms. Brooks decorates her bulletin board each month. She covers the 3-foot-by-12-foot board with construction paper. She outlines the board with a border.

3. What is the total area of the bulletin board? (Lesson 8-6)

4. How many feet of border will she need each month? (Lesson 8-4)

5. **FOOD SERVICE** Elan has two containers for storing popcorn. (Lesson 8-7)

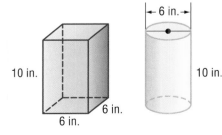

How much more will the rectangular prism hold? Round to the nearest tenth.

6. **INTERIOR DESIGN** Suzanne is painting the walls and door of a room with no windows. The room is 14 feet long, 9 feet high, and 12 feet wide. The table shows her paint choices.

Buckets of Paint	Area of Coverage	Price
Super Bucket	1000 ft²	$55.99
Monster Bucket	700 ft²	$41.99
Regular Bucket	400 ft²	$24.99
Mini Bucket	250 ft²	$17.99

Which bucket(s) of paint should she buy? Explain. (Lesson 8-8)

7. **RETAIL** Delmar bought half of the cards at a trading-card store. Next, Millie bought one-third of the cards that were left. Then Marilyn bought half of the remaining cards.

   If only three cards remained at the store, how many cards did Delmar buy? Using the Make-a-Model problem-solving method. Show your work. (Lesson 8-6)

8. **MEASUREMENT** To find the height of a tree, Rekha placed a mirror on the ground. She moved until she could look at the mirror and see the top of the tree in the reflection. The picture shows the measurements she made.

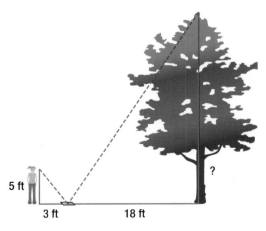

Using similar triangles, determine the height of the tree. (Lesson 8-3)

PART 1 Multiple Choice

Choose the *best* answer.

1 When 3 is added to the product of a number and 2, the result is 14. Which equation represents this situation?

   A $3 + 2x = 14$

   B $3(2 + x) = 14$

   C $3x + 2 = 14$

   D $3 + x + 2 = 14$

2 What is the value of $2^2 \cdot 3 + 20 \div (4 + 1)$?

   F 9   **H** 18

   G 16   **J** 28

3 Simplify: $3x + 2(x + 4)$.

   A $4x + 8$ **C** $5x + 8$

   B $5x + 4$ **D** $11x$

4 Which of the following is *not* a true mathematical statement?

   F $2(x + 3) = 2x + 6$

   G $2(3 + x) = 2x + 6$

   H $(x + 6) + x = 2x + 6$

   J $2 \cdot x + 3 = 2x + 6$

5 Which expression does *not* simplify to $3x + 21$?

   A $(2x + 21) + (x + 21)$

   B $3(x + 7)$

   C $3 \cdot x + 3 \cdot 7$

   D $x + x + x + 7 + 7 + 7$

6 Which property does this picture illustrate?

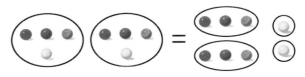

   F Associative Property of Addition

   G Associative Property of Multiplication

   H Identity Property of Multiplication

   J Distributive Property

7 J'ana bought two packs of trading cards. Each pack had the same number of cards. She gave 11 of the cards to Rey. But he already had 14 trading cards.

Now, Rey and J'ana have the same number of trading cards. Which equation represents this situation?

   A $2x - 11 = 14$

   B $2x - 11 = 25$

   C $2x + 11 = 14$

   D $2x + 11 = 25$

8 Which property states that $153 + 0 = 153$?

   F Identity Property of Multiplication

   G Multiplicative Property of Zero

   H Associative Property of Addition

   J Identity Property of Addition

Record your answers on the answer sheet provided by your teacher or on a separate sheet of paper.

9 Mallory volunteers at a local animal shelter. The table shows how many hours she volunteered during an eight-week period. How many total hours did Mallory volunteer during the eight weeks?

Mallory's Volunteer Hours			
Week	**Hours**	**Week**	**Hours**
March 3–9	15	March 31–April 6	7
March 10–16	8	April 7–13	13
March 17–23	12	April 14–20	14
March 24–30	5	April 21–27	16

10 What is the value of this expression?
$$\frac{9 \div 3 - 2}{4 + 2 \cdot 6 - 5}$$

11 Find the missing number.
$$3 + 5(\underline{\quad?\quad} - 4) = 38$$

12 Which property is illustrated by
$$6 + 7 = 7 + 6?$$

13 Allen is driving to his friend's house, which is 150 miles away. The speedometer shows his car's speed.

At this rate, how long will it take Allen to get to his friend's house?

14 Sasha is making the program for a school play. Each program will be made from four sheets of paper. A program will be given to every person who attends, and 1,125 tickets have been sold.

The paper that Sasha will use is sold in reams of 500 sheets. How many reams of paper does Sasha need to buy?

Record your answers on the answer sheet provided by your teacher or on a separate sheet of paper.

15 A telemarketer earns $10 per hour plus $2 for each magazine subscription that he sells.

  a. Write an expression that shows the amount he would earn in one hour.

  b. Write an expression that shows the amount he would earn in x hours.

  c. Find the amount he would earn if he sold eight subscriptions in an hour.

  d. Find the number of subscriptions he must have sold if he earned $34 in one hour.

  e. Find the number of subscriptions he must have sold if he earned $108 during a four-hour shift.

Standards Practice

NEED EXTRA HELP?															
If You Missed Question...	1	2	3	4	5	6	7	8	9	10	11	12	13	14	15
Go to Lesson...	1-2	1-3	1-3	1-8	1-3	1-5	1-2	1-7	1-1	1-3	1-5	1-4	1-1	1-1	1-2
For Help with Algebra Readiness Standard...	7AF1.1	7NS1.2 7AF2.1	ALG4.0	7AF1.3	ALG4.0	5AF1.3	7AF1.1	7AF1.3	MR2.0	7NS1.2	7AF1.3	7AF1.3	MR2.0	MR1.0	7AF1.1

PART 1   Multiple Choice

Choose the *best* answer.

1 What is the value of this expression?

$$\frac{30 + 2 \cdot 5}{2(3 - 1)}$$

A 8   **C** 32

B 10   **D** 40

2 Which expression simplifies to $7x + 14$?

F $2(x + 7)$

G $2(7x + 7)$

H $7(x + 2)$

J $7(2x + 1)$

3 Which property allows us to say that $153 \cdot 1 = 153$?

A Associative Property of Addition

B Associative Property of Multiplication

C Identity Property of Addition

D Identity Property of Multiplication

4 Which integer is less than 1 but greater than -5?

F -6 **H** 2

G 0   **J** 6

5 Solve for n: $4n + 8 = 36$.

A 3   **C** 9

B 7   **D** 28

6 Which expression does *not* have the same value as 2^4?

F $(-2)^4$

G $(-4)^2$

H $(-2)^3(2)$

J $(2)(-2)(2)(-2)$

7 The table below compares the length of a square's sides with its perimeter.

Length of a Square's Sides (inches)	Perimeter (inches)
2	8
3	12
4	16
5	20

If the pattern continues, what is the perimeter of a square with sides that are 96 inches long?

A 23   **C** 25

B 24   **D** 384

8 Which set of steps can be used to solve the equation $3f - 4 = 41$?

F Add 4 to both sides of the equation; then, divide both sides by 3.

G Add 4 to both sides of the equation; then, multiply both sides by 3.

H Subtract 4 from both sides of the equation; then, multiply both sides by 3.

J Subtract 4 from both sides of the equation; then, divide both sides by 3.

PART 2   Short Answer

Record your answers on the answer sheet provided by your teacher or on a separate sheet of paper.

9 What is the value of the following expression?

$$12 + 5(2 + 4) \div 10$$

Standards Practice

10 The week following Hurricane Katrina, a group of volunteers wanted to help with the relief effort. They drove at an average rate of 60 miles per hour.

Approximately how many hours did it take to make the trip?

11 When running long distances, Andrea holds a pace of s meters per minute. Ichiro holds a pace that is 15 meters per minute faster than Andrea's. Write an expression to show how many meters Ichiro will travel in 8 minutes.

12 The cost for using a calling card is 25 cents for the first minute and 15 cents for each additional minute.

  a. Suppose that Donald used this card for an 11-minute phone call. How much would he have to pay?

  b. Using this card, Donald paid $1.30 for a call he made. In minutes, how long was Donald's call?

13 For what value of h is the following equation true?
$$\frac{h}{3} = 12$$

14 What is the value of the following expression?
$$-3 + 3(-4 + 2) + \frac{-12}{-2}$$

Record your answers on the answer sheet provided by your teacher or on a separate sheet of paper.

15 Seventeen toothpicks were used to create the 3 · 2 arrangement of squares below.

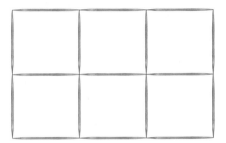

  a. How many more toothpicks would be needed to build a 3 × 4 arrangement of squares?

  b. How many toothpicks total would be needed to build a 5 × 7 arrangement of squares?

  c. In general, how can you determine the number of toothpicks needed to build an arrangement of squares?

Standards Practice

NEED EXTRA HELP?															
If You Missed Question...	1	2	3	4	5	6	7	8	9	10	11	12	13	14	15
Go to Lesson...	1-3	1-8	1-7	2-2	2-7	2-4	2-6	2-7	1-5	1-1	1-2	2-7	2-1	2-3 2-4 2-5	2-6
For Help with Algebra Readiness Standard...	7NS1.2	7AF1.3	7AF1.3	ALG2.0	7AF4.1	7NS1.2 7AF1.3	7MR1.0 MR2.0	7AF4.1	7AF1.3	MR1.0	7AF1.1	ALG5.0 7AF4.1	6AF1.1	7NS1.2 7AF1.3	7MR1.1

PART 1 Multiple Choice

Choose the *best* answer.

1 After 7 is added to a number, the entire result is divided by 4. Which expression represents this situation?

   A $x + 7 \div 4$ **C** $\dfrac{x}{4 + 7}$

   B $\dfrac{x + 7}{4}$ **D** $\dfrac{4}{x + 7}$

2 Which property is illustrated by the following picture?

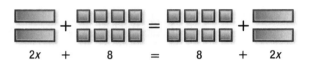

   F Associative Property of Addition

   G Associative Property of Multiplication

   H Commutative Property of Multiplication

   J Commutative Property of Addition

3 There are 3 feet in a yard. Which equation could be used to determine the number of feet represented by 7 yards?

   A $\dfrac{f}{3} = 7$ **C** $3f = 7$

   B $f + 3 = 7$ **D** $7f = 3$

4 The temperature at 8:00 A.M. was 2°F. If the temperature drops 3° every hour, what will the temperature be at 11:00 A.M.?

   F −11°F   **H** 1°F

   G −7°F   **J** 11°F

5 On a cold winter day, weather services reported the following temperatures:

   • Bishop—low −8°F; high 26°F

   • Mammoth Lakes—low −9°F; high 22°F

   • Merced—low 15°F; high 36°F

Which of the following statements is *true*?

   A Merced recorded the greatest change in temperature.

   B Mammoth Lakes recorded the greatest change in temperature.

   C The temperature changed more in Bishop than it did in Merced.

   D The temperature changed less in Bishop than it did in Mammoth Lakes.

6 What is the process for dividing one fraction by another fraction?

   F Multiply the first fraction by the second fraction.

   G Multiply the reciprocal of the first fraction by the second fraction.

   H Multiply the reciprocal of the first fraction by the reciprocal of the second fraction.

   J Multiply the first fraction by the reciprocal of the second fraction.

7 Solve: $\dfrac{3}{7}x + \dfrac{1}{2} = \dfrac{13}{14}$.

   A $\dfrac{9}{49}$ **C** 1

   B $\dfrac{3}{7}$ **D** $\dfrac{49}{9}$

8 What is $\dfrac{13}{4}$ written as a mixed number?

   F $\dfrac{4}{13}$ **H** $3\dfrac{1}{4}$

   G $1\dfrac{3}{4}$ **J** $3\dfrac{3}{4}$

Record your answers on the answer sheet provided by your teacher or on a separate sheet of paper.

9 Federico has quarters, dimes, and pennies in his pocket. What is the fewest number of coins he could use to total $1.05?

10 Order the expressions from least to greatest.

$$3(-4) \quad (-2)(-6) \quad 5(2) \quad -3(5)$$

11 List the prime whole numbers from 10 to 20.

12 When Wendi left home, her gas tank was $\frac{7}{8}$ full. The picture shows the level in her gas tank at the end of her trip.

What portion of the gas in her tank did she use during the trip?

13 What is the least common denominator of $\frac{2}{3}$ and $\frac{4}{5}$?

14 Three-fourths of a cup of sugar is needed to make one batch of cookies. How many cups of sugar are needed to make eight batches?

PART 3 Extended Response

Record your answers on the answer sheet provided by your teacher or on a separate sheet of paper.

15 Coach Larson schedules practice for two hours. Half of the time is spent on offense, $\frac{1}{3}$ is spent on defense, and the remaining time is used for conditioning.

a. How much time at practice is spent on defense?

b. What fraction of practice is spent on offense and defense combined?

c. How much time is spent on conditioning?

Standards Practice

NEED EXTRA HELP?															
If You Missed Question...	1	2	3	4	5	6	7	8	9	10	11	12	13	14	15
Go to Lesson...	1-2	1-4	2-1	2-2	2-3	3-6	3-10	3-2	1-1	2-4	3-3	3-3	3-9	3-5	3-8 3-9
For Help with Algebra Readiness Standard...	7AF1.1	7AF1.3	6AF1.1	7NS1.2 7AF1.3	ALG2.0 7NS1.1	6NS2.2	ALG5.0 7AF1.2	6NS1.1	MR1.0 MR2.0	7NS1.2 7AF1.3	5NS1.4	7NS1.2 6NS2.1	5NS1.4	7NS1.2 6NS2.1	7NS1.2 6NS2.1

PART 1 Multiple Choice

Choose the *best* answer.

1 The students in Ms. Kilmeyer's class were asked how many pets they have. The bar graph below shows the results.

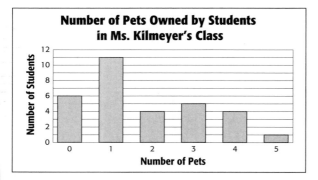

Number of Pets Owned by Students in Ms. Kilmeyer's Class

Which expression could *not* be used to find the total number of pets owned by all students in the class?

A $6 + 11 + 2(4) + 5 + 1$

B $0 + 11 + 8 + 15 + 16 + 5$

C $(6 \cdot 0) + (11 \cdot 1) + 4(2 + 4) + (5 \cdot 3) + (1 \cdot 5)$

D $(6 \cdot 0) + (11 \cdot 1) + (4 \cdot 2) + (5 \cdot 3) + (4 \cdot 4) + (1 \cdot 5)$

2 One-hundred ninth graders at Founder High School are enrolled in either French or Geometry. There are 68 students enrolled in French and 52 enrolled in Geometry. How many are enrolled in both?

F 16   **H** 32

G 20   **J** 48

3 For what value of x is this equation true?

$$\frac{x}{3} + 6 = 21$$

A 5   **C** 45

B 15   **D** 63

4 Which is a simplified form of $\frac{-39}{-26}$?

F $-\frac{3}{2}$ **H** $\frac{2}{3}$

G $-\frac{2}{3}$ **J** $\frac{3}{2}$

5 An $8\frac{1}{4}$-acre piece of land is divided into $1\frac{3}{8}$-acre plots. How many $1\frac{3}{8}$-acre plots are there?

A 6   **C** 13

B 12   **D** 33

6 The graph compares the probability of getting all heads with the number of coins flipped.

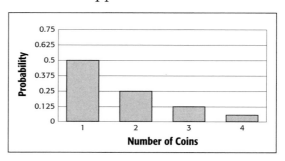

Which bar shows a value equivalent to $\frac{1}{8}$?

F 1   **H** 3

G 2   **J** 4

7 What is $13.1 \cdot 8.1$?

A 11.79   **C** 117.9

B 106.11   **D** 1061.1

8 A jet traveled 2,470 miles in 6.75 hours. What was its average speed in miles per hour, rounded to the nearest tenth?

F 36.59   **H** 366.0

G 365.9   **J** 370.0

Record your answers on the answer sheet provided by your teacher or on a separate sheet of paper.

9 The manager at Spokes Bike Shop took an inventory of the parts in his warehouse. The inventory is shown below.

Bike Shop Inventory					
Items	**Tires**	**Seats**	**Handlebars**	**Pedals**	**Frames**
Number of Items per Case	24	16	16	22	16
Number of Cases	4	1	8	11	9

Use the Distributive Property to write an expression that summarizes the information in the table.

10 When Tanika began a trip, her odometer read 11,030.4 miles. When she reached her grandmother's house, her odometer read 11,124.9 miles. How many miles did Tanika travel?

11 Eva brought a bag of carrot sticks to school. She ate half of them during the morning break. She ate a third of what was left at lunch. And, she ate the last four during the afternoon break. How many carrot sticks did she bring to school?

12 What is the sum of $\frac{1}{3} + \frac{5}{6}$?

13 Solve for x: $\frac{1}{3}x + \frac{1}{4} = \frac{7}{8}$.

14 As part of a camp project, Ben and his friends are shaping wet adobe into bricks. The volume of adobe in each brick is found by multiplying its length, width, and height. Each brick is to be 18 centimeters long, 6.2 centimeters wide, and 1.5 centimeters high. How much adobe will be needed to make each brick in cubic centimeters?

Record your answers on the answer sheet provided by your teacher or on a separate sheet of paper.

15 At the Snack Shack, a cheese pizza costs $8.99. A cheese pizza with four additional toppings costs $14.39. Suppose each topping costs the same amount.

   a. Write an equation that can be used to find the cost of each additional topping.

   b. Solve your equation from part (a) to find the cost per topping.

   c. What is the cost of a pizza with two toppings?

Standards Practice

NEED EXTRA HELP?															
If You Missed Question...	1	2	3	4	5	6	7	8	9	10	11	12	13	14	15
Go to Lesson...	1-2	1-1	2-7	2-5	3-6	4-1	4-3	4-4	1-4	4-5	4-5	3-3 3-8	3-10	4-3	4-6
For Help with Algebra Readiness Standard...	7AF1.1	MR2.0	ALG5.0 7AF4.1	7NS1.2 7AF1.3	7NS1.2 6NS2.1	7NS1.3	7NS1.2	7NS1.2 7NS1.3	7AF1.3	7NS1.2 7NS1.3	MR2.5	5NS1.4 6NS2.1 7NS1.2	7AF4.1 6NS2.1 7NS1.2	7NS1.2 7NS1.3	ALG5.0 7AF4.1

Standards Practice

PART 1 Multiple Choice

Choose the *best* answer.

1 The table below shows the number of skateboards that the Board Store has in each of its warehouses.

Skateboard Inventory	
Warehouse	**Number of Skateboards**
La Plata	87
Springfield	115
San Jose	34
Wheaton	241

Which is the *best* estimate for the number of skateboards in all of their warehouses combined?

A 140

B 300

C 500

D 1,400

2 On a number line, how many units apart are -8 and 4?

F 2

G 4

H 12

J 32

3 What value of n makes the equation below true?

$$3n + 15 = 24$$

A 3

B 9

C 13

D 39

4 Which of the following fractions is in simplest form?

F $\frac{3}{1}$

G $\frac{2}{4}$

H $\frac{3}{6}$

J $\frac{2}{9}$

5 Which is the *best* estimate of $49.1 \cdot 8.14$?

A 320

B 400

C 450

D 500

6 Which of the following is not equivalent to 5^5?

F $5 \cdot 5 \cdot 5 \cdot 5 \cdot 5$

G $5^0 \cdot 5^5$

H $5^1 \cdot 5^5$

J $5^2 \cdot 5^3$

7 Simplify:

$$\frac{32g^3h^4}{24g^2h^6}$$

A $\frac{4g}{3h^2}$

B $\frac{4}{3gh^2}$

C $\frac{4gh^2}{3}$

D $\frac{16g}{12h^2}$

8 If $s = -2$, what is the value of $2s^{-4}$?

F -16

G $-\frac{1}{8}$

H $\frac{1}{8}$

J 16

Record your answers on the answer sheet provided by your teacher or on a separate sheet of paper.

9 Of all voters in a certain district, one-half are registered as Democrats. In a recent election, $\frac{3}{4}$ of the registered Democratic voters voted for the Democratic candidate. What fraction of the population do these voters represent?

10 Simplify $16.8x - 12.45x$.

11 Eliza flips a coin 5 times in a row. How many different patterns of heads and tails could she get?

12 Order the following values from least to greatest.

$$8^{-1}, \sqrt{0}, -\frac{1}{4}, -\sqrt{9}$$

13 The formula for the radius of a ball is $r = \sqrt{\frac{SA}{12}}$, where SA is the surface area of the ball. What is the radius of a beach ball with a surface area of 588 square inches?

Record your answers on the answer sheet provided by your teacher or on a separate sheet of paper.

14 A developer is drawing plans for a parking garage. Each level will be a square. City code requires that a parking garage allow at least 16 square yards for each parked car.

Number of cars	9	16	25	36
Number of square yards	144	256	400	576

a. Suppose each level must hold a minimum of 25 cars. How long should each side of the garage be?

b. Suppose each level is expected to hold at least 144 cars. Write an expression that could be used to determine how long each side of the parking garage must be.

c. Simplify your expression to find the length of each side of the parking garage.

Standards Practice

NEED EXTRA HELP?														
If You Missed Question...	1	2	3	4	5	6	7	8	9	10	11	12	13	14
Go to Lesson...	1-1	2-3	2-7	3-3	4-3	5-1	5-1	5-2	3-5	4-2	5-3	5-6	5-5	5-4
For Help with Algebra Readiness Standard...	MR2.0	7NS1.2	7AF4.1 ALG5.0	5NS1.4	7NS1.2	7AF2.1	7AF2.1	7AF2.1	7NS1.2	7NS1.2	7MR2.2	6NS1.1	ALG2.0	ALG2.0

PART 1 Multiple Choice

Choose the *best* answer.

1 At his part-time job, Romeo earns $120 per month plus $2 for every t-shirt he sells. His total earnings last month were $488.

Which equation represents this situation?

A $120 + 2n = 488$

B $120n + 2 = 488$

C $122n = 488$

D $488 + 2n = 120$

2 What is the sum when 16 is added to its additive inverse?

F 0

G 16

H 17

J 32

3 Twenty-four freshman are divided into groups. Forty sophomores are also divided into groups. Every group has the same number of people.

What is the greatest number of students that could be in each group?

A 2

B 4

C 8

D 20

4 An online store sells campaign buttons. The price per button depends on the quantity ordered.

Button Pricing	
Number of buttons purchased	Cost per button
1–10	$1.35
11–30	$1.25
31 or more	$1.10

How much more does it cost to buy 30 buttons than to buy 33 buttons?

F $1.20

G $3.00

H $3.40

J $8.25

5 How is $4 \cdot 4 \cdot 4 \cdot 4 \cdot 4$ expressed in using exponents?

A $5 \cdot 4$ **C** 5^4

B 4^4 **D** 4^5

6 Suppose 1 pound of bananas costs $0.59 and 3 pounds cost $1.77. How much will 2 pounds cost?

F $0.59   **H** $2.36

G $1.18   **J** $3.00

7 What is $\frac{3}{5}$ expressed as a percent?

A 30%   **C** 60%

B 35%   **D** 167%

8 Nasir buys a $350 stereo system at a 25% discount. How much will he save?

F $8.75   **H** $87.50

G $26.25   **J** $262.50

Standards Practice

PART 2 Short Answer

Record your answers on the answer sheet provided by your teacher or on a separate sheet of paper.

9 For what value of x is
$3.3x - 1.26 = 4.68$?

10 What is the value of the expression below if $a = 4$ and $b = -2$?

$$\frac{-2b + \sqrt{a}}{a^b}$$

11 What is a 15% tip on a dinner bill of $48.63?

12 Are the fractions $\frac{15}{20}$ and $\frac{42}{56}$ equivalent? Explain.

13 A rectangular pool is shown below. To find the area of the pool, multiply the length and the width. What is the area of the pool?

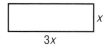

x

$3x$

PART 3 Extended Response

Record your answers on the answer sheet provided by your teacher or on a separate sheet of paper.

14 To raise money, the chess club sells fruit drinks at the football games. The drink's recipe uses 3 cups of cranberry juice for every 5 cups of ginger ale.

   a. Write the ratio of cranberry juice to ginger ale as a fraction in simplest form.

   b. What percent of the fruit drink is cranberry juice? What percent is ginger ale?

   c. Suppose they expect to sell 320 cups of fruit drink at the next game. How many cups of ginger ale should they buy?

Standards Practice

NEED EXTRA HELP?														
If You Missed Question...	1	2	3	4	5	6	7	8	9	10	11	12	13	**14**
Go to Lesson...	1-2	2-2	3-3	4-3	5-1	6-6	6-2	6-5	4-6	5-4	6-5	6-3	6-7	6-1 6-3
For Help with Algebra Readiness Standard...	7AF1.1	ALG2.0	5NS1.4	7NS1.2 7AF1.3	7NS2.1	7AF4.2 ALG5.0	7N.S1.3 ALG5.0	6NS1.4	ALG5.0 7AF4.1	ALG2.0	6NS1.4	ALG5.0	7MR2.5 6NS1.2	7NS1.3 ALG5.0

PART 1 Multiple Choice

Choose the *best* answer.

1 A theater sold 310 regular seats for $15. A small number of special tickets were sold for $25. The theater collected a total of $5,275 from the sale of tickets. Let t be the number of special tickets.

Which equation represents this situation?

A $15t + 5,275 = 7,750$

B $25t + 4,650 = 5,275$

C $25t + 5,275 = 4,650$

D $310t + 40 = 5,275$

2 Solve for h.

$$\frac{-h}{15} = 3$$

F -45 **H** 5

G -5 **J** 45

3 How many cups of sugar are needed to make this recipe?

Coated Cookies	
For Cookies	$2\frac{1}{2}$ cups all purpose flour
	1 teaspoon baking powder
	$\frac{1}{2}$ teaspoon ground nutmeg
	$\frac{1}{2}$ teaspoon salt
	$\frac{1}{2}$ cup butter
	$1\frac{2}{3}$ cups sugar
	1 egg
For Topping	$2\frac{1}{4}$ cups sugar
	3 tablespoon milk
	1 tablespoon butter

A $1\frac{2}{3}$ **C** $3\frac{3}{7}$

B $2\frac{1}{4}$ **D** $3\frac{11}{12}$

4 What is the difference of $14s$ and $3.2s$?

F $10.8s$ **H** $11.8s$

G $11.2s$ **J** $18s$

5 Which is the *best* estimate of $1.9d \cdot 39.1$?

A $39d$ **C** $80d$

B $40d$ **D** $800d$

6 Which expression is equivalent to $n^4 \cdot n^2$?

F n^2 **H** n^8

G n^6 **J** n^{42}

7 The formula for the area of a circle is $A = \pi r^2$. To the nearest tenth, what is the area of a circle when $r = 3$? Use $\pi = 3.1$.

A 9.3 **C** 27.9

B 18.6 **D** 28.8

8 A jar contains 150 marbles, and 33 are red. Which proportion could be used to find the percent of marbles that are red?

F $\frac{x}{100} = \frac{33}{150}$

G $\frac{33}{x} = \frac{100}{150}$

H $\frac{33}{100} = \frac{x}{150}$

J $\frac{150}{33} = \frac{x}{100}$

9 In simplest form, $55\% =$

A $\frac{11}{100}$

B $\frac{1}{2}$

C $\frac{11}{20}$

D $\frac{55}{100}$

10 The price of a ski jacket was $130. Allan bought the jacket at a 20% discount. How much was the discounted price?

F $26   H $110

G $104   J $156

11 Nicolas has $1,000 in a savings account that earns 4.5% interest per year. How much interest will he earn in the next year?

A $45

B $54

C $450

D $540

12 A section of a hiking trail rises 10 feet for every horizontal change of 50 feet. What is the slope for this section of the trail?

F -5

G $-\frac{1}{5}$

H $\frac{1}{5}$

J 5

13 There are 2.2 pounds per kilogram. Kono has a mass of 33 kilograms. What is his weight in pounds?

A 0.07

B 15

C 72.6

D 150

14 Find the slope of the line that includes these x and y coordinates.

x	0	1	2	3
y	−6	4	14	24

F -10

G $-\frac{1}{10}$

H $\frac{1}{10}$

J 10

15 The following are perimeters and side lengths for several equilateral triangles.

Perimeter	27	36	48	51
Side Length	9	12	16	17

What is the ratio between an equilateral triangle's perimeter and its side length?

A $\frac{1}{3}$

B $\frac{9}{27}$

C $\frac{3}{1}$

D $\frac{9}{1}$

NEED EXTRA HELP?

If You Missed Question...	1	2	3	4	5	6	7	8	9	10	11	12	13	**14**	**15**
Go to Lesson...	1-2	2-1 2-5 2-7	3-8	4-2	4-3	5-1	5-5	6-4	6-2	6-5	6-5	7-5	7-4	7-5	7-3
For Help with Algebra Readiness Standard...	7AF1.1	6AF1.1 7NS1.2 ALG5.0	7NS1.2	7NS1.2	7NS1.2	7AF2.1	ALG2.0	7NS1.3	7NS1.3	6NS1.4	ALG5.0	7AF3.3	7MG1.3	7AF3.3	7AF3.4

Standards Practice

16 Which graph shows the point $(2, -3)$?

F

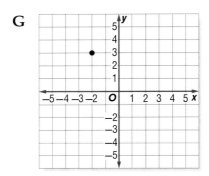

G

H

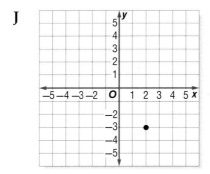

J

Record your answers on the answer sheet provided by your teacher or on a separate sheet of paper.

17 Simplify.

$$5(x + 2) - 3x + 4$$

18 Solve for p.

$$22p + 19 = 151$$

19 Solve for g.

$$\frac{2}{3}g - \frac{1}{6} = \frac{2}{3}$$

20 Dianna drove at an average speed of 53 miles per hour for 1.3 hours. Her car used 2.6 gallons of gas. What was the average number of miles her car traveled per gallon of gas?

21 Simplify.

$$\frac{3gh^2}{9g^2h}$$

22 Graph the following set of numbers on a number line.

$$\left\{ \frac{2}{3}, 1.3, 2^{-2}, \sqrt{4} \right\}$$

23 Write a fraction in simplest form that is equivalent to 0.14? Write an equivalent percent.

24 Ronni sells used CDs at a price of $25 for 4 CDs. Draw a graph that shows the relationship between the number of CDs and their total cost.

Standards Practice

25 Determine if the triangle shown below is a right triangle.

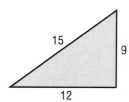

26 Describe the relationship between height and shoe size.

Height (inches)	54	43	45	51
Shoe Size	7	6	4	8

PART 3 Extended Response

Record your answers on the answer sheet provided by your teacher or on a separate sheet of paper.

27 Ramona drove at a constant speed. After 1.5 hours, she had driven 90 miles. After 2 hours, she had driven 120 miles.

  a. Draw a graph to represent this situation.

  b. How many miles would she drive in 4 hours?

  c. What is her unit rate of speed, in miles per hour?

28 One hundred twenty math teachers were asked the following. *On what day do you prefer to have students take tests?* The results of the survey are shown below.

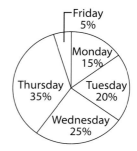

  a. Write a proportion that could be used to determine the number of teachers who chose Friday.

  b. How many more teachers prefer to give tests on Thursday than on Tuesday?

  c. Solve the proportion from part **a** to determine the number of teachers who chose Friday.

NEED EXTRA HELP?												
If You Missed Question...	16	17	18	19	20	21	22	23	24	25	26	27	28
Go to Lesson...	7-1	1-8	2-7	3-10	4-3 4-4	5-1	5-6	6-2	7-2 7-3	7-7	7-3	7-2	6-4
For Help with Algebra Readiness Standard...	5AF1.4	7AF1.3	ALG5.0	7AF4.1	7NS1.2 ALG5.0	7AF2.1	6NS1.1	7NS1.3	7AF3.4 7MR2.5	7MG3.3	7AF3.4	7MR2.5	7NS1.3 ALG5.0

Chapter 8 Standards Practice

PART 1 Multiple Choice

Choose the *best* answer.

1 What is the value of the expression below?

$$\frac{9 + 3(9 - 3) - 3}{2 + 2^2}$$

A $\frac{3}{2}$ **C** 4

B $\frac{15}{8}$ **D** 5

2 For each step of a staircase, the tread length (t) plus twice the riser height (h) of each step should be 24.

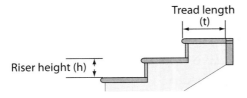

Which equation shows this relationship?

F $t + h = 24$ **H** $2t + h = 24$

G $t + 2h = 24$ **J** $2t + 2h = 24$

3 What value of x makes the equation below true?

$$\frac{2}{3}x + \frac{1}{2} = \frac{4}{5}$$

A $x = \frac{1}{5}$ **C** $x = \frac{13}{15}$

B $x = \frac{9}{20}$ **D** $x = \frac{39}{20}$

4 What is the value of $8.5 \cdot 4.7$?

F 9.35 **H** 39.95

G 32.35 **J** 399.5

5 In May 2006, Justin Gatlin broke the 100-meter world record with a time of 9.76 seconds. What was his speed in meters per second?

A 9.76 m/s **C** 97.6 m/s

B 10.25 m/s **D** 102.5 m/s

6 Which of the following shows the lowest unit rate?

7 Which is *not* equivalent to $6^3 \cdot 6^2$?

A 6^5 **C** $2^5 \cdot 3^5$

B $6^1 \cdot 6^4$ **D** 6^6

8 A bookstore is having a 20% off sale on novels by local writers. Suppose you want to buy a book that normally sells for $24. How much money would you save?

F $4.00 **H** $6.00

G $4.80 **J** $19.20

9 For a fuel pump to meet EPA regulations, it must take at least 12 seconds to dispense 2 gallons of fuel. At that rate, how long will the pump take to fill a 20-gallon tank?

A 1.2 minutes   C $3\frac{1}{3}$ minutes

B 2 minutes   D 8 minutes

10 The Hat Shack sells 40 different hats with team logos. Of these hats, 14 are black. Which proportion could be used to find the percent of hats that are black?

F $\dfrac{14}{100} = \dfrac{r}{40}$ H $\dfrac{100}{r} = \dfrac{14}{40}$

G $\dfrac{40}{100} = \dfrac{r}{14}$ J $\dfrac{r}{100} = \dfrac{14}{40}$

11 What is the length of the missing side to the nearest tenth?

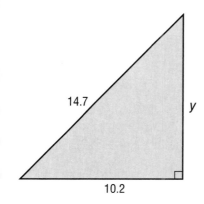

A 4.5

B 10.6

C 17.9

D 24.9

12 What are the coordinates of point *G*?

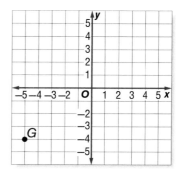

F (−5, −4)

G (−5, 4)

H (5, −4)

J (5, 4)

13 The speed of four different Internet providers is shown in the table below.

Swift Systems	10.2 megabits in 1 second
Speedy Bits	49.2 megabits in 5 seconds
PDQ Online	30.7 megabits in 3 seconds
Mega-Fast	616 megabits in 1 minute

Which provider is the fastest?

A Swift Systems

B PDQ Online

C Speedy Bits

D Mega-Fast

14 A parallelogram has adjacent side lengths of 9 centimeters and 5 centimeters. What is the perimeter?

F 14 cm   H 28 cm

G 23 cm   J 45 cm

NEED EXTRA HELP?														
If You Missed Question...	1	2	3	4	5	6	7	8	9	10	11	12	13	14
Go to Lesson...	1-3	2-6	3-10	4-3	4-4	5-6	5-1	6-5	6-6	6-4	7-7	7-1	7-4	8-4
For Help with Algebra Readiness Standard...	7NS1.2	7MR1.1	ALG5.0 7AF4.1	7NS1.2 7AF1.3	7NS1.2 7NS1.3	6NS1.1	7AF2.1	7AF4.2 6NS1.4	7AF4.2 ALG5.0	ALG5.0 7NS1.3	7MG3.3	5AF1.4	7MG1.3	7MR1.1 GEO8.0

15 Trees are to be planted along both sides of a walkway that is 150 feet long. The trees will be placed at 10-foot intervals as shown in the partial diagram below.

How many trees will be needed?

A 15

B 16

C 30

D 32

16 What is the difference in the volume of the two figures shown below?

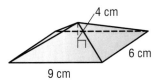

Rectangular prism

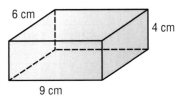

Rectangular pyramid

F 72 cm^3

G 144 cm^3

H 216 cm^3

J 288 cm^3

PART 2 Short Answer

Record your answers on the answer sheet provided by your teacher or on a separate sheet of paper.

17 For what value of n is the following equation true?

$$n - 13 = -5$$

18 What is $\frac{16}{5}$ written as a mixed number?

19 A number is decreased by 3^2, and then divided by 5. The resulting number is 13. What was the original number?

20 From Jane's house to school, how many different paths are 6 blocks long?

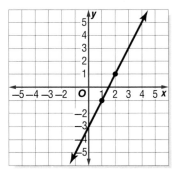

21 Mohamed travels 282.2 miles in 6.8 hours. Express his speed as a unit rate.

22 What is the slope of the line below?

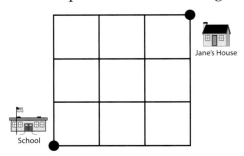

23 Trading cards come in packs of 4 cards for $1.50. Draw a graph that shows the relationship between the number of packs of cards and the total cost.

24 Four identical coins can be arranged to form a clover. The centers of the four coins form a rhombus.

If the radius of each coin is 2 cm, what is the perimeter of the rhombus?

25 Lavish Rent-a-Car charges $22 per day to rent a car. Entertain Rentals charges $19 per day plus a fee of $27. Graph these two linear functions to find the number of days for which it will cost the same amount to rent a car from either company.

26 For the quadrilateral below, what is the value of x?

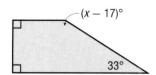

Record your answers on the answer sheet provided by your teacher or on a separate sheet of paper.

27 Lorena makes and sells picture frames. The materials cost $120. She sells the frames for $16 each.

   a. Write a linear equation that gives the profit, P, that Lorena will earn if she sells f picture frames.

   b. Use your equation to determine the profit if Lorena sells 11 frames.

28 Wendell wants to find the height of a tree in his backyard. Standing next to the tree on a sunny day, he notices that both he and the tree cast a shadow. Wendell is 5 feet tall.

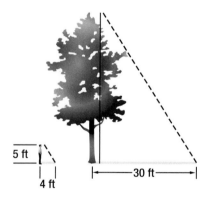

   a. Are the triangles similar, congruent, or neither?

   b. Write an equation that could be used to find the height of the tree.

   c. Use your equation from (b) to find the height of the tree.

Standards Practice

NEED EXTRA HELP?														
If You Missed Question...	15	16	17	18	19	20	21	22	23	24	25	26	27	28
Go to Lesson...	8-6	8-5	2-1	3-2	4-5	5-3	6-1	7-5	7-3	8-4	7-2 7-6	8-1	7-6	8-2
For Help with Algebra Readiness Standard...	7MR2.5	7MG2.1 7MR2.3	6AF1.1	6NS1.1	7NS1.2 7MR2.5	MR1.0 7MR2.2	7NS1.3	7AF3.3	7AF3.4	7MR1.1 GEO8.0	7MR2.5 7AF3.3	7MR1.1 5MG2.2	7AF3.3	7MR1.1 7MG4.4

Glossary/Glosario

English Español

absolute value (p. 119)
The distance between a number and zero on a number line.

acute angle (p. 46)
An angle that is less than 90 degrees.

additive inverse (pp. 129, 165)
The opposite of a number that when added to the number is zero.

add (p. 14)
An operation on two or more numbers to find a total or sum.

addend (p. 14)
A number that is to be added to another.

algebraic expression (p. 65)
A mathematical statement that contains at least one variable and may contain operations.

altitude (p. 442)
The height of a geometric figure measured by the line segment that is perpendicular to the base.

angle (p. 46)
Two rays with a common endpoint.

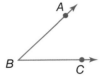

area (p. 440)
The space inside a plane figure measured in square units.

Associative Property of Addition (pp. 79, 245)
The property that states that the way numbers are grouped does not change their sum.

Associative Property of Multiplication (pp. 79, 253)
The property that states that the way numbers are grouped does not change their product.

valor absoluto (pág. 119)
La distancia entre un número y cero en la recta numérica.

ángulo agudo (pág. 46)
Ángulo que mide menos de 90 grados.

inverso aditivo (págs. 129, 165)
El opuesto de un número que cuando se suma a ese número da como resultado cero.

adición (pág. 14)
Operación que se realiza sobre dos o más números para calcular un total o una suma.

sumando (pág. 14)
Número que le sumará a otro.

expresión algebraica (pág. 65)
Enunciado matemático que contiene por lo menos una variable y puede contener operaciones.

altura (pág. 442)
La altura de una figura geométrica medida por el segmento de recta perpendicular a la base.

ángulo (pág. 46)
Dos rayos con un extremo común.

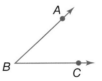

área (pág. 440)
El espacio dentro de una figura plana medido en unidades cuadradas.

propiedad asociativa de la adición (págs. 79, 245)
Propiedad que establece que la forma en que se agrupan los números no cambia su suma.

propiedad asociativa de la multiplicación (págs. 79, 253)
Propiedad que establece que la forma en que se agrupan los números no cambia su producto.

base (pp. 22, 72)
The number being multiplied in an exponent.

base (p. 336)
In a percent proportion, the whole quantity, or the number to which the part is being compared.

base (p. 452)
The side or face on which a three-dimensional shape stands.

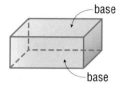

base (págs. 22, 72)
El número que se multiplica en una potencia.

base (pág. 336)
En una proporción percentual, toda la cantidad o número al que se compara la parte.

base (pág. 452)
El lado o el area sobre la acual reposa una figura tridimensional.

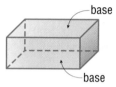

C

certain (p. 50)
The likelihood that an event will definitely happen.

check (p. 60)
A look back at work to make sure the answer makes sense. The fourth step in the four-step problem solving plan.

circle (pp. 415, 436)
A closed plane figure in which all points are the same distance from a fixed point called the center.

certeza (pág. 50)
La posibilidad de que un evento ocurrirá definitivamente.

verificar (pág. 60)
Revisión del trabajo para asegurarse que las respuestas tienen sentido. El cuarto paso en el plan de cuatro pasos para resolver problemas.

círculo (págs. 415, 436)
Figura plana cerrada en la cual todos los puntos están a la misma distancia de un punto fijo llamado centro.

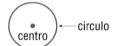

circumference (p. 436)
The distance around a circle.

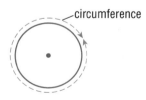

circunferencia (pág. 436)
La distancia alrededor de un círculo.

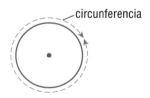

coefficient (p. 101)
A number that is multiplied by a variable in an algebraic expression.

commission (p. 345)
Money earned based on a percent of sales.

common factor (p. 179)
A whole number that is a factor of two or more numbers.

coeficiente (pág. 101)
Número que se multiplica por una variable en una expresión algebraica.

comisión (pág. 345)
Dinero que se gana en base a un porcentaje de las ventas.

factor común (pág. 179)
Número entero que es factor de dos o más números.

common multiple (p. 38)
A whole number that is a multiple of two or more numbers.

Commutative Property of Addition (pp. 77, 245)
The property that states that the order in which two numbers are added does not change the sum.

Commutative Property of Multiplication (pp. 78, 253)
The property that states that the order in which two numbers are multiplied does not change their product.

compatible numbers (p. 43)
Numbers in a problem that are easy to work with mentally.

composite number (p. 177)
A whole number that is greater than one.

congruent (p. 420)
Figures or angles that have the same size and same shape.

constant of variation (p. 349)
A constant ratio in a direct variation.

converse (p. 403)
An if/then statement in which terms are expressed in reverse order.

coordinate geometry (p. 429)
The study of geometry using the principles of algebra.

coordinate plane (pp. 364, 366)
A plane in which a horizontal number line and a vertical number line intersect at the point where each line is zero.

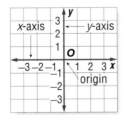

coordinate (p. 366)
One of the numbers in an ordered pair.

múltiplo común (pág. 38)
Número entero que es múltiplo de dos o más números.

propiedad conmutativa de la adición (págs. 77, 245)
Propiedad que establece que el orden en que se suman dos números no altera la suma.

propiedad conmutativa de la multiplicación (págs. 78, 253)
Propiedad que establece que el orden en que se multiplican dos números no altera su producto.

números compatibles (pág. 43)
Números de un problema con los cuales es fácil trabajar mentalmente.

número compuesto (pág. 177)
Número que tiene más factores comunes que él mismo y uno.

congruentes (pág. 420)
Figuras o ángulos que tienen el mismo tamaño y la misma forma.

constante de variación (pág. 349)
Razón constante en una variación directa.

contrario (pág. 403)
Enunciado si/entonces en el cual los términos se expresan en orden inverso.

geometría analítica (pág. 429)
El estudio de la geometría usando los principles del álgebra.

plano de coordenadas (págs. 364, 366)
Plano en el cual una recta numérica horizontal y una recta numérica vertical se intersecan en el punto donde cada recta es cero.

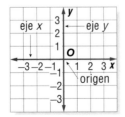

coordenadas (pág. 366)
Los números en un par ordenado.

corresponding parts (p. 420)
Parts of congruent figures that have the same measure or length.

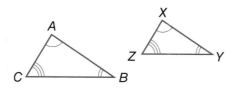

$\overline{AB}$ and $\overline{XY}$ are corresponding sides.
∠C and ∠Z are corresponding angles.

counting number (p. 6)
One of the set of whole numbers, not including 0.

cross product (p. 330)
A product of the numerator of one fraction and the denominator of another fraction.

cube (p. 452)
A rectangular prism with six faces that are congruent squares.

cylinder (p. 453)
A three-dimensional figure having two parallel congruent circular bases and a curved surface connecting the two bases.

partes correspondientes (pág. 420)
Partes de figuras congruentes que tienen la misma medida o longitud.

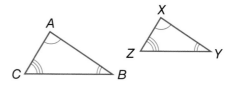

$\overline{AB}$ y $\overline{XY}$ son lados correspondientes.
∠C y ∠Z son ángulos correspondientes.

números de conteo (pág. 6)
El conjunto de números enteros sin incluir el 0.

producto cruzado (pág. 330)
El producto entre el numerador de una fracción y el denominador de otra fracción.

cubo (pág. 452)
Prisma rectangular con seis caras que son cuadrados congruentes.

cilindro (pág. 453)
Figura de tres dimensiones que tiene dos bases circulares paralelas y congruentes y una superficie curva que conecta las dos bases.

D

decimal (pp. 34, 234, 236)
A number with one or more digits to the right of the decimal point, such as 8.37 or 0.05.

decimal point (pp. 34, 236)
A period separating the ones and the tenths of a number.

define (p. 116)
Identifying the unknown and choosing a variable to represent it.

denominator (p. 30)
The number on the bottom of a fraction. It tells the total number of equal parts.

density (p. 383)
A measure of how much mass is in a given space (density = mass/volume).

difference (p. 15)
The result of subtracting one number from another.

decimal (págs. 34, 234, 236)
Número con uno o más dígitos a la derecha del punto decimal, como 8.37 ó 0.05.

punto decimal (págs. 34, 236)
Punto que separa las unidades y las décimas de un número.

definir (pág. 116)
Identificar la incógnita y elegir una variable que la represente.

denominador (pág. 30)
Término inferior de una fracción que indica en cuántas partes se divide un todo o unidad.

densidad (pág. 383)
Medida de la cantidad de masa que hay en un espacio dado (densidad = masa/volumen).

diferencia (pág. 15)
El resultado de restar un número de otro.

digit (p. 10)
One of the symbols used to write numbers. The ten digits are 0, 1, 2, 3, 4, 5, 6, 7, 8, 9.

dimensional analysis (p. 383)
A process where units of measurement are included when computing an answer.

direct variation (p. 349)
A relationship between two variable quantities that have a constant ratio.

discount (p. 342)
A decrease in the regular price.

Distance Formula (p. 429)
A formula used to find the distance between any two points in the coordinate plane.

Distributive Property (pp. 84, 131, 253)
The property that states that multiplying the sum of two or more numbers by another number is the same as multiplying each addend by the same number and adding the products.

dividend (p. 19)
A number that is being divided.

divide (p. 19)
An operation to separate a group into equal-sized groups.

divisor (p. 19)
The number by which the dividend is being divided.

draw a diagram (p. 184)
A strategy of creating a sketch, graph, chart, or picture to help understand information.

draw a graph (p. 372)
A strategy of creating an organized drawing that shows sets of data and how they are related to each other.

dígitos (pág. 10)
Símbolos que se usan para escribir números. Los diez dígitos son 0, 1, 2, 3, 4, 5, 6, 7, 8, 9.

análisis dimensional (pág. 383)
Proceso que incluyen unidades de medidas al calcular un resultado.

variación directa (pág. 349)
Relación entre dos cantidades variables que tiene una razón constante.

descuento (pág. 342)
Disminución en el precio regular.

fórmula de distancia (pág. 429)
Fórmula que se usa para calcular la distancia entre dos puntos cualesquiera en un plano de coordenadas.

propiedad distributiva (págs. 84, 131, 253)
Propiedad que establece que multiplicar la suma de dos o más números por otro número es igual a multiplicar cada sumando por el mismo número y sumar los productos.

dividendo (pág. 19)
Número que se divide entre otro.

dividir (pág. 19)
Operación para separar un grupo en grupos de igual tamaño.

divisor (pág. 19)
Número entre el cual se divide el dividendo.

dibujar un diagrama (pág. 184)
Estrategia de crear bosquejos, gráficas, tablas o dibujos como ayuda para comprender información.

trazar una gráfica (pág. 372)
Estrategia de crear un dibujo organizado que muestre conjuntos de datos y como se relacionan entre sí.

E

equation (pp. 67, 112, 114, 221, 264)
A mathematical sentence stating that two expressions are equal.

equivalent fraction (p. 172)
One of two or more fractions that are different, but have the same value, such as $\frac{1}{2}$ and $\frac{2}{4}$.

estimate (p. 42)
An answer that is close to the exact answer.

ecuación (págs. 67, 112, 114, 221, 264)
Enunciado matemático que establece la igualdad de dos expresiones.

fracciones equivalentes (pág. 172)
Fracciones que son diferentes pero que tienen el mismo valor, como $\frac{1}{2}$ y $\frac{2}{4}$.

estimación (pág. 42)
Respuesta cercana a la respuesta exacta.

evaluate (p. 103)
Finding the value of an expression by replacing variables with numbers.

expanded form (p. 10)
A form showing the sum of the values represented by each digit.

exponent (pp. 22, 26, 72, 276)
The number that tells how many times the base is multiplied.

expression (pp. 58, 220, 264)
A combination of numbers, variables, and operation symbols.

evaluar (pág. 103)
Calcular el valor de una expresión reemplazando las variables con números.

forma desarrollada (pág. 10)
Forma que muestra la suma de los valores representados por cada dígito.

exponente (págs. 22, 26, 72, 276)
El número que indica la cantidad de veces que se multiplica la base.

expresión (págs. 58, 220, 264)
Combinación de números, variables y símbolos operacionales.

F

factor (pp. 18, 26, 135, 177, 251)
Number or expression that is multiplied by another.

factor tree (p. 26)
A diagram used to find the prime factor of any number.

formula (p. 299)
An equation that describes a relationship among certain quantities.

fraction (pp. 30, 162, 164, 236)
A number that represents equal parts of a whole or group.

function (p. 393)
A relationship in which exactly one output is assigned to each input.

factor (págs. 18, 26, 135, 177, 251)
Número o expresión que se multiplica por otro número o expresión.

árbol de factores (pág. 26)
Diagrama que se usa para calcular los factores primos de cualquier número.

fórmulas (pág. 299)
Ecuaciones que describen una relación entre ciertas cantidades.

fracción (págs. 30, 162, 164, 236)
Número que representa partes iguales de un todo o un grupo.

función (pág. 393)
Relación en la cual se le asigna exactamente una salida a cada entrada.

G

greatest common factor (GCF) (p. 179)
The largest number that divides evenly into two or more numbers.

guess and check (p. 91)
A strategy of guessing the answer to a problem, then checking to *see* if the answer makes sense.

máximo común divisor (MCD) (pág. 179)
El mayor número que divide exactamente a dos o más números.

hacer una conjetura y verificar (pág. 91)
Estrategia de adivinar la respuesta a un problema y luego verificar para *ver* si tiene sentido.

Identity Property of Addition (pp. 93, 245)
The property that states that the sum of any number and zero is the original number.

Identity Property of Multiplication (pp. 94, 253)
The property that states that the product of any number and one is the original number.

impossible (p. 50)
The likelihood of an event that cannot happen.

improper fraction (p. 169)
A fraction in which the numerator is greater than or equal to the denominator.

integer (pp. 112, 119)
The set of whole numbers and their opposites, including zero.

interest (p. 342)
Money you earn on a bank account or money you pay towards a bank loan.

inverse operations (pp. 149, 221)
Operations that undo each other, such as addition and subtraction or multiplication and division.

Inverse Property of Multiplication (p. 192)
The property that states that the product of a number and its multiplicative inverse is one.

propiedad de identidad de la adición (págs. 93, 245)
Propiedad que establece que la suma de cualquier número más cero es el número original.

propiedad de identidad de la multiplicación (págs. 94, 253)
Propiedad que establece que el producto de cualquier número y uno es el número original.

imposible (pág. 50)
La posibilidad de un evento que no puede suceder.

fracción impropia (pág. 169)
Fracción en la cual el numerador es mayor que o igual al denominador.

enteros (págs. 112, 119)
El conjunto de números enteros y sus opuestos, incluyendo el cero.

interés (pág. 342)
El dinero que ganas en una cuenta bancaria o el dinero que pagas por un préstamo bancario.

operaciones inversas (págs. 149, 221)
Operaciones que se anulan entre sí, como la adición y la sustracción o la multiplicación y la división.

propiedad del inverso de la multiplicación (pág. 192)
Propiedad que establece que el producto de un número y su inverso multiplicativo es uno.

least common denominator (LCD) (p. 207)
The least common multiple of the denominators of two or more fractions, used as a denominator.

least common multiple (LCM) (pp. 38, 206)
The smallest multiple two or more numbers have in common.

like denominators (pp. 199, 213)
Fraction denominators that are the same.

likely (p. 50)
The likelihood that an event will probably happen.

mínimo común denominador (mcd) (pág. 207)
El mínimo común múltiplo de los denominadores de dos o más fracciones que se usa como un denominador.

mínimo común múltiplo (mcm) (págs. 38, 206)
El menor múltiplo que dos o más números tienen en común.

denominadores iguales (págs. 199, 213)
Denominadores iguales de fracciones.

posibilidad (pág. 50)
La posibilidad de que un evento suceda probablemente.

Glossary/Glosario

like terms (p. 100)
Terms that have the same variables raised to the same powers.

linear relationship (p. 376)
A relationship that has a straight-line graph.

look for a pattern (p. 146)
A strategy of finding repeated orders or sequences of numbers.

términos semejantes (pág. 100)
Términos que tienen las mismas variables elevadas a los mismos exponentes.

relación lineal (pág. 376)
Relación cuya gráfica es una línea recta.

hallar un patrón (pág. 146)
Estrategia de hallar órdenes repetidas o sucesiones de números.

M

make a model (p. 448)
A strategy of creating a model to represent a problem situation.

make a table (p. 355)
A strategy of creating an organized display of data.

mixed number (p. 169)
A number that combines an integer and a fraction.

multiple (p. 38)
The product of a number and any other whole number.

multiplicative inverse (p. 192)
The reciprocal of a number that when multiplied by that number has a product of one.

Multiplicative Property of Zero (pp. 94, 253)
A property that states that the product of a number and zero is zero.

multiply (p. 18)
An operation to combine equal-sized groups into one larger group.

hacer un modelo (pág. 448)
Estrategia de crear un modelo para representar un problema.

hacer una tabla (pág. 355)
Estrategia de crear una representación organizada de datos.

número mixto (pág. 169)
Número que combina un entero y una fracción.

múltiplos (pág. 38)
El producto de un número por cualquier otro número entero.

inverso multiplicativo (pág. 192)
El recíproco de un número que, cuando se multiplica por ese número, da como producto uno.

propiedad multiplicativa del cero (págs. 94, 253)
Propiedad que establece que el producto de un número por cero es cero.

multiplicación (pág. 18)
Operación que combina grupos de igual tamaño en un grupo más grande.

N

negative exponent (p. 285)
Exponent used to express a number less than one.

nonlinear function (p. 396)
A function that does not have a constant rate of change. The graph of a nonlinear function is not a straight line.

exponentes negativos (pág. 285)
Exponentes que se usan para expresar números menores que la unidad.

función no lineal (pág. 396)
Función que no tiene una tasa constante de combio. La gráfica de una función no lineal no es uno recta.

numerator (p. 30)
The number on top of a fraction. It tells the part of the whole or group.

numerical expression (p. 65)
A statement that contains only numbers and may contain operations.

numerador (pág. 30)
El número en la parte superior de una fracción. Indica las partes que se toman del todo o del grupo.

expresión numérica (pág. 65)
Enunciado que sólo contiene números y puede contener operaciones.

O

obtuse angle (p. 46)
An angle that is greater than 90 degrees and less than 180 degrees.

ángulo obtuso (pág. 46)
Ángulo mayor que 90 grados y menor que 180 grados.

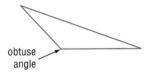

obtuse angle

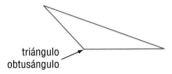

triángulo obtusángulo

one-step equation (p. 150)
An equation that contains only one operation.

ecuación de un paso (pág. 150)
Ecuación que contiene sólo una operación.

opposite (p. 119)
A positive rational number or its negative.

opuestos (pág. 119)
Todo número racional positivo y su par negativo.

ordered pair (p. 366)
A pair of numbers that gives the location of a point on a map or grid.

par ordenado (pág. 366)
Un par de números que da la ubicación de un punto en un mapa o una cuadrícula.

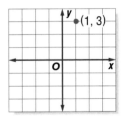

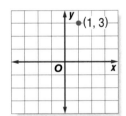

order of operations (p. 70)
Rules that are followed to find the correct value of an expression.

orden de operaciones (pág. 70)
Reglas que se siguen para calcular el valor correcto de una expresión.

origin (p. 366)
The point at which the two axes of a coordinate plane intersect.

origin (pág. 366)
El punto en el cual se intersecan los dos ejes de un plano de coordenadas.

P

percent (p. 323)
A number expressed in relation to 100.

por ciento (pág. 323)
Número expresado en relación con 100.

percentage (p. 336)
A number divided by 100.

porcentaje (pág. 336)
Número dividido entre 100.

percent proportion (p. 336)
An equality between a fraction and a percentage.

perimeter (p. 434)
The distance around a plane figure.

place value (p. 10)
The value given to a digit by its position in a numeral.

plane figure (pp. 413, 414)
A flat, two-dimensional figure that lies entirely within one plane.

plan (p. 60)
Selecting the best strategy for solving the problem. The second step in the four-step problem solving plan.

polygon (p. 414)
A closed plane figure formed by three or more line segments called sides.

Power of Powers (p. 281)
The exponent rule that states that exponents are multiplied to raise a power to a power.

power (pp. 22, 278)
A base raised to an exponent.

prime factorization (p. 26)
Expressing a composite number as a product of its prime factors.

prime number (pp. 26, 177)
Any whole number whose only factors are one and itself.

principal (p. 342)
The starting amount to which interest is added.

prism (p. 452)
A solid figure with two parallel, congruent polygonal faces called bases.

rectangular prism   triangular prism

proporción porcentual (pág. 336)
Igualdad entre una fracción y un porcentaje.

perímetro (pág. 434)
La distancia alrededor de una figura plana.

valor de posición (pág. 10)
El valor dado a un dígito según su posición en el número.

figura plana (págs. 413, 414)
Figura plana bidimensional que se halla completamente dentro de un plano.

planificar (pág. 60)
Elegir la mejor estrategia para resolver un problema. El segundo paso en el plan de cuatro pasos para resolver problemas.

polígono (pág. 414)
Figura plana y cerrada formada por tres segmentos de recta, o más, llamados lados.

potencia de potencias (pág. 281)
La regla de los exponentes que establece que los exponentes se multiplican para elevar una potencia a una potencia.

potencia (págs. 22, 278)
Base elevada a un exponente.

factorización prima (pág. 26)
Expresar un número compuesto como el producto de sus factores primos.

número primo (págs. 26, 177)
Cualquier número entero cuyos únicos factores sean uno y él mismo.

capital (pág. 342)
La cantidad inicial a la cual se le suma el interés.

prisma (pág. 452)
Figura sólida con dos caras poligonales, paralelas y congruentes llamadas bases.

prisma rectangular   prisma triangular

probability (p. 50)
The likelihood that an event will occur.

Product of Powers (p. 278)
The exponent rule that states that exponents are added to multiply powers that have the same base.

product (pp. 18, 135, 186)
The result of multiplying numbers or expressions together.

proper fraction (p. 169)
A fraction in which the numerator is less than the denominator.

property (pp. 58, 77)
A statement that is true for all numbers.

proportion (p. 330)
An equation that shows two equivalent ratios.

pyramid (p. 452)
A solid figure with a polygonal base and triangular faces.

Pythagorean Theorem (p. 400)
A theorem that states, for any right triangle, the square of the length of the hypotenuse is equal to the sum of the squares of the lengths of the legs.

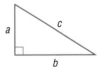

probabilidad (pág. 50)
La posibilidad de que ocurra un evento.

producto de potencias (pág. 278)
La regla de los exponentes que establece que para multiplicar potencias que tienen la misma base, se suman los exponentes.

producto (págs. 18, 135, 186)
El resultado de multiplicar números o expresiones entre sí.

fracción propia (pág. 169)
Fracción en la cual el numerador es menor que el denominador.

propiedad (págs. 58, 77)
Enunciado que es verdadero para todo número.

proporción (pág. 330)
Ecuación que muestra dos razones equivalentes.

pirámide (pág. 452)
Figura sólida con una base poligonal y caras triangulares.

teorema de Pitágoras (pág. 400)
Teorema que establece, para cualquier triángulo rectángulo, el cuadrado de la longitud de la hipotenusa es igual a la suma de los cuadrados de las longitudes de los catetos.

Glossary/Glosario

quadrant (p. 366)
One of the sections of a coordinate graph formed by the intersection of the *x*-axis and *y*-axis.

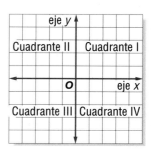

cuadrantes (pág. 366)
Las secciones de una gráfica de coordenadas formada por la intersección del eje *x* y el eje *y*.

quadrilateral (p. 416)
A four-sided polygon.

cuadrilátero (pág. 416)
Polígono de cuatro lados.

Quotient of Powers (p. 280)
The exponent rule that states that exponents are subtracted to divide powers that have the same bases.

cociente de potencias (pág. 280)
La regla de los exponentes que establece que los exponentes se restan para dividir potencias cuyas bases son las mismas.

quotient (pp. 19, 141)
The result from dividing two numbers or expressions.

cociente (págs. 19, 141)
El resultado de dividir dos números o expresiones.

radical sign ($\sqrt{\ }$) (pp. 276, 293)
A symbol indicating a nonnegative (positive or 0) square root.

signo radical ($\sqrt{\ }$) (págs. 276, 293)
Símbolo que indica una raíz cuadrada no negativa (positiva ó 0).

radicand (p. 295)
Any number or expression inside a radical sign.

radicando (pág. 295)
Cualquier número o expresión dentro de un signo radical.

rate (p. 319)
A ratio of two measurements or quantities with different units.

tasa (pág. 319)
Razón de dos medidas o cantidades con diferentes unidades.

rational number (pp. 160, 164, 305)
A number that takes the form a/b, where b is not equal to 0.

número racional (págs. 160, 164, 305)
Número que toma la forma de a/b, donde b es distinta de 0.

ratio (p. 318)
A comparison of two quantities by division.

razón (pág. 318)
Comparación de dos cantidades mediante división.

reciprocal (p. 192)
A number found by switching the numerator and denominator of the given number.

recíproco (pág. 192)
Número que se calcula intercambiando el numerador y el denominador del número dado.

repeating decimal (pp. 34, 237)
A repeating pattern of digits to the right of the decimal point.

decimal periódico (págs. 34, 237)
Patrón repetitivo de dígitos a la derecha del punto decimal.

right angle (p. 46)
An angle that equals 90 degrees.

rise (p. 388)
The vertical change along the *y*-axis of a graph.

round (pp. 42, 258)
Approximating a number to a given decimal place.

run (p. 388)
The horizontal change along the *x*-axis of a graph.

ángulo recto (pág. 46)
Ángulo que mide 90 grados.

run (pág. 388)
El cambio vertical a lo largo del eje *y* de una gráfica.

redondear (págs. 42, 258)
Aproximar un número a un lugar decimal dado.

carrera (pág. 388)
El cambio horizontal a lo largo del eje *x* de una grafica.

S

scatter plot (p. 374)
A graph that shows the general relationship, or correlation, between two sets of data.

diagrama de dispersión (pág. 374)
Gráfica que muestra la relación general o correlación, entre dos conjuntos de datos.

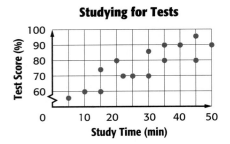

similar (p. 422)
Figures that have the same shapes but different, proportional sizes.

simplest form (pp. 102, 179)
A fraction in which the greatest common factor of the numerator and denominator is one.

slope (pp. 364, 388)
The ratio of the change in *y* to the change in *x* of a linear relationship.

solid figure (p. 452)
A shape that has three dimensions: length, width, and height.

semejantes (pág. 422)
Figuras que tienen la misma forma pero distintos tamaños proporcionales.

forma reducida (págs. 102, 179)
Fracción en la cual el máximo común divisor del numerador y el denominador, es uno.

pediente (págs. 364, 388)
Tasa constante de cambio en una relación lineal.

figuras sólidas (pág. 452)
Forma que tiene las tres dimensiones: largo, ancho y altura.

solution (p. 114)
Any number that makes an equation true.

solución (pág. 114)
Cualquier número que hace que se cumpla una ecuación.

solve a simpler problem (p. 291)
A strategy for solving part of a problem at a time.

solve (pp. 60, 114, 221)
Finding the answer to a problem. The third step in the four-step problem solving plan.

square root (pp. 276, 293)
One of two equal factors of a number.

subtract (p. 15)
An operation on two numbers that tells how many are left (difference), when some or all are taken away.

sum (p. 14)
The result of adding two or more numbers.

surface area (p. 457)
The sum of the areas of a solid figure's faces.

resolver un problema más simple (pág. 291)
Estrategia para resolver parte de un problema a la vez.

resolver (págs. 60, 114, 221)
Calcular la respuesta a un problema. El tercer paso en el plan de cuatro pasos para resolver problemas.

raíz cuadrada (págs. 276, 293)
Uno de los dos factores iguales de un número.

sustracción (pág. 15)
Operación sobre dos números que indica cuánto queda (diferencia), cuando algo o todo se retira.

suma (pág. 14)
El resultado de sumar dos o más números.

área de superficie (pág. 457)
La suma de las áreas de las caras de una figura sólida.

T

terminating decimal (pp. 34, 234, 237)
A decimal that has a finite (certain) number of digits to the right of the decimal point.

term (p. 100)
A number, a variable, or a product of numbers and variables.

tip (p. 345)
Money given for a job well done.

trend (p. 374)
Consistent change over time.

triangle (p. 414)
A polygon with three sides and three angles.

two-step equation (p. 150)
An equation that contains two operations.

decimal terminal (págs. 34, 234, 237)
Decimal que tiene un número finito (cierto número) de dígitos a la derecha del punto decimal.

término (pág. 100)
Número, variable o producto de números y variables.

propina (pág. 345)
Dinero que se da por un trabajo bien hecho.

tendencias (pág. 374)
Cambios consistentes en el tiempo.

triángulo (pág. 414)
Polígono con tres lados y tres ángulos.

ecuación de dos pasos (pág. 150)
Ecuación que contiene dos operaciones.

U

understand (p. 60)
Identifying the given information and problem to be solved. The first step in the four-step problem solving plan.

unit rate (p. 319)
The rate in lowest terms.

entender (pág. 60)
Identificar la información dada y el problema a resolver. El primer paso en el plan de cuatro pasos para resolver un problema.

tasa unitaria (pág. 319)
La tasa en los menores términos.

unlike denominators (pp. 206, 213)
Fraction denominators that are not the same.

unlikely (p. 50)
The likelihood that an event will probably not happen.

unlike term (p. 100)
One of two or more terms in which the variables are not alike.

distintos denominadores (págs. 206, 213)
Denominadores diferentes de fracciones.

improbable (pág. 50)
La posibilidad de que un evento no ocurra probablemente.

términos diferentes (pág. 100)
Uno o más términos en los cuales las variables y/o los exponentes no son iguales.

variable (pp. 58, 65)
A symbol or letter used to represent a number.

vertex (pp. 46, 414)
The common endpoint where two rays meet.

volume (p. 452)
The space inside a solid figure.

variable (págs. 58, 65)
Símbolo o letra que se usa para representar un número.

vértice (págs. 46, 414)
El extremo común donde se encuentran dos rayos.

volumen (pág. 452)
El espacio dentro de una figura sólida.

whole number (p. 6)
One of the set of numbers that includes the counting numbers and zero.

work backward (p. 262)
A strategy of using final answer to find the original information.

números enteros (pág. 6)
Los números {0, 1, 2, 3, 4, . . .}

trabajar al revés (pág. 262)
Estrategia en la que se usa la respuesta final para calcular la información original.

x-axis (p. 366)
The horizontal number line in a coordinate plane.

y-axis (p. 366)
The vertical number line in a coordinate plane.

eje x (pág. 366)
La recta numérica horizontal en un plano de coordenadas.

eje y (pág. 366)
La recta numérica vertical en un plano de coordenadas.

Glossary/Glosario

Selected Answers

Prerequisite Skills

Page 8   **Prerequisite Lesson 1**
1. All the counting numbers are whole numbers. The whole numbers include zero, and the counting numbers do not.   **3.** 8   **5.** 9   **7.** 6 > 3   **9.** 14 > 9
11. 19 < 25   **13.** 4, 5, 9   **15.** 7, 10, 12   **17.** 98, 195, 214
19. Answers will vary.

Page 12   **Prerequisite Lesson 2**
1. The place value tells the value of the digit because of its place in the number.   **3.** 900   **5.** 500
7. 3   **9.** 30   **11.** 60   **13.** 400 + 80 + 7
15. 1,000 + 400 + 2   **17.** 900 + 80 + 4
19. 2,000 + 900 + 50 + 4   **21.** 4,000 + 3   **23.** $493
25. Sample answers: 1,548 (1,000 and 8); or 8,541 (8,000 and 1)

Page 16   **Prerequisite Lesson 3**
1. Sample answer: Add to find the total amount of money in an account after making a deposit.
3. 556   **5.** 5,698   **7.** 2,902   **9.** 232   **11.** 6,389
13. 2,473   **15.** 3,725 hours   **17.** Sample answer: Stack numbers. Add 7 and 9. Write 6 in ones place, and carry 10. Rename 80 as 90, and add 90 + 0. Write 9 in tens place. Write 1 in hundreds place.

Page 20   **Prerequisite Lesson 4**
1. Multiply to find the total number of items in a number of equal-sized groups. Sample answer: Find the total number of paper clips in a number of boxes.   **3.** 1,073   **5.** 8,692   **7.** 15,045   **9.** 234
11. 628   **13.** 529   **15.** 7   **17.** Sample answer: when you multiply 6 in the tens place by 385

Page 24   **Prerequisite Lesson 5**
1. the base   **3.** 16   **5.** 64   **7.** 64   **9.** 8   **11.** 4^3
13. 7^5 **15.** 1^7 **17.** k^3 **19.** 64 squares
21. No; because $3^4 = 81$ and $4^3 = 64$.

Page 28   **Prerequisite Lesson 6**
1. A prime number has only itself and 1 as factors.
3. Write the number as the product of any two factors. Then, on the next line, break the factors that are not prime numbers into two factors, using leader lines to connect each number to its factors. Continue until you have only prime numbers.
5. 3 · 5   **7.** 3 · 7   **9.** 3^3 **11.** 5 · 7   **13.** 2 · 3 · 7
15. 2 · 5^2 **17.** 2, 3, and 5 inches   **19.** Sample answer: 4 ways, starting with 8 · 6, 12 · 4, 16 · 3, and 24 · 2

Page 32   **Prerequisite Lesson 7**
1. Sample answer: 3 of 5 equal parts   **3.** $\frac{1}{4}$ **5.** $\frac{5}{5}$
7. $\frac{2}{3}$ **9.** Sample answer: $\frac{3}{5}$ is less than 1; $\frac{5}{3}$ is more than 1.

Page 36   **Prerequisite Lesson 8**
1. 32 is a counting number and 32.4 is a decimal

number. The decimal point indicates that the number is a decimal because it has digits to the right of the decimal point.   **3.** Sample answer: 0.96
5. repeating   **7.** terminating   **9.** repeating
11. $1\frac{59}{100}$ **13.** $\frac{231}{1,000}$ **15.** $\frac{51}{100}$
17. Both are decimals, which have digits to the right of the decimal point. Terminating decimals have a certain number of digits to the right of the decimal point. Repeating decimals have an endless number of digits to the right of the decimal point that repeat in a certain pattern.

Page 40   **Prerequisite Lesson 9**
1. 5, 10, 15, and 20 are the products of 5 and the first four counting numbers.   **3.** 6   **5.** 12   **7.** 40   **9.** 24
11. 30   **13.** six sets of paper plates, five sets of paper cups   **15.** Sample answer: 112 is the LCM of 14 and 16. 14: 14, 28, 42, 56, 70, 84, 98, (112) 126; 16: 16, 32, 48, 64, 80, 96, (112) 128

Page 44   **Prerequisite Lesson 10**
1. Sample answer: rounding and using compatible numbers; Rounding numbers that are to be added, subtracted, multiplied, or divided makes the operation easier. Using compatible numbers makes division easier by changing the dividend to the nearest number that can be divided evenly by the divisor.   **3.** 640   **5.** 5   **7.** 240   **9.** 3,500   **11.** 470
13. 60   **15.** 50   **17.** 70   **19.** about 2,400 students
21. about 160 more cars   **23.** Sample answer: The estimated product will be less than the exact product because both factors were rounded down.

Page 48   **Prerequisite Lesson 11**
1. The vertex is the point where two rays come together to form an angle.   **3.** Both are formed where two rays meet at a vertex. An acute angle is less than a right angle. An obtuse angle is greater than a right angle.   **5.** acute   **7.** obtuse
9. obtuse   **11.** $\frac{3}{4}$ turn   **13.** Sample answer: right

Page 52   **Prerequisite Lesson 12**
1. Probability measures the likelihood that an event will occur.   **3.** likely   **5.** certain   **7.** $\frac{2}{8}$, or $\frac{1}{4}$
9. $\frac{0}{8}$, or 0   **11.** $\frac{0}{10}$, or 0   **13.** $\frac{6}{10}$, or $\frac{3}{5}$ **15.** certain
17. $\frac{1}{4}$ **19.** Remove all the blue marbles.

Chapter 1 Variables, Expressions, and Properties

Page 63   **Lesson 1-1**
1. plan   **3.** check   **5.** 12 students play basketball, 9 students play soccer, 18 students play at least one sport   **7.** 3   **9.** 2 notebooks, 4 pencils, 8 markers

11. how many hours a month Marcus needs to work, Sample answer: work backward   **13.** Sample answer: Yes. The answer checks.   **15.** how many hours Hannah worked; Sample answer: guess and check   **17.** Sample answer: Yes. No other combination equals $38.
19. 2:00 P.M.   **21.** five 2-point baskets   **23.** Sample answer: Show 240 divided into 8 equal parts.
25. F   **27.** 169   **29.** 73   **31.** 6,708   **33.** 16,142

Page 68   Lesson 1-2
1. Sample answers: 4; 5 + 3; $\frac{4}{5}$ (cannot include a variable)   **3.** Sample answers: $k = 2$; $4 + y = 7$; $x - 3 = 9$ (must include an equal sign)   **5.** $7m$
7. 4 more than the product of 20 and m
9. $4r - 6 = 10$ **11.** The product of 4 and n is 80.   **13.** $S = f - 26$ **15.** $35 + z$ **17.** $16p$
19. $3d + 18$ **21.** the quotient of b and 11   **23.** 1 less than the product of m and 2   **25.** $3 + w = 15$
27. $5r = 7$ **29.** Sample answer: Ten decreased by k is h.   **31.** Sample answer: Three times r is 18.
33. $(2 \cdot 15) + (3 \cdot 25)$ **35.** Sample answers: $2h - 7 = h + 15$; the difference between 2 times a number and 7 is the same as the number increased by 15.   **37.** $0.10   **39.** C   **41.** 29   **43.** 9   **45.** 2

Page 73   Lesson 1-3
1. division   **3.** base   **5.** 12   **7.** 12   **9.** 48
11. 30   **13.** 3   **15.** 140   **17.** 35   **19.** 33   **21.** 45
23. 47   **25.** $48   **27.** $50 \cdot 6 + 8 \cdot 12$; $396
29. Sample answer: $775   **31.** 90   **33.** 35
35. 89   **37.** 37   **39.** 6   **41.** 38   **43.** 46   **45.** 36
47. Emily is correct. She followed the order of operations: division and multiplication from left to right.   **49.** Sample answers:
$\frac{4 + 4 + 4}{4} = 3$, $4 + \frac{4 - 4}{4} = 4$, $\frac{4 \cdot 4 + 4}{4} = 5$,
$4 + \frac{4 + 4}{4} = 6$, $4 + 4 - \frac{4}{4} = 7$, $4 + \frac{4 \cdot 4}{4} = 8$,
$4 + 4 + \frac{4}{4} = 9$ **51.** A   **53.** B   **55.** $\frac{m}{n}$
57. $8y + 5$ **59.** the difference in d and 2
61. 10 more than the quotient of m and n

Page 81   Lesson 1-4
1. Sample answer: $3 + n = n + 3$ **3.** 9   **5.** 7
7. 10   **9.** 12, 2   **11.** Comm. (+)   **13.** Assoc. (×)
15. $12 + (39 + 8) = 12 + (8 + 39)$ Comm. (+); $= (12 + 8) + 39$ Assoc. (+); $= 20 + 39$; $= 59$
17. Sample answer: $2.95 + (1.29 + 6.05) = 2.95 + 7.34 = 10.29; $(2.95 + 1.29) + 6.05 = 4.24 + 6.05 = 10.29 **19.** Answers will vary. For example, $(5 + 2) + 8 = 5 + (2 + 8)$ and $5 + (2 + 8) = 5 + (8 + 2)$ **21.** 15   **23.** y, x **25.** 4   **27.** d, c
29. 3   **31.** 11, 25   **33.** 3, 10   **35.** y, z **37.** Comm. (×)   **39.** Assoc. (+)   **41.** Assoc. (+)   **43.** Comm. (×)   **45.** $5 \times (39 \times 2) = 5 \times (2 \times 39)$ Comm. (×); $= (5 \times 2) \times 39$ Assoc. (×); $= 10 \times 39$; $= 390$
47. $2 \times (71 \times 5) = 2 \times (5 \times 71)$ Comm. (×); $= (2 \times 5) \times 71$ Assoc. (×); $= 10 \times 71$; $= 710$ **49.** $(5 + 6) + 4 = 5 + (6 + 4)$ **51.** Comm. (×)   **53.** Sample answer: $12 \cdot 2 = 2 \cdot 12$ **55.** $(17 + 19) + 22 = (22 + 19) + 17$; $36 + 22 = 41 + 17$; $58 = 58$ **57.** D   **59.** B   **61.** 19   **63.** 38   **65.** mn **67.** $7 - t$

Page 87   Lesson 1-5
1. Sample answer: $(7 \cdot 3) + (7 \cdot 4) = 7(3 + 4)$
3. 3, 10, 7, 10   **5.** 4, 2   **7.** $52   **9.** $3.90   **11.** 1
13. 4, 8, 1, 8   **15.** 3   **17.** 2, 5   **19.** $150
21. $300   **23.** $25.00   **25.** $2,300   **27.** $4(0.50 + 1.00 + 2.00) + 2(3.50 + 2.50)$ **29.** $40   **31.** H
33. F   **35.** Comm. (+)   **37.** Assoc. (×)   **39.** 13
41. 1   **43.** $1.65   **45.** The difference between n and 20 is the same as the quotient of 100 and p.

Page 92   Lesson 1-6
1. 30 adult tickets   **3.** 35 peanut butter sandwiches
5. five 2-point baskets   **7.** $105   **9.** 30 students

Page 96   Lesson 1-7
1. Sample answers: $4 + 0 = 4$; $t + 0 = t$
3. $14 \times 0 = 0$; $0 \times 23 = 0$ **5.** 523   **7.** 409   **9.** 0
11. 0   **13.** Identity (+)   **15.** 0   **17.** 57   **19.** 1
21. 88   **23.** 0   **25.** 0   **27.** 0   **29.** 0   **31.** 0
33. $S = 3 \cdot 5 + 2 \cdot 0 + 1 \cdot 4$ **35.** $98.6 + 0 = 98.6$
37. Sample answer: The expression that shows her earnings is $125 + 1 \cdot x$. The Identity Property of Multiplication is used to find the earnings from sales. If she sells no CDs, the Multiplicative Property of Zero shows that she makes no money from sales. If this 0 is added to $125, the Identity Property of Addition is used to find her total earnings.   **39.** A   **41.** Comm. (+)   **43.** Assoc. (+)
45. 6

Page 104   Lesson 1-8
1. Sample answer: $4k$ and k, $9av$ and $12av$, $9x$ and $2x$ **3.** $6y$, $3y$ **5.** unlike terms   **7.** $13f + 3f = (13 + 3)f$ Distributive; $= 16f$ Order of Operations;
9. $+ 4 + 3y = 6y + (3y + 4)$ Comm. (+); $= (6y + 3y) + 4$ Assoc. (+); $= (6 + 3)y + 4$ Distributive; $= 9y + 4$ Order of Operations   **11.** $8(4g + 7n) = 32g + 56n$ Distributive   **13.** $150n$
15. $100c + 50b$ **17.** $9a$, $8a$ **19.** unlike terms
21. unlike terms   **23.** unlike terms   **25.** $15n + 3n = (15 + 3)n$ Distributive; $= 18n$ Order of Operations
27. $3n + n = 3n + 1n$ Identity (×); $= (3 + 1)n$ Distributive; $= 4n$ Order of Operations   **29.** $2n + 5 + 4n = 2n + 4n + 5$ Comm. (+); $= (2 + 4)n + 5$ Distributive; $= 6n + 5$ Order of Operations
31. $4(8j + 5) = 32j + 20$ Distributive   **33.** $8(2v + 9w) = 16v + 72w$ Distributive   **35.** $10c + 8(2cd + 3c) = 10c + 8 \cdot 2cd + 8 \cdot 3c$ Distributive; $= 10c + 16cd + 24c$ Order of Operations; $= 10c + 24c + 16cd$ Comm. (+); $= (10 + 24)c + 16cd$ Distributive; $= 34c + 16cd$; Order of Operations   **37.** $124.80   **39.** 138 pages   **41.** Gwen   **43.** 0   **45.** 0   **47.** $4k$
49. $50 + (30 \cdot 5)$

Chapter 2 Integers and Equations

Page 117   Lesson 2-1
1. Sample answer: $x + 3 = 11$ **3.** 4   **5.** 20
7. 7   **9.** 22   **11.** $4p = 24$; $p = 6$ **13.** 3   **15.** 6
17. 9   **19.** 12   **21.** 11   **23.** 14   **25.** 5   **27.** 21
29. $20p = 60$; $p = 3$ **31.** $x + 3 = 15$; $x = 12$

33. Sample answer: Find the value of the variable that makes the equation true.   **35.** A   **37.** $5x$
39. $6p + 6 + 9 + 4p = 6p + 4p + 6 + 9$ Comm. (+);
$= (6 + 4)p + 6 + 9$ Distributive; $10p + 15$ Order of Operations   **41.** $5 \times 8 + 16$; $56

Page 122   **Lesson 2-2**
1. Sample answer: 4 and -4. The numbers are the same distance from 0 on a number line but have different signs.   **3.** -5 **5.** -32 **7.** -3; 0; 2   **9.** >
11. $5 > -2$ or $-2 < 5$ **13.** Sample answer: $-8 < -2$; Which is colder, -8 degrees or -2 degrees?
15. 2   **17.** 11   **19.** 8   **21.** -5; -1; 3
23.
$\begin{array}{cccccccccccccc} -7 & -6 & -5 & -4 & -3 & -2 & -1 & 0 & 1 & 2 & 3 & 4 & 5 & 6 & 7 \end{array}$
25. <   **27.** >   **29.** -282; Use -282 because it is below sea level.   **31.** Sample answer: $4 < -6$ or $-6 > 4$; On the first play the team gained 4 yards. On the second play the team lost 6 yards. Show using < or >.   **33.** 3; -5 **35.** C   **37.** 3   **39.** 5
41. $53 = h + 30$; $h = 23$ **43.** 6   **45.** 4; 3

Page 132   **Lesson 2-3**
1. Sample answer: 6 and -6. The numbers are opposites and their sum is zero.   **3.** 4   **5.** -4
7. -11 **9.** 0   **11.** -7 **13.** 2   **15.** $-5n$ **17.** $5°C$
19. -1 **21.** 6   **23.** -3 **25.** -4 **27.** -6
29. -6 **31.** -16 **33.** -12 **35.** 0   **37.** 0
39. -3 **41.** 4   **43.** -8 **45.** 2   **47.** $4k$ **49.** $-5r$
51. 57 feet   **53.** Sample answer: $-6m + (-2m)$ and $4m - 12m$ **55.** C   **57.** <   **59.** >   **61.** $2n$

Page 138   **Lesson 2-4**
1. Sample answer: $(-2)(-5)(-2)$, $(2)^2(-5)$, and $(5)(-4)$ **3.** -12 **5.** 28   **7.** -6 **9.** -24 **11.** 18
13. -48 **15.** -20 **17.** -90 cm   **19.** -18
21. -40 **23.** 56   **25.** 20   **27.** 16   **29.** -30
31. -48 **33.** 18   **35.** 48   **37.** 10   **39.** 28
41. -20 **43.** 54   **45.** $-\$120$ **47.** -50
49. 1, -1, 1, -1; Sample answer: When -1 is used as a factor an even number of times, the product is 1, and when used an odd number of times, the product is -1; -1 is multiplied by itself an odd number of times. Each pair of $-1 \cdot -1$ is even. The last -1 makes the product negative.   **51.** G   **53.** 3
55. -5 **57.** 14 years old

Page 144   **Lesson 2-5**
1. Sample answers: $-27 \div 9$, $30 \div (-10)$, $-15 \div 5$
3. -6 **5.** 5   **7.** 625   **9.** -3 **11.** $-\$3$ **13.** -3
15. -11 **17.** 7   **19.** 9   **21.** -6 **23.** -8 **25.** -1
27. 81   **29.** -7 **31.** 9   **33.** -6 **35.** -10 yards
37. -3 meters   **39.** Sample answer: $-36 \div 9$, $12 \div (-3)$, and $16 \div (-4)$ **41.** C   **43.** 0   **45.** 0

Page 147   **Lesson 2-6**
1. 14 minutes   **3.** 2 P.M.   **5.** 7 cars
7. 10 handshakes

Page 152   **Lesson 2-7**
1. Sample answer: $x - 3 = 5$ **3.** 29   **5.** -9 **7.** 3
9. 8   **11.** $70°$ **13.** 0   **15.** 16   **17.** -8 **19.** 14
21. 3   **23.** 1   **25.** 3   **27.** 4   **29.** 12 feet
31. 450 miles   **33.** Sample answer: $3x + 15 = 24$ **35.** A   **37.** -5 **39.** -4 **41.** -28 **43.** 27

Chapter 3 Fractions

Page 167   **Lesson 3-1**
1. Both fractions are the same distance from 0 on the number line. The negative fraction is left of 0 and has a negative sign.
3.
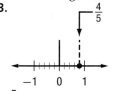
5. $\frac{5}{9}$
7. $\frac{4}{5}$; $-\frac{4}{5}$
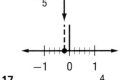
9. ; $-\frac{2}{5} > -\frac{3}{5}$
11. $-\frac{7}{9} < -\frac{5}{9} < -\frac{4}{9}$ **13.** $\frac{1}{5} < \frac{3}{5}$
15.
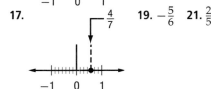
17.
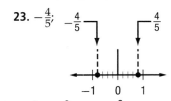
19. $-\frac{5}{6}$ **21.** $\frac{2}{5}$
23. $-\frac{4}{5}$;

25. $\frac{2}{7}$; $-\frac{2}{7}$

27. ; $\frac{4}{5} > \frac{2}{5}$
29. ; $-\frac{1}{8} > -\frac{3}{8}$
31. $-\frac{6}{7} < -\frac{5}{7} < -\frac{1}{7}$
33. $-\frac{3}{4}$ $\frac{3}{4}$; $\frac{3}{4} > -\frac{3}{4}$

35. Sample answer: $-\frac{1}{5} > -\frac{3}{5}$ **37.** B   **39.** 4
41. 18   **43.** 14 points

Page 173 **Lesson 3-2**

1. Sample answer: $-\frac{2}{3}$ and $-\frac{7}{4}$ **3.** improper

5. $1\frac{3}{5}$ **7.** $1\frac{1}{5}$

9.

11.

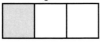

13. $2\frac{3}{10} > 1\frac{9}{10}$; He walked farther on Saturday.

15. Sample answer:

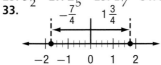

17. Sample answer:

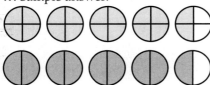

19. Answers will vary. **21.** proper **23.** improper

25. $5\frac{1}{2}$ **27.** $2\frac{2}{5}$ **29.** $1\frac{2}{7}$ **31.** $2\frac{1}{6}$

33.

35.

37.

39.

41. $\frac{37}{8} > 4\frac{1}{8}$; He has more tea.

43. Sample answer:

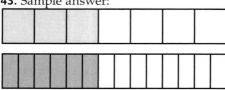

45. Sample answer:

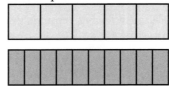

47. $-1\frac{1}{2}°F$ **49.** Answers will vary. **51.** B

53. ; $-\frac{3}{6} > -\frac{5}{6}$

55. ; $\frac{6}{7} > \frac{5}{7}$ **57.** 7 **59.** 4 inches

Page 181 **Lesson 3-3**

1. 11; It only has two factors: 1 and 11; 9; It has more than two factors: 1, 3, and 9. **3.** The greatest common factor of the numerator and denominator is 1. **5.** prime **7.** composite **9.** 1, 2, 5, 10
11. 1, 3, 9, 27 **13.** 1 **15.** 4 **17.** 3 **19.** $\frac{5}{8}$ **21.** $\frac{1}{3}$
23. $\frac{1}{6}$ **25.** prime **27.** composite
29. composite **31.** prime **33.** 1, 2, 4, 8
35. 1, 2, 3, 6, 9, 18 **37.** 1, 23
39. 1, 2, 3, 5, 6, 10, 15, 30 **41.** 1, 3 **43.** 1, 5 **45.** 6
47. 15 **49.** $\frac{2}{5}$ **51.** $\frac{1}{3}$ **53.** $\frac{2}{9}$ **55.** 6 beads
57. 1; A prime number has only 1 and itself as factors, so the GCF of any two prime numbers can only be 1. **59.** $-2\frac{2}{3} > -3\frac{1}{8}$ **61.** 4 **63.** $-\$2$

Page 185 **Lesson 3-4**

1. 15 games **3.** $\frac{1}{8}$ **5.** 30 people; $\frac{4}{15}$ **7.** 72 dots
9. 3 cakes

Page 190 **Lesson 3-5**

1. Multiply the numerators and multiply the denominators, then simplify. **3.** $\frac{4}{45}$ **5.** $\frac{1}{3}$ **7.** $\frac{2}{15}$

9. $2\frac{1}{2}$

11. Sample answer:

The diagram shows $\frac{3}{4}$ of the basement carpeted. The dashed line shows half of the carpeted portion. The fraction $\frac{3}{8}$ represents how much of the whole basement will be carpeted. **13.** $\frac{16}{81}$
15. Multiply the numerators and the denominators separately, then simplify: $\frac{a}{b} \cdot \frac{c}{d} = \frac{ac}{bd}$. **17.** $\frac{8}{15}$ **19.** $\frac{9}{32}$
21. $\frac{1}{3}$ **23.** $\frac{11}{16}$ **25.** $-\frac{15}{32}$ **27.** $\frac{21}{40}$ **29.** $\frac{5}{6}$
31. Sample answer:

The diagram shows $\frac{1}{4}$ of the bar's Calories are fat. The dashed line shows $\frac{2}{3}$ of the fat Calories are unsaturated. The fraction $\frac{1}{6}$ represents how much of the total calories are unsaturated fat.
33. $\frac{1}{3}$ cup; Sample answer: Multiply $\frac{2}{3}$ cup of apples by $\frac{1}{2}$ to get the new apple measurement, and simplify. $\frac{2}{3} \times \frac{1}{2} = \frac{1}{3}$ **35.** $\frac{16}{81}$ **37.** $\frac{27}{1000}$

39. Sample answer: If the numerator in the other factor is 1, the numerator in the product will be two; $2 \times 1 = 2$. The product $\frac{2}{6} = \frac{1}{3}$. So, the denominator must be 2 since $3 \times 2 = 6$. Therefore, the other factor is $\frac{1}{2}$. $\frac{2}{3} \times \frac{1}{2} = \frac{2}{6} = \frac{1}{3}$ **41.** G
43. 4 **45.** 6

Page 196   Lesson 3-6
1. A multiplicative inverse is the reciprocal of a number. **3.** $\frac{10}{7}$ **5.** $\frac{9}{14}$ **7.** $\frac{15}{14}$ or $1\frac{1}{14}$ **9.** $\frac{33}{16}$ or $2\frac{1}{16}$
11. Sample answer:

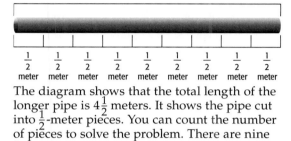

$4\frac{1}{2}$ meters

| $\frac{1}{2}$ meter | $\frac{1}{2}$ meter | $\frac{1}{2}$ meter | $\frac{1}{2}$ meter | $\frac{1}{2}$ meter | $\frac{1}{2}$ meter | $\frac{1}{2}$ meter | $\frac{1}{2}$ meter | $\frac{1}{2}$ meter |

The diagram shows that the total length of the longer pipe is $4\frac{1}{2}$ meters. It shows the pipe cut into $\frac{1}{2}$-meter pieces. You can count the number of pieces to solve the problem. There are nine $\frac{1}{2}$-meter pipes. **13.** Answers will vary.
15. $-\frac{2}{4}$ or $-\frac{1}{2}$ **17.** $\frac{1}{19}$ **19.** $\frac{2}{3}$ **21.** $\frac{2}{3}$ **23.** $-\frac{4}{3}$ or $-1\frac{1}{3}$
25. $-\frac{3}{5}$ **27.** $\frac{9}{4}$ or $2\frac{1}{4}$ **29.** $\frac{29}{14}$ or $2\frac{1}{14}$
31. Sample answer:

$2\frac{2}{5}$ meters

| $\frac{3}{5}$ meter | $\frac{3}{5}$ meter | $\frac{3}{5}$ meter | $\frac{3}{5}$ meter |

The diagram shows that the total length of the pipe is $2\frac{2}{5}$ meters. It shows the pipe cut into 4 equal pieces. You can use the measurement of each piece to solve the problem. Each piece is $\frac{3}{5}$ meter long.
33. $2\frac{1}{2}$; Sample answer: Divide the total $\frac{25}{4}$ gallons of water by the $\frac{5}{2}$ gallons the watering can holds. To divide, multiply $\frac{25}{4}$ and the reciprocal of $\frac{5}{2}$. Change the product to a mixed number. She fills the watering can $2\frac{1}{2}$ times. $\frac{25}{4} \times \frac{2}{5} = \frac{5}{2} = 2\frac{1}{2}$
35. quotient; When you multiply by a rational number between 0 and 1, the product is smaller than the original number because you are dividing a part of the original. When you divide by a rational number between 0 and 1, the result is larger because you are finding how many number parts are in the original. For example, $2 \times \frac{1}{3} = \frac{1}{6}$ which is less than $2 \div \frac{1}{3} = 6$ **37.** H **39.** $\frac{1}{6}$
41. $\frac{7}{2}$ or $3\frac{1}{2}$

Page 203   Lesson 3-7
1. Sample answer: $\frac{5}{8}$ and $\frac{7}{8}$ **3.** $\frac{1}{2}$ **5.** $\frac{1}{9}$ **7.** 4 **9.** $2\frac{1}{2}$
11. subtraction; Katie is taking fabric away from the roll. **13.** 1 whole note **15.** $\frac{8}{9}$ **17.** 1 **19.** $\frac{3}{4}$
21. $\frac{2}{7}$ **23.** $\frac{5}{6}$ **25.** $\frac{2}{3}$ **27.** $4\frac{1}{2}$ **29.** 4 **31.** $1\frac{2}{3}$

33. $2\frac{1}{2}$ **35.** Addition. Mia is adding additional trim to her scarf. **37.** $\frac{1}{2}$ inch **39.** $1\frac{1}{2}$ minutes
41. Sample answer: $\frac{1}{4} + \frac{1}{4}$; $\frac{3}{4} - \frac{1}{4}$ **43.** C **45.** $\frac{20}{27}$
47. $\frac{1}{2}$ **49.** 8 bars

Page 210   Lesson 3-8
1. Sample answer: $\frac{3}{4}$ and $\frac{5}{7}$ **3.** 12 **5.** $\frac{6}{8}$ **7.** 8
9. $\frac{7}{8}$ **11.** $\frac{7}{10}$ **13.** $2\frac{2}{3}$ **15.** addition; Sample answer: Adding the thickness of the base and the thickness of the tile on top gives the total thickness.
17. $\frac{3}{4}$ page **19.** 6 **21.** 20 **23.** 3 **25.** 16 **27.** 12
29. $\frac{5}{8}$ **31.** $\frac{5}{6}$ **33.** $\frac{19}{20}$ **35.** $\frac{19}{12}$ or $1\frac{7}{12}$ **37.** $\frac{39}{35}$ or $1\frac{4}{35}$
39. $3\frac{4}{5}$ miles **41.** $\frac{3}{16}$ inch **43.** $2\frac{3}{4}$ **45.** $5\frac{7}{12}$
47. $4\frac{1}{8}$ inches **49.** No; Sample answer: The product is always a common multiple, but not always the LCM. For example, the LCM of 5 and 10 is 10, not $5 \cdot 10 = 50$. **51.** J **53.** $\frac{1}{2}$ **55.** $\frac{8}{5}$ or $1\frac{3}{5}$

Page 216   Lesson 3-9
1. The fractions will have different denominators.
3. $\frac{5}{8}$ **5.** $\frac{1}{3}$ **7.** $1\frac{5}{8}$ **9.** $1\frac{1}{2}$ **11.** subtraction; Sample answer: You subtract the hours he has worked from the total hours to find how many hours are left. **13.** $\frac{1}{12}$ inch **15.** $\frac{1}{8}$ **17.** $\frac{1}{6}$ **19.** $\frac{1}{24}$ **21.** $\frac{3}{14}$
23. $\frac{3}{10}$ **25.** $1\frac{1}{4}$ **27.** $2\frac{1}{2}$ **29.** $1\frac{2}{3}$ **31.** $2\frac{5}{8}$ **33.** subtraction; Subtract the measure of the bracelet beads from the measure of the necklace beads to find the difference. **35.** $1\frac{1}{4}$ miles **37.** $\frac{5}{8}$ acre
39. Sample answers: $\frac{2}{3} - \frac{1}{3}$; $\frac{5}{6} - \frac{1}{2}$ **41.** $1\frac{3}{4}$ cups
43. C **45.** $\frac{5}{6}$ **47.** $\frac{7}{8}$ **49.** $\frac{3}{4}$

Page 224   Lesson 3-10
1. Sample answer: An equation is two expressions on either side of an equal sign. **3.** $\frac{3}{4}f + \frac{11}{12}$
5. $\frac{5}{8}b + \frac{5}{6}$ **7.** $z = \frac{3}{8}$ **9.** $x = \frac{4}{9}$ **11.** 36 red roses
13. $w = \frac{1}{6}$ **15.** $t = 3\frac{2}{3}$ **17.** $\frac{1}{2}d + \frac{1}{4}$ **19.** $\frac{11}{12}t - \frac{2}{3}$
21. $\frac{1}{4}m + \frac{1}{2}$ **23.** $x = \frac{1}{2}$ **25.** $y = \frac{1}{12}$ **27.** $d = 1\frac{5}{6}$
29. $b = \frac{9}{10}$ **31.** $m = \frac{2}{3}$ **33.** $n = 1\frac{1}{5}$ **35.** $12\frac{3}{4}$ acres
37. $5 **39.** $k = \frac{3}{4}$ **41.** $m = \frac{1}{4}$ **43.** $b = 1\frac{2}{3}$
45. $\frac{1}{2}m + \frac{1}{2}c$; $\frac{1}{2}$ **47.** Sample answer: I disagree with her solution. The answer should be $\frac{4}{9}$. **49.** G
51. $\frac{3}{8}$ **53.** $\frac{11}{24}$

Chapter 4 Decimals

Page 241   Lesson 4-1
1. Sample answer: Both represent part of a whole. A fraction is written as a quotient; decimals are easier to compare. **3.** $-\frac{101}{200}$ **5.** $-3\frac{2}{5}$ **7.** $-0.4\overline{6}$

9. 5.775 **11.** Fabio **13.** $\frac{14}{25}$ **15.** $-\frac{22}{25}$ **17.** $-2\frac{12}{25}$
19. $-17\frac{3}{4}$ **21.** $-0.\overline{45}$ **23.** $0.7\overline{3}$ **25.** $16.3\overline{18}$
27. 28.2 **29.** No; $7\frac{5}{8} = 7.625$, so $7.5 < 7\frac{5}{8}$.
31. The bush grew $\frac{1}{8}$ cm each time; $51\frac{3}{8}$ cm.
33. B **35.** $y = 1\frac{1}{4}$ **37.** $t = \frac{13}{20}$ **39.** \$20 **41.** Identity
Property (×) **43.** Commutative Property (×)

Page 246   Lesson 4-2
1. Identity Property of Addition **3.** Commutative
Property of Addition **5.** 74.023 **7.** 2.728 **9.** 21.5
km **11.** -5.36 **13.** $3.67x + (6.78 - 5.87x); = 3.67x +$
$(5.87x + 6.78)$ Commutative Prop. (+); $= (3.67x -$
$5.87x) + 6.78$ Associative Prop. (+); $= -2.20x + 6.78$
Subtract **15.** 154.46 **17.** 28.97 **19.** 15.56 **21.** 7.02
23. 45.3 inches **25.** \$30.88 **27.** -7.7 **29.** -35.2
31. $14.5k + (5.4 + 4.5k); = 14.5k + (4.5k + 5.4)$
Commutative Prop. (+); $= (14.5k + 4.5k) + 5.4$
Associative Prop. (+); $= 19.0k + 5.4$ Add
33. $8.0n + (12.3 - 7.8n); = 8.0n \, (-7.8n + 12.3)$
Commutative Prop. (+); $= (8.0n - 7.8n) + 12.3$
Associative Prop. (+); $= 0.2n + 12.3$ Subtract
35. $(4.2m + 6.4p) - (4.2p + 6.7m) = (4.2m + 6.4p) +$
$(-4.2p - 6.7m)$ Distributive Prop.; $= (4.2m + 6.4p) +$
$(-6.7m - 4.2p)$ Commutative Prop. (+); $= (4.2m +$
$-6.7m) + (6.4p - 4.2p)$ Commutative Prop. (+);
Associative Prop. (+); $= (-2.5m) + (6.4p - 4.2p)$
Add; $= -2.5m + 2.2p$ Subtract **37.** Sample answer:
If zeros are not used, mistakes might be made. For
example, $42 - 34.8$ might be calculated as 8.8 instead
of 7.2. **39.** B **41.** 0.22 **43.** 0.45 **45.** $2\frac{19}{20}$ **47.** $17\frac{2}{3}$

Page 254   Lesson 4-3
1. Distributive Property **3.** Commutative Property
(×) **5.** 0.03366 **7.** 541.35 gal **9.** 12.96 **11.** 357.911
13. 0 **15.** $5.2p + 11.96$ **17.** 342.0 **19.** 3.564
21. -17.16 **23.** \$240.19 **25.** 120 min **27.** 50.41
29. 26.2144 **31.** 118.5921 **33.** $7.1y$ **35.** $26.13m$
37. $4.1\ell + 19.27$ **39.** $9.0f + 65.70$ **41.** Sample
answer: Nestor bought 2.3 pounds of steak at \$4.89
a pound and 1.2 pounds of chicken at \$2.99 a
pound. How much did he spend? **43.** H
45. 54.525 **47.** 3.413 **49.** $m = 37$ **51.** $p = 13$

Page 259   Lesson 4-4
1. Locate the digit in hundredths place. If this digit
is 5 or greater, round up the digit in tenths place.
If this digit is 4 or less, the tenths digit stays the
same. **3.** 12 **5.** 4.6 **7.** 0.37 **9.** 20 servings
11. 13 **13.** 5 **15.** 20 **17.** 1.3 **19.** 10.1 **21.** 620.88
23. 7.88 **25.** 47.7 kilometers per hour
27. 19 pillows (She would not cover parts of a
pillow. The answer must be a whole number.)
29. Orlando correctly multiplied speed and time.
Mishka divided speed by time and misplaced
the decimal by one place. **31.** B **33.** -394.68
35. 0.0135 **37.** 60 **39.** 36

Page 263   Lesson 4-5
1. \$3.05 per pound **3.** 13.2 pounds
5. 108 cards **7.** 24 outfits

Page 268   Lesson 4-6
1. equation **3.** 1.32
5. $3 \times 3.6 \times 3.6 + 2 \times 6.8 \times 6.8$; 131.36 square
inches **7.** $6.4t - 1.9s$ **9.** $j = 4.2$ **11.** 16.28
13. 10.27 **15.** $4.7 \times 2.19 + 4.9 \times 0.79$; \$14.16
17. $0.8y$ **19.** $7.73q - 7.234$ **21.** $s = 3.7$ **23.** $x = 1.5$
25. $45 = 1.875n$; $n = 24$ table tops **27.** Sample
answer: Mo works four hours a day. She gets paid
\$13.95 for each shift and \$4.75 for each DVD she
sells. If she makes \$28.20 on one shift, how many
DVDs did she sell? Answer: 3 DVDs **29.** B
31. 120 bacteria **33.** $3\frac{1}{7}$

Chapter 5 Exponents and Roots

Page 282   Lesson 5-1
1. product of powers **3.** quotient of powers
5. t^{11} **7.** $-70g^9$ **9.** $12t^{11}$ **11.** 2^6 **13.** $9r^2s^7$
15. 4^{12} **17.** $27f^6g^{12}$ **19.** 3^{10} **21.** z^{14} **23.** d^9
25. $15z^{15}$ **27.** $15d^{17}f^2$ **29.** $3x^4y^9$ **31.** 3^5 panes
33. 8 **35.** c^8 **37.** $-2h^{11}$ **39.** 4^4 campers
41. 5^{12} **43.** $16d^{20}$ **45.** $r^{48}s^8$ **47.** The expressions
are alike because they all have exponents and all
simplify to x^{12}. They are different because one is a
quotient, one is a product, and one is a power.
49. C **51.** B **53.** $x = 34.84$ **55.** 14

Page 289   Lesson 5-2
1. Sample answer: $x^{-3} = \frac{1}{x^3}$ **3.** 169 **5.** $\frac{k^9}{m^5}$ **7.** $\frac{1}{b^3}$
9. s^7 **11.** 10^3 millimeters **13.** $\frac{16h^{16}}{k^4}$ **15.** $\frac{1}{512}$
17. 1 **19.** $\frac{1}{y^{34}}$ **21.** $\frac{q^{76}}{p^{50}}$ **23.** $\frac{32}{q^8}$ **25.** r^5
27. $-12u^5v^4$ **29.** $\frac{21}{pq^2}$ **31.** $\frac{1}{t^2}$ **33.** $\frac{j}{3k^2}$ **35.** $\frac{3n^4}{2m^3}$
37. 2^8 grains of powder **39.** $\frac{4x^8}{y^4}$ **41.** $\frac{2c^9}{y^{15}}$
43. -1 **45.** x^n means x multiplied n times. x^{-n}
means 1 divided by (x multiplied n times).
47. H **49.** $63g^{10}h^6$ **51.** $27m^4n^9$

Page 292   Lesson 5-3
1. 30 squares **3.** 20,100 **5.** 27 triangles
7. 77 pieces **9.** 34567

Page 296   Lesson 5-4
1. $5 \cdot 5 = 25$ **3.** 4 **5.** -8 **7.** $\sqrt{81}, 10.4, \sqrt{121},$
$11\frac{3}{5}$ **9.** 9 cm **11.** 7 **13.** 5 **15.** 1 **17.** 13 **19.** -5
21. -3 **23.** $\sqrt{49}, 7\frac{3}{8}, \sqrt{64}, \sqrt{100}$ **25.** $4\frac{5}{8}, 4.8, \sqrt{25},$
$5\frac{1}{5}$ **27.** $\sqrt{144}, 12\frac{1}{8}, \sqrt{169}, 13\frac{3}{4}$ **29.** $\sqrt{25}, 5.25, 5\frac{5}{8},$
$\sqrt{36}$ **31.** 10 in. **33.** 3 **35.** 55 **37.** No number
multiplied by itself equals -16. **39.** H **41.** 10
43. 20 **45.** $6s^2$

Page 301   Lesson 5-5
1. to find how fast water leaves the nozzle of a fire
hose **3.** 25, 25 **5.** 50, 50 **7.** 4 **9.** 240 gallons per
minute **11.** 10 in. **13.** 16, 16 **15.** 64, 64

17. 16, 16 **19.** 160, 160 **21.** −72 **23.** 83
25. 4 in. **27.** 480 gal/min **29.** No; $(5 − 3)^2 = 4$,
but $5^2 − 3^2 = 25 − 9 = 16$. **31.** A **33.** 11 in.
35. −3 **37.** −36 **39.** −16

Page 308   **Lesson 5-6**
1. Answers will vary. **3.** > **5.** < **7.** < **9.** Yes;
$0.375 = \frac{3}{8}$ **11.** $−\sqrt{36}, −5\frac{1}{4}, 8^{−2}$ **13.** > **15.** <
17. > **19.** < **21.** > **23.** > **25.** middle school
team **27.** Yes. $2 × 0.05 = 0.1 = \frac{1}{10}$ **29.** $−3^2, −\frac{9}{9}$,
$\sqrt{9}$ **31.** $4^5 = (2^2)^5 = 2^{10} > 2^9$ so 4^5 is greater.
33. J **35.** 0.8 **37.** $2\frac{4}{7}$

Chapter 6 Ratios, Rates, Proportion, and Percent

Page 321   **Lesson 6-1**
1. ratio **3.** $\frac{3}{5}$ **5.** 0.8 **7.** 44 feet per second
9. $\frac{1}{5}$ **11.** $\frac{2}{3}$ **13.** 6:11 **15.** 0.8 **17.** 33 miles per
gallon **19.** $250 per employee **21.** 7.0
23. 9 minutes per mile **25.** 90 miles **27.** D
29. B **31.** > **33.** 351 square yards larger

Page 327   **Lesson 6-2**
1. Sample answer: Percent means *hundredths* or *out
of 100*. It is a ratio that compares a number to 100.
3. $\frac{9}{100}$; 0.09 **5.** $\frac{17}{20}$; 0.85 **7.** $\frac{53}{20}$; 2.65 **9.** 0.52; 52%
11. 1.3; 130% **13.** 0.93; 93% **15.** 0.35 **17.** $\frac{1}{50}$; 0.02
19. $\frac{1}{100}$; 0.01 **21.** $\frac{9}{50}$; 0.18 **23.** $\frac{13}{50}$; 0.26 **25.** $\frac{57}{20}$; 2.85
27. $\frac{22}{5}$; 4.4 **29.** 0.8; 80% **31.** 0.68; 68% **33.** 1.2;
120% **35.** 1.75; 175% **37.** 1.5; 150% **39.** 1.04;
104% **41.** 28% **43.** 31% = $\frac{31}{100}$; 26% = $\frac{13}{50}$; 8% = $\frac{2}{25}$;
35% = $\frac{35}{100}$ **45.** Sample answer: Find the total
students by adding boys and girls: $9 + 15 = 24$.
Then find the percent of 15 girls out of 24 students:
$\frac{15}{24} = 0.625 = 62.5\%$ **47.** H **49.** 0.8 **51.** 0.75
53. 2 **55.** 6 **57.** $28x + 51y$

Page 333   **Lesson 6-3**
1. cross product **3.** $\frac{4}{6} \stackrel{?}{=} \frac{30}{45}$; 180 = 180; It is a
proportion. **5.** $\frac{15}{24} \stackrel{?}{=} \frac{10}{16}$; 240 = 240; It is a
proportion. **7.** $r = 15$ **9.** $c = 3$ **11.** $h = 0.36$
13. 7.8 meters **15.** $\frac{11}{13} \stackrel{?}{=} \frac{4}{7}$; 77 ≠ 52; It is not a
proportion. **17.** $\frac{6}{7} \stackrel{?}{=} \frac{8}{12}$; 72 ≠ 56; It is not a
proportion. **19.** $\frac{12}{18} \stackrel{?}{=} \frac{8}{12}$; 144 = 144; It is a
proportion. **21.** $n = 2$ **23.** $m = 12$ **25.** $x = 60$
27. $s = 3$ **29.** $c = 0.72$ **31.** $k = 18$ **33.** 324 lamps
35. $2.00 **37.** Answers will vary. **39.** C **41.** 0.18
43. 1.43 **45.** $\frac{1}{20}$ **47.** 16

Page 339   **Lesson 6-4**
1. percentage **3.** 60% **5.** 375% **7.** 28 **9.** 45
11. 20% **13.** $70 **15.** Sample answer: Write the
information you know as a percent proportion
equation. Solve for the unknown. **17.** 80%
19. 62.5% **21.** 260% **23.** 125% **25.** 3 **27.** 4
29. 60 **31.** 625 **33.** 70% **35.** 288% **37.** $144
39. about 5,250 **41.** 15% **43.** G **45.** $n = 6$
47. $0.\overline{8}$ **49.** $0.958\overline{3}$ **51.** $7x$ **53.** $−72ab$

Page 346   **Lesson 6-5**
1. discount **3.** commission **5.** $1,350 **7.** $10.50
9. Answers will vary. **11.** $6.40 **13.** $72
15. $31.50 **17.** $39 **19.** $4,950 **21.** 8.72 MHz
23. Jan is correct. She subtracted 10% of the
previous week's price. Patrick incorrectly
subtracted 10% of the original price each
week. **25.** F **27.** 120 **29.** 22

Page 352   **Lesson 6-6**
1. Sample answer: $y = 3x$; $y = 5x$; $y = 7x$ **3.** yes; 43
5. $t = 6.5$ **7.** 50 pounds **9.** yes; −150 **11.** −30
13. $180 **15.** 28 feet **17.** $x = 5$; $y = 75$; $y = 12.5x$
19. F **21.** $7.38 **23.** 2.75

Page 356   **Lesson 6-7**
1. 8 hours **3.** 12 years **5.** 12 **7.** 31.5 cups **9.** $78

Chapter 7 Algebra on the Coordinate Plane

Page 369   **Lesson 7-1**
1–4.

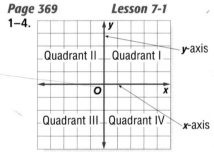

5. A (4, −1) **7.** C (3, 0)
8–10.

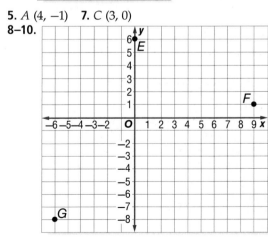

11.

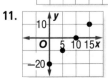

13. $J(-7, -6)$ **15.** $L(4, 4)$ **17.** $N(-5, 3)$
18–23.

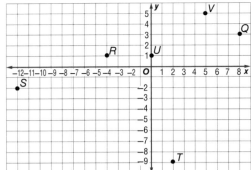

25.

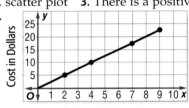

27. The taxi would not drive for -2 miles and charge a passenger.   **29.** H   **31.** yes   **33.** $\frac{1}{e^2}$

Page 373   Lesson 7-2
1. \$6   **3.** 400 miles   **5.** $5\frac{1}{2}$ feet   **7.** $\frac{1}{2}$ in.   **9.** 8 and 4

Page 376   Lesson 7-3
1. scatter plot   **3.** There is a positive relationship.
5.

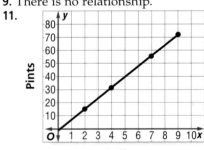

Boxes of Cereal

A box of cereal costs \$2.50.
7. There is a positive relationship.
9. There is no relationship.
11.

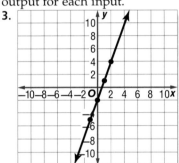

Gallons

There are 8 pints in a gallon.
13.

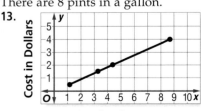

Pounds of Bananas
One pound of bananas costs \$2.20.

15. If the points tend to go up, there is a positive relationship. If they tend to go down, it is a negative relationship. If they show no pattern, there is no relationship.   **17.** 300 miles   **19.** $x = -4$
21. $x = 4$

Page 385   Lesson 7-4
1. change units or check your answer.   **3.** 3,520 strides   **5.** 220 passenger-miles   **7.** 12 ft/sec
9. 80 m/min   **11.** 350 gal/wk   **13.** 2 people per square mile   **15.** 990 passenger-miles
17. 260 person-hours   **19.** Pedro — 2 people travel 100 miles (200 passenger-miles) and 3 travel 50 miles (150 passenger-miles): $200 + 150 = 350$ passenger-miles.   **21.** G   **23.** positive relationship
25. -6

Page 391   Lesson 7-5
1. slope   **3.** run   **5.** $\frac{1}{20}$ **7.** $\frac{3}{2}$ **9.** 3   **11.** $-\frac{3}{2}$ **13.** $-\frac{4}{5}$
15. 1   **17.** 1   **19.** 5   **21.** -2 **23.** No; a change down is a negative change and a change left is also a negative change. That produces a positive slope, but the given slope is negative.   **25.** C   **27.** 48 passenger-hours   **29.** $\frac{1}{bd^3}$

Page 397   Lesson 7-6
1. A function is a relation that assigns exactly one output for each input.
3.

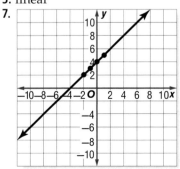

Sample answer:

x	$3x - 2$	y	(x, y)
-1	-5	-5	$(-1, -5)$
0	-2	-2	$(0, -2)$
1	1	1	$(1, 1)$
2	4	4	$(2, 4)$

5. linear
7.

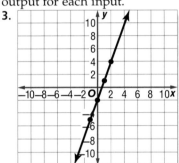

Sample answer:

x	$x + 4$	y	(x, y)
−2	2	2	(−2, 2)
−1	3	3	(−1, 3)
0	4	4	(0, 4)
1	5	5	(1, 5)

9.

Sample answer:

x	$2x + 4$	y	(x, y)
0	4	4	(0, 4)
1	6	6	(1, 6)
2	8	8	(2, 8)
3	10	10	(3, 10)

11. linear **13.** nonlinear

15.

x	y
50	1
100	2
150	3
200	4

The car does travel at a linear rate.

17. See students' work. Sample answer: Linear functions are ones in which the rate of change in the x-values and the y-values are constant. Nonlinear functions are functions in which the rate of change in the x-values and y-values are not constant. **19.** D **21.** $-\frac{2}{5}$ **23.** 10 **25.** 0.32 **27.** 9°C

Page 404   **Lesson 7-7**
1. In a right triangle with legs a and b and hypotenuse c, $a^2 + b^2 = c^2$. **3.** 2^2 cm + 1.5^2 cm = 2.5^2 cm **5.** 8 in. **7.** 3^2 cm + 4.2^2 cm = 5.2^2 cm **9.** 34 ft **11.** 5 ft **13.** 5 units **15.** Yes. **17.** 3 ft **19.** Yes. Using the Pythagorean Theorem, the hypotenuse of a triangle with legs 8 feet and 22 feet is 23.4 feet. This means the ladder must be at least that long. The 24-foot ladder is longer, so it meets the safety requirements.
21. 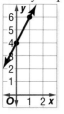   **23.** 4

Page 418   **Lesson 8-1**
1. quadrilateral **3.** triangle **5.** 12° **7.** 95°
9. triangle **11.** 79° **13.** 40° **15.** 36°
17. rectangle **19.** trapezoid **21.** Both figures are triangles, both have three sides and three angles, and both have angles that total 180 degrees. One triangle has an angle of 90 degrees and the other triangle does not. In the triangle with the 90-degree angle, the sum of the square of the sides equals the square of the hypotenuse. **23.** B **25.** 25 meters **27.** 11 **29.** 46.627

Page 423   **Lesson 8-2**
1. congruent **3.** $\angle A \cong \angle L$, $\angle B \cong \angle M$, $\angle C \cong \angle N$, $\overline{AB} \cong \overline{LM}$, $\overline{BC} \cong \overline{MN}$, $\overline{CA} \cong \overline{NL}$ **5.** 20 in. **7.** $\angle C \cong \angle F$, $\angle D \cong \angle G$, $\angle E \cong \angle H$, $\overline{CD} \cong \overline{FG}$, $\overline{DE} \cong \overline{GH}$, $\overline{EC} \cong \overline{HF}$ **9.** $\angle W \cong \angle T$, $\angle X \cong \angle U$, $\angle Y \cong \angle V$, $\overline{WX} \cong \overline{TU}$, $\overline{XY} \cong \overline{UV}$, $\overline{YW} \cong \overline{VT}$ **11.** $\angle K \cong \angle A$; $\angle L \cong \angle C$; $\angle M \cong \angle B$; $\overline{KL} \cong \overline{AC}$; $\overline{LM} \cong \overline{CB}$; $\overline{MK} \cong \overline{BA}$; BCA **13.** 4 cm **15.** Sample answers: 3.0 cm by 4.4 cm; 4.5 cm by 6.6 cm; 15 cm by 22 cm **17.** C **19.** J **21.** 110° **23.** $y = -\frac{4}{3}x + 4$

Page 432   **Lesson 8-3**
1. Distance Formula **3.** 17 **5.** 5 **7.** $\sqrt{338} \approx 18.4$ miles **9.** Use the Distance Formula to find the length of each side. If opposite sides are congruent, $LMNO$ is a parallelogram. **11.** B **13.** 5 in. **15.** $x = 50°$

Page 437   **Lesson 8-4**
1. perimeter **3.** 105 ft **5.** 18.84 mm **7.** 17 cm **9.** 11 ft **11.** 31.4 in. **13.** 81.64 in. **15.** The circle has a greater circumference (37.68 cm), compared with the perimeter of the triangle (36 cm). **17.** B **19.** 15 **21.** $(3 \cdot x) + (3 \cdot 6)$

Page 445   **Lesson 8-5**
1. Area is the measure of the surface enclosed by a geometric figure. **3.** 555 mm² **5.** 24 in² **7.** 12.56 m² **9.** 544 m² **11.** 17 mm² **13.** 85 m² **15.** 803.84 mm² **17.** about 51,852 mi² **19.** 1,450 ft² **21.** about 37.1 mi **23.** C **25.** 48 yd **27.** 75.36 in. **29.** 33.3%

Page 449   **Lesson 8-6**
1. 100 cm **3.** 15 inches by 15 inches **5.** 36 square meters **7.** $438.26

Page 455   **Lesson 8-7**
1. a and b **3.** b **5.** 125 cm³ **7.** 226.08 cm³ **9.** 216 mm³ **11.** 64 ft³ **13.** 1,884 yd³ **15.** 1,004.8 in³ **17.** 160 in³ **19.** Doubling each demension results in a volume eight times the original. **21.** D **23.** 12.56 ft² **25.** 6 m

Page 460   **Lesson 8-8**
1. Surface area is the sum of the areas of all faces of a figure. **3.** 664 in² **5.** 10,048 ft² **7.** 1,350 m² **9.** 28 mm² **11.** 151 ft² **13.** 384 in² **15.** 24 ft² **17.** Answers will vary. **19.** D **21.** 35.25 in³ **23.** $4\frac{5}{14}$ **25.** $3\frac{7}{12}$

Photo Credits

Cover: (t)Knott's Silver Bullet photo courtesy of Knott's Berry Farm, Buena Park, CA, (b)Created by Michael Trott ; **iv** (tl)Doug Martin, (tr)(b)The McGraw-Hill Companies; **v** The McGraw-Hill Companies; **vi** The McGraw-Hill Companies; **viii** Digital Vision/PunchStock; **ix** Duomo/Corbis; **x** Pierre Holtz/epa/Corbis; **xi** Robert E Daemmrich/Getty Images; **xii** Yellow Dog Productions/Getty Images; **xiii, xiv** Corbis; **xv** Doug Menuez/Getty Images; **xvi** Davis Barber/PhotoEdit; **xvii** Digital Vision/PunchStock; **xxii** Jeremy Woodhouse/Getty Images; **1** Barros & Barros/Getty Images; **24** Corbis; **33** United States coin images from the United States Mint; **58–59** Duomo/Corbis; **61** Tom Stewart/Corbis; **62, 67, 71** Corbis; **73** PunchStock; **80** Comstock Images/Alamy; **86** PhotoLink/Photodisc/Getty Images; **88** Bob Daemmrich/PhotoEdit; **95** SSC/SuperStock; **103** Paul Barton/Corbis; **112–113** Pierre Holtz/epa/Corbis; **118** Brad Wrobleski/Masterfile; **123** Wonderfile/Masterfile; **129** Masterfile; **131** Myrleen Ferguson Cate/Photo Edit; **133** (l)Ed Freeman/Getty Images, (r)F. Lukasseck/Masterfile; **140** Stuart Westmorland/Getty Images; **147** Masterfile; **153** Randy Faris/Corbis; **160–161** Robert E Daemmrich/Getty Images; **168** B.S.P.I./Corbis; **174** Dwayne Newton/PhotoEdit; **182** (l)Amos Morgan/Getty Images, (r)Tim Pannell/Corbis; **184** Ed Murray/Star Ledger/Corbis; **191** David Wells/The Image Works; **195** Corbis; **197** Michael Newman/PhotoEdit; **204** Ariel Skelley/Getty Images; **209** Stockbyte/PunchStock; **210** Eddie Brady/Lonely Planet Images; **215** Ingram Publishing/Alamy; **218** Jeff Greenberg/PhotoEdit; **222** Creatas/PunchStock; **225** Ariel Skelley/Corbis; **234–235** Yellow Dog Productions/Getty Images; **241** (l)Corbis, (tr)Isabelle Rozebaum/PhotoAlto, (cr)C Squared Studios/Getty Images, (br)Envision/Corbis; **252** Jeremy Hoare/Life File/Getty Images; **262** Ariel Skelley/Corbis; **263** (l)Michael Houghton/StudiOhio, (r)United States coin images from the United States Mint; **267** The Palma Collection/Photodisc/Getty Images; **276–277** Corbis; **278** Brand X Pictures; **280** BananaStock/PunchStock; **281** Frank Krahmer/Masterfile; **288** Ambient Images Inc./Alamy; **291** United States coin images from the United States Mint; **295** Photodisc/Getty Images; **297** David Madison/Getty Images; **301** Philip Rostron/Masterfile; **302** David Hume Kennerly/Getty Images; **309** Jon Feingersh/Masterfile; **316–317** Corbis; **319** Dynamic Graphics/Jupiter Images; **324** Comstock/PictureQuest; **332** Richard Price/Getty Images; **338** Corbis; **340** Masterfile; **347** Ronnie Kaufman/Corbis; **351** Brand X Pictures/PunchStock; **355** John Rowley/Getty Images; **364–365** Doug Menuez/Getty Images; **366** Corbis; **367** Image Courtesy of U.S. Geological Survey; **372** Harald Sund/Getty Images; **376** Jack Hollingsworth/Getty Images; **393** George Tiedemann/GT Images/Corbis; **396** Corbis; **412–413** Davis Barber/PhotoEdit; **414** Photodisc/Getty Images; **448** (inset)Images.com/Corbis, (frame)Squared Studios/Getty Images; **476** PNC/Getty Images; **527** United States coin images from the United States Mint.

Index

Index

Index

Rounding, 42, 257–258

Run, 388

Scalene triangles, 414

Scatter plot, 374

Set model, 162

Similarity, 422, 426–427

Simplest form, 102, 160, 179–180

Simplification

    of algebraic expressions, 131, 266

    of expressions, 100–103

    of expressions using tiles, 131

    of fractions, 177–180, 228

    of numerical expressions, 265–266

Slope, 388–390

    finding, 389

    and graphs, 389

    and tables, 390

Solid figures, 452–454

Solutions, 114

    by substitution, 115

    with mental math, 115

Solve (problem-solving step), 60

Solving equations, 149–151, 221–223, 230

Solving simpler problems, 291

Speed equation, 256

Square, area of, 442

Square pyramid, 452

Square roots, 293–295, 401

Substitution, 115, 155

Subtracting

    decimals, 243–245, 271

    improper fractions, 214

    integers, 127, 128–131, 155

    like denominators, 229

    negative numbers, 431

    unlike denominators, 213–215, 230

Subtraction, 15, 70

Sum, 14

Surface area, 457–459

    and volume, 451

    of cubes, 458

    of cylinders, 459

    formulas for, 458

    of rectangular prisms, 458

Tables

    and slope, 390

    making, 355

Term, 100

Terminating decimals, 34, 35, 234, 237–238

Tile, algebra, 98–99

Tips, calculating, 345

Trapezoids, 416, 442

Trends, 374

Triangles, 414

    angle measures in, 415–417

    area of, 442, 443

    congruent and similar, 426

    rotating, 420

    types of, 414

Triangular models, 163

Triangular prisms, 453

Two-digit numbers, multiplying by, 18

Two-step equations, 150–151, 223

Understand (problem-solving step), 60

Unit rates, 319, 320, 382

Unlike denominators, 206–209, 213

    adding, 206–209, 230

    subtracting, 213–215, 230

Unlikely (unlikelihood), 50

Unlike terms, 100

Variables, 58, 65

Vertex, 46, 414, 421

Volume, 412

    and solid figures, 452–454

    and surface area, 451

    formulas for, 452

    of cubes, 453

    of cylinders, 454

    of prisms, 453

Whole numbers, 6–8

    adding, 14

    comparing, 7

    identifying, 7

    ordering, 8

    subtracting, 15

Whole thing, modeling the, 245–246

Working backward, 61, 262

x–axis, 366

y–axis, 366

Zero

    as integer, 119

    Multiplicative Property of, 94–95, 253

Zero pairs, 126